MONOGRAPHS ON STATISTICS AND APPLIED PROBABILITY

General Editors

V. Isham, N. Keiding, T. Louis, N. Reid, R. Tibshirani, and H. Tong

Monographs on Statistics and Applied Probability 101

Hierarchical Modeling and Analysis for Spatial Data

Sudipto Banerjee
Bradley P. Carlin
Alan E. Gelfand

CHAPMAN & HALL/CRC

A CRC Press Company
Boca Raton London New York Washington, D.C.

Library of Congress Cataloging-in-Publication Data

Banerjee, Sudipto.
  Hierarchical modeling and analysis for spatial data / Sudipto Banerjee, Bradley P. Carlin,
Alan E. Gelfand.
      p. cm. — (Monographs on statistics and applied probability : 101)
      Includes bibliographical references and index.
      ISBN 1-58488-410-X (alk. paper)
      1. Spatial analysis (Statistics)—Mathematical models. I. Carlin, Bradley P. II. Gelfand,
Alan E., 1945- III. Title. IV. Series.

  QA278.2.B36 2004
  519.5—dc22   2003062652

Visit the CRC Press Web site at www.crcpress.com

© 2004 by Chapman & Hall/CRC

No claim to original U.S. Government works
International Standard Book Number 1-58488-410-X
Library of Congress Card Number 2003062652
Printed in the United States of America   6 7 8 9 0
Printed on acid-free paper

TO SHARBANI, CAROLINE, AND MARY ELLEN

Contents

Preface

As recently as two decades ago, the impact of hierarchical Bayesian methods outside of a small group of theoretical probabilists and statisticians was minimal at best. Realistic models for challenging data sets were easy enough to write down, but the computations associated with these models required integrations over hundreds or even thousands of unknown parameters, far too complex for existing computing technology. Suddenly, around 1990, the "Markov chain Monte Carlo (MCMC) revolution" in Bayesian computing took place. Methods like the Gibbs sampler and the Metropolis algorithm, when coupled with ever-faster workstations and personal computers, enabled evaluation of the integrals that had long thwarted applied Bayesians. Almost overnight, Bayesian methods became not only feasible, but the method of choice for almost any model involving multiple levels incorporating random effects or complicated dependence structures. The growth in applications has also been phenomenal, with a particularly interesting recent example being a Bayesian program to delete spam from your incoming email (see `popfile.sourceforge.net`).

Our purpose in writing this book is to describe hierarchical Bayesian methods for one class of applications in which they can pay substantial dividends: spatial (and spatiotemporal) statistics. While all three of us have been working in this area for some time, our motivation for writing the book really came from our experiences teaching courses on the subject (two of us at the University of Minnesota, and the other at the University of Connecticut). In teaching we naturally began with the textbook by Cressie (1993), long considered the standard as both text and reference in the field. But we found the book somewhat uneven in its presentation, and written at a mathematical level that is perhaps a bit high, especially for the many epidemiologists, environmental health researchers, foresters, computer scientists, GIS experts, and other users of spatial methods who lacked significant background in mathematical statistics. Now a decade old, the book also lacks a current view of hierarchical modeling approaches for spatial data.

But the problem with the traditional teaching approach went beyond the mere need for a less formal presentation. Time and again, as we presented the traditional material, we found it wanting in terms of its flexibility to deal with realistic assumptions. Traditional Gaussian kriging is obviously the most important method of point-to-point spatial interpolation, but extending the paradigm beyond this was awkward. For areal (block-level) data, the problem seemed even more acute: CAR models should most naturally appear as priors for the parameters in a model, not as a model for the observations themselves.

This book, then, attempts to remedy the situation by providing a fully Bayesian treatment of spatial methods. We begin in Chapter 1 by outlining and providing illustrative examples of the three types of spatial data: point-level (geostatistical), areal (lattice), and spatial point process. We also provide a brief introduction to map projection and the proper calculation of distance on the earth's surface (which, since the earth is round, can differ markedly from answers obtained using the familiar notion of Euclidean distance). Our statistical presentation begins in earnest in Chapter 2, where we describe both exploratory data analysis tools and traditional modeling approaches for point-referenced data. Modeling approaches from traditional geostatistics (variogram fitting, kriging, and so forth) are covered here. Chapter 3 offers a similar presentation for areal data models, again starting with choropleth maps and other displays and progressing toward more formal statistical models. This chapter also presents Brook's Lemma and Markov random fields, topics that underlie the conditional, intrinsic, and simultaneous autoregressive (CAR, IAR, and SAR) models so often used in areal data settings.

Chapter 4 provides a review of the hierarchical Bayesian approach in a fairly generic setting, for readers previously unfamiliar with these methods and related computing and software. (The penultimate sections of Chapters 2, 3, and 4 offer tutorials in several popular software packages.) This chapter is not intended as a replacement for a full course in Bayesian methods (as covered, for example, by Carlin and Louis, 2000, or Gelman et al., 2004), but should be sufficient for readers having at least some familiarity with the ideas. In Chapter 5 then we are ready to cover hierarchical modeling for univariate spatial response data, including Bayesian kriging and lattice modeling. The issue of nonstationarity (and how to model it) also arises here.

Chapter 6 considers the problem of spatially misaligned data. Here, Bayesian methods are particularly well suited to sorting out complex interrelationships and constraints and providing a coherent answer that properly accounts for all spatial correlation and uncertainty. Methods for handling multivariate spatial responses (for both point- and block-level data) are discussed in Chapter 7. Spatiotemporal models are considered in Chapter 8, while Chapter 9 presents an extended application of areal unit data modeling in the context of survival analysis methods. Chapter 10 considers novel methodology associated with spatial process modeling, including spatial directional derivatives, spatially varying coefficient models, and spatial cumulative distribution functions (SCDFs). Finally, the book also features two useful appendices. Appendix A reviews elements of matrix theory and important related computational techniques, while Appendix B contains solutions to several of the exercises in each of the book's chapters.

Our book is intended as a research monograph, presenting the "state of the art" in hierarchical modeling for spatial data, and as such we hope readers will find it useful as a desk reference. However, we also hope it will be of benefit to instructors (or self-directed students) wishing to use it as a textbook. Here we see several options. Students wanting an introduction to methods for point-referenced data (traditional geostatistics and its extensions) may begin with Chapter 1, Chapter 2, Chapter 4, and Section 5.1 to Section 5.3. If areal data models are of greater interest, we suggest beginning with Chapter 1, Chapter 3, Chapter 4, Section 5.4, and Section 5.5. In addition, for students wishing to minimize the mathematical presentation, we have also marked sections containing more advanced material with a star ($\star$). These sections may be skipped (at least initially) at little cost to the intelligibility of the subsequent narrative. In our course in the Division of Biostatistics at the University of Minnesota, we are able to cover much of the book in a 3-credit-hour, single-semester (15-week) course. We encourage the reader to check http://www.biostat.umn.edu/~brad/ on the web for many of our data sets and other teaching-related information.

We owe a debt of gratitude to those who helped us make this book a reality. Kirsty Stroud and Bob Stern took us to lunch and said encouraging things (and more importantly, picked up the check) whenever we needed it. Cathy Brown, Alex Zirpoli, and Desdamona Racheli prepared significant portions of the text and figures. Many of our current and former graduate and postdoctoral students, including Yue Cui, Xu Guo, Murali Haran, Xiaoping Jin, Andy Mugglin, Margaret Short, Amy Xia, and Li Zhu at Minnesota, and Deepak Agarwal, Mark Ecker, Sujit Ghosh, Hyon-Jung Kim, Ananda Majumdar, Alexandra Schmidt, and Shanshan Wu at the University of Connecticut, played a big role. We are also grateful to the Spring 2003 *Spatial Biostatistics* class in the School of Public Health at the University of Minnesota for taking our draft for a serious "test drive." Colleagues Jarrett Barber, Nicky Best, Montserrat Fuentes, David Higdon, Jim Hodges, Oli Schabenberger, John Silander, Jon Wakefield, Melanie Wall, Lance Waller, and many others provided valuable input and assistance. Finally, we thank our families, whose ongoing love and support made all of this possible.

SUDIPTO BANERJEE   Minneapolis, Minnesota
BRADLEY P. CARLIN   Durham, North Carolina
ALAN E. GELFAND   October 2003

CHAPTER 1

Overview of spatial data problems

1.1 Introduction to spatial data and models

Researchers in diverse areas such as climatology, ecology, environmental health, and real estate marketing are increasingly faced with the task of analyzing data that are

- highly multivariate, with many important predictors and response variables,
- geographically referenced, and often presented as maps, and
- temporally correlated, as in longitudinal or other time series structures.

For example, for an epidemiological investigation, we might wish to analyze lung, breast, colorectal, and cervical cancer rates by county and year in a particular state, with smoking, mammography, and other important screening and staging information also available at some level. Public health professionals who collect such data are charged not only with surveillance, but also statistical *inference* tasks, such as *modeling* of trends and correlation structures, *estimation* of underlying model parameters, *hypothesis testing* (or comparison of competing models), and *prediction* of observations at unobserved times or locations.

In this text we seek to present a practical, self-contained treatment of hierarchical modeling and data analysis for complex spatial (and spatiotemporal) data sets. Spatial statistics methods have been around for some time, with the landmark work by Cressie (1993) providing arguably the only comprehensive book in the area. However, recent developments in Markov chain Monte Carlo (MCMC) computing now allow fully Bayesian analyses of sophisticated multilevel models for complex geographically referenced data. This approach also offers full inference for non-Gaussian spatial data, multivariate spatial data, spatiotemporal data, and, for the first time, solutions to problems such as geographic and temporal misalignment of spatial data layers.

This book does not attempt to be fully comprehensive, but does attempt to present a fairly thorough treatment of hierarchical Bayesian approaches for handling all of these problems. The book's mathematical level is roughly comparable to that of Carlin and Louis (2000). That is, we sometimes state

2   OVERVIEW OF SPATIAL DATA PROBLEMS

results rather formally, but spend little time on theorems and proofs. For
more mathematical treatments of spatial statistics (at least on the geostatistical side), the reader is referred to Cressie (1993), Wackernagel (1998),
Chiles and Delfiner (1999), and Stein (1999a). For more descriptive presen-
tations the reader might consult Bailey and Gattrell (1995), Fotheringham
and Rogerson (1994), or Haining (1990). Our primary focus is on the issues
of *modeling* (where we offer rich, flexible classes of hierarchical structures
to accommodate both static and dynamic spatial data), *computing* (both
in terms of MCMC algorithms and methods for handling very large matrices), and *data analysis* (to illustrate the first two items in terms of inferential summaries and graphical displays). Reviews of both traditional spatial
methods (Chapters 2 and 3) and Bayesian methods (Chapter 4) attempt to
ensure that previous exposure to either of these two areas is not required
(though it will of course be helpful if available).

Following convention, we classify spatial data sets into one of three basic
types:

- *point-referenced data*, where $Y(\mathbf{s})$ is a random vector at a location $\mathbf{s} \in \Re^r$,
  where $\mathbf{s}$ varies *continuously* over D, a fixed subset of $\Re^r$ that contains
  an r-dimensional rectangle of positive volume;

- *areal data*, where D is again a fixed subset (of regular or irregular shape),
  but now partitioned into a finite number of areal units with well-defined
  boundaries;

- *point pattern data*, where now D is itself random; its index set gives the
  locations of random events that are the spatial point pattern. $Y(\mathbf{s})$ itself
  can simply equal 1 for all $\mathbf{s} \in D$ (indicating occurrence of the event), or
  possibly give some additional covariate information (producing a *marked
  point pattern process*).

The first case is often referred to as *geocoded* or *geostatistical* data, names
apparently arising from the long history of these types of problems in mining and other geological sciences. Figure 1.1 offers an example of this case,
showing the locations of 114 air-pollution monitoring sites in three midwestern U.S. states (Illinois, Indiana, and Ohio). The plotting character
indicates the 2001 annual average PM2.5 level (measured in ppb) at each
site. PM2.5 stands for particulate matter less than 2.5 microns in diame-
ter, and is a measure of the density of very small particles that can travel
through the nose and windpipe and into the lungs, potentially damaging a
person's health. Here we might be interested in a model of the geographic
distribution of these levels that account for spatial correlation and perhaps underlying covariates (regional industrialization, traffic density, and
the like). The use of symbols to denote the approximate level is convenient
for this black-and-white map, but the color version in Figure C.1 (located
in this book's color insert) is somewhat easier to read, since the color allows the categories to be ordered more naturally, and helps sharpen the

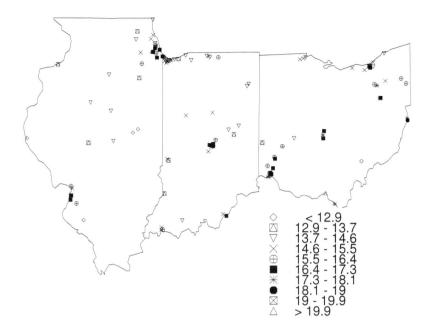

◇	< 12.9
▱	12.9 - 13.7
▽	13.7 - 14.6
×	14.6 - 15.5
⊕	15.5 - 16.4
■	16.4 - 17.3
✳	17.3 - 18.1
●	18.1 - 19
⊠	19 - 19.9
△	> 19.9

Figure 1.1 *Map of PM2.5 sampling sites over three midwestern U.S. states; plotting character indicates range of average monitored PM2.5 level over the year 2001 (see also color insert).*

contrast between the urban and rural areas. Again, traditional analysis methods for point level data like this are described in Chapter 2, while Chapter 5 introduces the corresponding hierarchical modeling approach.

The second case above (areal data) is often referred to as *lattice* data, a term we find misleading since it connotes observations corresponding to "corners" of a checkerboard-like grid. Of course, there *are* data sets of this type; for example, as arising from agricultural field trials (where the plots cultivated form a regular lattice) or image restoration (where the data correspond to pixels on a screen, again in a regular lattice). However, in practice most areal data are summaries over an *irregular* lattice, like a collection of county or other regional boundaries, as in Figure 1.2 (see also color insert Figure C.2). Here we have information on the percent of a surveyed population with household income falling below 200% of the federal poverty limit, for a collection of regions comprising Hennepin County, MN. Note that we have no information on any single household in the study area, only regional summaries for each region. Figure 1.2 is an example of a *choropleth map*, meaning that it uses shades of color (or greyscale) to classify values into a few broad classes (six in this case), like

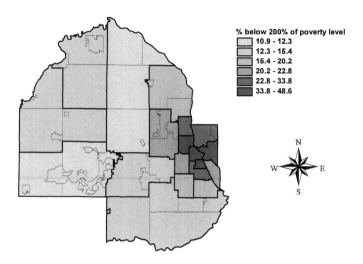

Figure 1.2 **ArcView** *map of percent of surveyed population with household income below 200% of the federal poverty limit, regional survey units in Hennepin County, MN (see also color insert).*

a histogram (bar chart) for nonspatial data. Choropleth maps are visually appealing (and therefore, also common), but of course provide a rather crude summary of the data, and one that can be easily altered simply by manipulating the class cutoffs.

As with any map of the areal units, choropleth maps *do* show reasonably precise *boundaries* between the regions (i.e., a series of exact spatial coordinates that when connected in the proper order will trace out each region), and thus we also know which regions are adjacent to (touch) which other regions. Thus the "sites" $s \in D$ in this case are actually the regions (or *blocks*) themselves, which in this text we will denote not by s_i but by B_i, $i = 1, \ldots, n$, to avoid confusion between points s_i and blocks B_i. It may also be illuminating to think of the county centroids as forming the vertices of an irregular lattice, with two lattice points being connected if and only if the counties are "neighbors" in the spatial map, with physical adjacency being the most obvious (but not the only) way to define a region's neighbors.

Some spatial data sets feature *both* point- and areal-level data, and require their simultaneous display and analysis. Figure 1.3 (see also color insert Figure C.3) offers an example of this case. The first component of this data set is a collection of eight-hour maximum ozone levels at 10 monitoring sites in the greater Atlanta, GA, area for a particular day in July 1995. Like the observations in Figure 1.1, these were made at fixed monitoring stations for which exact spatial coordinates (say, latitude and longitude)

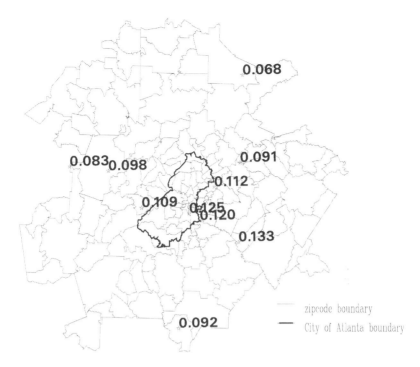

Figure 1.3 *Zip code boundaries in the Atlanta metropolitan area and 8-hour maximum ozone levels (ppm) at 10 monitoring sites for July 15, 1995 (see also color insert).*

are known. (That is, we assume the $Y(\mathbf{s}_i)$, $i = 1, \ldots, 10$ are random, but the $\mathbf{s}_i$ are not.) The second component of this data set is the number of children in the area's zip codes (shown using the irregular subboundaries on the map) that reported at local emergency rooms (ERs) with acute asthma symptoms on the following day; confidentiality of health records precludes us from learning the precise address of any of the children. These are areal summaries that could be indicated by shading the zip codes, as in Figure 1.2. An obvious question here is whether we can establish a connection between high ozone and subsequent high pediatric ER asthma visits. Since the data are misaligned (point-level ozone but block-level ER counts), a formal statistical investigation of this question requires a preliminary *realignment* of the data; this is the subject of Chapter 6.

The third case above (spatial point pattern data) could be exemplified by residences of persons suffering from a particular disease, or by locations of a certain species of tree in a forest. Here the response Y is often fixed (occurrence of the event), and only the locations $\mathbf{s}_i$ are thought of as random. In some cases this information might be supplemented by age or other covariate information, producing a *marked* point pattern). Such data are often of interest in studies of event *clustering*, where the goal is to determine whether an observed spatial point pattern is an example of a clustered process (where points tend to be spatially close to other points), or merely the result of a random event process operating independently and homogeneously over space. Note that in contrast to areal data, where no individual points in the data set could be identified, here (and in point-referenced data as well) precise locations are known, and so must often be protected to protect the privacy of the persons in the set.

In the remainder of this initial section, we give a brief outline of the basic models most often used for each of these three data types. Here we only intend to give a flavor of the models and techniques to be fully described in the remainder of this book. However, in subsequent chapters we confine ourselves to the case where the locations (or areal units) are fixed, and the only randomness is in the measurements at these locations or units.

Even though our preferred inferential outlook is Bayesian, the statistical inference tools discussed in Chapters 2 and 3 are entirely classical. While all subsequent chapters adopt the Bayesian point of view, our objective here is to acquaint the reader with the classical techniques first, since they are more often implemented in standard software packages. Moreover, as in other fields of data analysis, classical methods can be easier to compute, and produce perfectly acceptable results in relatively simple settings. Classical methods often have interpretations as limiting cases of Bayesian methods under increasingly vague prior assumptions. Finally, classical methods can provide insight for formulating and fitting hiearchical models.

1.1.1 Point-level models

In the case of point-level data, the location index $\mathbf{s}$ varies *continuously* over D, a fixed subset of $\Re^d$. Suppose we assume that the covariance between the random variables at two locations depends on the *distance* between the locations. One frequently used association specification is the exponential model. Here the covariance between measurements at two locations is an exponential function of the interlocation distance, i.e., $Cov(Y(\mathbf{s}_i), Y(\mathbf{s}_{i'})) \equiv C(d_{ii'}) = \sigma^2 e^{-\phi d_{ii'}}$ for $i \neq i'$, where $d_{ii'}$ is the distance between sites s_i and $s_{i'}$, and σ^2 and ϕ are positive parameters called the *partial sill* and the *decay parameter*, respectively ($1/\phi$ is called the *range parameter*). A plot of the covariance versus distance is called the *covariogram*. When $i = i'$, $d_{ii'}$ is of course 0, and $C(d_{ii'}) = Var(Y(\mathbf{s}_i))$ is often expanded to $\tau^2 + \sigma^2$, where $\tau^2 > 0$ is called a *nugget effect*, and $\tau^2 + \sigma^2$ is called the *sill*. Of course, while the exponential model is convenient and has some desirable properties, many other parametric models are commonly used; see Section 2.1 for further discussion of these and their relative merits.

Adding a joint distributional model to these variance and covariance assumptions then enables likelihood inference in the usual way. The most convenient approach would be to assume a multivariate *normal* (or *Gaussian*) distribution for the data. That is, suppose we are given observations $\mathbf{Y} \equiv \{Y(\mathbf{s}_i)\}$ at known locations $\mathbf{s}_i$, $i = 1, \ldots, n$. We then assume that

$$\mathbf{Y} \mid \mu, \boldsymbol{\theta} \sim N_n(\mu \,, \, \Sigma(\boldsymbol{\theta})) \,, \tag{1.1}$$

where N_n denotes the n-dimensional normal distribution, μ is the (constant) mean level, and $(\Sigma(\boldsymbol{\theta}))_{ii'}$ gives the covariance between $Y(\mathbf{s}_i)$ and $Y(\mathbf{s}_{i'})$. For the variance-covariance specification of the previous paragraph, we have $\boldsymbol{\theta} = (\tau^2, \sigma^2, \phi)^T$, since the covariance matrix depends on the nugget, sill, and range.

In fact, the simplest choices for Σ are those corresponding to *isotropic* covariance functions, where we assume that the spatial correlation is a function solely of the distance $d_{ii'}$ between $\mathbf{s}_i$ and $\mathbf{s}_{i'}$. As mentioned above, exponential forms are particularly intuitive examples. Here,

$$(\Sigma(\boldsymbol{\theta}))_{ii'} = \sigma^2 \exp(-\phi d_{ii'}) + \tau^2 I(i = i'), \ \sigma^2 > 0, \ \phi > 0, \ \tau^2 > 0 \,, \tag{1.2}$$

where I denotes the indicator function (i.e., $I(i = i') = 1$ if $i = i'$, and 0 otherwise). Many other choices are possible for $Cov(Y(\mathbf{s}_i), Y(\mathbf{s}_{i'}))$, including for example the powered exponential,

$$(\Sigma(\boldsymbol{\theta}))_{ii'} = \sigma^2 \exp(-\phi d_{ii'}^{\kappa}) + \tau^2 I(i = i'), \ \sigma^2 > 0, \ \phi > 0, \ \tau^2 > 0, \ \kappa \in (0, 2] \,,$$

the spherical, the Gaussian, and the Matérn (see Subsection 2.1.3 for a full discussion). In particular, while the latter requires calculation of a modified Bessel function, Stein (1999a, p. 51) illustrates its ability to capture a broader range of local correlation behavior despite having no more parameters than the powered exponential. Again, we shall say much more about point-level spatial methods and models in Section 2.1.

1.1.2 Areal models

In models for areal data, the geographic regions or *blocks* (zip codes, counties, etc.) are denoted by B_i, and the data are typically sums or averages of variables over these blocks. To introduce spatial association, we define a *neighborhood* structure based on the arrangement of the blocks in the map. Once the neighborhood structure is defined, models resembling autoregressive time series models are considered. Two very popular models that incorporate such neighborhood information are the *simultaneously* and *conditionally autoregressive* models (abbreviated SAR and CAR), originally developed by Whittle (1954) and Besag (1974), respectively. The SAR model is computationally convenient for use with likelihood methods. By contrast, the CAR model is computationally convenient for Gibbs sampling used in conjunction with Bayesian model fitting, and in this regard is often used to incorporate spatial correlation through a vector of spatially varying random effects $\phi = (\phi_1, \ldots, \phi_n)^T$. For example, writing $Y_i \equiv Y(B_i)$, we might assume $Y_i \overset{ind}{\sim} N(\phi_i, \sigma^2)$, and then impose the CAR model

$$\phi_i | \phi_{(-i)} \sim N\left(\mu + \sum_{j=1}^{n} a_{ij}(\phi_j - \mu), \tau_i^2\right), \qquad (1.3)$$

where $\phi_{(-i)} = \{\phi_j : j \neq i\}$, τ_i^2 is the conditional variance, and the a_{ij} are known or unknown constants such that $a_{ii} = 0$ for $i = 1, \ldots, n$. Letting $A = (a_{ij})$ and $M = Diag(\tau_1^2, \ldots, \tau_n^2)$, by Brook's Lemma (c.f. Section 3.2), we can show that

$$p(\phi) \propto \exp\{-(\phi - \mu\mathbf{1})^T M^{-1}(I - A)(\phi - \mu\mathbf{1})/2\}, \qquad (1.4)$$

where $\mathbf{1}$ is an n-vector of 1's, and I is a $n \times n$ identity matrix.

A common way to construct A and M is to let $A = \rho\, Diag(1/w_{i+})W$ and $M^{-1} = \tau^{-2}Diag(w_{i+})$. Here ρ is referred to as the *spatial correlation* parameter, and $W = (w_{ij})$ is a neighborhood matrix for the areal units, which can be defined as

$$w_{ij} = \begin{cases} 1 & \text{if subregions } i \text{ and } j \text{ share a common boundary, } i \neq j \\ 0 & \text{otherwise} \end{cases}.$$

$$(1.5)$$

Thus $Diag(w_{i+})$ is a diagonal matrix with (i, i) entry equal to $w_{i+} = \sum_j w_{ij}$. Letting $\alpha \equiv (\rho, \tau^2)$, the covariance matrix of ϕ then becomes $C(\alpha) = \tau^2[Diag(w_{i+}) - \rho W]^{-1}$, where the inverse exists for an appropriate range of ρ values; see Subsection 3.3.1.

In the context of Bayesian hierarchical areal modeling, when choosing a prior distribution $\pi(\phi)$ for a vector of spatial random effects ϕ, the CAR distribution (1.3) is often used with the 0–1 *weight* (or *adjacency*) *matrix* W in (1.5) and $\rho = 1$. While this results in an *improper* (nonintegrable) prior distribution, this problem is remedied by imposing a sum-to-zero constraint on the ϕ_i (which turns out to be easy to implement numerically using Gibbs sampling). In this case the more general conditional form (1.3) is replaced by

$$\phi_i | \phi_{(-i)} \sim N(\bar{\phi}_i, \tau^2/m_i), \qquad (1.6)$$

where $\bar{\phi}_i$ is the average of the $\phi_{j \neq i}$ that are adjacent to ϕ_i, and m_i is the number of these adjacencies (see, e.g., Besag, York, and Mollié, 1991).

1.1.3 Point process models

In the point process model, the spatial domain D is itself random, so that the elements of the index set D are the locations of random events that constitute the spatial point pattern. $Y(\mathbf{s})$ then normally equals the constant 1 for all $\mathbf{s} \in D$ (indicating occurrence of the event), but it may also provide additional covariate information, in which case the data constitute a marked point process.

Questions of interest with data of this sort typically center on whether the data are *clustered* more or less than would be expected if the locations were determined completely by chance. Stochastically, such uniformity is often described through a *homogeneous Poisson process*, which implies that the expected number of occurrences in region A is $\lambda|A|$, where λ is the *intensity* parameter of the process and $|A|$ is the area of A. To investigate this in practice, plots of the data are typically a good place to start, but the tendency of the human eye to see clustering or other structure in virtually every point pattern renders a strictly graphical approach unreliable. Instead, statistics that measure clustering, and perhaps even associated significance tests, are often used. The most common of these is *Ripley's K function*, given by

$$K(d) = \frac{1}{\lambda} E[\text{number of points within } d \text{ of an arbitrary point}] , \qquad (1.7)$$

where again λ is the intensity of the process, i.e., the mean number of points per unit area.

The theoretical value of K is known for certain spatial point process models. For instance, for point processes that have no spatial dependence at all, we would have $K(d) = \pi d^2$, since in this case the number of points within d of an arbitrary point should be proportional to the area of a circle of radius d; the K function then divides out the average intensity λ. However, if the data are clustered we might expect $K(d) > \pi d^2$, while if the points follow some regularly spaced pattern we would expect $K(d) < \pi d^2$. This suggests a potential inferential use for K; namely, comparing an estimate of it from a data set to some theoretical quantities, which in turn suggests if clustering is present, and if so, which model might be most plausible. The usual estimator for K is given by

$$\widehat{K}(d) = n^{-2}|A| \sum_{i \neq j} \sum p_{ij}^{-1} I_d(d_{ij}) , \qquad (1.8)$$

where n is the number of points in A, d_{ij} is the distance between points i and j, p_{ij} is the proportion of the circle with center i and passing through j that lies within A, and $I_d(d_{ij})$ equals 1 if $d_{ij} < d$, and 0 otherwise.

A popular spatial add-on to the S+ package, S+SpatialStats, allows computation of K for any data set, as well as approximate 95% intervals for it so the significance of departure from some theoretical model may be judged. However, full inference likely requires use of the Splancs software (www.maths.lancs.ac.uk/~rowlings/Splancs/), or perhaps a fully Bayesian approach along the lines of Wakefield and Morris (2001). Again, in the remainder of this book we confine ourselves to the case of a fixed index set D, i.e., random observations at either fixed points $\mathbf{s}_i$ or areal units

B_i. The reader may wish to consult the recent books by Diggle (2003), Lawson and Denison (2002), and Møller and Waagepetersen (2004) for recent treatments of spatial point processes and related methods in spatial cluster detection and modeling.

1.2 Fundamentals of cartography

In this section we provide a brief introduction to how geographers and spatial statisticians understand the geometry of (and determine distances on) the surface of the earth. This requires a bit of thought regarding cartography (mapmaking), especially map projections, and the meaning of latitude and longitude, which are often understood informally (but incorrectly) by laypersons and even some spatial modelers as being equivalent to Cartesian x and y coordinates.

1.2.1 Map projections

A map projection is a systematic representation of all or part of the surface of the earth on a plane. This typically comprises lines delineating meridians (longitudes) and parallels (latitudes), as required by some definitions of the projection. A well-known fact from topology is that it is impossible to prepare a distortion-free flat map of a surface curving in all directions. Thus, the cartographer must choose the characteristic (or characteristics) that are to be shown accurately in the map. In fact, it cannot be said that there is a "best" projection for mapping. The purpose of the projection and the application at hand lead to projections that are appropriate. Even for a single application, there may be several appropriate projections, and choosing the "best" projection can be subjective. Indeed there are an infinite number of projections that can be devised, and several hundred have been published.

Since the sphere cannot be flattened onto a plane without distortion, the general strategy for map projections is to use an intermediate surface that can be flattened. This intermediate surface is called a *developable surface* and the sphere is first projected onto the this surface, which is then laid out as a plane. The three most commonly used surfaces are the cylinder, the cone and the plane itself. Using different orientations of these surfaces lead to different classes of map projections. Some examples are given in Figure 1.4. The points on the globe are projected onto the wrapping (or tangential) surface, which is then laid out to form the map. These projections may be performed in several ways, giving rise to different projections.

Before the availability of computers, the above orientations were used by cartographers in the physical construction of maps. With computational advances and digitizing of cartography, analytical formulae for projections were desired. Here we briefly outline the underlying theory for equal-area

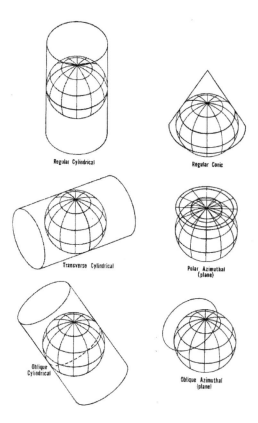

Figure 1.4 *The geometric constructions of projections using developable surfaces (figure courtesy of the U.S. Geological Survey).*

and conformal (locally shape-preserving) maps. A much more detailed and rigorous treatment may be found in Pearson (1990).

The basic idea behind deriving equations for map projections is to consider a sphere with the geographical coordinate system (λ, ϕ) for longitude and latitude and to construct an appropriate (rectangular or polar) coordinate system (x, y) so that

$$x = f(\lambda, \phi), \ y = g(\lambda, \phi) \,,$$

where f and g are appropriate functions to be determined, based upon the properties we want our map to possess. We will study map projections using differential geometry concepts, looking at infinitesimal patches on the sphere (so that curvature may be neglected and the patches are closely approximated by planes) and deriving a set of (partial) differential equations whose solution will yield f and g. Suitable initial conditions are set to create projections with desired geometric properties.

Thus, consider a small patch on the sphere formed by the infinitesimal quadrilateral, $ABCD$, given by the vertices,

$$A = (\lambda, \phi), \ B = (\lambda, \phi + d\phi), \ C = (\lambda + d\lambda, \phi), \ D = (\lambda + d\lambda, \phi + d\phi).$$

So, with R being the radius of the earth, the horizontal differential component along an arc of latitude is given by $|AC| = (R \cos \phi) d\lambda$ and the vertical component along a great circle of longitude is given by $|AB| = Rd\phi$. Note that since AC and AB are arcs along the latitude and longitude of the globe, they intersect each other at right angles. Therefore, the area of the patch $ABCD$ is given by $|AC||AB|$. Let $A'B'C'D'$ be the (infinitesimal) image of the patch $ABCD$ on the map. Then, we see that

$$A' = (f(\lambda, \phi), g(\lambda, \phi)),$$
$$C' = (f(\lambda + d\lambda, \phi), g(\lambda + d\lambda, \phi)),$$
$$B' = (f(\lambda, \phi + d\phi), g(\lambda, \phi + d\phi)),$$
$$\text{and } D' = (f(\lambda + d\lambda, \phi + d\phi), g(\lambda + d\lambda, \phi + d\phi)) .$$

This in turn implies that

$$\overrightarrow{A'C'} = \left(\frac{\partial f}{\partial \lambda}, \frac{\partial g}{\partial \lambda} \right) d\lambda \text{ and } \overrightarrow{A'B'} = \left(\frac{\partial f}{\partial \phi}, \frac{\partial g}{\partial \phi} \right) d\phi .$$

If we desire an equal-area projection we need to equate the area of the patches $ABCD$ and $A'B'C'D'$. But note that the area of $A'B'C'D'$ is given by the area of parallelogram formed by vectors $\overrightarrow{A'C'}$ and $\overrightarrow{A'B'}$. Treating them as vectors in the xy plane of an xyz system, we see that the area of $A'B'C'D'$ is the cross-product,

$$(\overrightarrow{A'C'}, 0) \times (\overrightarrow{A'B'}, 0) = \left(\frac{\partial f}{\partial \lambda} \frac{\partial g}{\partial \phi} - \frac{\partial f}{\partial \phi} \frac{\partial g}{\partial \lambda} \right) d\lambda d\phi .$$

Therefore, we equate the above to $|AC||AB|$, leading to the following partial differential equation in f and g:

$$\left(\frac{\partial f}{\partial \lambda} \frac{\partial g}{\partial \phi} - \frac{\partial f}{\partial \phi} \frac{\partial g}{\partial \lambda} \right) = R^2 \cos \phi .$$

Note that this is the equation that must be satisfied by any equal-area projection. It is an underdetermined system, and further conditions need to be imposed (that ensure other specific properties of the projection) to arrive at f and g.

Example 1.1 Equal-area maps are used for statistical displays of areal-referenced data. An easily derived equal-area projection is the sinusoidal projection, shown in Figure 1.5. This is obtained by specifying $\partial g / \partial \phi = R$, which yields equally spaced straight lines for the parallels, and results in

Figure 1.5 *The sinusoidal projection.*

the following analytical expressions for f and g (with the 0 degree meridian as the central meridian):

$$f(\lambda, \phi) = R\lambda \cos \phi; g(\lambda, \phi) = R\phi.$$

Another popular equal-area projection (with equally spaced straight lines for the meridians) is the Lambert cylindrical projection given by

$$f(\lambda, \phi) = R\lambda; \; g(\lambda, \phi) = R \sin \phi.$$

■

For conformal (angle-preserving) projections we set the angle $\angle(AC, AB)$ equal to $\angle(A'C', A'B')$. Since $\angle(AC, AB) = \pi/2$, $\cos(\angle(AC, AB)) = 0$, leading to

$$\frac{\partial f}{\partial \lambda} \frac{\partial f}{\partial \phi} + \frac{\partial g}{\partial \lambda} \frac{\partial g}{\partial \phi} = 0$$

or, equivalently, the Cauchy-Riemann equations of complex analysis,

$$\left(\frac{\partial f}{\partial \lambda} + i \frac{\partial g}{\partial \lambda} \right) \left(\frac{\partial f}{\partial \phi} - i \frac{\partial g}{\partial \phi} \right) = 0.$$

A sufficient partial differential equation system for conformal mappings of the Cauchy-Riemman equations that is simpler to use is

$$\frac{\partial f}{\partial \lambda} = \frac{\partial g}{\partial \phi} \cos \phi; \; \frac{\partial g}{\partial \lambda} = \frac{\partial f}{\partial \phi} \cos \phi.$$

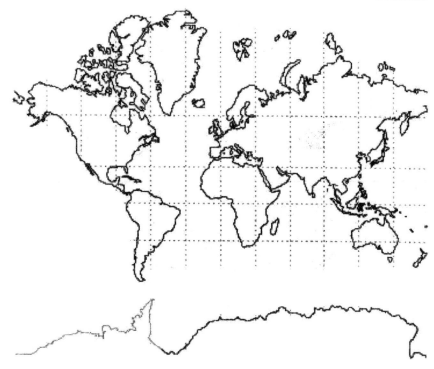

Figure 1.6 *The Mercator projection.*

Example 1.2 The Mercator projection shown in Figure 1.6 is a classical example of a conformal projection. It has the interesting property that rhumb lines (curves that intersect the meridians at a constant angle) are shown as straight lines on the map. This is particularly useful for navigation purposes. The Mercator projection is derived by letting $\partial g/\partial \phi = R \sec \phi$. After suitable integration, this leads to the analytical equations (with the 0 degree meridian as the central meridian),

$$f(\lambda, \phi) = R\lambda; \ g(\lambda, \phi) = R \ln \tan \left(\frac{\pi}{4} + \frac{\phi}{2} \right) .$$

■

As is seen above, even the simplest map projections lead to complex transcendental equations relating latitude and longitude to positions of points on a given map. Therefore, rectangular grids have been developed for use by surveyors. In this way, each point may be designated merely by its distance from two perpendicular axes on a flat map. The y-axis usually coincides with a chosen central meridian, y increasing north, and the x-axis is perpendicular to the y-axis at a latitude of origin on the central meridian, with x increasing east. Frequently, the x and y coordinates

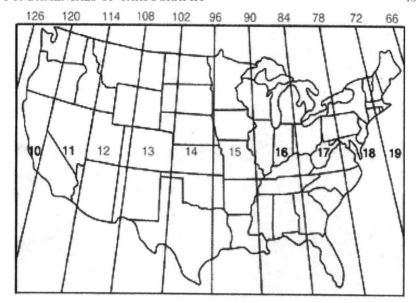

Figure 1.7 *Example of a UTM grid over the United States (figure courtesy of the U.S. Geological Survey).*

are called "eastings" and "northings," respectively, and to avoid negative coordinates, may have "false eastings" and "false northings" added to them. The grid lines usually do not coincide with any meridians and parallels except for the central meridian and the equator.

One such popular grid, adopted by The National Imagery and Mapping Agency (NIMA) (formerly known as the Defense Mapping Agency) and used especially for military use throughout the world, is the Universal Transverse Mercator (UTM) grid; see Figure 1.7. The UTM divides the world into 60 north-south zones, each of width six degrees longitude. Starting with Zone 1 (between 180 degrees and 174 degrees west longitude), these are numbered consecutively as they progress eastward to Zone 60, between 174 degrees and 180 degrees east longitude. Within each zone, coordinates are measured north and east in meters, with northing values being measured continuously from zero at the Equator, in a northerly direction. Negative numbers for locations south of the Equator are avoided by assigning an arbitrary false northing value of 10,000,000 meters (as done by NIMA's cartographers). A central meridian cutting through the center of each 6 degree zone is assigned an easting value of 500,000 meters, so that values to the west of the central meridian are less than 500,000 while those to the east are greater than 500,000. In particular, the conterminous

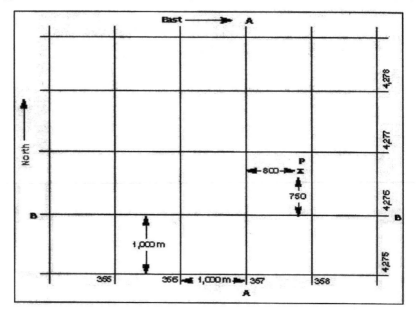

Figure 1.8 *Finding the easting and northing of a point in a UTM projection (figure courtesy of the U.S. Geological Survey).*

48 states of the United States are covered by 10 zones, from Zone 10 on the west coast through Zone 19 in New England.

In practice, the UTM is used by overlaying a transparent grid on the map, allowing distances to be measured in meters at the map scale between any map point and the nearest grid lines to the south and west. The northing of the point is calculated as the sum of the value of the nearest grid line south of it and its distance north of that line. Similarly, its easting is the value of the nearest grid line west of it added to its distance east of that line. For instance, in Figure 1.8, the grid value of line A-A is 357,000 meters east, while that of line B-B is 4,276,000 meters north. Point P is 800 meters east and 750 meters north of the grid lines resulting in the grid coordinates of point P as north 4,276,750 and east 357,800.

Finally, since spatial modeling of point-level data often requires computing distances between points on the earth's surface, one might wonder about a *planar* map projection, which would preserve distances between points. Unfortunately, the existence of such a map is precluded by Gauss' Theorema Eggregium in differential geometry (see, e.g., Guggenheimer, 1977, pp. 240–242). Thus, while we have seen projections that preserve area and shapes, distances are always distorted. The *gnomonic* projection (Snyder, 1987, pp. 164–168) gives the correct distance from a single reference point, but is less useful for the practicing spatial analyst who needs to obtain complete intersite distance matrices (since this would require not one but many such maps). See Banerjee (2005) for more details.

1.2.2 Calculating distance on the earth's surface

As we have seen, the most common approach in spatial statistics is to model spatial dependence between two variables as a function of the distance between them. For data sets covering relatively small spatial domains, ordinary Euclidean distance is fine for this purpose. However, for larger domains (say, the entire continental U.S.) we must account for the curvature of the earth when computing such distances.

Suppose we have two points on the surface of the earth, $P_1 = (\theta_1, \lambda_1)$ and $P_2 = (\theta_2, \lambda_2)$. We assume both points are represented in terms of latitude and longitude. That is, let θ_1 and λ_1 be the latitude and longitude, respectively, of the point P_1, while θ_2 and λ_2 are those for the point P_2. The main problem is to find the shortest distance (*geodesic*) between the points. The solution is obtained via the following formulae:

$$D = R\phi$$

where R is the radius of the earth and ϕ is an angle (measured in *radians*) satisfying

$$\cos\phi = \sin\theta_1 \sin\theta_2 + \cos\theta_1 \cos\theta_2 \cos(\lambda_1 - \lambda_2) . \qquad (1.9)$$

These formulae are derived as follows. The geodesic is actually the arc of the great circle joining the two points. Thus the distance will be the length of the arc of a *great circle* (i.e., a circle with radius equal to the radius of the earth). Recall that the length of the arc of a circle equals the angle subtended by the arc at the center multiplied by the radius of the circle. Therefore it suffices to find the angle subtended by the arc; denote this angle by ϕ.

Let us form a three-dimensional Cartesian coordinate system (x, y, z), with the origin at the center of the earth, the z-axis along the North and South Poles, and the x-axis on the plane of the equator joining the center of the earth and the Greenwich meridian. Using the left panel of Figure 1.9 as a guide, elementary trigonometry provides the following relationships between (x, y, z) and the latitude-longitude (θ, λ):

$$
\begin{aligned}
x &= R\cos\theta\cos\lambda, \\
y &= R\cos\theta\sin\lambda, \\
\text{and } z &= R\sin\theta .
\end{aligned}
$$

Now form the vectors $\mathbf{u}_1 = (x_1, y_1, z_1)$ and $\mathbf{u}_2 = (x_2, y_2, z_2)$ as the Cartesian coordinates corresponding to points P_1 and P_2. Hence ϕ is the angle

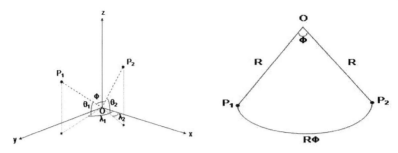

Figure 1.9 *Diagrams illustrating the geometry underlying the calculation of great circle (geodesic) distance.*

between $\mathbf{u}_1$ and $\mathbf{u}_2$. From standard analytic geometry, the easiest way to find this angle is therefore to use the following relationship between the cosine of this angle and the dot product of $\mathbf{u}_1$ and $\mathbf{u}_2$:

$$\cos\phi = \frac{\langle \mathbf{u}_1, \mathbf{u}_2 \rangle}{||\mathbf{u}_1||\,||\mathbf{u}_2||}\;.$$

We then compute $\langle \mathbf{u}_1, \mathbf{u}_2 \rangle$ as

$$R^2 \left[\cos\theta_1 \cos\lambda_1 \cos\theta_2 \cos\lambda_2 + \cos\theta_1 \sin\lambda_1 \cos\theta_2 \sin\lambda_2 + \sin\theta_1 \sin\theta_2 \right]$$
$$= R^2 \left[\cos\theta_1 \cos\theta_2 \cos(\lambda_1 - \lambda_2) + \sin\theta_1 \sin\theta_2 \right]\;.$$

But $||\mathbf{u}_1|| = ||\mathbf{u}_2|| = R$, so the result in (1.9) follows. Looking at the right panel of Figure 1.9, our final answer is thus

$$D = R\phi = R \arccos[\sin\theta_1 \sin\theta_2 + \cos\theta_1 \cos\theta_2 \cos(\lambda_1 - \lambda_2)]\;. \qquad (1.10)$$

1.3 Exercises

1. What sorts of areal unit variables can you envision that could be viewed as arising from point-referenced variables? What sorts of areal unit variables can you envision whose mean could be viewed as arising from a point-referenced surface? What sorts of areal unit variables fit neither of these scenarios?

2. What sorts of sensible properties should characterize association between point-referenced measurements? What sorts of sensible properties should characterize association between areal unit measurements?

3. Suggest some regional-level covariates that might help explain the spatial pattern evident in Figure 1.2. (*Hint:* The roughly rectangular group of regions located on the map's eastern side is the city of Minneapolis, MN.)

4.(a) Suppose you recorded elevation and average daily temperature on a particular day for a sample of locations in a region. If you were given the elevation at a new location, how would you make a plausible estimate of the average daily temperature for that location?

 (b) Why might you expect spatial association between selling prices of single-family homes in this region to be weaker than that between the observed temperature measurements?

5. For what sorts of point-referenced spatial data would you expect measurements across time to be essentially independent? For what sorts of point-referenced data would you expect measurements across time to be strongly dependent?

6. For point-referenced data, suppose the means of the variables are spatially associated. Would you expect the association between the variables themselves to be weaker than, stronger than, or the same as the association between the means?

7.(a) Write an S-plus or R function that will compute the distance between 2 points P_1 and P_2 on the surface of the earth. The function should take the latitude and longitude of the P_i as input, and output the geodesic distance D given in (1.10). Use $R = 6371$ km.

 (b) Use your program to obtain the geodesic distance between Chicago (87.63W, 41.88N) and Minneapolis (93.22W, 44.89N), and between New York (73.97W, 40.78N) and New Orleans (90.25W, 29.98N).

8. A "naive Euclidean" distance may be computed between two points by simply applying the Euclidean distance formula to the longitude-latitude coordinates, and then multiplying by $(R\pi/180)$ to convert to kilometers. Find the naive Euclidean distance between Chicago and Minneapolis, and between New York and New Orleans, comparing your results to the geodesic ones in the previous problem.

9. The *chordal* ("burrowing through the earth") distance separating two points is given by the Euclidean distance applied to the cartesian spherical coordinate system given in Subsection 1.2.2. Find the chordal distance between Chicago and Minneapolis, and between New York and New Orleans, comparing your results to the geodesic and naive Euclidean ones above.

10. A two-dimensional projection, often used to approximate geodesic distances by applying Euclidean metrics, sets up rectangular axes along the centroid of the observed locations, and scales the points according to these axes. Thus, with N locations having geographical coordinates $(\lambda_i, \theta_i)_{i=1}^N$, we first compute the centroid $(\bar{\lambda}, \bar{\theta})$ (the mean longitude and latitude). Next, two distances are computed. The first, d_X, is the geodesic distance (computed using (1.10)) between $(\bar{\lambda}, \theta_{\min})$ and $(\bar{\lambda}, \theta_{\max})$, where $\theta_{\min}$ and $\theta_{\max}$ are the minimum and maximum of the

observed latitudes. Analogously, d_Y is the geodesic distance computed between $(\lambda_{\min}, \bar{\theta})$ and $(\lambda_{\max}, \bar{\theta})$. These actually scale the axes in terms of true geodesic distances. The projection is then given by

$$x = \frac{\lambda - \bar{\lambda}}{\lambda_{\max} - \lambda_{\min}} d_X; \text{ and } y = \frac{\theta - \bar{\theta}}{\theta_{\max} - \theta_{\min}} d_Y .$$

Applying the Euclidean metric to the projected coordinates yields a good approximation to the intersite geodesic distances. This projection is useful for entering coordinates in spatial statistics software packages that require two-dimensional coordinate input and uses Euclidean metrics to compute distances (e.g., the variogram functions in S+SpatialStats, the spatial.exp function in WinBUGS, etc.).

(a) Compute the above projection for Chicago and Minneapolis ($N = 2$) and find the Euclidean distance between the projected coordinates. Compare with the geodesic distance. Repeat this exercise for New York and New Orleans.

(b) When will the above projection fail to work?

## CHAPTER 2

Basics of point-referenced data models

In this chapter we present the essential elements of spatial models and classical analysis for point-referenced data. As mentioned in Chapter 1, the fundamental concept underlying the theory is a stochastic process $\{Y(\mathbf{s}) : \mathbf{s} \in D\}$, where D is a fixed subset of r-dimensional Euclidean space. Note that such stochastic processes have a rich presence in the time series literature, where $r = 1$. In the spatial context, usually we encounter r to be 2 (say, northings and eastings) or 3 (e.g., northings, eastings, and altitude above sea level). For situations where $r > 1$, the process is often referred to as a *spatial process*. For example, $Y(\mathbf{s})$ may represent the level of a pollutant at site $\mathbf{s}$. While it is conceptually sensible to assume the existence of a pollutant level at all possible sites in the domain, in practice the data will be a partial realization of that spatial process. That is, it will consist of measurements at a finite set of locations, say $\{\mathbf{s}_1, \ldots, \mathbf{s}_n\}$, where there are monitoring stations. The problem facing the statistician is inference about the spatial process $Y(\mathbf{s})$ and prediction at new locations, based upon this partial realization.

This chapter is organized as follows. We begin with a survey of the building blocks of point-level data modeling, including stationarity, isotropy, and variograms (and their fitting via traditional moment-matching methods). We then add the spatial (typically Gaussian) process modeling that enables likelihood (and Bayesian) inference in these settings. We also illustrate helpful exploratory data analysis tools, as well as more formal classical methods, especially kriging (point-level spatial prediction). We close with short tutorials in S+SpatialStats and geoR, two easy to use and widely available point-level spatial statistical analysis packages.

The material we cover in this chapter is traditionally known as *geostatistics*, and could easily fill many more pages than we devote to it here. While we prefer the more descriptive term "point-level spatial modeling," we will at times still use "geostatistics" for brevity and perhaps consistency when referencing the literature.

2.1 Elements of point-referenced modeling

2.1.1 Stationarity

For our discussion we assume that our spatial process has a mean, say $\mu(\mathbf{s}) = E(Y(\mathbf{s}))$, associated with it and that the variance of $Y(\mathbf{s})$ exists for all $\mathbf{s} \in D$. The process $Y(\mathbf{s})$ is said to be *Gaussian* if, for any $n \geq 1$ and any set of sites $\{\mathbf{s}_1, \ldots, \mathbf{s}_n\}$, $\mathbf{Y} = (Y(\mathbf{s}_1), \ldots, Y(\mathbf{s}_n))^T$ has a multivariate normal distribution. The process is said to be *strictly stationary* if, for any given $n \geq 1$, any set of n sites $\{\mathbf{s}_1, \ldots, \mathbf{s}_n\}$ and any $\mathbf{h} \in \Re^r$, the distribution of $(Y(\mathbf{s}_1), \ldots, Y(\mathbf{s}_n))$ is the same as that of $(Y(\mathbf{s}_1 + \mathbf{h}), \ldots, Y(\mathbf{s}_n + \mathbf{h}))$. Here D is envisioned as $\Re^r$ as well.

A less restrictive condition is given by *weak stationarity* (also called second-order stationarity). Cressie (1993, p. 53) defines a spatial process to be weakly stationary if $\mu(\mathbf{s}) \equiv \mu$ (i.e., the process has a constant mean) and $Cov(Y(\mathbf{s}), Y(\mathbf{s} + \mathbf{h})) = C(\mathbf{h})$ for all $\mathbf{h} \in \Re^r$ such that $\mathbf{s}$ and $\mathbf{s} + \mathbf{h}$ both lie within D. (We note that, strictly speaking, for stationarity as a second-order property we will need only the second property; $E(Y(\mathbf{s}))$ need not equal $E(Y(\mathbf{s}+\mathbf{h}))$. But since we will apply the definition only to a mean 0 spatial residual term, this distinction is unimportant for us.) Weak stationarity implies that the covariance relationship between the values of the process at any two locations can be summarized by a covariance function $C(\mathbf{h})$, and this function depends only on the separation vector $\mathbf{h}$. Note that with all variances assumed to exist, strong stationarity implies weak stationarity. The converse is not true in general, but it *does* hold for Gaussian processes; see Exercise 2.

2.1.2 Variograms

There is a third type of stationarity called *intrinsic* stationarity. Here we assume $E[Y(\mathbf{s} + \mathbf{h}) - Y(\mathbf{s})] = 0$ and define

$$E[Y(\mathbf{s} + \mathbf{h}) - Y(\mathbf{s})]^2 = Var(Y(\mathbf{s} + \mathbf{h}) - Y(\mathbf{s})) = 2\gamma(\mathbf{h}) . \qquad (2.1)$$

Equation (2.1) makes sense only if the left-hand side depends *only* on $\mathbf{h}$ (so that the right-hand side can be written at all), and not the particular choice of $\mathbf{s}$. If this is the case, we say the process is *intrinsically stationary*. The function $2\gamma(\mathbf{h})$ is then called the *variogram*, and $\gamma(\mathbf{h})$ is called the *semivariogram*. (The covariance function $C(\mathbf{h})$ is sometimes referred to as the *covariogram*, especially when plotted graphically.) Note that intrinsic stationarity defines only the first and second moments of the differences $Y(\mathbf{s} + \mathbf{h}) - Y(\mathbf{s})$. It says nothing about the joint distribution of a collection of variables $Y(\mathbf{s}_1), \ldots, Y(\mathbf{s}_n)$, and thus provides no likelihood.

It is easy to see the relationship between the variogram and the covariance function:

$$
\begin{aligned}
2\gamma(\mathbf{h}) &= Var\left(Y\left(\mathbf{s}+\mathbf{h}\right)-Y\left(\mathbf{s}\right)\right) \\
&= Var(Y(\mathbf{s}+\mathbf{h}))+Var(Y(\mathbf{s}))-2Cov(Y(\mathbf{s}+\mathbf{h}),Y(\mathbf{s})) \\
&= C(\mathbf{0})+C(\mathbf{0})-2C\left(\mathbf{h}\right) \\
&= 2\left[C\left(\mathbf{0}\right)-C\left(\mathbf{h}\right)\right] .
\end{aligned}
$$

Thus,

$$
\gamma\left(\mathbf{h}\right)=C\left(\mathbf{0}\right)-C\left(\mathbf{h}\right) . \tag{2.2}
$$

From (2.2) we see that given C, we are able to recover γ easily. But what about the converse; in general, can we recover C from γ? Here it turns out we need to assume a bit more: if the spatial process is *ergodic*, then $C\left(\mathbf{h}\right) \to 0$ as $||\mathbf{h}|| \to \infty$, where $||\mathbf{h}||$ denotes the length of the $\mathbf{h}$ vector. This is an intuitively sensible condition, since it means that the covariance between the values at two points vanishes as the points become further separated in space. But taking the limit of both sides of (2.2) as $||\mathbf{h}|| \to \infty$, we then have that $lim_{||\mathbf{h}|| \to \infty}\gamma\left(\mathbf{h}\right)=C\left(\mathbf{0}\right)$. Thus, using the dummy variable $\mathbf{u}$ to avoid confusion, we have

$$
C\left(\mathbf{h}\right)=C(\mathbf{0})-\gamma(\mathbf{h})=lim_{||\mathbf{u}|| \to \infty}\gamma\left(\mathbf{u}\right)-\gamma\left(\mathbf{h}\right) . \tag{2.3}
$$

In general, the limit on the right-hand side need not exist, but if it does, then the process is weakly (second-order) stationary with $C\left(\mathbf{h}\right)$ as given in (2.3). We then have a way to determine the covariance function C from the semivariogram γ. Thus weak stationarity implies intrinsic stationarity, but the converse is not true; indeed, the next section offers examples of processes that are intrinsically stationary but not weakly stationary.

A valid variogram necessarily satisfies a negative definiteness condition. In fact, for any set of locations $\mathbf{s}_1,\ldots,\mathbf{s}_n$ and any set of constants $a_1,\ldots,a_n$ such that $\sum_i a_i = 0$, if $\gamma(\mathbf{h})$ is valid, then

$$
\sum_i \sum_j a_i a_j \gamma(\mathbf{s}_i - \mathbf{s}_j) \le 0 . \tag{2.4}
$$

To see this, note that

$$
\begin{aligned}
\sum_i \sum_j a_i a_j \gamma(\mathbf{s}_i - \mathbf{s}_j) &= \frac{1}{2}E\sum_i \sum_j a_i a_j (Y(\mathbf{s}_i)-Y(\mathbf{s}_j))^2 \\
&= -E\sum_i \sum_j a_i a_j Y(\mathbf{s}_i)Y(\mathbf{s}_j) \\
&= -E\left[\sum_i a_i Y(\mathbf{s}_i)\right]^2 \le 0 .
\end{aligned}
$$

Note that, despite the suggestion of expression (2.2), there is no relationship between this result and the positive definiteness condition for

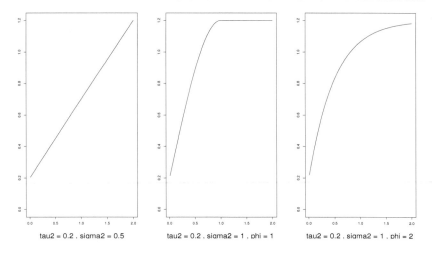

tau2 = 0.2 . sigma2 = 0.5   tau2 = 0.2 . sigma2 = 1 . phi = 1   tau2 = 0.2 . sigma2 = 1 . phi = 2

Figure 2.1 *Theoretical semivariograms for three models: (a) linear, (b) spherical, and (c) exponential.*

covariance functions (see Subsection 2.2.2). Cressie (1993) discusses further necessary conditions for a valid variogram. Lastly, the condition (2.4) emerges naturally in ordinary kriging (see Section 2.4).

2.1.3 Isotropy

Another important related concept is that of isotropy (as mentioned in Subsection 1.1.1). If the semivariogram function $\gamma(\mathbf{h})$ depends upon the separation vector only through its length $||\mathbf{h}||$, then we say that the process is *isotropic*; if not, we say it is *anisotropic*. Thus for an isotropic process, $\gamma(\mathbf{h})$ is a real-valued function of a univariate argument, and can be written as $\gamma(||\mathbf{h}||)$. If the process is intrinsically stationary and isotropic, it is also called *homogeneous*.

Isotropic processes are popular because of their simplicity, interpretability, and, in particular, because a number of relatively simple parametric forms are available as candidates for the semivariogram. Denoting $||\mathbf{h}||$ by t for notational simplicity, we now consider a few of the more important such forms.

    1. *Linear:*

$$\gamma(t) = \begin{cases} \tau^2 + \sigma^2 t & \text{if } t > 0, \ \tau^2 > 0, \ \sigma^2 > 0 \\ 0 & \text{otherwise} \end{cases}.$$

Note that $\gamma(t) \to \infty$ as $t \to \infty$, and so this semivariogram does not correspond to a weakly stationary process (although it is intrinsically stationary). This semivariogram is plotted in Figure 2.1(a) using the parameter values $\tau^2 = 0.2$ and $\sigma^2 = 0.5$.

2. *Spherical:*

$$\gamma(t) = \begin{cases} \tau^2 + \sigma^2 & \text{if } t \geq 1/\phi, \\ \tau^2 + \sigma^2 \left\{ \frac{3\phi t}{2} - \frac{1}{2} (\phi t)^3 \right\} & \text{if } 0 < t \leq 1/\phi, \\ 0 & \text{otherwise} \end{cases} .$$

The spherical semivariogram is valid in $r = 1, 2$, or 3 dimensions, but for $r \geq 4$ it fails to correspond to a spatial variance matrix that is positive definite (as required to specify a valid joint probability distribution). The spherical form does give rise to a stationary process and so the corresponding covariance function is easily computed (see the exercises that follow).

This variogram owes its popularity largely to the fact that it offers clear illustrations of the *nugget*, *sill*, and *range*, three characteristics traditionally associated with variograms. Specifically, consider Figure 2.1(b), which plots the spherical semivariogram using the parameter values $\tau^2 = 0.2$, $\sigma^2 = 1$, and $\phi = 1$. While $\gamma(0) = 0$ by definition, $\gamma(0^+) \equiv lim_{t \to 0^+} \gamma(t) = \tau^2$; this quantity is the *nugget*. Next, $lim_{t \to \infty} \gamma(t) = \tau^2 + \sigma^2$; this asymptotic value of the semivariogram is called the *sill*. (The sill minus the nugget, which is simply σ^2 in this case, is called the *partial sill*.) Finally, the value $t = 1/\phi$ at which $\gamma(t)$ first reaches its ultimate level (the sill) is called the *range*. It is for this reason that many of the variogram models of this subsection are often parametrized through $R \equiv 1/\phi$. Confusingly, both R and ϕ are sometimes referred to as the *range* parameter, although ϕ is often more accurately referred to as the *decay* parameter.

Note that for the linear semivariogram, the nugget is τ^2 but the sill and range are both infinite. For other variograms (such as the next one we consider), the sill is finite, but only reached asymptotically.

3. *Exponential:*

$$\gamma(t) = \begin{cases} \tau^2 + \sigma^2 \left(1 - \exp\left(-\phi t \right) \right) & \text{if } t > 0, \\ 0 & \text{otherwise} \end{cases} .$$

The exponential has an advantage over the spherical in that it is simpler in functional form while still being a valid variogram in all dimensions (and without the spherical's finite range requirement). However, note from Figure 2.1(c), which plots this semivariogram assuming $\tau^2 = 0.2$, $\sigma^2 = 1$, and $\phi = 2$, that the sill is only reached asymptotically, meaning that strictly speaking, the range $R = 1/\phi$ is infinite. In cases like this, the notion of an *effective range* is often used, i.e., the distance at which there is essentially no lingering spatial correlation. To make this notion precise, we must convert from γ scale to C scale (possible here since $lim_{t \to \infty} \gamma(t)$ exists; the exponential is not only intrinsically but also weakly stationary).

From (2.3) we have

$$
\begin{aligned}
C(t) &= lim_{u \to \infty} \gamma(u) - \gamma(t) \\
&= \tau^2 + \sigma^2 - \left[\tau^2 + \sigma^2 (1 - \exp(-\phi t)) \right] \\
&= \sigma^2 \exp(-\phi t) \ .
\end{aligned}
$$

Hence

$$
C(t) = \left\{ \begin{array}{ll} \tau^2 + \sigma^2 & \text{if } t = 0 \\ \sigma^2 \exp(-\phi t) & \text{if } t > 0 \end{array} \right. \ . \tag{2.5}
$$

If the nugget $\tau^2 = 0$, then this expression reveals that the correlation between two points t units apart is $\exp(-\phi t)$; note that $\exp(-\phi t) = 1^-$ for $t = 0^+$ and $\exp(-\phi t) = 0$ for $t = \infty$, both in concert with this interpretation.

A common definition of the *effective range*, t_0, is the distance at which this correlation has dropped to only 0.05. Setting $\exp(-\phi t_0)$ equal to this value we obtain $t_0 \approx 3/\phi$, since $\log(0.05) \approx -3$. The range will be discussed in more detail in Subsection 2.2.2.

Finally, the form of (2.5) gives a clear example of why the nugget (τ^2 in this case) is often viewed as a "nonspatial effect variance," and the partial sill (σ^2) is viewed as a "spatial effect variance." Along with ϕ, a statistician would likely view fitting this model to a spatial data set as an exercise in estimating these three parameters. We shall return to variogram model fitting in Subsection 2.1.4.

4. *Gaussian:*

$$
\gamma(t) = \left\{ \begin{array}{ll} \tau^2 + \sigma^2 \left(1 - \exp\left(-\phi^2 t^2 \right) \right) & \text{if } t > 0 \\ 0 & \text{otherwise} \end{array} \right. \ . \tag{2.6}
$$

The Gaussian variogram is an analytic function and yields very smooth realizations of the spatial process. We shall say much more about process smoothness in Subsection 2.2.3.

5. *Powered exponential:*

$$
\gamma(t) = \left\{ \begin{array}{ll} \tau^2 + \sigma^2 \left(1 - \exp\left(-|\phi t|^p \right) \right) & \text{if } t > 0 \\ 0 & \text{otherwise} \end{array} \right. \ . \tag{2.7}
$$

Here $0 < p \le 2$ yields a family of valid variograms. Note that both the Gaussian and the exponential forms are special cases of this one.

6. *Rational quadratic:*

$$
\gamma(t) = \left\{ \begin{array}{ll} \tau^2 + \frac{\sigma^2 t^2}{(\phi + t^2)} & \text{if } t > 0 \\ 0 & \text{otherwise} \end{array} \right. \ .
$$

7. *Wave:*

$$
\gamma(t) = \left\{ \begin{array}{ll} \tau^2 + \sigma^2 \left(1 - \frac{\sin(\phi t)}{\phi t} \right) & \text{if } t > 0 \\ 0 & \text{otherwise} \end{array} \right. \ .
$$

Model	Covariance function, $C(t)$
Linear	$C(t)$ does not exist
Spherical	$C(t) = \begin{cases} 0 & \text{if } t \geq 1/\phi \\ \sigma^2 \left[1 - \frac{3}{2}\phi t + \frac{1}{2}(\phi t)^3\right] & \text{if } 0 < t \leq 1/\phi \\ \tau^2 + \sigma^2 & \text{otherwise} \end{cases}$
Exponential	$C(t) = \begin{cases} \sigma^2 \exp(-\phi t) & \text{if } t > 0 \\ \tau^2 + \sigma^2 & \text{otherwise} \end{cases}$
Powered exponential	$C(t) = \begin{cases} \sigma^2 \exp(-\|\phi t\|^p) & \text{if } t > 0 \\ \tau^2 + \sigma^2 & \text{otherwise} \end{cases}$
Gaussian	$C(t) = \begin{cases} \sigma^2 \exp(-\phi^2 t^2) & \text{if } t > 0 \\ \tau^2 + \sigma^2 & \text{otherwise} \end{cases}$
Rational quadratic	$C(t) = \begin{cases} \sigma^2 \left(1 - \frac{t^2}{(\phi + t^2)}\right) & \text{if } t > 0 \\ \tau^2 + \sigma^2 & \text{otherwise} \end{cases}$
Wave	$C(t) = \begin{cases} \sigma^2 \frac{\sin(\phi t)}{\phi t} & \text{if } t > 0 \\ \tau^2 + \sigma^2 & \text{otherwise} \end{cases}$
Power law	$C(t)$ does not exist
Matérn	$C(t) = \begin{cases} \frac{\sigma^2}{2^{\nu-1}\Gamma(\nu)} \left(2\sqrt{\nu}t\phi\right)^\nu K_\nu(2\sqrt{\nu}t\phi) & \text{if } t > 0 \\ \tau^2 + \sigma^2 & \text{otherwise} \end{cases}$
Matérn at $\nu = 3/2$	$C(t) = \begin{cases} \sigma^2 \left(1 + \phi t\right) \exp\left(-\phi t\right) & \text{if } t > 0 \\ \tau^2 + \sigma^2 & \text{otherwise} \end{cases}$

Table 2.1 *Summary of covariance functions (covariograms) for common parametric isotropic models.*

Note this is an example of a variogram that is not monotonically increasing. The associated covariance function is $C(t) = \sigma^2 \sin(\phi t)/(\phi t)$. Bessel functions of the first kind include the wave covariance function and are discussed in detail in Subsections 2.2.2 and 5.1.3.

8. *Power law*

$$\gamma(t) = \begin{cases} \tau^2 + \sigma^2 t^\lambda & \text{of } t > 0 \\ 0 & \text{otherwise} \end{cases} .$$

This generalizes the linear case and produces valid intrinsic (albeit not weakly) stationary semivariograms provided $0 \leq \lambda < 2$.

9. *Matérn :* The variogram for the Matérn class is given by

$$\gamma(t) = \begin{cases} \tau^2 + \sigma^2 \left[1 - \frac{(2\sqrt{\nu}t\phi)^\nu}{2^{\nu-1}\Gamma(\nu)} K_\nu(2\sqrt{\nu}t\phi)\right] & \text{if } t > 0 \\ \tau^2 & \text{otherwise} \end{cases} . \qquad (2.8)$$

This class was originally suggested by Matérn (1960, 1986). Interest in it was revived by Handcock and Stein (1993) and Handcock and Wallis (1994),

model	Variogram, $\gamma(t)$
Linear	$\gamma(t) = \begin{cases} \tau^2 + \sigma^2 t & \text{if } t > 0 \\ 0 & \text{otherwise} \end{cases}$
Spherical	$\gamma(t) = \begin{cases} \tau^2 + \sigma^2 & \text{if } t \geq 1/\phi \\ \tau^2 + \sigma^2 \left[\frac{3}{2}\phi t - \frac{1}{2}(\phi t)^3\right] & \text{if } 0 < t \leq 1/\phi \\ 0 & \text{otherwise} \end{cases}$
Exponential	$\gamma(t) = \begin{cases} \tau^2 + \sigma^2(1 - \exp(-\phi t)) & \text{if } t > 0 \\ 0 & \text{otherwise} \end{cases}$
Powered exponential	$\gamma(t) = \begin{cases} \tau^2 + \sigma^2(1 - \exp(-|\phi t|^p)) & \text{if } t > 0 \\ 0 & \text{otherwise} \end{cases}$
Gaussian	$\gamma(t) = \begin{cases} \tau^2 + \sigma^2(1 - \exp(-\phi^2 t^2)) & \text{if } t > 0 \\ 0 & \text{otherwise} \end{cases}$
Rational quadratic	$\gamma(t) = \begin{cases} \tau^2 + \frac{\sigma^2 t^2}{(\phi + t^2)} & \text{if } t > 0 \\ 0 & \text{otherwise} \end{cases}$
Wave	$\gamma(t) = \begin{cases} \tau^2 + \sigma^2(1 - \frac{\sin(\phi t)}{\phi t}) & \text{if } t > 0 \\ 0 & \text{otherwise} \end{cases}$
Power law	$\gamma(t) = \begin{cases} \tau^2 + \sigma^2 t^\lambda & \text{if } t > 0 \\ 0 & \text{otherwise} \end{cases}$
Matérn	$\gamma(t) = \begin{cases} \tau^2 + \sigma^2 \left[1 - \frac{(2\sqrt{\nu}t\phi)^\nu}{2^{\nu-1}\Gamma(\nu)} K_\nu(2\sqrt{\nu}t\phi)\right] & \text{if } t > 0 \\ 0 & \text{otherwise} \end{cases}$
Matérn at $\nu = 3/2$	$\gamma(t) = \begin{cases} \tau^2 + \sigma^2 \left[1 - (1 + \phi t)\exp(-\phi t)\right] & \text{if } t > 0 \\ 0 & \text{otherwise} \end{cases}$

Table 2.2 *Summary of variograms for common parametric isotropic models.*

who demonstrated attractive interpretations for ν as well as ϕ. Here $\nu > 0$ is a parameter controlling the smoothness of the realized random field (see Subsection 2.2.3) while ϕ is a spatial scale parameter. The function $\Gamma(\cdot)$ is the usual gamma function while K_ν is the modified Bessel function of order ν (see, e.g., Abramowitz and Stegun, 1965, Chapter 9). Implementations of this function are available in several C/C++ libraries and also in the R package geoR. Note that special cases of the above are the exponential ($\nu = 1/2$) and the Gaussian ($\nu \to \infty$). At $\nu = 3/2$ we obtain a closed form as well, namely $\gamma(t) = \tau^2 + \sigma^2 \left[1 - (1 + \phi t)\exp(-\phi t)\right]$ for $t > 0$, and τ^2 otherwise.

The covariance functions and variograms we have described in this subsection are conveniently summarized in Tables 2.1 and 2.2, respectively.

2.1.4 Variogram model fitting

Having seen a fairly large selection of models for the variogram, one might well wonder how we choose one of them for a given data set, or whether the data can really distinguish them (see Subsection 5.1.3 in this latter regard). Historically, a variogram model is chosen by plotting the *empirical semivariogram* (Matheron, 1963), a simple nonparametric estimate of the semivariogram, and then comparing it to the various theoretical shapes available from the choices in the previous subsection. The customary empirical semivariogram is

$$\hat{\gamma}(t) = \frac{1}{2N(t)} \sum_{(\mathbf{s}_i, \mathbf{s}_j) \in N(t)} [Y(\mathbf{s}_i) - Y(\mathbf{s}_j)]^2 , \qquad (2.9)$$

where $N(t)$ is the set of pairs of points such that $||\mathbf{s}_i - \mathbf{s}_j|| = t$, and $|N(t)|$ is the number of pairs in this set. Notice that, unless the observations fall on a regular grid, the distances between the pairs will all be different, so this will not be a useful estimate as it stands. Instead we would "grid up" the t-space into intervals $I_1 = (0, t_1), I_2 = (t_1, t_2)$, and so forth, up to $I_K = (t_{K-1}, t_K)$ for some (possibly regular) grid $0 < t_1 < \cdots < t_K$. Representing the t values in each interval by its midpoint, we then alter our definition of $N(t)$ to

$$N(t_k) = \{(\mathbf{s}_i, \mathbf{s}_j) : ||\mathbf{s}_i - \mathbf{s}_j|| \in I_k\} , \ k = 1, \ldots, K .$$

Selection of an appropriate number of intervals K and location of the upper endpoint t_K is reminiscent of similar issues in histogram construction. Journel and Huijbregts (1979) recommend bins wide enough to capture at least 30 pairs per bin.

Clearly (2.9) is nothing but a method of moments (MOM) estimate, the semivariogram analogue of the usual sample variance estimate s^2. While very natural, there is reason to doubt that this is the best estimate of the semivariogram. Certainly it will be sensitive to outliers, and the sample average of the squared differences may be rather badly behaved since under a Gaussian distributional assumption for the $Y(\mathbf{s}_i)$, the squared differences will have a distribution that is a scale multiple of the heavily skewed χ_1^2 distribution. In this regard, Cressie and Hawkins (1980) proposed a robustified estimate that uses sample averages of $|Y(\mathbf{s}_i) - Y(\mathbf{s}_j)|^{1/2}$; this estimate is available in several software packages (see Section 2.5.1 below). Perhaps more uncomfortable is that (2.9) uses data differences, rather than the data itself. Also of concern is the fact that the components of the sum in (2.9) will be dependent within and across bins, and that $N(t_k)$ will vary across bins.

In any case, an empirical semivariogram estimate can be plotted, viewed, and an appropriately shaped theoretical variogram model can be fit to this "data." Since any empirical estimate naturally carries with it a signifi-

cant amount of noise in addition to its signal, this fitting of a theoretical
model has traditionally been as much art as science: in any given real data
setting, any number of different models (exponential, Gaussian, spherical,
etc.) may seem equally appropriate. Indeed, fitting has historically been
done "by eye," or at best by using trial and error to choose values of
nugget, sill, and range parameters that provide a good match to the empirical semivariogram (where the "goodness" can be judged visually or by
using some least squares or similar criterion); again see Section 2.5.1. More
formally, we could treat this as a statistical estimation problem, and use
nonlinear maximization routines to find nugget, sill, and range parameters
that minimize some goodness-of-fit criterion.

If we also have a distributional model for the data, we could use maximum likelihood (or restricted maximum likelihood, REML) to obtain sensible parameter estimates; see, e.g., Smith (2001) for details in the case
of Gaussian data modeled with the various parametric variogram families
outlined in Subsection 2.1.3. In Chapter 4 and Chapter 5 we shall see that
the hierarchical Bayesian approach is broadly similar to this latter method,
although it will often be easier and more intuitive to work directly with
the covariance model $C(t)$, rather than changing to a partial likelihood in
order to introduce the semivariogram.

2.2 Spatial process models ⋆

2.2.1 Formal modeling theory for spatial processes

When we write the collection of random variables $\{Y(\mathbf{s}) : \mathbf{s} \in D\}$ for some
region of interest D or more generally $\{Y(\mathbf{s}) : \mathbf{s} \in \Re^r\}$, it is evident that
we are envisioning a stochastic process indexed by $\mathbf{s}$. To capture spatial
association it is also evident that these variables will be pairwise dependent
with strength of dependence that is specified by their locations.

So, in fact, we have to determine the joint distribution for an uncountable number of random variables. In fact, we do this through specification
of arbitrary finite dimensional distributions, i.e., for an arbitrary number
of and choice of locations. Consistency of such specifications in terms of
ensuring a unique joint distribution will rarely hold and will be difficult to
establish. We avoid such technical concerns here by confining ourselves to
Gaussian processes or to mixtures of such processes. In this case, all that
is required is a valid correlation function, as we discuss below.

Again, to clarify the inference setting, in practice we will only observe
$Y(\mathbf{s})$ at a finite set of locations, $\mathbf{s}_1, \mathbf{s}_2, \ldots, \mathbf{s}_n$. Based upon $\{Y(\mathbf{s}_i), i = 1, \ldots, n\}$, we seek to infer about the mean, variability, and association
structure of the process. We also seek to predict $Y(\mathbf{s})$ at arbitrary un-
observed locations. Since our focus is on hierarchical modeling, often the
spatial process is introduced through random effects at the second stage of

the modeling specification. In this case, we still have the same inferential questions but now the process is never actually observed. It is latent and the data, modeled at the first stage, helps us to learn about the process.

In this sense, we can make intuitive connections with familiar dynamic models (e.g., West and Harrison, 1997) where there is a latent state space model that is temporally updated. In fact, this reminds us of a critical difference between the one-dimensional time domain and the two-dimensional spatial domain: we have full order in the former, but only partial order in two or more dimensions.

The implications of this remark are substantial. Large sample analysis for time series usually lets time go to ∞. Asymptotics envision an increasing time domain. By contrast, large sample analysis for spatial process data usually envisions a fixed region with more and more points filling in this domain (so-called infill asymptotics). When applying increasing domain asymptotic results, we can assume that, as we collect more and more data, we can learn about temporal association at increasing distance in time. When applying infill asymptotic results for a fixed domain we can learn more and more about association as distance between points tends to 0. However, with a maximum distance fixed by the domain we cannot learn about association (in terms of consistent inference) at increasing distance. The former remark indicates that we may be able to do an increasingly better job with regard to spatial prediction at a given location. However, we need not be doing better in terms of inferring about other features of the process. See the work of Stein (1999a, 1999b) for a full technical discussion regarding such asymptotic results. Here, we view such concerns as providing encouragement for using a Bayesian framework for inference, since then we need not rely on any asymptotic theory for inference, but rather obtain exact inference given whatever data we have observed.

Before we turn to some technical discussion regarding covariance and correlation functions, we note that the above restriction to Gaussian processes enables several advantages. First, it allows very convenient distribution theory. Joint marginal and conditional distributions are all immediately obtained from standard theory once the mean and covariance structure have been specified. In fact, this is all we need to specify in order to determine all distributions. Also, as we shall see, in the context of hierarchical modeling, a Gaussian process assumption for spatial random effects introduced at the second stage of the model is very natural in the same way that independent random effects with variance components are customarily introduced in linear or generalized linear mixed models. From a technical point of view, as noted in Subsection 2.1.1, if we work with Gaussian processes and stationary models, strong stationarity is equivalent to weak stationarity. We will clarify these notions in the next subsection. Lastly, in most applications, it is difficult to criticize a Gaussian assumption. To argue this as simply as possible, in the absence of replication we have $\mathbf{Y} = (Y(s_1), \ldots, Y(s_n))$, a single realization from an n-dimensional distribution. With a sample size of one, how can we criticize *any* multivariate distributional specification (Gaussian or otherwise)?

Strictly speaking this last assertion is not quite true with a Gaussian process model. That is, the joint distribution is a multivariate normal with mean say $\mathbf{0}$, and a covariance matrix that is a parametric function of the parameters in the covariance function. When n is large enough, the effective sample size will also be large. We can obtain by linear transformation a set of approximately uncorrelated variables through which the adequacy of the normal assumption can be studied. We omit details.

2.2.2 Covariance functions and spectra

In order to specify a stationary process we must provide a valid covariance function. Here "valid" means that $c(\mathbf{h}) \equiv cov(Y(\mathbf{s}), Y(\mathbf{s} + \mathbf{h}))$ is such that for any finite set of sites $\mathbf{s}_1, \ldots, \mathbf{s}_n$ and for any $a_1, \ldots, a_n$,

$$Var\left[\sum_i a_i Y(s_i)\right] = \sum_{i,j} a_i a_j Cov(Y(\mathbf{s}_i), Y(\mathbf{s}_j)) = \sum_{i,j} a_i a_j c(\mathbf{s}_i - \mathbf{s}_j) \geq 0 ,$$

with strict inequality if not all the a_i are 0. That is, we need $c(\mathbf{h})$ to be a positive definite function.

Verifying the positive definiteness condition is evidently not routine. Fortunately, we have *Bochner's Theorem* (see, e.g., Gikhman and Skorokhod, 1974, p. 208), which provides a necessary and sufficient condition for $c(\mathbf{h})$ to be positive definite. This theorem is applicable for $\mathbf{h}$ in arbitrary r-dimensional Euclidean space, although our primary interest is in $r = 2$.

In general, for real-valued processes, Bochner's Theorem states that $c(\mathbf{h})$ is positive definite if and only if

$$c(\mathbf{h}) = \int \cos(\mathbf{w}^T \mathbf{h})\, G(d\mathbf{w}) , \tag{2.10}$$

where G is a bounded, positive, symmetric about 0 measure in $\Re^r$. Then $c(\mathbf{0}) = \int G d(\mathbf{w})$ becomes a normalizing constant, and $G(d\mathbf{w})/c(\mathbf{0})$ is referred to as the *spectral distribution* that induces $c(\mathbf{h})$. If $G(d\mathbf{w})$ has a density with respect to Lebesgue measure, i.e., $G(d\mathbf{w}) = g(\mathbf{w})d\mathbf{w}$, then $g(\mathbf{w})/c(\mathbf{0})$ is referred to as the *spectral density*. Evidently, (2.10) can be used to generate valid covariance functions; see (2.12) below. Of course, the behavioral implications associated with c arising from a given G will only be clear in special cases, and (2.10) will be integrable in closed form only in cases that are even more special.

Since $e^{i\mathbf{w}^T \mathbf{h}} = \cos(\mathbf{w}^T \mathbf{h}) + i \sin(\mathbf{w}^T \mathbf{h})$, we have $c(\mathbf{h}) = \int e^{i\mathbf{w}^T \mathbf{h}} G(d\mathbf{w})$. That is, the imaginary term disappears due to the symmetry of G around 0. In other words, $c(\mathbf{h})$ is a valid covariance function if and only if it is the characteristic function of an r-dimensional symmetric random variable (random variable with a symmetric distribution). We note that if G is not assumed to be symmetric about $\mathbf{0}$, $c(\mathbf{h}) = \int e^{i\mathbf{w}^T \mathbf{h}} G(d\mathbf{w})$ still provides a valid covariance function (i.e., positive definite) but now for a complex-valued random process on $\Re^r$.

The Fourier transform of $c(\mathbf{h})$ is

$$\widehat{c}(\mathbf{w}) = \int e^{-i\mathbf{w}^T \mathbf{h}}\, c(\mathbf{h})d\mathbf{h} \; . \tag{2.11}$$

Applying the inversion formula, $c(\mathbf{h}) = (2\pi)^{-r} \int e^{i\mathbf{w}^T \mathbf{h}}\widehat{c}(\mathbf{w})d\mathbf{w}$, we see that $(2\pi)^{-r}\widehat{c}(\mathbf{w})/c(0) = g(\mathbf{w})$, the spectral density. Explicit computation of (2.11) is usually not possible except in special cases. However, approximate calculation is available through the fast Fourier transform (FFT); see Appendix A, Section A.4. Expression (2.11) can be used to check whether a given $c(\mathbf{h})$ is valid: we simply compute $\widehat{c}(\mathbf{w})$ and check whether it is positive and integrable (so it is indeed a density up to normalization).

The one-to-one relationship between $c(\mathbf{h})$ and $g(\mathbf{w})$ enables examination of spatial processes in the spectral domain rather than in the observational domain. Computation of $g(\mathbf{w})$ can often be expedited through fast Fourier transforms; g can be estimated using the so-called *periodogram*. Likelihoods can be obtained approximately in the spectral domain enabling inference to be carried out in this domain. See, e.g., Guyon (1995) or Stein (1999a) for a full development. Likelihood evaluation is much faster in the spectral domain. However, in this book we confine ourselves to the observational domain because of concerns regarding the accuracy associated with approximation in the spectral domain (e.g., the likelihood of Whittle, 1954), and with the ad hoc creation of the periodogram (e.g., how many low frequencies are ignored). We do however note that the spectral domain may afford the best potential for handling computation associated with large data sets.

Isotropic covariance functions, i.e., $c(\|\mathbf{h}\|)$, where $\|\mathbf{h}\|$ denotes the length of $\mathbf{h}$, are the most frequently adopted choice within the stationary class. There are various direct methods for checking the permissibility of isotropic covariance and variogram specifications. See, e.g., Armstrong and Diamond (1984), Christakos (1984), and McBratney and Webster (1986). Again denoting $\|\mathbf{h}\|$ by t for notational simplicity, recall that Tables 2.1 and 2.2 provide the covariance function $C(t)$ and variogram $\gamma(t)$, respectively, for the widely encountered parametric istropic choices that were initially presented in Subsection 2.1.3.

It is noteworthy that an isotropic covariance function that is valid in dimension r need not be valid in dimension $r + 1$. The intuition may be gleaned by considering $r = 1$ versus $r = 2$. For three points, in one-dimensional space, given the distances separating points 1 and 2 (d_{12}) and points 2 and 3 (d_{23}), then the distance separating points 1 and 3 d_{13} is either $d_{12} + d_{23}$ or $|d_{12} - d_{23}|$. But in two-dimensional space, given d_{12} and d_{23}, d_{13} can take any value in $\Re^+$ (subject to triangle inequality). With increasing dimension more sets of interlocation distances are possible for a given number of locations; it will be more difficult for a function to satisfy the positive definiteness condition. Armstrong and Jabin (1981) provide an explicit example that we defer to Exercise 3.

There are isotropic correlation functions that are valid in all dimensions. The Gaussian correlation function, $k(\|h\|) = \exp(-\phi \|h\|^2)$ is an example. It is the characteristic function associated with r i.i.d. normal random variables, each with variance $1/(2\phi)$ for any r. More generally, the powered exponential, $\exp(-\phi \|h\|^\alpha)$, $0 < \alpha \le 2$ (and hence the exponential correlation function) is valid for any r.

Rather than seeking isotropic correlation functions that are valid in all dimensions, we might seek all valid isotropic correlation function in a particular dimension r. Matérn (1960, 1986) provides the general result. The set of $c(\|h\|)$ of the form

$$c(\|h\|) = \int_0^\infty \left(\frac{2}{w \|\mathbf{h}\|} \right)^\alpha \Gamma(\nu + 1) J_\nu(w \|\mathbf{h}\|) G(dw) , \qquad (2.12)$$

where G is nondecreasing and integrable on $\Re^+$, J_ν is the Bessel function of the first kind of order ν, and $\nu = (r - 2)/2$ provides all valid isotropic correlation functions on $\Re^r$.

When $r = 2$, $v = 0$ so that arbitrary correlation functions in two-dimensional space arise as scale mixtures of Bessel functions of order 0. In particular, $J_0(d) = \sum_{k=0}^\infty \frac{(-1)^k}{(k!)^2} \left(\frac{d}{2} \right)^{k/2}$. J_0 decreases from 1 at $d = 0$ and will oscillate above and below 0 with amplitudes and frequencies that are diminishing as d increases (see Figure 5.1 in Section 5.1). Typically, correlation functions that are monotonic and decreasing to 0 are chosen but, apparently, valid correlation functions can permit negative associations with w determining the scale in distance space. Such behavior might be appropriate in certain applications.

The form in (2.12) at $\nu = 0$ was exploited in Shapiro and Botha (1981) and Ver Hoef and Barry (1998) to develop "nonparametric" variogram models and "black box" kriging. It was employed in Ecker and Gelfand (1997) to obtain flexible spatial process models within which to do inference from a Bayesian perspective (see Subsection 5.1.3).

If we confine ourselves to strictly monotonic isotropic covariance functions then we can introduce the notion of a range. As described above, the range is conceptualized as the distance beyond which association becomes negligible. If the covariance function reaches 0 in a finite distance, then we refer to this distance as the range. However, as Table 2.1 reveals, we customarily work with covariance functions that attain 0 asymptotically as $\|\mathbf{h}\| \to \infty$. In this case, it is common to define the range as the distance beyond which correlation is less than .05, and this is the definition we employ in the sequel. So if ρ is the correlation function, then writing the range as R we solve $\rho(R; \boldsymbol{\theta}) = .05$, where $\boldsymbol{\theta}$ denotes the parameters in the correlation function. Therefore, R is an implicit function of the parameter $\boldsymbol{\theta}$.

We do note that some authors define the range through the variogram, i.e., the distance at which the variogram reaches .95 of its sill. That is, we would solve $\gamma(R) = .95(\sigma^2 + \tau^2)$. Note, however, that if we rewrite this equation in terms of the correlation function we obtain $\tau^2 + \sigma^2(1 - \rho(R; \boldsymbol{\theta})) = .95(\tau^2 + \sigma^2)$, so that $\rho(R; \boldsymbol{\theta}) = .05\left(\frac{\sigma^2 + \tau^2}{\sigma^2}\right)$. Evidently, the solution to this equation is quite different from the solution to the above equation. In fact, this latter equation may not be solvable, e.g., if $\sigma^2/(\sigma^2 + \tau^2) \leq .05$, the case of very weak "spatial story" in the model. As such, one might argue that a spatial model is inappropriate in this case. However, with σ^2 and τ^2 unknown, it seems safer to work with the former definition.

We note that one can offer constructive strategies to build larger classes of correlation functions. Three approaches are mixing, products, and convolution. Mixing notes simply that if $C_1, \ldots, C_m$ are valid correlation functions in $\Re^r$ and if $\sum_{i=1}^m p_i = 1$, $p_i > 0$, then $C(\mathbf{h}) = \sum_{i=1}^m p_i C_i(\mathbf{h})$ is also a valid correlation function in $\Re^r$. This follows since $C(\mathbf{h})$ is the characteristic function associated with $\sum p_i f_i(\mathbf{x})$, where $f_i(\mathbf{x})$ is the symmetric about 0 density in r-dimensional space associated with $C_i(\mathbf{h})$.

Using products simply notes that again if $c_1, \ldots, c_n$ are valid in $\Re^r$, then $\prod_{i=1}^m c_i$ is a valid correlation function in $\Re^r$. This follows since $\prod_{i=1}^m c_i(n)$ is the characteristic function associated with $V = \sum_{i=1}^m V_i$ where the V_i are independent with V_i having characteristic function $c_i(\mathbf{h})$.

Convolution simply recognizes that if c_1 and c_2 are valid correlation functions in $\Re^r$, then $c_{12}(\mathbf{h}) = \int c_1(\mathbf{h} - \mathbf{t})c_2(\mathbf{t})d\mathbf{t}$ is a valid correlation function in $\Re^r$. The argument here is to look at the Laplace transform of $c_{12}(\mathbf{h})$. That is,

$$\begin{aligned}
\widehat{c}_{12}(\mathbf{w}) &= \int e^{-i\mathbf{w}^T\mathbf{h}}c_{12}(\mathbf{h})d\mathbf{h} \\
&= \int e^{-i\mathbf{w}^T\mathbf{h}}\int c_1(\mathbf{h} - \mathbf{t})c_2(\mathbf{t})d\mathbf{t}d\mathbf{h} \\
&= \widehat{c}_1(\mathbf{w}) \cdot \widehat{c}_2(\mathbf{w}),
\end{aligned}$$

where $\widehat{c}_i(\mathbf{w})$ is the Laplace transform of $c_i(\mathbf{h})$ for $i = 1, 2$. But then $c_{12}(\mathbf{h}) = (2\pi)^{-2}\int e^{i\mathbf{w}^T\mathbf{h}}\widehat{c}_1(\mathbf{w})\widehat{c}_2(\mathbf{w})d\mathbf{w}$. Now $\widehat{c}_1(\mathbf{w})$ and $\widehat{c}_2(\mathbf{w})$ are both symmetric about $\mathbf{0}$ since, up to a constant, they are the spectral densities associated with $c_1(\mathbf{h})$ and $c_2(\mathbf{h})$, respectively. Hence, $c_{12}(\mathbf{h}) = \int \cos\mathbf{w}^T\mathbf{h}G(d\mathbf{w})$ where $G(d\mathbf{w}) = (2\pi)^{-2}\widehat{c}_1(\mathbf{w})^2\widehat{c}_2(\mathbf{w})d\mathbf{w}$.

Thus, from (2.10), $c_{12}(\mathbf{h})$ is a valid correlation function, i.e., G is a bounded, positive, symmetric about 0 measure on $\Re^2$. In fact, if c_1 and c_2 are isotropic then c_{12} is as well; we leave this verification as Exercise 5.

2.2.3 Smoothness of process realizations

How does one select among the various choices of correlation functions? Usual model selection criteria will typically find it difficult to distinguish, say, among one-parameter isotropic scale choices such as the exponential, Gaussian, or Cauchy. Ecker and Gelfand (1997) provide some graphical illustration showing that, through suitable alignment of parameters, the correlation curves will be very close to each other. Of course, in comparing choices with parametrizations of differing dimensions (e.g., correlation functions developed using results from the previous section), we will need to employ a selection criterion that penalizes complexity and rewards parsimony (see Section 4.2.3).

An alternative perspective is to make the selection based upon theoretical considerations. This possibility arises from the powerful fact that the choice of correlation function determines the smoothness of realizations from the spatial process. More precisely, a process realization is viewed as a random surface over the region. By choice of c we can ensure that these realizations will be almost surely continuous, or mean square continuous, or mean square differentiable, and so on. Of course, at best the process is only observed at finitely many locations. (At worst, it is never observed, e.g., when the spatial process is used to model random spatial effects.) So, it is not possible to "see" the smoothness of the process realization. Elegant theory, developed in Kent (1989), Stein (1999a) and extended in Banerjee and Gelfand (2003), clarifies the relationship between the choice of correlation function and such smoothness. We provide a bit of this theory below, with further discussion in Section 10.1. For now, the key point is that, according to the process being modeled, we may, for instance, anticipate surfaces not be continuous (as with digital elevation models in the presence of gorges, escarpments, or other topographic features), or to be differentiable (as in studying land value gradients or temperature gradients). We can choose a correlation function to essentially ensure such behavior.

Of particular interest in this regard is the Matérn class of covariance functions. The parameter v (see Table 2.1) is, in fact, a smoothness parameter. In two-dimensional space, the greatest integer in v indicates the number of times process realizations will be mean square differentiable. In particular, since $v = \infty$ corresponds to the Gaussian correlation function, the implication is that use of the Gaussian correlation function results in process realizations that are mean square analytic, which may be too smooth to be appropriate in practice. That is, it is possible to predict $Y(\mathbf{s})$ perfectly for all $\mathbf{s} \in \Re^2$ based upon observing $Y(\mathbf{s})$ in an arbitrarily small neighborhood. Expressed in a different way, use of the Matérn covariance function as a model enables the data to inform about v; we can learn about process smoothness despite observing the process at only a finite number of locations.

Hence, we follow Stein (1999a) in recommending the Matérn class as a general tool for building spatial models. The computation of this function requires evaluation of a modified Bessel function. In fact, evaluation will be done repeatedly to obtain a covariance matrix associated with n locations, and then iteratively if a model is fit via MCMC methods. This may appear off-putting but, in fact, such computation can be done efficiently using expansions to approximate $K_v(\cdot)$ (Abramowitz and Stegun, p. 435), or working through the inversion formula below (2.11), which in this case becomes

$$2 \left(\frac{\phi \|\mathbf{h}\|}{2} \right)^{\nu} \frac{K_v(\phi(\|\mathbf{h}\|))}{\phi^{2\nu}\Gamma(v + \frac{r}{2})} = \int_{\Re^r} e^{i\mathbf{w}^T\mathbf{h}}(\phi^2 + \|\mathbf{w}\|^2)^{-(v+r/2)}d\mathbf{w} \,, \quad (2.13)$$

where K_v is the modified Bessel function of order ν.

Computation of (2.13) is discussed further in Appendix Section A.4. In particular, the right side of (2.13) is readily approximated using fast Fourier transforms. Again, we revisit process smoothness in Section 10.1.

2.2.4 Directional derivative processes

The previous section offered discussion intended to clarify, for a spatial process, the connection between correlation function and smoothness of process realizations. When realizations are mean square differentiable, we can think about a directional derivative process. That is, for a given direction, at each location we can define a random variable that is the directional derivative of the original process at that location in the given direction. The entire collection of random variables can again be shown to be a spatial process. We offer brief development below but note that, intuitively, such variables would be created through limits of finite differences. In other words, we can also formalize a finite difference process in a given direction. The value of formalizing such processes lies in the possibility of assessing where, in a region of interest, there are sharp gradients and in which directions. They also enable us to work at different scales of resolution. Application could involve land-value gradients away from a central business district, temperature gradients in a north-south direction as mentioned above, or perhaps the maximum gradient at a location and the direction of that gradient, in order to identify zones of rapid change (boundary analysis). Some detail on the development of directional derivative processes appears in Subsection 10.1.2.

2.2.5 Anisotropy

Geometric anisotropy

Stationary correlation functions extend the class of correlation functions from isotropy where association only depends upon distance to association that depends upon the separation vector between locations. As a result, association depends upon direction. An illustrative example is the class of geometric anisotropic correlation functions where we set

$$c(\mathbf{s} - \mathbf{s}') = \sigma^2 \rho((\mathbf{s} - \mathbf{s}')^T B(\mathbf{s} - \mathbf{s}')) . \qquad (2.14)$$

In (2.14), B is positive definite with ρ a valid correlation function in $\Re^r$ (say, from Table 2.1). We would omit the range/decay parameter since it can be incorporated into B. When $r = 2$ we obtain a specification with three parameters rather than one. Contours of constant association arising from c in (2.14) are elliptical. In particular, the contour corresponding to $\rho = .05$ provides the range in each spatial direction. Ecker and Gelfand (1997) provide the details for Bayesian modeling and inference incorporating (2.14); see also Subsection 5.1.4.

Following the discussion in Subsection 2.2.2, we can extend geometric anisotropy to *product* geometric anisotropy. In the simplest case, we would set

$$c(\mathbf{s} - \mathbf{s}') = \sigma^2 \, \rho_1((\mathbf{s} - \mathbf{s}')^T B_1(\mathbf{s} - \mathbf{s}')) \, \rho_2((\mathbf{s} - \mathbf{s}')^T B_2(\mathbf{s} - \mathbf{s}')) ,$$

noting that c is valid since it arises as a product of valid covariance functions. See Ecker and Gelfand (2003) for further details and examples.

Other notions of anisotropy

In a more general discussion, Zimmerman (1993) suggests three different notions of anisotropy: *sill* anisotropy, *nugget* anisotropy, and *range* anisotropy. More precisely, working with a variogram $\gamma(\mathbf{h})$, let $\mathbf{h}$ be an arbitrary separation vector so that $\mathbf{h}/\|\mathbf{h}\|$ is a unit vector in $\mathbf{h}$'s direction. Consider $\gamma(c\mathbf{h}/\|\mathbf{h}\|)$. Let $c \to \infty$ and suppose $\lim_{c \to \infty} \gamma(c\mathbf{h}/\|\mathbf{h}\|)$ depends upon $\mathbf{h}$. This situation is naturally referred to as sill anisotropy. If we work with the usual relationship $\gamma(c\mathbf{h}/\|\mathbf{h}\|) = \tau^2 + \sigma^2 \left(1 - \rho\left(c\frac{\mathbf{h}}{\|\mathbf{h}\|}\right)\right)$, then, in some directions, ρ must not go to 0 as $c \to \infty$. If this can be the case, then ergodicity assumptions (i.e., convergence assumptions associated with averaging) will be violated. If this can be the case, then perhaps the constant mean assumption, implicit for the variogram, does not hold. Alternatively, it is also possible that the constant nugget assumption fails.

Instead, let $c \to 0$ and suppose $\lim_{c \to 0} \gamma(c\mathbf{h}/\|\mathbf{h}\|)$ depends upon $\mathbf{h}$. This situation is referred to as nugget anisotropy. Since, by definition, ρ must go to 1 as $c \to 0$, this says that the measurement errors that are assumed uncorrelated with common variance may be correlated. More generally, a simple white noise process model for the nonspatial errors is not appropriate.

A third type of anisotropy is range anisotropy where the range depends upon direction. Zimmerman (1993) asserts that "this is the form most often seen in practice." Geometric anisotropy and the more general product geometric anisotropy from the previous subsections are illustrative cases. However, given the various constructive strategies offered in Subsection 2.2.2 to create more general stationary covariance functions, we can envision nongeometric range anisotropy, implying general correlation function or variogram contours in $\Re^2$. However, due to the positive definiteness restriction on the correlation function, the extent of possible contour shapes is still rather limited.

Lastly, motivated by directional variograms (see Subsection 2.3.2), some authors propose the idea of nested models (see Zimmerman, 1993, and the references therein). That is, for each separation vector there is an associated angle with, say, the x-axis, which by symmetry considerations can be restricted to $[0, \pi)$. Partitioning this interval into a set of angle classes, a different variogram model is assumed to operate for each class. In terms of correlations, this would imply a different covariance function is operating for each angle class. But evidently this does not define a valid process model: the resulting covariance matrix for an arbitrary set of locations need not be positive definite.

This can be seen with as few as three points and two angle classes. Let $(\mathbf{s}_1, \mathbf{s}_2)$ belong to one angle class with $(\mathbf{s}_1, \mathbf{s}_3)$ and $(\mathbf{s}_2, \mathbf{s}_3)$ in the other. With exponential isotropic correlation functions in each class by choosing ϕ_1 and ϕ_2 appropriately we can make $\rho(\mathbf{s}_1 - \mathbf{s}_2) \approx 0$ while $\rho(\mathbf{s}_1 - \mathbf{s}_3) = \rho(\mathbf{s}_2 - \mathbf{s}_3) \approx 0.8$. A quick calculation shows that the resulting 3×3 covariance (correlation) matrix is not positive definite. So, in terms of being able to write proper joint distributions for the resulting data, nested models are inappropriate; they do not provide an extension of isotropy that allows for likelihood based inference.

2.3 Exploratory approaches for point-referenced data

2.3.1 Basic techniques

Exploratory data analysis (EDA) tools are routinely implemented in the process of analyzing one- and two-sample data sets, regression studies, generalized linear models, etc. (see, e.g., Chambers et al., 1983; Hoaglin, Mosteller, and Tukey, 1983, 1985; Aiktin et al., 1989). Similarly, such tools are appropriate for analyzing point-referenced spatial data.

For continuous data, the starting point is the so-called "first law of geostatistics." Figure 2.2 illustrates this "law" in a one-dimensional setting. The data is partitioned into a mean term and an error term. The mean corresponds to global (or *first-order*) behavior, while the error captures local (or *second-order*) behavior through a covariance function. EDA tools examine both first- and second-order behavior.

The law also clarifies that spatial association in the data, $Y(\mathbf{s})$, need not resemble spatial association in the residuals, $\epsilon(\mathbf{s})$. That is, spatial as-

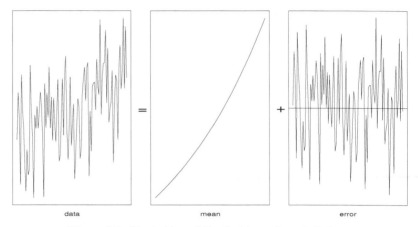

Figure 2.2 *Illustration of the first law of geostatistics.*

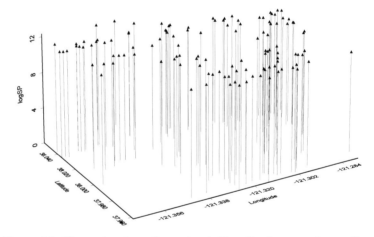

Figure 2.3 *Illustrative three-dimensional "drop line" scatterplot, scallop data.*

sociation in the $Y(\mathbf{s})$ corresponds to looking at $E(Y(\mathbf{s}) - \mu)(Y(\mathbf{s}') - \mu)$, while spatial structure in the $\epsilon(\mathbf{s})$ corresponds to looking at $E(Y(\mathbf{s}) - \mu(\mathbf{s}))(Y(\mathbf{s}') - \mu(\mathbf{s}'))$. The difference between the former and the latter is $(\mu - \mu(\mathbf{s}))(\mu - \mu(\mathbf{s}'))$, which need not be negligible.

Certainly an initial exploratory display should be a simple map of the locations themselves. We need to assess how *regular* the arrangement of the points is. Next, some authors would recommend a stem-and-leaf display of the $Y(\mathbf{s})$. This plot is evidently nonspatial and is customarily for observations which are i.i.d. We expect both nonconstant mean and spatial dependence, but such a plot may at least suggest potential outliers. Next we might develop a three-dimensional "drop line" scatterplot of $Y(\mathbf{s}_i)$ versus

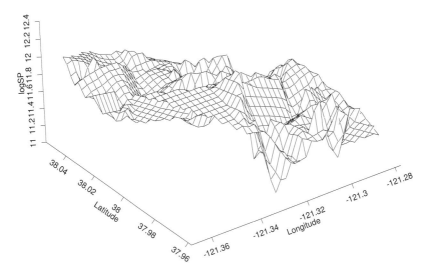

Figure 2.4 *Illustrative three-dimensional surface ("perspective") plot, Stockton real estate data.*

$\mathbf{s}_i$, which we could convert to a three-dimensional surface plot or perhaps a contour plot as a *smoothed* summary. Examples of these three plots are shown for a sample of 120 log-transformed home selling prices in Stockton, CA, in Figures 2.3, 2.4, and 2.5, respectively. However, as the preceding paragraph clarifies, such displays may be deceiving. They may show spatial pattern that will disappear after $\mu(\mathbf{s})$ is fitted, or perhaps vice versa. It seems more sensible to study spatial pattern in the residuals.

In exploring $\mu(\mathbf{s})$ we may have two types of information at location $\mathbf{s}$. One is the purely geographic information, i.e., the geocoded location expressed in latitude and longitude or as projected coordinates such as eastings and northings (Subsection 1.2.1 above). The other will be features relevant for explaining the $Y(\mathbf{s})$ at $\mathbf{s}$. For instance, if $Y(\mathbf{s})$ is a pollution concentration, then elevation, temperature, and wind information at $\mathbf{s}$ could well be useful and important. If instead $Y(\mathbf{s})$ is the selling price of a single-family home at $\mathbf{s}$, then characteristics of the home (square feet, age, number of bathrooms, etc.) would be useful.

When the mean is described purely through geographic information, $\mu(\mathbf{s})$ is referred to as a *trend surface*. When $\mathbf{s} \in \Re^2$, the surface is usually developed as a bivariate polynomial. For data that is roughly gridded (or can be assigned to row and column bins by overlaying a regular lattice on the

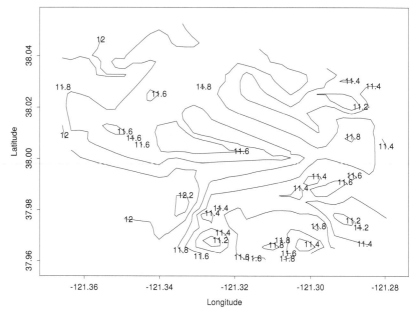

Figure 2.5 *Illustrative contour plot, Stockton real estate data.*

points), we can make row and column boxplots looking for trend. Plotting these boxplots versus their center could clarify the existence and nature of such trend. In fact, median polishing (see, e.g., Hoaglin, Mosteller, and Tukey, 1985) could be used to extract row and column effects, and also to see if a multiplicative trend surface term is useful; see Cressie (1983, pp. 46–48) in this regard.

Figures 2.6 and 2.7 illustrate the row and column boxplot approach for a data set previously considered by Diggle and Ribeiro (2002). The response variable is the surface elevation ("height") at 52 locations on a regular grid within a 310-foot square (and where the mesh of the grid is 50 feet). The plots reveals some evidence of spatial pattern as we move along the rows, but not along the columns of the regular grid.

To assess small-scale behavior, some authors recommend creating the *semivariogram cloud*, i.e., a plot of $(Y(\mathbf{s}_i) - Y(\mathbf{s}_j))^2$ versus $||\mathbf{s}_i - \mathbf{s}_j||$. Usually this cloud is too "noisy" to reveal very much; see, e.g., Figure 5.2. The empirical semivariogram (2.9) is preferable in terms of reducing some of the noise, and can be a helpful tool in seeing the presence of spatial structure. Again, the caveat above suggests employing it for residuals (not the data itself) unless a constant mean is appropriate.

An empirical (nonparametric) covariance estimate, analogous to (2.9), is

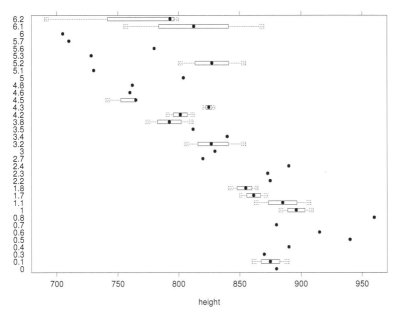

Figure 2.6 *Illustrative row box plots, Diggle and Ribeiro (2002) surface elevation data.*

also available. Creating bins as in this earlier approach, define

$$\widehat{c}(t_k) = \frac{1}{N_k} \sum_{(\mathbf{s}_i, \mathbf{s}_j) \in N(t_k)} (Y(\mathbf{s}_i) - \bar{Y})(Y(\mathbf{s}_j) - \bar{Y}) , \qquad (2.15)$$

where again $N(t_k) = \{(\mathbf{s}_i, \mathbf{s}_j) : ||\mathbf{s}_i - \mathbf{s}_j|| \in I_k\}$ for $k = 1, \ldots, K$, I_k indexes the kth bin, and there are N_k pairs of points falling in this bin. Equation (2.15) is a spatial generalization of a lagged autocorrelation in time series analysis. Since $\widehat{c}$ uses a common $\bar{Y}$ for all $Y(\mathbf{s}_i)$, it may be safer to employ (2.15) on the residuals. Two further issues arise: first, what should we define $\widehat{c}(0)$ to be, and second, regardless of this choice, the fact that $\widehat{\gamma}(t_k)$ does *not* equal $\widehat{c}(0) - \widehat{c}(t_k)$, $k = 1, \ldots, K$. Details for both of these issues are left to Exercise 6.

Again, with a regular grid or binning we can create "same-lag" scatterplots. These are plots of $Y(\mathbf{s}_i + h\mathbf{e})$ versus $Y(\mathbf{s}_i)$ for a fixed h and a fixed unit vector $\mathbf{e}$. Comparisons among such plots may reveal the presence of anisotropy and perhaps nonstationarity.

Lastly, suppose we attach a neighborhood to each point. We can then compute the sample mean and variance for the points in the neighborhood, and even a sample correlation coefficient using all pairs of data in the neighborhood. Plots of each of them versus location can be informative.

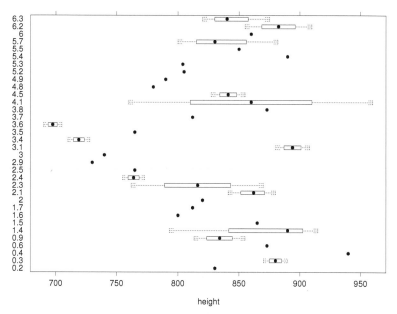

Figure 2.7 *Illustrative column box plots, Diggle and Ribeiro (2002) surface elevation data.*

The first may give some idea regarding how the mean structure changes across the study region. Plots of the second and third may provide evidence of nonstationarity. Implicit in extracting useful information from these plots is a roughly constant local mean. If $\mu(\mathbf{s})$ is to be a trend surface, this is plausible. But if $\mu(\mathbf{s})$ is a function of some geographic variables at $\mathbf{s}$ (say, home characteristics), then use of residuals would be preferable.

2.3.2 Assessing anisotropy

We illustrate various EDA techniques to assess anisotropy using sampling of scallop abundance on the continental shelf off the coastline of the northeastern U.S. The data from this survey, conducted by the Northeast Fisheries Science Center of the National Marine Fisheries Service, is available within the S+SpatialStats package; see Subsection 2.5.1. Figure 2.8 shows the sampling sites for 1990 and 1993.

Directional semivariograms and rose diagrams

The most common EDA technique for assessing anisotropy involves use of directional semivariograms. Typically, one chooses angle classes $\eta_i \pm \epsilon$, $i = 1, \ldots, L$ where ϵ is the halfwidth of the angle class and L is the

1990 Scallop Sites

1993 Scallop Sites

Figure 2.8 *Sites sampled in the Atlantic Ocean for 1990 and 1993 scallop catch data.*

number of angle classes. For example, a common choice of angle classes involves the four cardinal directions measured counterclockwise from the x-axis ($0°$, $45°$, $90°$, and $135°$) where ϵ is $22.5°$. Journel and Froidevaux (1982) display directional semivariograms at angles $35°$, $60°$, $125°$, and $150°$ in deducing anistropy for a tungsten deposit. While knowledge of the underlying spatial characteristics of region D is invaluable in choosing directions, often the choice of the number of angle classes and the directions seems to be arbitrary.

For a given angle class, the Matheron empirical semivariogram (2.9) can be used to provide a directional semivariogram for angle η_i. Theoretically, all types of anisotropy can be assessed from these directional semivariograms; however, in practice determining whether the sill, nugget, and/or range varies with direction can be difficult. Figure 2.9(a) illustrates directional semivariograms for the 1990 scallop data in the four cardinal directions. Note that the semivariogram points are connected only to aid

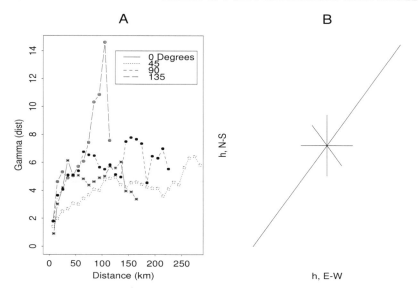

Figure 2.9 *Directional semivariograms (a) and a rose diagram (b) for the 1990 scallop data.*

comparison. Possible conclusions are: the variability in the 45° direction (parallel to the coastline) is significantly less than in the other three directions and the variability perpendicular to the coastline (135°) is very erratic, possibly exhibiting sill anisotropy. We caution however that it is dangerous to read too much significance and interpretation into directional variograms. No sample sizes (and thus no assessments of variability) are attached to these pictures. Directional variograms from data generated under a simple isotropic model will routinely exhibit differences of the magnitudes seen in Figure 2.9(a). Furthermore, it seems difficult to draw any conclusions regarding the presence of geometric anisotropy from this figure.

A rose diagram (Isaaks and Srivastava, 1989, pp. 151–154) can be created from the directional semivariograms to evaluate geometric anisotropy. At an arbitrarily selected γ^*, for a directional semivariogram at angle η, the distance d^* at which the directional semivariogram attains γ^* can be interpolated. Then, the rose diagram is a plot of angle η and corresponding distance d^* in polar coordinates. If an elliptical contour describes the extremities of the rose diagram reasonably well, then the process exhibits geometric anisotropy. For instance, the rose diagram for the 1990 scallop data is presented in Figure 2.9(b) using the γ^* contour of 4.5. It is approximately elliptical, oriented parallel to the coastline ($\approx 45°$) with a ratio of major to minor ellipse axes of about 4.

Empirical semivariogram contour (ESC) plots

A more informative method for assessing anisotropy is a contour plot of the empirical semivariogram surface in $\Re^2$. Such plots are mentioned informally in Isaaks and Srivastava (1989, pp. 149–151) and in Haining (1990, pp. 284–286); the former call them contour maps of the grouped variogram values, the latter an isarithmic plot of the semivariogram. Following Ecker and Gelfand (1999), we formalize such a plot here calling it an *empirical semivariogram contour* (ESC) plot. For each of the $\frac{N(N-1)}{2}$ pairs of sites in $\Re^2$, calculate h_x and h_y, the separation distances along each axis. Since the sign of h_y depends upon the arbitrary order in which the two sites are compared, we demand that $h_y \geq 0$. (We could alternatively demand that $h_x \geq 0$.) That is, we take $(-h_x, -h_y)$ when $h_y < 0$. These separation distances are then aggregated into rectangular bins B_{ij} where the empirical semivariogram values for the (i,j)th bin are calculated by

$$\gamma_{ij}^* = \frac{1}{2N_{B_{ij}}} \sum_{\{(k,l):(\mathbf{s}_k - \mathbf{s}_l) \in B_{ij}\}} (Y(\mathbf{s}_k) - Y(\mathbf{s}_l))^2, \qquad (2.16)$$

where $N_{B_{ij}}$ equals the number of sites in bin B_{ij}. Because we force $h_y \geq 0$ with h_x unrestricted, we make the bin width on the y-axis half of that for the x-axis. We also force the middle class on the x-axis to be centered around zero. Upon labeling the center of the (i,j)th bin by (x_i, y_j), a three dimensional plot of γ_{ij}^* versus (x_i, y_j) yields an empirical semivariogram surface. Smoothing this surface using, for example, the algorithm of Akima (1978) available in the S-plus software package (see Subsection 2.5.1) produces a contour plot that we call the ESC plot. A symmetrized version of the ESC plot can be created by reflecting the upper left quadrant to the lower right and the upper right quadrant to the lower left.

The ESC plot can be used to assess departures from isotropy; isotropy is depicted by circular contours while elliptical contours capture geometric anisotropy. A rose diagram traces only one arbitrarily selected contour of this plot. A possible drawback to the ESC plot is the occurrence of sparse counts in extreme bins. However, these bins may be trimmed before smoothing if desired. Concerned that use of geographic coordinates could introduce artificial anisotropy (since $1°$ latitude $\neq 1°$ longitude in the northeastern United States), we have employed a Universal Transverse Mercator (UTM) projection to kilometers in the E-W and N-S axes (see Subsection 1.2.1).

Figure 2.10 is the empirical semivariogram contour plot constructed using x-axis width of 30 kilometers for the 1993 scallop data. We have overlaid this contour plot on the bin centers with their respective counts. Note that using empirical semivariogram values in the row of the ESC plot for which $h_y \approx 0$ provides an alternative to the usual $0°$ directional semivariogram. The latter directional semivariograms are based on a polar representation

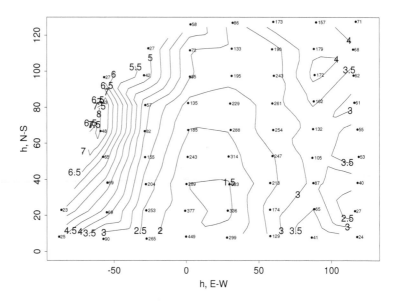

Figure 2.10 *ESC plot for the 1993 scallop data.*

of the angle and distance. For a chosen direction η and tolerance ϵ, the area for a class fans out as distance increases (see Figure 7.1 of Isaaks and Srivastava, 1989, p. 142). Attractively, a directional semivariogram based on the rectangular bins associated with the empirical semivariogram in $\Re^2$ has bin area remaining constant as distance increases. In Figure 2.11, we present the four customary directional (polar representation) semivariograms for the 1993 scallop data. Clearly, the ESC plot is more informative, particularly in suggesting evidence of geometric anisotropy.

2.4 Classical spatial prediction

In this section we describe the classical (i.e., minimum mean-squared error) approach to spatial prediction in the point-referenced data setting. The approach is commonly referred to as *kriging*, so named by Matheron (1963) in honor of D.G. Krige, a South African mining engineer whose seminal work on empirical methods for geostatistical data (Krige, 1951) inspired the general approach (and indeed, inspired the convention of using the terms "point-level spatial" and "geostatistical" interchangeably!). The problem is one of optimal spatial prediction: given observations of a random field $\mathbf{Y} = (Y(\mathbf{s}_1), \dots, Y(\mathbf{s}_n))'$, how do we predict the variable Y at a site $\mathbf{s}_0$ where it has not been observed? In other words, what is the best predictor of the value of $Y(\mathbf{s}_0)$ based upon the data $\mathbf{y}$?

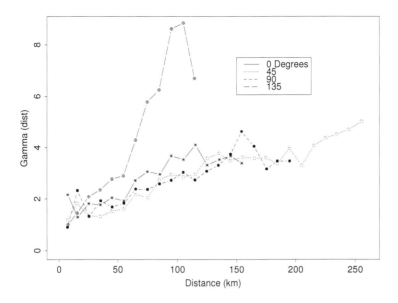

Figure 2.11 *Directional semivariograms for the 1993 scallop data.*

A linear predictor for $Y(\mathbf{s}_0)$ based on $\mathbf{y}$ would take the form $\sum \ell_i Y(\mathbf{s}_i) + \delta_0$. Using squared error loss, the best linear prediction would minimize $E[Y(\mathbf{s}_0) - (\sum \ell_i Y(\mathbf{s}_i) + \delta_0)]^2$ over δ_0 and the ℓ_i. For a constant mean process we would take $\sum \ell_i = 1$, in which case we would minimize $E[Y(\mathbf{s}_0) - \sum \ell_i Y(\mathbf{s}_i)]^2 + \delta_0^2$, and clearly δ_0 would be set to 0. Now letting $a_0 = 1$ and $a_i = -\ell_i$ we see that the criterion becomes $E[\sum_{i=0}^{n} a_i Y(\mathbf{s}_i)]^2$ with $\sum a_i = 0$. But from (2.4) this expectation becomes $-\sum_i \sum_j a_i a_j \gamma(\mathbf{s}_i - \mathbf{s}_j)$, revealing how, historically, the variogram arose in kriging within the geostatistical framework. Indeed, the optimal ℓ's can be obtained by solving this constrained optimization (e.g., using Lagrange multipliers), and will be functions of $\gamma(\mathbf{h})$ (see, e.g., Cressie, 1983, Sec. 3.2). With an estimate of γ, one immediately obtains the so-called *ordinary kriging* estimate. Other than the intrinsic stationarity model (Subsection 2.1.2), no further distributional assumptions are required for the $Y(\mathbf{s})$'s.

Let us take a more formal look at kriging in the context of Gaussian processes. Consider first the case where we have no covariates, but only the responses $Y(\mathbf{s}_i)$. This is developed by means of the following model for the observed data:

$$\mathbf{Y} = \mu \mathbf{1} + \boldsymbol{\epsilon}, \text{ where } \boldsymbol{\epsilon} \sim N(\mathbf{0}, \Sigma) .$$

For a spatial covariance structure having no nugget effect, we specify Σ as

$$\Sigma = \sigma^2 H(\phi) \text{ where } (H(\phi))_{ij} = \rho(\phi; d_{ij}) ,$$

where $d_{ij} = ||\mathbf{s}_i - \mathbf{s}_j||$, the distance between $\mathbf{s}_i$ and $\mathbf{s}_j$ and ρ is a valid correlation function on $\Re^r$ such as those in Table 2.1. For a model having a nugget effect, we instead set

$$\Sigma = \sigma^2 H (\phi) + \tau^2 I ,$$

where τ^2 is the nugget effect variance.

When covariate values $\mathbf{x} = (x(\mathbf{s}_1), \ldots, x(\mathbf{s}_n))'$ and $x(\mathbf{s}_0)$ are available for incorporation into the analysis, the procedure is often referred to as *universal kriging*, though we caution that some authors (e.g., Kaluzny et al., 1998) use the term "universal" in reference to the case where only latitude and longitude are available as covariates. The model now takes the more general form

$$\mathbf{Y} = X\boldsymbol{\beta} + \boldsymbol{\epsilon}, \text{ where } \boldsymbol{\epsilon} \sim N(\mathbf{0}, \Sigma) ,$$

with Σ being specified as above, either with or without the nugget effect. Note that ordinary kriging may be looked upon as a particular case of universal kriging with X being the $n \times 1$ matrix (i.e., column vector) $\mathbf{1}$, and $\boldsymbol{\beta}$ the scalar μ.

We now pose our prediction problem as follows: we seek the function $f(\mathbf{y})$ that minimizes the mean-squared prediction error,

$$E\left[(Y(\mathbf{s}_0) - f(\mathbf{y}))^2 \,\Big|\, \mathbf{y}\right] . \tag{2.17}$$

By adding and subtracting the conditional mean $E[Y(\mathbf{s}_0)|\mathbf{y}]$ inside the square, grouping terms, and squaring we obtain

$$E\left[(Y(\mathbf{s}_0) - f(\mathbf{y}))^2 \,\Big|\, \mathbf{y}\right]$$
$$= E\left\{(Y(\mathbf{s}_0) - E[Y(\mathbf{s}_0)|\mathbf{y}])^2 \,\Big|\, \mathbf{y}\right\} + \{E[Y(\mathbf{s}_0)|\mathbf{y}] - f(\mathbf{y})\}^2 ,$$

since (as often happens in statistical derivations like this) the expectation of the cross-product term equals zero. But since the second term on the right-hand side is nonnegative, we have

$$E\left[(Y(\mathbf{s}_0) - f(\mathbf{y}))^2 \,\Big|\, \mathbf{y}\right] \geq E\left\{(Y(\mathbf{s}_0) - E[Y(\mathbf{s}_0)|\mathbf{y}])^2 \,\Big|\, \mathbf{y}\right\}$$

for any function $f(\mathbf{y})$. Equality holds if and only if $f(\mathbf{y}) = E[Y(\mathbf{s}_0)|\mathbf{y}]$, so it must be that the predictor $f(\mathbf{y})$ that minimizes the error is the conditional expectation of $Y(\mathbf{s}_0)$ given the data. This result is quite intuitive from a Bayesian point of view, since this $f(\mathbf{y})$ is just the *posterior mean* of $Y(\mathbf{s}_0)$, and it is well known that the posterior mean is the Bayes rule (i.e., the minimizer of posterior risk) under squared error loss functions of the sort adopted in (2.17) above as our scoring rule.

Having identified the form of the best predictor we now turn to its estimation. Consider first the wildly unrealistic situation in which all the population parameters $(\boldsymbol{\beta}, \sigma^2, \phi$, and $\tau^2)$ are known. From standard multivariate normal theory we have the following general result: If

$$\begin{pmatrix} \mathbf{Y}_1 \\ \mathbf{Y}_2 \end{pmatrix} \sim N \left(\begin{pmatrix} \boldsymbol{\mu}_1 \\ \boldsymbol{\mu}_2 \end{pmatrix}, \begin{pmatrix} \Omega_{11} & \Omega_{12} \\ \Omega_{21} & \Omega_{22} \end{pmatrix} \right),$$

where $\Omega_{21} = \Omega_{12}^T$, then the conditional distribution $p(\mathbf{Y}_1 | \mathbf{Y}_2)$ is normal with mean and variance:

$$\begin{aligned} E[\mathbf{Y}_1 | \mathbf{Y}_2] &= \boldsymbol{\mu}_1 + \Omega_{12}\Omega_{22}^{-1}(\mathbf{Y}_2 - \boldsymbol{\mu}_2); \\ Var[\mathbf{Y}_1 | \mathbf{Y}_2] &= \Omega_{11} - \Omega_{12}\Omega_{22}^{-1}\Omega_{21}. \end{aligned}$$

In our framework, we have $\mathbf{Y}_1 = Y(\mathbf{s}_0)$ and $\mathbf{Y}_2 = \mathbf{y}$. It then follows that

$$\Omega_{11} = \sigma^2 + \tau^2, \quad \Omega_{12} = \boldsymbol{\gamma}^T, \quad \text{and } \Omega_{22} = \Sigma = \sigma^2 H(\phi) + \tau^2 I,$$

where $\boldsymbol{\gamma}^T = \left(\sigma^2 \rho(\phi; d_{01}), \dots, \sigma^2 \rho(\phi; d_{0n}) \right)$. Substituting these values into the mean and variance formulae above, we obtain

$$\begin{aligned} E[Y(\mathbf{s}_0) | \mathbf{y}] &= \mathbf{x}_0^T \boldsymbol{\beta} + \boldsymbol{\gamma}^T \Sigma^{-1} (\mathbf{y} - X\boldsymbol{\beta}), & (2.18) \\ \text{and } Var[Y(\mathbf{s}_0) | \mathbf{y}] &= \sigma^2 + \tau^2 - \boldsymbol{\gamma}^T \Sigma^{-1} \boldsymbol{\gamma}. & (2.19) \end{aligned}$$

We remark that this solution assumes we have actually observed the covariate value $\mathbf{x}_0 = \mathbf{x}(\mathbf{s}_0)$ at the "new" site $\mathbf{s}_0$; we defer the issue of missing $\mathbf{x}_0$ for the time being.

Note that one could consider prediction *not* at a new location, but at one of the already observed locations. In this case one can ask whether or not the predictor in (2.18) will equal the observed value at that location. We leave it as an exercise to verify that if $\tau^2 = 0$ (i.e., the no-nugget case, or so-called noiseless prediction) then the answer is yes, while if $\tau^2 > 0$ then the answer is no.

Next, consider how these answers are modified in the more realistic scenario where the model parameters are unknown and so must be estimated from the data. Here we would modify $f(\mathbf{y})$ to

$$\widehat{f(\mathbf{y})} = \mathbf{x}_0^T \widehat{\boldsymbol{\beta}} + \widehat{\boldsymbol{\gamma}}^T \widehat{\Sigma}^{-1} \left(\mathbf{y} - X\widehat{\boldsymbol{\beta}} \right),$$

where $\widehat{\boldsymbol{\gamma}} = \left(\hat{\sigma}^2 \rho(\hat{\phi}; d_{01}), \dots, \hat{\sigma}^2 \rho(\hat{\phi}; d_{0n}) \right)^T$, $\widehat{\boldsymbol{\beta}} = \left(X^T \widehat{\Sigma}^{-1} X \right)^{-1} X^T \widehat{\Sigma}^{-1} \mathbf{y}$, the usual weighted least squares estimator of $\boldsymbol{\beta}$, and $\widehat{\Sigma} = \hat{\sigma}^2 H(\hat{\phi})$. Thus $\widehat{f(\mathbf{y})}$ can be written as $\boldsymbol{\lambda}^T \mathbf{y}$, where

$$\boldsymbol{\lambda} = \widehat{\Sigma}^{-1} \widehat{\boldsymbol{\gamma}} + \widehat{\Sigma}^{-1} X \left(X^T \widehat{\Sigma}^{-1} X \right)^{-1} \left(\mathbf{x}_0 - X^T \widehat{\Sigma}^{-1} \widehat{\boldsymbol{\gamma}} \right). \quad (2.20)$$

If $\mathbf{x}_0$ is unobserved, we can estimate it and $Y(\mathbf{s}_0)$ jointly by iterating between this formula and a corresponding one for $\hat{\mathbf{x}}_0$, namely

$$\hat{\mathbf{x}}_0 = X^T \boldsymbol{\lambda},$$

which arises simply by multiplying both sides of (2.20) by X^T and simplifying. This is essentially an EM (expectation-maximization) algorithm (Dempster, Laird, and Rubin, 1977), with the calculation of $\hat{\mathbf{x}}_0$ being the E step and (2.20) being the M step.

In the classical framework a lot of energy is devoted to the determination of the optimal estimates to plug into the above equations. Typically, restricted maximum likelihood (REML) estimates are selected and shown to have certain optimal properties. However, as we shall see in Chapter 5, how to perform the estimation is not an issue in the Bayesian setting. There, we instead impose prior distributions on the parameters and produce the full posterior predictive distribution $p(Y(\mathbf{s}_0)|\mathbf{y})$. Any desired point or interval estimate (the latter to express our uncertainty in such prediction) may then be computed with respect to this distribution.

2.5 Computer tutorials

2.5.1 EDA and variogram fitting in S+SpatialStats

In this section we outline the use of the S+SpatialStats package in performing exploratory analysis on spatially referenced data. Throughout we use a "computer tutorial" style, as follows.

First, we need to load the spatial module into the S-plus environment:

```
>module(spatial)
```

The scallops data, giving locations and scallop catches in the Atlantic waters off the coasts of New Jersey and Long Island, New York, is preloaded as a data frame in S-plus, and can therefore be accessed directly. For example, a descriptive summary of the data can be obtained by typing

```
>summary(scallops)
```

In order to present graphs and maps in S-plus we will need to open a graphing device. The best such device is called trellis.device(). Here we draw a histogram of the variable tcatch in the dataframe scallops. Note the generic notation a$b for accessing a member b of a dataframe a. Thus the member tcatch of dataframe scallops is accessed as scallops$tcatch. We also print the histogram to a .ps file:

```
>trellis.device()
>hist(scallops$tcatch)
>printgraph(file=''histogram.tcatch.ps'')
```

Noticing the data to be highly skewed, we feel the need to create a new variable log(tcatch). But since tcatch contains a number of 0's, we instead compute $log(tcatch + 1)$. For that it is best to create our own dataframe since it is not a good idea to "spoil" S-plus' own dataframe. As such we assign scallops to myscallops. Then we append the variable lgcatch (which is actually $log(tcatch + 1)$) to `myscallops`. We then draw the histogram of the variable `lgcatch`:

```
>myscallops <- scallops
>myscallops[,''lgcatch''] <- log(scallops$tcatch+1)
>summary(myscallops$lgcatch)
>hist(myscallops$lgcatch)
```

This histogram exhibits much more symmetry than the earlier one, suggesting a normality assumption we might make later when kriging will be easier to accept.

We next plot the locations as an ordinary line plot:

```
>plot(myscallops$long, myscallops$lat)
```

For spatial purposes, we actually need a *geographic information system (GIS)* interface. This is offered by the `S-plus` library `maps`. We next invoke this library and extract a map of the U.S. from it.

```
>library(maps)
>map(''usa'')
```

For our data, however, we do not need the map of the entire U.S. Looking at the earlier summary of scallops, we note the range of the latitude and longitude variables and decide upon the following limits. Note that `xlim` sets the x-axis limits and `ylim` sets the y-axis limits; the `c()` function creates vectors.

```
>map(''usa'', xlim=c(-74, -71), ylim=c(38.2, 41.5))
```

The observed sites may be embedded on the map, reducing their size somewhat using the `cex` ("character expansion") option:

```
>points(myscallops$long, myscallops$lat, cex=0.75)
```

It is often helpful to add contour lines to the plot. In order to add such lines it is necessary to carry out an interpolation. This essentially fills in the gaps in the data over a regular grid (where there are no actual observerd data) using a bivariate linear interpolation. This is done in `S-plus` using the `interp` function. The contour lines may then be added to the plot using the `contour` command:

```
>int.scp <- interp(myscallops$long,
     myscallops$lat, myscallops$lgcatch)
>contour(int.scp, add=T)
```

Figure 2.12 shows the result of the last four commands, i.e., the map of the scallop locations and log catch contours arising from the linear interpolation.

Two other useful ways of looking at the data may be through `image` and `perspective` (three-dimensional surface) plots. Remember that they will use the interpolated object so a preexisting interpolation is also compulsory here.

Figure 2.12 *Map of observed scallop sites and contours of (linearly interpolated) raw log catch data, scallop data.*

```
>image(int.scp)
>persp(int.scp)
```

The empirical variogram can be estimated in both the standard and "robust" (Cressie and Hawkins) way with built-in functions. We first demonstrate the standard approach. After a variogram object is created, typing that object yields the actual values of the variogram function with the distances at which they are computed. A summary of the object may be invoked to see information for each lag, the total number of lags, and the maximum intersite distance.

```
>scallops.var <- variogram(lgcatch~loc(long,lat),
    data=myscallops)
>scallops.var
>summary(scallops.var)
```

In scallops.var, distance corresponds to the spatial lag (h in our usual notation), gamma is the variogram $\gamma(h)$, and np is the number of points in each bin. In the output of the summary command, maxdist is the largest distance on the map, nlag is the number of lags (variogram bins), and lag is maxdist/nlag, which is the width of each variogram bin.

By contrast, the robust method is obtained simply by specifying "robust" in the method option:

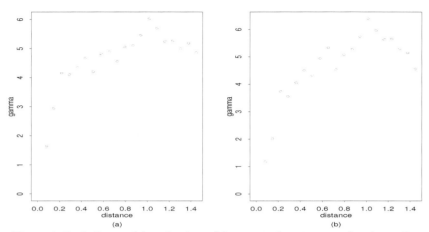

Figure 2.13 *Ordinary (a) and robust (b) empirical variograms for the scallops data.*

```
>scallops.var.robust <- variogram(lgcatch~loc(long,lat),
    data=myscallops, method = ''robust'')
```

Plotting is usually done by just calling the `plot` function with the variogram object as its argument. It may be useful to compare the plots one below the other. Setting up these plots is done as follows:

```
>par(mfrow=c(1,2))
>plot(scallops.var)
>plot(scallops.var.robust)
>printgraph(file=''scallops.empvario.ps'')
```

The output from this picture is shown in Figure 2.13.

The *covariogram* (a plot of an isotropic empirical covariance function (2.15) versus distance) and *correlogram* (a plot of (2.15) divided by $\hat{C}(\mathbf{0})$ versus distance) may be created using the `covariogram` and `correlogram` functions. (When we are through here, we set the graphics device back to having one plot per page using the `par` command.)

```
>scallops.cov <- covariogram(lgcatch~loc(long,lat),
    data=myscallops)
>plot(scallops.cov)
>scallops.corr <- correlogram(lgcatch~loc(long,lat),
    data=myscallops)
>plot(scallops.corr)
>printgraph(file=''scallops.covariograms.ps'')
> par(mfrow=c(1,1))
```

Theoretical variograms may also be computed and compared to the observed data as follows. Invoke the `model.variogram` function and choose an initial theoretical model; say, range=0.8, sill=1.25, and nugget=0.50. Note that the fun option specifies the variogram type we want to work with. Below we choose the spherical (spher.vgram); other options include exponential (exp.vgram), Gaussian (gauss.vgram), linear (linear.vgram), and power (power.vgram).

```
>model.variogram(scallops.var.robust, fun=spher.vgram,
    range=0.80, sill=1.25, nugget = 0.50)
```

We remark that this particular model provides relatively poor fit to the data; the objective function takes a relatively high value (roughly 213). (You are asked to find a better-fitting model in Exercise 7.)

Formal estimation procedures for variograms may also be carried out by invoking the nls function on the spher.fun function that we can create:

```
>spher.fun <- function(gamma,distance,range,sill,nugget){
    gamma - spher.vgram(distance, range=range,
    sill=sill, nugget=nugget)}
>scallops.nl1 <- nls(~spher.fun(gamma, distance, range,
    sill, nugget), data = scallops.var.robust,
    start=list(range=0.8, sill=1.05, nugget=0.7))
>coef(scallops.nl1)
```

Thus we are using nls to minimize the squared distance between the theoretical and empirical variograms. Note there is nothing to the left of the "~" character at the beginning of the nls statement.

Many times our interest lies in spatial *residuals*, or what remains after detrending the response from the effects of latitude and longitude. An easy way to do that is by using the gam function in S-plus. Here we plot the residuals of the scallops lgcatch variable after the effects of latitude and longitude have been accounted for:

```
>gam.scp <- gam(lgcatch~lo(long)+lo(lat), data= myscallops)
>par(mfrow=c(2,1))
>plot(gam.scp, residuals=T, rug=F)
```

Finally, at the end of the session we unload the spatial module, after which we can either do other work, or quit.

```
>module(spatial, unload=T)
>q()
```

2.5.2 *Kriging in* S+SpatialStats

We now present a tutorial in using S+SpatialStats to do basic kriging. At the command prompt type S-plus to start the software, and load the spatial module into the environment:

```
>module(spatial)
```

Recall that the `scallops` data is preloaded as a data frame in `S-plus`, and a descriptive summary of this data set can be obtained by typing

```
>summary(scallops)
```

while the first row of the data may be seen by typing

```
>scallops[1,]
```

Recall from our Section 2.5.1 tutorial that the data on `tcatch` was highly skewed, so we needed to create another dataframe called `myscallops`, which includes the log transform of `tcatch` (or actually $\log(tcatch+1)$). We called this new variable `lgcatch`. We then computed both the regular empirical variogram and the "robust" (Cressie and Hawkins) version, and compared both to potential theoretical models using the `variogram` command.

```
>scallops.var.robust <- variogram(lgcatch~loc(long,lat),
    data=myscallops, method = ''robust'')
```

Plotting is usually done using the `plot` function on the variogram object:

```
>trellis.device()
>plot(scallops.var.robust)
```

Next we recall `S-plus`' ability to compute theoretical variograms. We invoke the `model.variogram` function, choosing a theoretical starting model (here, range=0.8, sill=4.05, and nugget=0.80), and using `fun` to specify the variogram type.

```
>model.variogram(scallops.var.robust, fun=spher.vgram,
    range=0.80, sill=4.05, nugget = 0.80)
>printgraph(file=''scallops.variograms.ps'')
```

The output from this command (robust empirical semivariogram with this theoretical variogram overlaid) is shown in Figure 2.14. Note again the `model.variogram` command allows the user to alter the theoretical model and continually recheck the value of the objective function (where smaller values indicate better fit of the theoretical to the empirical).

Formal estimation procedures for variograms may also be carried out by invoking the `nls` function on the `spher.fun` function that we create:

```
>spher.fun <- function(gamma,distance,range,sill,nugget){
    gamma - spher.vgram(distance, range=range,
    sill=sill, nugget=nugget)}
>scallops.nl1 <- nls(~spher.fun(gamma, distance, range,
    sill, nugget), data = scallops.var.robust,
    start=list(range=0.8, sill=4.05, nugget=0.8))
>summary(scallops.nl1)
```

We now call the kriging function `krige` on the variogram object to produce estimates of the parameters for ordinary kriging:

```
>scallops.krige <- krige(lgcatch~loc(long,lat),
    data=myscallops, covfun=spher.cov, range=0.71,
```

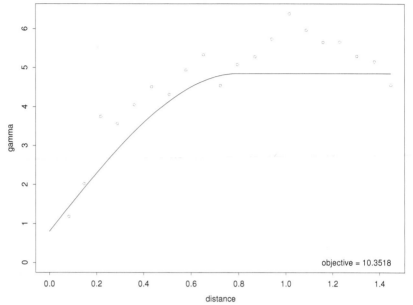

Figure 2.14 *Robust empirical and theoretical (spherical) variograms for the scallops data.*

nugget=0.84, sill=4.53)

Note that the covfun option here specifies an intrinsic spherical covariance function. Now suppose we want to predict the response at a small collection of new locations. We need to create a new text file containing the latitudes and longitudes for these new locations. We *must* label the coordinates as lat and long, exactly matching the names in our original dataframe myscallops. Download the file

www.biostat.umn.edu/~brad/data/newdata.txt

from the web, and save the file as newdata.txt. The file contains two new locations:

```
long  lat
-71.00 40.0
-72.75 39.5
```

Here the first site is far from the bulk of the observed data (so the predicted values should have high standard errors), while the second site is near the bulk of the observed data (so the predicted values should have low standard errors).

Next we create a data frame called newdata that reads in the new set of sites using the read.table function in S-plus, remembering to include the `header=T` option. Having done that we may call the `predict` function, specifying our `newdata` data frame using the `newdata` option:

```
>newdata <- read.table(''newdata.txt'', header=T)
>scallops.predicttwo <- predict(scallops.krige,
    newdata=newdata)
```

`scallops.predicttwo` then contains the predictions and associated standard errors for the two new sites, the latter of which are ordered as we anticipated.

Next, we consider the case where we wish to predict not at a few specific locations, but over a fine grid of sites, thus enabling a prediction *surface*. In such a situation one can use the `expand.grid` function, but the `interp` function seems to offer an easier and less error-prone approach. To do this, we first call the `predict` function *without* the `newdata` option. After checking the first row of the `scallops.predict` object, we collect the coordinates and the predicted values into three vectors `x`, `y`, and `z`. We then invoke the `interp` and `persp` functions:

```
>scallops.predict <- predict(scallops.krige)
>scallops.predict[1,]
>x <- scallops.predict[,1]
>y <- scallops.predict[,2]
>z <- scallops.predict[,3]
>scallops.predict.interp <- interp(x,y,z)
>persp(scallops.predict.interp)
```

It may be useful to recall the location of the sites and the surface plot of the raw data for comparison. We create these plots on a separate graphics device:

```
>trellis.device()
>plot(myscallops$long, myscallops$lat)
>int.scp <- interp(myscallops$long, myscallops$lat,
    myscallops$lgcatch)
>persp(int.scp)
```

Figure 2.15 shows these two perspective plots side by side for comparison. The predicted surface on the left is smoother, as expected.

It is also useful to have a surface plot of the standard errors, since we expect to see higher standard errors where there is less data. This is well illustrated by the following commands:

```
>z.se <- scallops.predict[,4]
>scallops.predict.interp.se <- interp(x,y,z.se)
>persp(scallops.predict.interp.se)
```

Other plots, such as image plots for the prediction surface with added contour lines, may be useful:

```
>image(scallops.predict.interp)
```

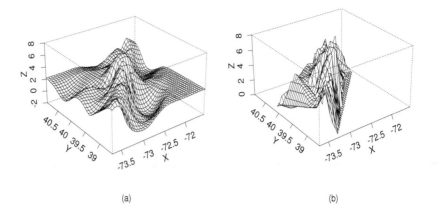

(a)   (b)

Figure 2.15 *Perspective plots of the kriged prediction surface (a) and interpolated raw data (b), log scallop catch data.*

```
>par(new=T, xaxs=''d'', yaxs=''d'')
>contour(scallops.predict.interp)
```

Turning to universal kriging, here we illustrate with the scallops data using latitude and longitude as the covariates (i.e., trend surface modeling). Our covariance matrix X is therefore $n \times 3$ ($n = 148$ here) with columns corresponding to the intercept, latitude and longitude.

```
>scallops.krige.universal <- krige(lgcatch~loc(long,lat)
    +long+lat, data=myscallops, covfun=spher.cov,
    range=0.71, nugget=0.84, sill=4.53)
```

Note that the `scallops.krige.universal` function gives point estimates, but not associated standard errors. You are asked to remedy this situation in Exercise 11.

Plots like those already seen for ordinary kriging may be done as well. It is also useful to produce a spatial surface of the standard errors of the fit.

```
>scallops.predict.universal
    <- predict(scallops.krige.universal)
>scallops.predict.universal[1,]
>x <- scallops.predict.universal[,1]
>y <- scallops.predict.universal[,2]
>z <- scallops.predict.universal[,3]
>scallops.predict.interp <- interp(x,y,z)
>persp(scallops.predict.interp)
>q()
```

2.5.3 EDA, variograms, and kriging in geoR

R is an increasingly popular freeware alternative to S-plus, available from the web at www.r-project.org. In this subsection we describe methods for kriging and related geostatistical operations available in geoR, a geostatistical data analysis package using R authored by Paulo Ribeiro Jr. and Peter Diggle, which is also freely available on the web at www.est.ufpr.br/geoR/.

Since the syntax of S-plus and R is virtually identical, we do not spend time here repeating the material of the past subsection. Rather, we only highlight a few differences in exploratory data analysis steps, before moving on to model fitting and kriging.

Consider again the scallop data. Our modified "myscallops" version can now be read into R directly from our website by starting the program and typing

```
>myscallops <- read.table(
  ''http://www.biostat.umn.edu/~brad/data/myscallops.txt'',
  header=T)
```

Recall that it is often helpful to create image plots and place contour lines on the plot. These provide a visual idea of the realized spatial surface. In order to do these, it is necessary to first carry out an interpolation. This essentially fills up the gaps (i.e., where there are no points) using a bivariate linear interpolation. This is done using the interp function in R, located in the library akima. Then the contour lines may be added to the plot using the contour command. The results are shown in Figure 2.16.

```
>library(akima)
>int.scp <- interp(myscallops$long, myscallops$lat,
  myscallops$lgcatch)
>image(int.scp, xlim=range(myscallops$long),
  ylim=range(myscallops$lat))
>contour(int.scp, add=T)
```

Another useful way of looking at the data is through surface plots (or perspective plots). This is done by invoking the persp function:

```
>persp(int.scp, xlim=range(myscallops$long),
  ylim=range(myscallops$lat))
```

The empirical variogram can be estimated in the classical way and in the robust way with in-built R functions. There are several packages in R that perform the above computations. We illustrate the geoR package, mainly because of its additional ability to fit Bayesian geostatistical models as well. Nevertheless, the reader might want to check out the CRAN website (http://cran.us.r-project.org/) for the latest updates and several other spatial packages. In particular, we mention fields, gstat, sgeostat, spatstat, and spatdep for exploratory work and some model fitting of spatial data, and GRASS and RArcInfo for interfaces to GIS software.

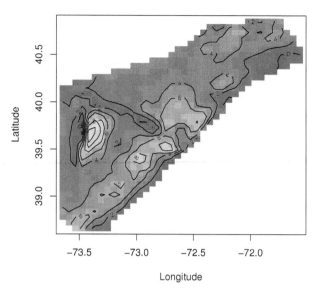

Figure 2.16 *An image plot of the scallops data, with contour lines super-imposed.*

Returning to the problem of empirical variogram fitting, we first invoke the geoR package. We will use the function variog in this package, which takes in a geodata object as input. To do this, we first create an object, obj, with only the coordinates and the response. We then create the geodata object using the as.geodata function, specifying the columns holding the coordinates, and the one holding the response.

```
>library(geoR)
>obj <- cbind(myscallops$long,myscallops$lat,
   myscallops$lgcatch)
>scallops.geo <- as.geodata(obj,coords.col=1:2,data.col=3)
```

Now, a variogram object is created.

```
>scallops.var <- variog(scallops.geo,
   estimator.type=''classical'')
>scallops.var
```

The robust estimator (see Cressie, 1993, p.75) can be obtained by typing

```
>scallops.var.robust <- variog(scallops.geo,
   estimator.type=''modulus'')
```

A plot of the two semivariograms (by both methods, one below the other, as in Figure 2.17) can be obtained as follows:

```
>par(mfrow=c(2,1))
>plot(scallops.var)
```

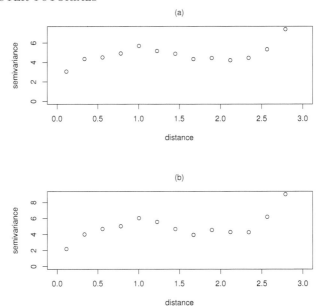

Figure 2.17 *Plots of the empirical semivariograms for the scallops data: (a) classical; (b) robust.*

```
>plot(scallops.var.robust)
```

Covariograms and correlograms are invoked using the `covariogram` and `correlogram` functions. The remaining syntax is the same as in `S-plus`.

The function `variofit` estimates the sill, the range, and the nugget parameters under a specified covariance model. A variogram object (typically an output from the `variog` function) is taken as input, together with initial values for the range and sill (in `ini.cov.pars`), and the covariance model is specified through `cov.model`. The covariance modeling options include `exponential`, `gaussian`, `spherical`, `circular`, `cubic`, `wave`, `power`, `powered.exponential`, `cauchy`, `gneiting`, `gneiting.matern`, and `pure.nugget` (no spatial covariance). Also, the initial values provided in `ini.cov.pars` do not include those for the nugget. It is concatenated with the value of the `nugget` option only if `fix.nugget=FALSE`. If the latter is `TRUE`, then the value in the `nugget` option is taken as the fixed true value.

Thus, with the exponential covariance function for the scallops data, we can estimate the parameters (including the nugget effect) using

```
>scallops.var.fit <- variofit(scallops.var.robust,
    ini.cov.pars = c(1.0,2.0), cov.model=''exponential'',
    fix.nugget=FALSE, nugget=1.0)
```

The output is given below. Notice that this is the weighted least squares approach for fitting the variogram:

```
variofit: model parameters estimated by WLS
         (weighted least squares):
covariance model is: matern with fixed kappa = 0.5
              (exponential)
          parameter estimates:
            tausq sigmasq phi
            0.0000 5.1289 0.2160
```

Likelihood model fitting

In the previous section we saw parameter estimation through weighted least squares of variograms. Now we introduce likelihood-based and Bayesian estimation functions in geoR.

Both maximum likelihood and REML methods are available through the geoR function likfit. To estimate the parameters for the scallops data, we invoke

```
>scallops.lik.fit <- likfit(scallops.geo,
   ini.cov.pars=c(1.0,2.0),cov.model = ''exponential'',
   trend = ''cte'', fix.nugget = FALSE, nugget = 1.0,
   nospatial = TRUE, method.lik = ''ML'')
```

The option trend = ''cte'' means a spatial regression model with constant mean. This yields the following output:

```
> scallops.lik.fit

        likfit: estimated model parameters:
            beta tausq sigmasq phi
            2.3748 0.0947 5.7675 0.2338
```

Changing method.lik = ''REML'' yields the restricted maximum likelihood estimation. Note that the variance of the estimate of beta is available by invoking scallops.lik.fit$beta.var, so calculating the confidence interval for the trend is easy. However, the variances of the estimates of the covariance parameters is not easily available within geoR.

Kriging in geoR

There are two in-built functions in geoR for kriging: one is for classical or conventional kriging, and is called krige.conv, while the other performs Bayesian kriging and is named krige.bayes. We now briefly look into these two types of functions. The krige.bayes function is not as versatile as WinBUGS in that it is more limited in the types of models it can handle, and also the updating is not through MCMC methods. Nevertheless, it is a handy tool and already improved upon the aforementioned likelihood methods by providing posterior samples of *all* the model parameters, which lead to estimation of their variability.

The `krige.bayes` function can be used to estimate parameters for spatial regression models. To fit a constant mean spatial regression model for the scallops data, without doing predictions, we invoke `krige.bayes` specifying a constant trend, an exponential covariance model, a flat prior for the constant trend level, the reciprocal prior for `sigmasq` (Jeffrey's), and a discrete uniform prior for `tausq`.

```
>scallops.bayes1 <- krige.bayes(scallops.geo,
   locations = ''no'', borders = NULL, model =
   model.control(trend.d = ''cte'',
   cov.model = ''exponential''),
   prior = prior.control(beta.prior = ''flat'',
   sigmasq.prior = ''reciprocal'',
   tausq.rel.prior = ''uniform'',
   tausq.rel.discrete=seq(from=0.0,to=1.0,by=0.01)))
```

We next form the quantiles in the following way:

```
> out <- scallops.bayes1$posterior
> out <- out$sample
> beta.qnt <- quantile(out$beta, c(0.50,0.025,0.975))
> phi.qnt <- quantile(out$phi, c(0.50,0.025,0.975))
> sigmasq.qnt <- quantile(out$sigmasq, c(0.50,0.025,0.975))
> tausq.rel.qnt <- quantile(out$tausq.rel,
   c(0.50,0.025,0.975))
> beta.qnt
```

```
              50%   2.5%   97.5%
          1.931822 -6.426464 7.786515
```

```
> phi.qnt
```

```
              50%   2.5%   97.5%
          0.5800106 0.2320042 4.9909913
```

```
sigmasq.qnt
```

```
              50%   2.5%   97.5%
          11.225002 4.147358 98.484722
```

```
> tausq.rel.qnt
```

```
              50%   2.5%   97.5%
              0.03  0.00  0.19
```

Note that `tausq.rel` refers to the ratio of the nugget variance to the spatial variance, and is seen to be negligible here, too. This is consistent with all the earlier analysis, showing that a purely spatial model (no nugget) would perhaps be more suitable for the scallops data.

2.6 Exercises

1. For semivariogram models #2, 4, 5, 6, 7, and 8 in Subsection 2.1.3,

   (a) identify the nugget, sill, and range (or effective range) for each;
   (b) find the covariance function $C(t)$ corresponding to each $\gamma(t)$, provided it exists.

2. Prove that for Gaussian processes, strong stationarity is equivalent to weak stationarity.

3. Consider the *triangular* (or "tent") covariance function,

$$C(\|h\|) = \begin{cases} \sigma^2(1 - \|h\| / \delta) & \text{if } \|h\| \leq \delta, \ \sigma^2 > 0, \ \delta > 0, \\ 0 & \text{if } \|h\| > \delta \end{cases}.$$

   It is valid in one dimension. (The reader can verify that it is the characteristic function of the density function $f(x)$ proportional to $[1 - \cos(\delta x)]/\delta x^2$.) Now in two dimensions, consider a 6×8 grid with locations $\mathbf{s}_{jk} = (j\delta/\sqrt{2}, k\delta/\sqrt{2})$, $j = 1, \ldots, 6$, $k = 1, \ldots, 8$. Assign a_{jk} to $\mathbf{s}_{jk}$ such that $a_{jk} = 1$ if $j + k$ is even, $a_{jk} = -1$ if $j + k$ is odd. Show that $Var[\Sigma a_{jk} Y(\mathbf{s}_{jk})] < 0$, and hence that the triangular covariance function is *invalid* in two dimensions.

4. The *turning bands method* (Christakos, 1984; Stein, 1999a) is a technique for creating stationary covariance functions on $\Re^r$. Let $\mathbf{u}$ be a random unit vector on $\Re^r$ (by random we mean that the coordinate vector that defines $\mathbf{u}$ is randomly chosen on the surface of the unit sphere in $\Re^r$). Let $c(\cdot)$ be a valid stationary covariance function on $\Re^1$, and let $W(t)$ be a mean 0 process on $\Re^1$ having $c(\cdot)$ as its covariance function. Then for any location $\mathbf{s} \in \Re^r$, define

$$Y(\mathbf{s}) = W(\mathbf{s}^T \mathbf{u}) .$$

   Note that we can think of the process either conditionally given $\mathbf{u}$, or marginally by integrating with respect to the uniform distribution for $\mathbf{u}$. Note also that $Y(\mathbf{s})$ has the possibly undesirable property that it is constant on planes (i.e., on $\mathbf{s}^T \mathbf{u} = k$).

   (a) If W is a Gaussian process, show that, given $\mathbf{u}$, $Y(\mathbf{s})$ is also a Gaussian process and is stationary.
   (b) Show that marginally $Y(\mathbf{s})$ is *not* a Gaussian process, but is isotropic. [*Hint:* Show that $Cov(Y(\mathbf{s}), Y(\mathbf{s}')) = E_{\mathbf{u}} c((\mathbf{s} - \mathbf{s}')^T \mathbf{u})$.]

5.(a) Based on (2.10), show that $c_{12}(\mathbf{h})$ is a valid correlation function; i.e., that G is a bounded, positive, symmetric about 0 measure on $\Re^2$.
   (b) Show further that if c_1 and c_2 are isotropic, then c_{12} is.

6.(a) What is the issue with regard to specifying $\hat{c}(0)$ in the covariance function estimate (2.15)?

(b) Show either algebraically or numerically that regardless of how $\widehat{c}(0)$ is obtained, $\widehat{\gamma}(t_k) \neq \widehat{c}(0) - \widehat{c}(t_k)$ for all t_k.

7. Carry out the steps outlined in Section 2.5.1 in S+SpatialStats. In addition:

(a) Provide a descriptive summary of the scallops data with the plots derived from the above session.

(b) Experiment with the model.variogram function to obtain rough estimates of the nugget, sill, and range; your final objective function should have a value less than 9.

(c) Repeat the theoretical variogram fitting with an exponential variogram, and report your results.

8. Consider the coal.ash data frame built into S+SpatialStats. This data comes from the Pittsburgh coal seam on the Robena Mine Property in Greene County, PA (Cressie, 1993, p. 32). This data frame contains 208 coal ash core samples (the variable coal in the data frame) collected on a grid given by x and y planar coordinates (*not* latitude and longitude).

Carry out the following tasks in S-plus:

(a) Plot the sampled sites embedded on a map of the region. Add contour lines to the plot.

(b) Provide a descriptive summary (histograms, stems, quantiles, means, range, etc.) of the variable coal in the data frame.

(c) Plot variograms and correlograms of the response and comment on the need for spatial analysis here.

(d) If you think that there is need for spatial analysis, use the interactive model.variogram method in S-plus to arrive at your best estimates of the range, nugget, and sill. Report your values of the objective functions.

(e) Try to estimate the above parameters using the nls procedure in S-plus.

Hint: You may wish to look at Section 3.2 in Kaluzny et al. (1998) for some insight into the coal.ash data.

9. Confirm expressions (2.18) and (2.19), and subsequently verify the form for $\boldsymbol{\lambda}$ given in equation (2.20).

10. Show that when using (2.18) to predict the value of the surface at one of the existing data locations $\mathbf{s}_i$, the predictor will equal the observed value at that location if and only if $\tau^2 = 0$. (That is, the usual Gaussian process is a spatial interpolator only in the "noiseless prediction" scenario.)

11. It is an unfortunate feature of S+SpatialStats that there is no intrinsic

routine to automatically obtain the standard errors of the estimated regression coefficients in the universal kriging model. Recall that

$$\mathbf{Y} \;=\; X\boldsymbol{\beta} + \boldsymbol{\epsilon}, \text{ where } \boldsymbol{\epsilon} \sim N\left(\mathbf{0}, \Sigma\right),$$

$$\text{and } \Sigma \;=\; \sigma^2 H\left(\phi\right) + \tau^2 I, \text{ where } \left(H\left(\phi\right)\right)_{ij} = \rho\left(\phi; d_{ij}\right).$$

Thus the dispersion matrix of $\widehat{\boldsymbol{\beta}}$ is given as $Var(\widehat{\boldsymbol{\beta}}) = \left(X^T \Sigma^{-1} X\right)^{-1}$. Thus $\widehat{Var(\boldsymbol{\beta})} = \left(X^T \widehat{\Sigma}^{-1} X\right)^{-1}$ where $\widehat{\Sigma} = \widehat{\sigma}^2 H(\widehat{\phi}) + \widehat{\tau}^2 I$ and $X = [\mathbf{1}, \texttt{long}, \texttt{lat}]$. Given the estimates of the sill, range, and nugget (from the $\texttt{nls}$ function), it is possible to estimate the covariance matrix $\widehat{\Sigma}$, and thereby get $\widehat{Var(\boldsymbol{\beta})}$. Develop an $\texttt{S-plus}$ or $\texttt{R}$ program to perform this exercise to obtain estimates of standard errors for $\widehat{\boldsymbol{\beta}}$ for the scallops data.

Hint: $\widehat{\tau}^2$ is the nugget; $\widehat{\sigma}^2$ is the partial sill (the sill minus the nugget). Finally, the correlation matrix $H(\widehat{\phi})$ can be obtained from the spherical covariance function, part of your solution to Exercise 1.

Note: It appears that $\texttt{S+SpatialStats}$ uses the ordinary Euclidean (not geodesic) metric when computing distance, so you may use this as well when computing $H(\widehat{\phi})$. However, you may also wish to experiment with geodesic distances here, perhaps using your solution to Chapter 1, Exercise 7.

## CHAPTER 3

Basics of areal data models

We now present a development of exploratory tools and modeling approaches that are customarily applied to data collected for areal units. We have in mind general, possibly irregular geographic units, but of course include the special case of regular grids of cells (pixels). Indeed, the ensuing models have been proposed for regular lattices of points and parameters, and sometimes even for point-referenced data (see Appendix A, Section A.5 on the problem of inverting very large matrices).

In the context of areal units the general inferential issues are the following:

(i) Is there spatial pattern? If so, how strong is it? Intuitively, "spatial pattern" suggests measurements for areal units that are near to each other will tend to take more similar values than those for units far from each other. Though you might "know it when you see it," this notion is evidently vague and in need of quantification. Indeed, with independent measurements for each unit we expect to see *no pattern*, i.e., a completely random arrangement of larger and smaller values. But again, randomness will inevitably produce some patches of similar values.

(ii) Do we want to smooth the data? If so, how much? Suppose, for example, that the measurement for each areal unit is a count, say, a number of cancers. Even if the counts were independent, and perhaps even after population adjustment, there would still be extreme values, as in any sample. Are the observed high counts more elevated than would be expected by chance? If we sought to present a surface of expected counts we might naturally expect that the high values would tend to be pulled down, the low values to be pushed up. This is the notion of smoothing. No smoothing would present a display using simply the observed counts. Maximal smoothing would result in a single common value for all units, clearly excessive. Suitable smoothing would fall somewhere in between, and take the spatial arrangement of the units into account.

Of course, how much smoothing is appropriate is not readily defined. In particular, for model-based smoothers such as we describe below, it is not evident what the extent of smoothing is, or how to control it.

Specification of a utility function for smoothing (as attempted in Stern and Cressie, 1999) would help to address these questions.

(iii) For a new areal unit or set of units, how can we infer about what data values we expect to be associated with these units? That is, if we modify the areal units to new units, e.g., from zip codes to census block groups, what can we say about the cancer counts we expect for the latter given those for the former? This is the so-called *modifiable areal unit problem (MAUP)*, which historically (and in most GIS software packages) is handled by crude areal allocation. Sections 6.2 and 6.3 propose model-based methodology for handling this problem.

As a matter of fact, in order to facilitate interpretation and better assess uncertainty, we will suggest model-based approaches to treat the above issues, as opposed to the more descriptive or algorithmic methods that have dominated the literature and are by now widely available in GIS software packages. We will also introduce further flexibility into these models by examining them in the context of regression. That is, we will assume that we have available potential covariates to explain the areal unit responses. These covariates may be available at the same or at different scales from the responses, but, regardless, we will now question whether there remains any spatial structure adjusted for these explanatory variables. This suggests that we may not try to model the data in a spatial way directly, but instead introduce spatial association through random effects. This will lead to versions of generalized linear mixed models (Breslow and Clayton, 1993). We will often view such models in the hierarchical fashion that is the primary theme of this text.

3.1 Exploratory approaches for areal data

We begin with the presentation of some tools that can be useful in the initial exploration of areal unit data. The primary concept here is a *proximity matrix*, W. Given measurements $Y_1, \ldots, Y_n$ associated with areal units $1, 2, \ldots, n$, the entries w_{ij} in W spatially connect units i and j in some fashion. (Customarily w_{ii} is set to 0.) Possibilities include binary choices, i.e., $w_{ij} = 1$ if i and j share some common boundary, perhaps a vertex (as in a regular grid). Alternatively, w_{ij} could reflect "distance" between units, e.g., a decreasing function of intercentroidal distance between the units (as in a county or other regional map). But distance can be returned to a binary determination. For example, we could set $w_{ij} = 1$ for all i and j within a specified distance. Or, for a given i, we could get $w_{ij} = 1$ if j is one of the K nearest (in distance) neighbors of i. The preceding choices suggest that W would be symmetric. However, for irregular areal units, this last example provides a setting where this need not be the case. Also, the w_{ij}'s may be standardized by $\sum_j w_{ij} = w_{i+}$. If $\widetilde{W}$ has entries $\widetilde{w}_{ij} = w_{ij}/w_{i+}$, then evidently $\widetilde{W}$ is row stochastic, i.e., $\widetilde{W}\mathbf{1} = \mathbf{1}$, but now $\widetilde{W}$ need not be symmetric.

As the notation suggests, the entries in W can be viewed as weights. More weight will be associated with j's closer (in some sense) to i than those farther away from i. In this exploratory context (but, as we shall see, more generally) W provides the mechanism for introducing spatial structure into our formal modeling.

Lastly, working with distance suggests that we can define distance bins, say, $(0, d_1], (d_1, d_2], (d_2, d_3]$, and so on. This enables the notion of *first-order neighbors* of unit i, i.e., all units within distance d_1 of i, *second-order neighbors*, i.e., all units more than d_1 but at most d_2 from i, *third-order neighbors*, and so on. Analogous to W we can define $W^{(1)}$ as the proximity matrix for first-order neighbors. That is, $w_{ij}^{(1)} = 1$ if i and j are first-order neighbors, and equal to 0 otherwise. Similarly we define $W^{(2)}$ as the proximity matrix for second-order neighbors; $w_{ij}^{(2)} = 1$ if i and j are second-order neighbors, and 0 otherwise, and so on to create $W^{(3)}$, $W^{(4)}$, etc.

Of course, the most obvious exploratory data analysis tool for lattice data is a map of the data values. Figure 3.1 gives the statewide average verbal SAT scores as reported by the College Board and initially analyzed by Wall (2004). Clearly these data exhibit strong spatial pattern, with midwestern states and Utah performing best, and coastal states and Indiana performing less well. Of course, before jumping to conclusions, we must realize there are any number of spatial covariates that may help to explain this pattern; the percentage of eligible students taking the exam, for instance (Midwestern colleges have historically relied on the ACT, not the SAT, and only the best and brightest students in these states would bother taking the latter exam). Still, the map of these raw data show significant spatial pattern.

3.1.1 Measures of spatial association

Two standard statistics that are used to measure strength of spatial association among areal units are Moran's I and Geary's C (see, e.g., Ripley, 1981, Sec. 5.4). These are spatial analogues of statistics for measuring association in time series, the lagged autocorrelation coefficient and the Durbin-Watson statistic, respectively. They can also be seen to be areal unit analogues of the empirical estimates for the correlation function and the variogram, respectively. Recall that, for point-referenced data, the empirical covariance function (2.15) and semivariogram (2.9), respectively, provide customary nonparametric estimates of these measures of association.

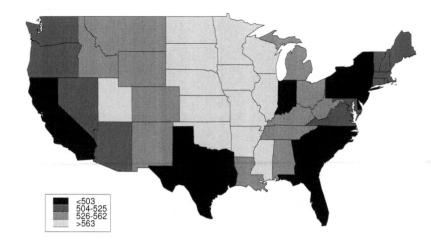

Figure 3.1 *Choropleth map of 1999 average verbal SAT scores, lower 48 U.S. states.*

Moran's I takes the form

$$I = \frac{n \sum_i \sum_j w_{ij} (Y_i - \overline{Y})(Y_j - \overline{Y})}{\left(\sum_{i \neq j} w_{ij} \right) \sum_i (Y_i - \overline{Y})^2} \, . \qquad (3.1)$$

I is not strictly supported on the interval $[-1, 1]$. It is evidently a ratio of quadratic forms in $\mathbf{Y}$ that provides the idea for obtaining approximate first and second moments through the delta method (see, e.g., Agresti, 2002, Ch. 14). Moran shows under the null model where the Y_i are i.i.d., I is asymptotically normally distributed with mean $-1/(n-1)$ and a rather unattractive variance of the form

$$Var(I) = \frac{n^2(n-1)S_1 - n(n-1)S_2 - 2S_0^2}{(n+1)(n-1)^2 S_0^2} \, . \qquad (3.2)$$

In (3.2), $S_0 = \sum_{i \neq j} w_{ij}$, $S_1 = \frac{1}{2} \sum_{i \neq j} (w_{ij} + w_{ji})^2$, and $S_2 = \sum_k (\sum_j w_{kj} + \sum_i w_{ik})^2$. We recommend the use of Moran's I as an exploratory measure of spatial association, rather than as a "test of spatial significance."

For the data mapped in Figure 3.1, we used the `spatial.cor` function in `S+SpatialStats` (see Section 2.5) to obtain a value for Moran's I of 0.5833, a reasonably large value. The associated standard error estimate of 0.0920 suggests very strong evidence against the null hypothesis of no spatial correlation in these data.

Geary's C takes the form

$$C = \frac{(n-1)\sum_i \sum_j w_{ij}(Y_i - Y_j)^2}{2\left(\sum_{i \neq j} w_{ij}\right) \sum_i (Y_i - \overline{Y})^2} . \tag{3.3}$$

C is never negative, and has mean 1 for the null model; *low* values (i.e., between 0 and 1) indicate *positive* spatial association. Also, C is a ratio of quadratic forms in $\mathbf{Y}$ and, like I, is asymptotically normal if the Y_i are i.i.d. We omit details of the distribution theory, recommending the interested reader to Cliff and Ord (1973), or Ripley (1981, p. 99).

Again using the `spatial.cor` function on the SAT verbal data in Figure 3.1, we obtained a value of 0.3775 for Geary's C, with an associated standard error estimate of 0.1008. Again, the marked departure from the mean of 1 indicates strong positive spatial correlation in the data.

If one truly seeks to run a significance test using (3.1) or (3.3), our recommendation is a Monte Carlo approach. Under the null model the distribution of I (or C) is invariant to permutation of the Y_i's. The exact null distribution of I (or C) requires computing its value under all $n!$ permutation of the Y_i's, infeasible for n in practice. However, a Monte Carlo sample of say 1000 permutations, including the observed one, will position the observed I (or C) relative to the remaining 999, to determine whether it is extreme (perhaps via an empirical p-value). Again using `spatial.cor` function on our SAT verbal data, we obtained empirical p-values of 0 using both Moran's I and Geary's C; *no* random permutation achieved I or C scores as extreme as those obtained for the actual data itself.

A further display that can be created in this spirit is the *correlogram*. Working with say I, in (3.1) we can replace w_{ij} with the previously defined $w_{ij}^{(1)}$ and compute say $I^{(1)}$. Similarly, we can replace w_{ij} with $w_{ij}^{(2)}$ and obtain $I^{(2)}$. A plot of $I^{(r)}$ vs. r is called a correlogram and, if spatial pattern is present, is expected to decline in r initially and then perhaps vary about 0. Evidently, this display is a spatial analogue of a temporal lag autocorrelation plot (e.g., see Carlin and Louis, 2000, p. 181). In practice, the correlogram tends to be very erratic and its information context is often not clear.

With large, regular grids of cells as we often obtain from remotely sensed imagery, it may be of interest to study spatial association in a particular direction (e.g., east-west, north-south, southwest-northeast, etc.). Now the spatial component reduces to one dimension and we can compute lagged autocorrelations (lagged appropriately to the size of the grid cells) in the specific direction. An analogue of this was proposed for the case where the Y_i are binary responses (e.g., presence or absence of forest in the cell) by Agarwal, Gelfand, and Silander (2002). In particular, Figure 3.2 shows rasterized maps of binary land use classifications for roughly 25,000 1 km

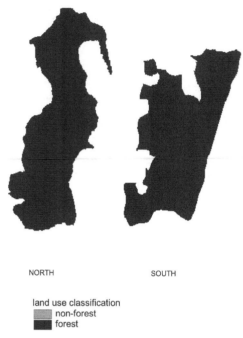

NORTH   SOUTH

land use classification
▒ non-forest
■ forest

Figure 3.2 *Rasterized north and south regions (1 km × 1 km) with binary land use classification overlaid.*

× 1 km pixels in eastern Madagascar; see Agarwal et al. (2002) as well as Section 6.4 for further discussion.

While the binary map in Figure 3.2 shows spatial pattern in land use, we develop an additional display to provide quantification. For data on a regular grid or lattice, we calculate binary analogues of the sample autocovariances, using the 1 km × 1 km resolution with four illustrative directions: East (E), Northeast (NE), North (N), and Northwest (NW). Relative to a given pixel, we can identify all pixels in the region in a specified direction from that pixel and associate with each a distance (Euclidean distance centroid to centroid) from the given pixel. Pairing the response at the given pixel (X) with the response at a directional neighbor (Y), we obtain a correlated binary pair. Collecting all such (X,Y) pairs at a given direction/distance combination yields a 2 × 2 table of counts. The resultant log-odds ratio measures the association between pairs in that direction at that distance. (Note that if we followed the same procedure but reversed direction, e.g., changed from E to W, the corresponding log odds ratio would be unchanged.)

In Figure 3.3, we plot log odds ratio against direction for each of the

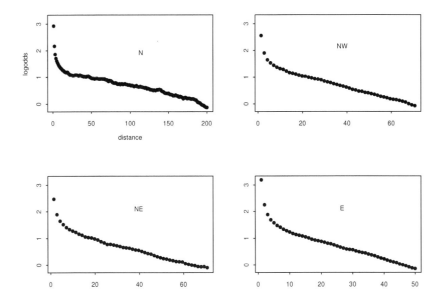

Figure 3.3 *Land use log-odds ratio versus distance in four directions.*

four directions. Note that the spatial association is quite strong, requiring a distance of at least 40 km before it drops to essentially 0. This suggests that we would not lose much spatial information if we work with the lower (4 km × 4 km) resolution. In exchange we obtain a richer response variable (17 ordered levels, indicating number of forested cells from 0 to 16) and a substantial reduction in number of pixels (from 26,432 to 1,652 in the north region, from 24,544 to 1,534 in the south region) to facilitate model fitting.

3.1.2 Spatial smoothers

Recall from the beginning of this chapter that often a goal for, say, a choropleth map of the Y_i's is *smoothing*. Depending upon the number of classes used to make the map, there is already some implicit smoothing in such a display (although this is not *spatial* smoothing, of course).

The W matrix directly provides a spatial smoother; that is, we can replace Y_i by $\widehat{Y}_i = \sum_j w_{ij} Y_j / w_{i+}$. This ensures that the value for areal unit i "looks like" its neighbors, and that the more neighbors we use in computing $\widehat{Y}_i$, the more smoothing we will achieve. In fact, $\widehat{Y}_i$ may be viewed as an unusual smoother in that it ignores the value actually observed for unit i. As such, we might revise the smoother to

$$\widehat{Y}_i^* = (1 - \alpha)Y_i + \alpha\widehat{Y}_i \,, \tag{3.4}$$

where $\alpha \in (0, 1)$. Working in an exploratory mode, various choices may be tried for α, but for any of these, (3.4) is a familiar *shrinkage* form. Thus, under a specific model with a suitable loss function, an optimal α could be sought. Finally, the form (3.4), viewed generally as a linear combination of the Y_j, is customarily referred to as a *filter* in the GIS literature. In fact, such software will typically provide choices of filters, and even a default filter to automatically smooth maps.

In Section 4.1 we will present a general discussion revealing how smoothing emerges as a byproduct of the hierarchical models we propose to use to explain the Y_i. In particular, when W is used in conjunction with a stochastic model (as in Section 3.3), the $\widehat{Y}_i$ are updated across i and across Monte Carlo iterations as well. So the observed Y_i will affect the eventual $\widehat{Y}_i$, and a "manual" inclusion of Y_i as in (3.4) is unnecessary.

3.2 Brook's Lemma and Markov random fields

A useful technical result for obtaining the joint distribution of the Y_i in some of the models we discuss below is *Brook's Lemma* (Brook, 1964). The usefulness of this lemma is exposed in Besag's (1974) seminal paper on conditionally autoregressive models.

It is clear that given $p(y_1, \ldots, y_n)$, the so-called *full conditional* distributions, $p(y_i|y_j, j \neq i)$, $i = 1, \ldots, n$, are uniquely determined. Brook's Lemma proves the converse and, in fact, enables us to constructively retrieve the unique joint distribution determined by these full conditionals. But first, it is also clear that we cannot write down an arbitrary set of full conditional distributions and assert that they uniquely determine the joint distribution. To see this, let $Y_1|Y_2 \sim N(\alpha_0 + \alpha_1 Y_2, \sigma_1^2)$ and let $Y_2|Y_1 \sim N(\beta_0 + \beta_1 Y_1^3, \sigma_2^2)$, where N denotes the normal (Gaussian) distribution. It is apparent that

$$E(Y_1) = E[E(Y_1|Y_2)] = E[\alpha_0 + \alpha_1 Y_2] = \alpha_0 + \alpha_1 E(Y_2) \,, \tag{3.5}$$

i.e., $E(Y_1)$ and $E(Y_2)$ are linearly related. But in fact, it must also be the case that

$$E(Y_2) = E[E(Y_2|Y_1)] = E[\beta_0 + \beta_1 Y_1] = \beta_0 + \beta_1 E(Y_1^3) \,. \tag{3.6}$$

Equations (3.5) and (3.6) could simultaneously hold only in trivial cases, so the two mean specifications are *incompatible*. Thus we can say that $f(y_1|y_2)$ and $f(y_2|y_1)$ are incompatible with regard to determining $p(y_1, y_2)$. We do not propose to examine conditions for compatibility here, although there has been considerable work in this area (see, e.g., Arnold and Strauss, 1991, and references therein).

Another point is that $p(y_1 \ldots, y_n)$ may be improper even if $p(y_i|y_j, j \neq i)$

is proper for all i. As an elementary illustration, consider $p(y_1, y_2) \propto \exp[-\frac{1}{2}(y_1 - y_2)^2]$. Evidently $p(y_1|y_2)$ is $N(y_2, 1)$ and $p(y_2|y_1)$ is $N(y_1, 1)$, but $p(y_1, y_2)$ is improper. Casella and George (1992) provide a similar example in a bivariate exponential (instead of normal) setting.

Brook's Lemma notes that

$$p(y_1, \ldots, y_n) = \frac{p(y_1|y_2, \ldots, y_n)}{p(y_{10}|y_2, \ldots, y_n)} \cdot \frac{p(y_2|y_{10}, y_3, \ldots, y_n)}{p(y_{20}|y_{10}, y_3, \ldots, y_n)} \qquad (3.7)$$
$$\ldots \frac{p(y_n|y_{10}, \ldots, y_{n-1,0})}{p(y_{n0}|y_{10}, \ldots, y_{n-1,0})} \cdot p(y_{10}, \ldots, y_{n0}),$$

an identity you are asked to check in Exercise 1. Here, $\mathbf{y}_0 = (y_{10}, \ldots, y_{n0})'$ is any fixed point in the support of $p(y_1, \ldots, y_n)$. Hence $p(y_1, \ldots, y_n)$ is determined by the full conditional distributions, since apart from the constant $p(y_{10}, \ldots, y_{n0})$ they are the only objects appearing on the right-hand side of (3.7). Hence the joint distribution is determined up to a proportionality constant. If $p(y_1, \ldots, y_n)$ is improper then this is, of course, the best we can do; if $p(y_1, \ldots, y_n)$ is proper then the fact that it integrates to 1 determines the constant. Perhaps most important is the constructive nature of (3.7): we can create $p(y_1, \ldots, y_n)$ simply by calculating the product of ratios. For more on this point see Exercise 2.

Usually, when the number of areal units is very large (say, a large number of small geographic regions, or a regular grid of pixels on a screen), we do not seek to write down the joint distribution of the Y_i. Rather we prefer to work (and model) exclusively with the n corresponding full conditional distributions. In fact, from a spatial perspective we would think that the full conditional distribution for Y_i should really depend only upon the neighbors of cell i. Adopting some definition of a neighbor structure (e.g., the one setting $W_{ij} = 1$ or 0 depending on whether i and j are adjacent or not), let ∂_i denote the set of neighbors of cell i.

Next suppose we specify a set of full conditional distributions for the Y_i such that

$$p(y_i|y_j, j \neq i) = p(y_i|y_j, j \in \partial_i) \qquad (3.8)$$

A critical question to ask is whether a specification such as (3.8) uniquely determines a joint distribution for $Y_1, \ldots Y_n$. That is, we do not need to see the explicit form of this distribution. We merely want to be assured that if, for example, we implement a Gibbs sampler (see Subsection 4.3.1) to simulate realizations from the joint distribution, that there is indeed a unique stationary distribution for this sampler.

The notion of using *local* specification to determine a joint (or global) distribution in the form (3.8) is referred to as a *Markov random field* (MRF). There is by now a substantial literature in this area, with Besag (1974) being a good place to start. Geman and Geman (1984) provide the next

critical step in the evolution, while Kaiser and Cressie (2000) offer a current
view and provide further references.

A critical definition in this regard is that of a *clique*. A clique is a set of
cells (equivalently, indices) such that each element is a neighbor of every
other element. With n cells, depending upon the definition of the neighbor
structure, cliques can possibly be of size 1, 2, and so on up to size n.
A *potential function* (or simply *potential*) of order k is a function of k
arguments that is exchangeable in these arguments. The arguments of the
potential would be the values taken by variables associated with the cells
for a clique of size k. For continuous Y_i, a customary potential when $k = 2$
is $Y_i Y_j$ if i and j are a clique of size 2. (We use the notation $i \sim j$ if i is
a neighbor of j and j is a neighbor of i.) For, say, binary Y_i, a potential
when $k = 2$ is

$$I(Y_i = Y_j) = Y_i Y_j + (1 - Y_i)(1 - Y_j) \,,$$

where again $i \sim j$ and I denotes the indicator function. Throughout this
book (and perhaps in most practical work as well), only cliques of order
less than or equal to 2 are considered.

Next, we define a *Gibbs distribution* as follows: $p(y_1, \ldots, y_n)$ is a Gibbs
distribution if it is a function of the Y_i only through potentials on cliques.
That is,

$$p(y_1, \ldots, y_n) \propto \exp \left\{ \gamma \sum_k \sum_{\alpha \in \mathcal{M}_k} \phi^{(k)}(y_{\alpha_1}, y_{\alpha_2}, \ldots, y_{\alpha_k}) \right\} \,. \qquad (3.9)$$

Here, $\phi^{(k)}$ is a potential of order k, $\mathcal{M}_k$ is the collection of all subsets of
size k from $\{1, 2, \ldots, n\}$, $\alpha = (\alpha_1, \ldots, \alpha_k)'$ indexes this set, and $\gamma > 0$ is a
scale (or "temperature") parameter.

Informally, the *Hammersley-Clifford Theorem* (see Besag, 1974; also Clifford, 1990) demonstrates that if we have an MRF, i.e., if (3.8) defines a
unique joint distribution, then this joint distribution is a Gibbs distribution. That is, it is of the form (3.9), with all of its "action" coming in the
form of potentials on cliques. Cressie (1993, pp. 417–18) offers a proof of
this theorem, and mentions that its importance for spatial modeling lies
in its limiting the complexity of the conditional distributions required, i.e.,
full conditional distributions can be specified locally.

Geman and Geman (1984) provided essentially the converse of the Hammersley-Clifford theorem. If we begin with (3.9) we have determined an
MRF. As a result, they argued that to sample a Markov random field, one
could sample from its associated Gibbs distribution, hence coining the term
"Gibbs sampler."

If we only use cliques of order 1, then the Y_i must be independent, as is
evidenced by (3.9). For continuous data on $\Re^1$, a common choice for the

joint distribution is a pairwise difference form

$$p(y_1,\ldots,y_n) \propto \exp\left\{-\frac{1}{2\tau^2}\sum_{i,j}(y_i - y_j)^2 I(i \sim j)\right\} . \quad (3.10)$$

Distributions such as (3.10) will be the focus of the next section. For the moment, we merely note that it is a Gibbs distribution on potentials of order 1 and 2 and that

$$p(y_i \mid y_j, j \neq i) = N\left(\sum_{j\in\partial_i} y_i/m_i , \ \tau^2/m_i\right) , \quad (3.11)$$

where m_i is the number of neighbors of cell i. The distribution in (3.11) is clearly of the form (3.8) and shows that the mean of Y_i is the average of its neighbors.

3.3 Conditionally autoregressive (CAR) models

Although they were introduced by Besag (1974) approximately 30 years ago, conditionally autoregressive (CAR) models have enjoyed a dramatic increase in usage only in the past decade or so. This resurgence arises from their convenient employment in the context of Gibbs sampling and more general Markov chain Monte Carlo (MCMC) methods for fitting certain classes of hierarchical spatial models (seen, e.g., in Section 5.4.3).

3.3.1 The Gaussian case

We begin with the Gaussian (or *autonormal*) case. Suppose we set

$$Y_i \mid y_j, j \neq i \sim N\left(\sum_j b_{ij}y_j , \ \tau_i^2\right) , \ i = 1,\ldots,n . \quad (3.12)$$

These full conditionals are compatible, so through Brook's Lemma we can obtain

$$p(y_1,\ldots,y_n) \propto \exp\left\{-\frac{1}{2}\mathbf{y}'D^{-1}(I - B)\mathbf{y}\right\} , \quad (3.13)$$

where $B = \{b_{ij}\}$ and D is diagonal with $D_{ii} = \tau_i^2$. Expression (3.13) suggests a joint multivariate normal distribution for $\mathbf{Y}$ with mean $\mathbf{0}$ and variance matrix $\Sigma_{\mathbf{y}} = (I - B)^{-1}D$.

But we are getting ahead of ourselves. First, we need to ensure that $D^{-1}(I - B)$ is symmetric. The simple resulting conditions are

$$\frac{b_{ij}}{\tau_i^2} = \frac{b_{ji}}{\tau_j^2} \quad \text{for all } i,j . \quad (3.14)$$

Evidently, from (3.14), B is not symmetric. Returning to our proximity matrix W (which we assume to be symmetric), suppose we set $b_{ij} = w_{ij}/w_{i+}$ and $\tau_i^2 = \tau^2/w_{i+}$. Then (3.14) is satisfied and (3.12) yields $p(y_i|y_j, j \neq i) = N\left(\sum_j w_{ij}y_j/w_{i+}, \; \tau^2/w_{i+}\right)$. Also, (3.13) becomes

$$p(y_1, \ldots, y_n) \propto \exp\left\{-\frac{1}{2\tau^2}\mathbf{y}'(D_w - W)\mathbf{y}\right\}, \tag{3.15}$$

where D_w is diagonal with $(D_w)_{ii} = w_{i+}$.

Now a second problem is noticed. $(D_w - W)\mathbf{1} = \mathbf{0}$, i.e., $\Sigma_{\mathbf{y}}^{-1}$ is singular, so that $\Sigma_{\mathbf{y}}$ does not exist and the distribution in (3.15) is improper. (The reader is encouraged to note the difference between the case of $\Sigma_{\mathbf{y}}^{-1}$ singular and the case of $\Sigma_{\mathbf{y}}$ singular. With the former we have a density function but one that is not integrable; effectively we have too many variables and we need a constraint on them to restore propriety. With the latter we have no density function but a proper distribution that resides in a lower dimensional space; effectively we have too *few* variables.) With a little algebra (3.15) can be rewritten as

$$p(y_1, \ldots, y_n) \propto \exp\left\{-\frac{1}{2\tau^2}\sum_{i \neq j} w_{ij}(y_i - y_j)^2\right\}. \tag{3.16}$$

This is a pairwise difference specification slightly more general than (3.10). But the impropriety of $p(\mathbf{y})$ is also evident from (3.16) since we can add any constant to all of the Y_i and (3.16) is unaffected; the Y_i are not "centered." A constraint such as $\sum_i Y_i = 0$ would provide the needed centering. Thus we have a more general illustration of a joint distribution that is improper, but has all full conditionals proper. The specification (3.16) is often referred to as an *intrinsically autoregressive* (IAR) model.

As a result, $p(\mathbf{y})$ in (3.15) cannot be used as a model for data; data could not arise under an improper stochastic mechanism, and we cannot impose a constant center on randomly realized measurements. Hence, the use of an improper autonormal model must be relegated to a *prior* distributional specification. That is, it will be attached to random spatial effects introduced at the second stage of a hierarchical specification (again, see e.g. Section 5.4.3).

The impropriety in (3.15) can be remedied in an obvious way. Redefine $\Sigma_{\mathbf{y}}^{-1} = D_w - \rho W$ and choose ρ to make $\Sigma_{\mathbf{y}}^{-1}$ nonsingular. This is guaranteed if $\rho \in \left(1/\lambda_{(1)}, 1/\lambda_{(n)}\right)$, where $\lambda_{(1)} < \lambda_{(2)} < \cdots < \lambda_{(n)}$ are the ordered eigenvalues of $D_w^{-1/2}WD_w^{-1/2}$; see Exercise 5. Moreover, since $tr(D_w^{-1/2}WD_w^{-1/2}) = 0 = \sum_{i=1}^n \lambda_{(i)}$, $\lambda_{(1)} < 0$, $\lambda_{(n)} > 0$, and 0 belongs to $\left(1/\lambda_{(1)}, 1/\lambda_{(n)}\right)$.

Simpler bounds than those given above for the propriety parameter ρ may be obtained if we replace the adjacency matrix W by the scaled adjacency matrix $\widetilde{W} \equiv Diag(1/w_{i+})W$; recall $\widetilde{W}$ is not symmetric, but it will be row stochastic (i.e., all of its rows sum to 1). $\Sigma_{\mathbf{y}}^{-1}$ can then be written as $M^{-1}(I - \alpha\widetilde{W})$ where M is diagonal. Then if $|\alpha| < 1$, $I - \alpha\widetilde{W}$ is nonsingular. (See the SAR model of the next section, as well as Exercise 7.) Carlin and Banerjee (2003) show that $\Sigma_{\mathbf{y}}^{-1}$ is diagonally dominant and symmetric. But diagonally dominant symmetric matrices are positive definite (Harville, 1997), providing an alternative argument for the propriety of the joint distribution.

Returning to the unscaled situation, ρ can be viewed as an additional parameter in the CAR specification, enriching this class of spatial models. Furthermore, $\rho = 0$ has an immediate interpretation: the Y_i become independent $N(0, \tau^2/w_{i+})$. If ρ is not included, independence cannot emerge as a limit of (3.15). (Incidentally, this suggests a clarification of the role of τ^2, the variance parameter associated with the full conditional distributions: the magnitude of τ^2 should *not* be viewed as in any way quantifying the strength of spatial association. Indeed if all Y_i are multiplied by c, τ^2 becomes $c\tau^2$ but the strength of spatial association among the Y_i is clearly unaffected.) Lastly, $\rho \sum_j w_{ij}Y_j/w_{i+}$ can be viewed as a *reaction function*, i.e., ρ is the expected proportional "reaction" of Y_i to $\sum_j w_{ij}Y_j/w_{i+}$.

With these advantages plus the fact that $p(\mathbf{y})$ (or the Bayesian posterior distribution, if the CAR specification is used to model constrained random effects) is now proper, is there any reason not to introduce the ρ parameter? In fact, the answer may be yes. Under $\Sigma_{\mathbf{y}}^{-1} = D_w - \rho W$, the full conditional $p(y_i|y_j, j \neq i)$ becomes $N\left(\rho \sum_j w_{ij}y_j/w_{i+}, \tau^2/w_{i+}\right)$. Hence we are modeling Y_i not to have mean that is an average of its neighbors, but some *proportion* of this average. Does this enable any sensible spatial interpretation for the CAR model? Moreover, does ρ calibrate very well with any familiar interpretation of "strength of spatial association?" Fixing $\tau^2 = 1$ without loss of generality, we can simulate CAR realizations for a given n, W, and ρ. We can also compute for these realizations a descriptive association measure such as Moran's I or Geary's C. Here we do not present explicit details of the range of simulations we have conducted. However, for a 10×10 grid using a first-order neighbor system, when $\rho = 0.8$, I is typically 0.1 to 0.15; when $\rho = 0.9$, I is typically 0.2 to 0.25; and even when $\rho = 0.99$, I is typically at most 0.5. It thus appears that ρ can mislead with regard to strength of association. Expressed in a different way, within a Bayesian framework, a prior on ρ that encourages a consequential amount of spatial association would place most of its mass near 1.

A related point is that if $p(\mathbf{y})$ is proper, the breadth of spatial pattern may be too limited. In the case where a CAR model is applied to random effects, an improper choice may actually enable wider scope for posterior spatial pattern. As a result, we do not take a position with regard to propriety or impropriety in employing CAR specifications (though in the remain-

der of this text we do sometimes attempt to illuminate relative advantages
and disadvantages).

Referring to (3.12), we may write the entire system of random variables
as

$$\mathbf{Y} = B\mathbf{Y} + \boldsymbol{\epsilon}, \quad \text{or equivalently,} \tag{3.17}$$

$$(I - B)\mathbf{Y} = \boldsymbol{\epsilon}. \tag{3.18}$$

In particular, the distribution for $\mathbf{Y}$ induces a distribution for $\boldsymbol{\epsilon}$. If $p(\mathbf{y})$ is
proper then $\mathbf{Y} \sim N(\mathbf{0}, (I - B)^{-1}D)$ whence $\boldsymbol{\epsilon} \sim N(\mathbf{0}, D(I - B)')$, i.e., the
components of $\boldsymbol{\epsilon}$ are not independent. Also, $Cov(\boldsymbol{\epsilon}, \mathbf{Y}) = D$.

When $p(\mathbf{y})$ is proper we can appeal to standard multivariate normal dis-
tribution theory to interpret the entries in $\Sigma_{\mathbf{y}}^{-1}$. For example, $1/(\Sigma_{\mathbf{y}}^{-1})_{ii} = Var(Y_i | Y_j, j \neq i)$. Of course with $\Sigma_{\mathbf{y}}^{-1} = D^{-1}(I - B)$, $(\Sigma_{\mathbf{y}}^{-1})_{ii} = 1/\tau_i^2$
providing immediate agreement with (3.12). But also, if $(\Sigma_{\mathbf{y}}^{-1})_{ij} = 0$, then
Y_i and Y_j are conditionally independent given $Y_k, k \neq i, j$, a fact you are
asked to show in Exercise 8. Hence if any $b_{ij} = 0$, we have conditional independence for that pair of variables. Connecting b_{ij} to w_{ij} shows that the
choice of neighbor structure implies an associated collection of conditional
independences. With first-order neighbor structure, all we are asserting is
a spatial illustration of the local Markov property (Whittaker, 1990, p. 68).

We conclude this subsection with three remarks. First, one can directly
introduce a regression component into (3.12), e.g., a term of the form $\mathbf{x}_i'\boldsymbol{\beta}$.
Conditional on $\boldsymbol{\beta}$, this does not affect the association structure that ensues
from (3.12); it only revises the mean structure. However, we omit details
here (the interested reader can consult Besag, 1974), since we will only use
the autonormal CAR as a distribution for spatial random effects. These
effects are added onto the regression structure for the mean on some transformed scale (again, see Section 5.4.3).

We also note that in suitable contexts it may be appropriate to think
of $\mathbf{Y}_i$ as a vector of dependent areal unit measurements or, in the context
of random effects, as a vector of dependent random effects associated with
an areal unit. This leads to the specification of multivariate conditionally
autoregressive (MCAR) models, which is the subject of Section 7.4. From
a somewhat different perspective, $\mathbf{Y}_i$ might arise as $(Y_{i1}, \ldots, Y_{iT})'$ where
Y_{it} is the measurement associated with areal unit i at time t, $t = 1, \ldots, T$.
Now we would of course think in terms of spatiotemporal modeling for Y_{it}.
This is the subject of Section 8.5.

Lastly, a (proper) CAR model can in principle be used for point-level
data, taking w_{ij} to be, say, an inverse distance between points i and j.
However, unlike the spatial prediction described in Section 2.4, now spatial
prediction becomes *ad hoc*. That is, to predict at a new site Y_0, we might
specify the distribution of Y_0 given $Y_1, \ldots, Y_n$ to be a normal distribution,
such as a $N\left(\rho \sum_j w_{0j} y_j / w_{0+}, \tau^2 / w_{0+}\right)$. Note that this determines the

joint distribution of $Y_0, Y_1, \ldots, Y_n$. However, this joint distribution is *not* the CAR distribution that would arise by specifying the full conditionals for $Y_0, Y_1, \ldots, Y_n$ and using Brook's Lemma, as in constructing (3.15).

3.3.2 The non-Gaussian case

If one seeks to model the data directly using a CAR specification then in many cases a normal distribution would not be appropriate. Binary response data and sparse count data are two examples. In fact, one can select any exponential family model as a first-stage distribution for the data and propose

$$p\left(y_i | y_j, j \neq i\right) \propto \exp\left(\{\psi\left(\theta_i y_i - \chi\left(\theta_i\right)\right)\}\right), \tag{3.19}$$

where, adopting a canonical link, $\theta_i = \sum_{j \neq i} b_{ij} y_j$ and ψ is a non-negative dispersion parameter. In fact (3.19) simplifies to

$$p\left(y_i | y_j, j \neq i\right) \propto \exp\left(\psi \sum_{j \neq i} b_{ij} y_i y_j\right). \tag{3.20}$$

Since the data are being modeled directly, it may be appropriate to introduce a nonautoregressive linear regression component to (3.20). That is, we can write $\theta_i = \mathbf{x}_i^T \boldsymbol{\gamma} + \sum_{j \neq i} b_{ij} y_j$, for some set of covariates $\mathbf{x}_i$. After obvious reparametrization (3.20) becomes

$$p\left(y_i | y_j, j \neq i\right) \propto \exp\left(\mathbf{x}_i^T \boldsymbol{\gamma} + \psi \sum_{j \neq i} b_{ij} y_j\right), \tag{3.21}$$

where on the left hand side we suppress the conditioning on $\boldsymbol{\gamma}$ and ψ.

In the case where the Y_i are binary, a version that has received attention in the literature is the *autologistic* model; see, e.g., Heikkinen and Hogmander (1994), Hogmander and Møller (1995), and Hoeting et al. (2000). Here,

$$\log \frac{P\left(Y_i = 1\right)}{P\left(Y_i = 0\right)} = \mathbf{x}_i^T \boldsymbol{\gamma} + \psi \sum w_{ij} y_j, \tag{3.22}$$

where $w_{ij} = 1$ if $i \sim j$, $= 0$ otherwise. Using Brook's Lemma the joint distribution of $Y_1, \ldots, Y_n$ can be shown to be

$$p\left(y_1, \ldots, y_n\right) \propto \exp\left(\boldsymbol{\gamma}^T \left(\sum_i y_i \mathbf{x}_i\right) + \psi \sum_{i,j} w_{ij} y_i y_j\right). \tag{3.23}$$

Expression (3.23) shows that f is indeed a Gibbs distribution and appears to be an attractive form. But for likelihood or Bayesian inference, the normalizing constant is required, since it is a function of $\boldsymbol{\gamma}$ and ψ. However,

computation of this constant requires summation over all of the 2^n possible values that $(Y_1, Y_2, ..., Y_n)$ can take on. Even for moderate sample sizes
this will present computational challenges. Hoeting et al. (2000) propose
approximations to the likelihood using a pseudo-likelihood and a normal
approximation.

The case where Y_i can take on one of several categorical values presents a
natural extension to the autologistic model. If we label the (say) L possible
outcomes as simply $1, 2, ..., L$, then we can define

$$P\left(Y_i = l \mid Y_j, j \neq i\right) \propto \exp\left(\psi \sum_{j \neq i} w_{ij} I\left(Y_j = l\right)\right), \qquad (3.24)$$

with w_{ij} as above. The distribution in (3.24) is referred to as a *Potts model*.
It obviously extends the binary case and encourages Y_i to be like its neighbors. It also suffers from the normalization problem. It can also be employed
as a random effects specification, as an alternative to an autonormal; see
the allocation model in Green and Richardson (2002) in this regard.

3.4 Simultaneous autoregressive (SAR) models

Returning to (3.17), suppose that instead of letting $\mathbf{Y}$ induce a distribution
for $\boldsymbol{\epsilon}$, we let $\boldsymbol{\epsilon}$ induce a distribution for $\mathbf{Y}$. Imitating usual autoregressive
time series modeling, suppose we take the ϵ_i to be independent innovations.
For a little added generality, assume that $\boldsymbol{\epsilon} \sim N\left(0, \tilde{D}\right)$ where $\tilde{D}$ is diagonal
with $\left(\tilde{D}\right)_{ii} = \sigma_i^2$. (Note $\tilde{D}$ has no connection with D in Section 3.3; the
B we use below may or may not be the same as the one we used in that
section.) Analogous to (3.12), now $Y_i = \sum_j b_{ij} Y_j + \epsilon_i$, $i = 1, 2, ..., n$, with
$\epsilon_i \sim N\left(0, \sigma_i^2\right)$. Therefore, if $(I - B)$ is full rank,

$$\mathbf{Y} \sim N\left(\mathbf{0}, (I - B)^{-1} \tilde{D}\left((I - B)^{-1}\right)'\right). \qquad (3.25)$$

Also, $Cov(\boldsymbol{\epsilon}, \mathbf{Y}) = \tilde{D}(I - B)^{-1}$. If $\tilde{D} = \sigma^2 I$ then (3.25) simplifies to $\mathbf{Y} \sim$
$N\left(\mathbf{0}, \sigma^2\left[(I - B)(I - B)'\right]^{-1}\right)$. In order that (3.25) be proper, $I - B$ must
be full rank. Two choices are most frequently discussed in the literature
(e.g., Griffith, 1988). The first assumes $B = \rho W$, where W is a so-called
contiguity matrix, i.e., W has entries that are 1 or 0 according to whether
or not unit i and unit j are direct neighbors (with $w_{ii} = 0$). So W is our
familiar first-order neighbor proximity matrix. Here ρ is called a *spatial
autoregression parameter* and, evidently, $Y_i = \rho \sum_j Y_j I(j \in \partial_i) + \epsilon_i$, where
∂_i denotes the set of neighbors of i. In fact, any proximity matrix can be
used and, paralleling the discussion below (3.15), $I - \rho W$ will be nonsingular

if $\rho \in \left(\frac{1}{\lambda_{(1)}}, \frac{1}{\lambda_{(n)}} \right)$ where now $\lambda_{(1)} < \cdots < \lambda_{(n)}$ are the ordered eigenvalues of W.

Alternatively, W can be replaced by $\widetilde{W}$ where now, for each i, the ith row has been normalized to sum to 1. That is, $\left(\tilde{W} \right)_{ij} = w_{ij}/w_{i+}$. Again, $\widetilde{W}$ is not symmetric, but it is row stochastic, i.e., $\widetilde{W}\mathbf{1} = \mathbf{1}$. If we set $B = \alpha\widetilde{W}$, α is called a *spatial autocorrelation parameter* and, were W a contiguity matrix, now $Y_i = \alpha \sum_j Y_i I(j \in \partial_i)/w_{i+} + \epsilon_i$. With a very regular grid the w_{i+} will all be essentially the same and thus α will be a multiple of ρ. But, perhaps more importantly, with $\widetilde{W}$ row stochastic the eigenvalues of $\widetilde{W}$ are all less than or equal to 1 (i.e., max $|\lambda_i| = 1$). Thus $I - \alpha\widetilde{W}$ will be nonsingular if $\alpha \in (-1, 1)$, justifying referring to α as an autocorrelation parameter; see Exercise 7.

A SAR model is customarily introduced in a regression context, i.e., the *residuals* $\mathbf{U} = \mathbf{Y} - X\boldsymbol{\beta}$ are assumed to follow a SAR model, rather than $\mathbf{Y}$ itself. But then, following (3.17), if $\mathbf{U} = B\mathbf{U} + \boldsymbol{\epsilon}$, we obtain the attractive form

$$\mathbf{Y} = B\mathbf{Y} + (I - B)X\boldsymbol{\beta} + \boldsymbol{\epsilon}. \tag{3.26}$$

Expression (3.26) shows that $\mathbf{Y}$ is modeled through a component that provides a spatial weighting of neighbors and a component that is a usual linear regression. If B is the zero matrix we obtain an OLS regression; if $B = I$ we obtain a purely spatial model.

We note that from (3.26) the SAR model does not introduce any spatial effects; the errors in (3.26) are independent. Expressed in a different way, if we modeled $\mathbf{Y} - X\boldsymbol{\beta}$ as $\mathbf{U} + \mathbf{e}$ with $\mathbf{e}$ independent errors, we would have $\mathbf{U} + \mathbf{e} = B\mathbf{U} + \boldsymbol{\epsilon} + \mathbf{e}$ and $\boldsymbol{\epsilon} + \mathbf{e}$ would result in a redundancy. As a result, in practice a SAR specification is not used in conjunction with a GLM. To introduce $\mathbf{U}$ as a vector of spatial adjustments to the mean vector, a transformed scale creates redundancy between the independent Gaussian error in the definition of the U_i and the stochastic mechanism associated with the conditionally independent Y_i.

We briefly note the somewhat related spatial modeling approach of Langford et al. (1999). Rather than modeling the residual vector $\mathbf{U} = B\mathbf{U} + \boldsymbol{\epsilon}$, they propose that $\mathbf{U} = \tilde{B}\boldsymbol{\epsilon}$ where $\boldsymbol{\epsilon} \sim N\left(\mathbf{0}, \sigma^2 I\right)$, i.e., that $\mathbf{U}$ be modeled as a spatially motivated linear combination of independent variables. This induces $\Sigma_U = \sigma^2 \tilde{B}\tilde{B}^T$. Thus, the U_i and hence the Y_i will be dependent and given $\tilde{B}$, $cov(Y_i, Y_{i'}) = \sigma^2 \sum_j b_{ij}b_{i'j}$. If B arises through some proximity matrix W, the more similar rows i and i' of W are, the stronger the association between Y_i and $Y_{i'}$. However, the difference in nature between this specification and that in (3.26) is evident. To align the two, we would set $(I - B)^{-1} = \tilde{B}$, i.e. $B = I - \tilde{B}^{-1}$ (assuming $\tilde{B}$ is of full rank). $I - \tilde{B}^{-1}$ would not appear to have any interpretation through a proximity matrix.

Perhaps the most important point to note with respect to SAR models is that they are well suited to maximum likelihood estimation but not at all for MCMC fitting of Bayesian models. That is, the log likelihood associated with (3.26) (assuming $\tilde{D} = \sigma^2 I$) is

$$\frac{1}{2} \log \left| \sigma^{-1} (I - B) \right| - \frac{1}{2\sigma^2} (\mathbf{Y} - X\boldsymbol{\beta})^T (I - B) (I - B)^T (\mathbf{Y} - X\boldsymbol{\beta}) . \quad (3.27)$$

Though B will introduce a regression or autocorrelation parameter, the quadratic form in (3.27) is quick to calculate (requiring no inverse) and the determinant can usually be calculated rapidly using diagonally dominant, sparse matrix approximations (see, e.g., Pace and Barry, 1997a,b). Thus maximization of (3.27) can be done iteratively but, in general, efficiently.

Also, note that while the form in (3.27) can certainly be extended to a full Bayesian model through appropriate prior specifications, the absence of a hierarchical form with random effects implies straightforward Bayesian model fitting as well. Indeed, the general spatial slice Gibbs sampler (see Appendix Section A.6, or Agarwal and Gelfand, 2002) can easily handle this model. However, suppose we attempt to introduce SAR random effects in some fashion. Unlike CAR random effects that are defined through full conditional distributions, the full conditional distributions for the SAR effects have no convenient form. For large n, computation of such distributions using a form such as (3.25) will be expensive.

SAR models as in (3.26) are frequently employed in the spatial econometrics literature. With point-referenced data, B is taken to be ρW where W is the matrix of interpoint distances. Likelihood-based inference can be implemented in S+SpatialStats as well as more specialized software, such as that from the Spatial Analysis Laboratory (sal.agecon.uiuc.edu)). Software for large data sets is supplied there, as well as through the website of Prof. Kelley Pace, www.spatial-statistics.com. An illustrative example is provided in Exercise 10.

CAR versus SAR models

Cressie (1993, pp. 408–10) credits Brook (1964) with being the first to make a distinction between the CAR and SAR models, and offers a comparison of the two. To begin with, we may note from (3.13) and (3.25) that the two forms are equivalent if and only if

$$(I - B)^{-1} D = (I - \tilde{B})^{-1} \tilde{D} ((I - \tilde{B})^{-1})' ,$$

where we use the tilde to indicate matrices in the SAR model. Cressie then shows that any SAR model can be represented as a CAR model (since D is diagonal), but gives a counterexample to prove that the converse is not true. For the "proper" CAR and SAR models that include spatial correlation parameters ρ, Wall (2004) shows that the correlations between neighboring regions implied by these two models can be rather different; in particular, the first-order neighbor correlations increase at a slower rate

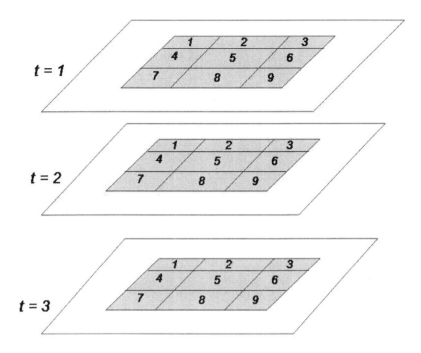

Figure 3.4 *Illustration of spatiotemporal areal unit setting for STAR model.*

as a function of ρ in the CAR model than they do for the SAR model. (As an aside, she notes that these correlations are not even monotone for $\rho <$ 0, another reason to avoid negative spatial correlation parameters.) Also, correlations among pairs can switch in nonintuitive ways. For example, when working with the adjacency relationships generated by the lower 48 contiguous U.S. states, she finds that when $\rho = .49$ in the CAR model, $Corr(Alabama, Florida) = .20$ and $Corr(Alabama, Georgia) = .16$. But when ρ increases to .975, we instead get $Corr(Alabama, Florida) = .65$ and $Corr(Alabama, Georgia) = .67$, a slight reversal in ordering.

STAR models

In the literature SAR models have frequently been extended to handle spatiotemporal data. The idea is that in working with proximity matrices, we can define neighbors in time as well as in space. Figure 3.4 shows a simple illustration with 9 areal units, 3 temporal units for each areal unit yielding $i = 1, \ldots, 9$, $t = 1, 2, 3$, labeled as indicated.

The measurements Y_{it} are spatially associated at each fixed t. But also, we might seek to associate, say, Y_{i2} with Y_{i1} and Y_{i3}. Suppose we write Y

as the 27×1 vector with the first nine entries at $t = 1$, the second nine at $t = 2$, and the last nine at $t = 3$. Also let $W_S = BlockDiag(W_1, W_1, W_1)$, where

$$W_1 = \begin{pmatrix} 0 & 1 & 0 & 1 & 0 & 0 & 0 & 0 & 0 \\ 1 & 0 & 1 & 0 & 1 & 0 & 0 & 0 & 0 \\ 0 & 1 & 0 & 0 & 0 & 1 & 0 & 0 & 0 \\ 1 & 0 & 0 & 0 & 1 & 0 & 1 & 0 & 0 \\ 0 & 1 & 0 & 1 & 0 & 1 & 0 & 1 & 0 \\ 0 & 0 & 1 & 0 & 1 & 0 & 0 & 0 & 1 \\ 0 & 0 & 0 & 1 & 0 & 0 & 0 & 1 & 0 \\ 0 & 0 & 0 & 0 & 1 & 0 & 1 & 0 & 1 \\ 0 & 0 & 0 & 0 & 0 & 1 & 0 & 1 & 0 \end{pmatrix}.$$

Then W_S provides a spatial contiguity matrix for the Y's. Similarly, let $W_T = \begin{pmatrix} 0 & W_2 & 0 \\ W_2 & 0 & W_2 \\ 0 & W_2 & 0 \end{pmatrix}$, where $W_2 = I_{3\times 3}$. Then W_T provides a *temporal* contiguity matrix for the Y's. But then, in our SAR model we can define $B = \rho_s W_S + \rho_t W_T$. In fact, we can also introduce $\rho_{ST} W_S W_T$ into B and note that

$$W_S W_T = \begin{pmatrix} 0 & W_1 & 0 \\ W_1 & 0 & W_1 \\ 0 & W_1 & 0 \end{pmatrix}.$$

In this way, we introduce association across both space and time. For instance Y_{21} and Y_{41} affect the mean of Y_{12} (as well as affecting Y_{11}) from W_S by itself. Many more possibilities exist. Models formulated through such more general definitions of B are referred to as *spatiotemporal autoregressive* (STAR) models. See Pace et al. (2000) for a full discussion and development. The interpretation of the ρ's in the above example measures the relative importance of first-order spatial neighbors, first order temporal neighbors, and first-order spatiotemporal neighbors.

3.5 Computer tutorials

In this section we outline the use of the S+SpatialStats package in constructing spatial neighborhood (adjacency) matrices, fitting CAR and SAR models using traditional maximum likelihood techniques, and mapping the results for certain classes of problems. Here we confine ourselves to the modeling of Gaussian data on areal units. As in Section 2.5, we adopt a tutorial style.

3.5.1 *Adjacency matrix construction in* S+SpatialStats

The most common specification for a SAR model is obtained by setting $B = \rho W$ and $\tilde{D} = Diag(\sigma_i^2)$, where W is some sort of spatial dependence matrix, and ρ measures the strength of spatial association. As such, of fundamental importance is the structure of W, which is often taken as an *adjacency* (or *contiguity*) matrix. It is therefore important to begin with the specification of such matrices in S+SpatialStats. Our discussion follows that outlined in Kaluzny et al. (1998, Ch.5), and we refer the reader to that text for further details.

One way of specifying neighborhood structures is through lists in an ordinary text (ASCII) file. For example,

```
1    2  4
2    1  3  5  6
3    2  4  5
4    1  3  6
5    2  3  7
6    2  4
7    5
```

is a typical text listing of adjacencies. Here we have 7 sites, where site 1 has sites 2 and 4 as neighbors, site 2 has sites 1, 3, 5, and 6 as neighbors, and so on. The function read.neighbor in S+SpatialStats reads such a text file and converts it to a spatial.neighbor object, the fundamental adjacency-storage object in the language. Thus, if we write the above matrix to a file called Neighbors.txt, we may create a spatial.neighbor object (say, ngb) as

>ngb <- read.neighbor(''Neighbors.txt'', keep=F)

By default, the spatial.neighbor object ngb is larger than required, since the symmetry of the neighbors is not accounted for. To correct this, the size can be reduced using the spatial.condense function:

>ngb <- spatial.condense(ngb, symmetry=T)

Another, perhaps more direct method of creating neighbor objects is by invoking the spatial.neighbor function directly on an $n \times n$ contiguity matrix. Note that the neighbor relations listed above are equivalent to the (symmetric, 0-1) contiguity matrix

```
0  1  0  1  0  0  0
1  0  1  0  1  1  0
0  1  0  1  1  0  0
1  0  1  0  0  1  0
0  1  1  0  0  0  1
0  1  0  1  0  0  0
0  0  0  0  1  0  0
```

Suppose these relations are stored in a file called Adjacency.txt. A spatial.neighbor object may be created from this contiguity matrix as follows:

>no.sites <- 7

```
>ngb.mat <- matrix(scan(''Adjacency.txt''),
    ncol=no.sites,byrow=T)
>ngb2 <- spatial.neighbor(neighbor.matrix=ngb.mat,
    nregion=no.sites, symmetric=T)
```

Note that the "symmetry" above refers to the spatial dependence matrix W, and may not always be appropriate. For example, recall that a common specification is to take W as the *row-normalized* adjacency matrix. In such cases, each element is scaled by the sum of the corresponding row, and the resulting W matrix is not symmetric. To form a row-normalized spatial dependence matrix W, we modify the above example to

```
>ngb2 <- spatial.neighbor(neighbor.matrix=ngb.mat,
    nregion=no.sites, weights=1/c(2,4,3,3,3,2,1))
```

Here, the weights are the number of neighbors (i.e., the number of elements in each row of Neighbors.txt).

3.5.2 SAR and CAR model fitting in S+SpatialStats

We next turn to fitting Gaussian linear spatial models using the slm (spatial linear model) function in S+SpatialStats. A convenient illustration is offered by the SIDS (sudden infant death syndrome) data, analyzed by Cressie (1993, Sec. 6.2) and Kaluzny et al. (1998, Sec. 5.3), and already loaded into the S+SpatialStats package. This data frame contains counts of SIDS deaths from 1974 to 1978 along with related covariate information for the 100 counties in the U.S. state of North Carolina. We fit two spatial autoregressive models with the dependent variable as sid.ft (a Freedman-Tukey transformation of the ratio of the number of SIDS cases to the total number of births in each county). Further information about the data frame can be obtained by typing

```
>help(sids)
```

We first fit a null model (no covariates). Note that sids.neighbor (built into S+SpatialStats) is a spatial.neighbor object containing the contiguity structure for the 100 North Carolina counties. Specifically, region.id is the variable that identifies the way the regions are numbered, while weights specifies the elements of the W matrix. We follow Cressie (1993) and assign the reciprocal of the births in the county as the weights. The resulting model fit (without covariates) is obtained by typing

```
>sids.nullslm.SAR <- slm(sid.ft~1, cov.family=SAR,
    data=sids, spatial.arglist=list(neighbor=sids.neighbor,
    region.id=1:100, weights=1/sids$births))
>null.SAR.summary <- summary(sids.nullslm.SAR)
```

To fit a Gaussian spatial regression model with a regressor (say, the ratio of non-white to total births in each county between 1974 and 1978), we simply modify the above to

```
>sids.raceslm.SAR <- slm(sid.ft~nwbirths.ft,
  cov.family=SAR, data=sids, spatial.arglist
  =list(neighbor=sids.neighbor, region.id=1:100,
  weights=1/sids$births))
>race.SAR.summary <- summary(sids.raceslm.SAR)
```

The output contained in `race.SAR.summary` is as follows:

```
Call:
slm(formula = sid.ft ~ nwbirths.ft, cov.family = SAR,
data = sids, spatial.arglist = list(neighbor = sids.neighbor,
region.id = 1:100, weights = 1/sids\$births))}


Residuals:
   Min    1Q  Median    3Q    Max
-106.9 -18.28   4.692 25.53 79.09


Coefficients:
            Value  Std. Error t value Pr(>|t|)
(Intercept) 1.6729    0.2480  6.7451   0.0000
nwbirths.ft 0.0337    0.0069  4.8998   0.0000


Residual standard error: 34.4053 on 96 degrees of freedom


Variance-Covariance Matrix of Coefficients
            (Intercept)     nwbirths.ft
(Intercept) 0.061513912   -1.633112e-03
nwbirths.ft -0.001633112    4.728317e-05


Correlation of Coefficient Estimates
            (Intercept)  nwbirths.ft
(Intercept ) 1.000000     -0.957582
nwbirths.ft -0.957582      1.000000
```

Note that a county's non-white birth rate does appear to be significantly associated with its SIDS rate, but this covariate is strongly negatively associated with the intercept. We also remark that the `slm` function can also fit a CAR (instead of SAR) model simply by specifying `cov.family=CAR` above.

Next, instead of defining a neighborhood structure completely in terms of spatial adjacency on the map, we may want to construct neighbors using a distance function. For example, given centroids of the various regions, we could identify regions as neighbors if and only if their intercentroidal distance is below a particular threshold.

We illustrate using `www.biostat.umn.edu/~brad/data/Columbus.dat`, a data set offering neighborhood-level information on crime, mean home value, mean income, and other variables for 49 neighborhoods in Columbus, OH, during 1980. More information on these data is available from Anselin (1988, p.189), or in Exercise 10.

We begin by creating the data frame:

```
>columbus <- read.table(''Columbus.dat'', header=T)
```

Suppose we would like to have regions with intercentroidal distances less than 2.5 units as neighbors. We first form an object, columbus.coords, that contain the centroids of the different regions. The function that we use is find.neighbor, but a required intermediate step is making a *quad tree*, which is a matrix providing the most efficient ordering for the nearest neighbor search. This is accomplished using the quad.tree function in S+SpatialStats. The following steps will create a spatial.neighbor object in this way:

```
>columbus.coords <- cbind(columbus$X, columbus$Y)
>columbus.quad <- quad.tree(columbus.coords)
>columbus.ngb <- find.neighbor(x=columbus.coords,
    quadtree=columbus.quad, max.dist=2.5)
>columbus.ngb <- spatial.neighbor(row.id=columbus.ngb[,1],
    col.id=columbus.ngb[,2])
```

Once our neighborhood structure is created, we proceed to fit a CAR model (having crime rate as the response and house value and income as covariates) as follows:

```
>columbus.CAR <- slm(CRIME ~ HOVAL + INC, cov.family=CAR,
    data=columbus, spatial.arglist=list(neighbor=
    columbus.ngb, region.id=1:49))
>columbus.CAR.summary <- summary(columbus.CAR)
```

The output from columbus.CAR.summary, similar to that given above for the SAR model, reveals both covariates to be significant (both p-values near .002).

3.5.3 Choropleth mapping using the maps library in S-plus

Finally, we describe the drawing of choropleth maps in S+SpatialStats. In fact, S-plus is all we need here, thanks to the maps library originally described by Becker and Wilks (1993). This map library, invoked using the command,

```
>library(maps)
```

contains the geographic boundary files for several maps, including county boundaries for every state in the U.S. However, other important regional boundary types (say, zip codes) and features (rivers, major roads, and railroads) are generally not available. As such, while S-plus is not nearly as versatile as `ArcView` or other GIS packages, it does offer a rare combination of GIS and statistical analysis capabilities.

We will now map the actual transformed SIDS rates along with their fitted values under the SAR model of the previous subsection. Before we can do this, however, a special feature of the polygon boundary file of the `S-plus` North Carolina county map must be accounted for. Specifically, county #27, Currituck county, is apparently comprised of not one but three separate regions. Thus, when mapping the raw SIDS rates, Kaluzny et al. (1998) propose the following solution: form a modified vector for the mapping variable (`sid.ft`), but with Currituck county appearing three times:

```
>sids.map <- c(sids$sid.ft[1:26],rep(sids$sid.ft[27],3),
    sids$sid.ft[28:100])
```

We next form a vector of the cutpoints that determine the different bins into which the rates will be classified, and assign the county rates to these bins:

```
>breaks.sids <- c(-0.001,2.0,3.0,3.5,7.0)
>sids.mapgrp <- cut(sids.map, breaks.sids)
```

We now must assign a color (or shade of gray) to each bin. An oddity in the default postscript color specification of `S-plus` is that color "1" is black, and then increasingly lighter shades are given by colors 3, 2, and 4 (not 2, 3, and 4, as you might expect). While this problem may be overcome by careful work with the `ps.options` command, here we simply use the nonintuitive 4-2-3-1 lightest to darkest grayscale ordering, which is obtained here simply by swapping categories 1 and 4:

```
>sids.mapgrp[(sids.mapgrp==1)] <- 0
>sids.mapgrp[(sids.mapgrp==4)] <- 1
>sids.mapgrp[(sids.mapgrp==0)] <- 4
```

Now the map of the actual (transformed) SIDS rates can be obtained as

```
>map("county", "north carolina", fill=T, color=sids.mapgrp)
>map("county", "north carolina", add=T)
>title(main="Actual Transformed SIDS Rates")
>legend(locator(1), legend=
    c("<2.0","2.0-3.0","3.0-3.5",">3.5"), fill=c(4,2,3,1))
```

In the first command, the modified mapping vector `sids.mapgrp` is specified as the grouping variable for the different colors. The `fill=T` option automates the shading of regions, while the next command (with `add=T`) adds the county boundaries. Finally, the `locator(1)` option within the `legend` command waits for the user to click on the position where the legend is desired; Figure 3.5(a) contains the result we obtained. We hasten to add that one can automate the placing of the legend by replacing the

a) actual transformed SIDS rates

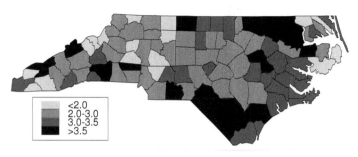

b) fitted SIDS rates from SAR model

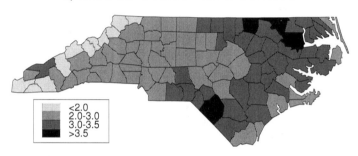

Figure 3.5 *Unsmoothed raw (a) and spatially smoothed fitted (b) rates, North Carolina SIDS data.*

locator(1) option with actual (x, y) coordinates for the upper left corner of the legend box.

To draw a corresponding map of the fitted values from our SAR model (using our parameter estimates in the mean structure), we must first create a modified vector of the fits (again due to the presence of Currituck county):

```
>sids.race.fit <- fitted(sids.raceslm.SAR)
>sids.race.fit.map <- c(sids.race.fit[1:26],
    rep(sids.race.fit[27],3), sids.race.fit[28:100])
>sids.race.fit.mapgrp <- cut(sids.race.fit.map,
    breaks.sids)
```

where breaks.sids is the same color cutoff vector as earlier. The map is then drawn as follows:

```
>sids.race.fit.mapgrp <- cut(sids.race.fit.map,
    breaks.sids)
```

```
>sids.race.fit.mapgrp[(sids.race.fit.mapgrp==1)] <- 0
>sids.race.fit.mapgrp[(sids.race.fit.mapgrp==4)] <- 1
>sids.race.fit.mapgrp[(sids.race.fit.mapgrp==0)] <- 4
>map("county", "north carolina", fill=T,
  color=sids.race.fit.mapgrp)
>map("county", "north carolina", add=T)
>title(main="Fitted SIDS Rates from SAR Model")
>legend(locator(1), legend=
  c("<2.0","2.0-3.0","3.0-3.5",">3.5"), fill=c(4,2,3,1))
```

Figure 3.5(b) contains the result. Note that the SAR model has resulted in significant smoothing of the observed rates, and clarified the generally increasing pattern as we move from west to east.

Finally, if a map of predicted (rather than fitted) values is desired, these values can be formed as

```
>noise <- 1/sqrt(sids$births)*resid(sids.raceslm.SAR)
>signal <- sids$sid.ft - sids.race.fit - noise
>sids.race.pred <- signal + sids.race.fit
>sids.race.pred.map <- c(sids.race.pred[1:26],
  rep(sids.race.pred[27],3), sids.race.pred[28:100])
>sids.race.pred.mapgrp <- cut(sids.race.pred.map,
  breaks.sids)
```

The actual drawing of the maps then proceeds exactly as before.

3.6 Exercises

1. Verify Brook's Lemma, equation (3.7).

2.(a) To appreciate how Brook's Lemma works, suppose Y_1 and Y_2 are both binary variables, and that their joint distribution is defined through conditional logit models. That is,

$$\log \frac{P(Y_1 = 1|Y_2)}{P(Y_1 = 0|Y_2)} = \alpha_0 + \alpha_1 Y_2 \quad \text{and} \quad \log \frac{P(Y_2 = 1|Y_1)}{P(Y_2 = 0|Y_1)} = \beta_0 + \beta_1 Y_1 .$$

Obtain the joint distribution of Y_1 and Y_2.

 (b) This result can be straightforwardly extended to the case of more than two variables, but the details become increasingly clumsy. Illustrate this issue in the case of *three* binary variables, Y_1, Y_2, and Y_3.

3. Returning to (3.13) and (3.14), let $B = ((b_{ij}))$ be an $n \times n$ matrix with positive elements; that is, $b_{ij} > 0$, $\sum_j b_{ij} \le 1$ for all i, and $\sum_j b_{ij} < 1$ for at least one i. Let $D = Diag\left(\tau_i^2\right)$ be a diagonal matrix with positive elements τ_i^2 such that $D^{-1}(I - B)$ is symmetric; that is, $b_{ij}/\tau_i^2 = b_{ji}/\tau_j^2$, for all i, j. Show that $D^{-1}(I - B)$ is positive definite.

4. Looking again at (3.13), obtain a simple sufficient condition on B such

that the CAR specification with precision matrix $D^{-1}(I - B)$ is a pairwise difference specification, as in (3.16).

5. Show that $\Sigma_{\mathbf{y}}^{-1} = D_w - \rho W$ is positive definite (thus resolving the impropriety in (3.15)) if $\rho \in \left(1/\lambda_{(1)}, 1/\lambda_{(n)}\right)$, where $\lambda_{(1)} < \lambda_{(2)} < \cdots < \lambda_{(n)}$ are the ordered eigenvalues of $D_w^{-1/2} W D_w^{-1/2}$.

6. Show that if all entries in W are nonnegative and $D_w - \rho W$ is positive definite with $0 < \rho < 1$, then all entries in $(D_w - \rho W)^{-1}$ are nonnegative.

7. Recalling the SAR formulation using the scaled adjacency matrix $\widetilde{W}$ just below (3.25), prove that $I - \alpha \widetilde{W}$ will be nonsingular if $\alpha \in (-1, 1)$, so that α may be sensibly referred to as an "autocorrelation parameter."

8. In the setting of Subsection 3.3.1, if $(\Sigma_{\mathbf{y}}^{-1})_{ij} = 0$, then show that Y_i and Y_j are conditionally independent given $Y_k, k \neq i, j$.

9. The file www.biostat.umn.edu/~brad/data/state-sat.dat gives the 1999 state average SAT data (part of which is mapped in Figure 3.1), while www.biostat.umn.edu/~brad/data/contig-lower48.dat gives the contiguity (adjacency) matrix for the lower 48 U.S. states (i.e., excluding Alaska and Hawaii, as well as the District of Columbia).

   (a) Use the S+SpatialStats software to construct a spatial.neighbor object from the contiguity file.

   (b) Use the slm function to fit the SAR model of Section 3.4, taking the verbal SAT score as the response Y and the percent of eligible students taking the exam in each state as the covariate X. Use row-normalized weights based on the contiguity information in spatial.neighbor object. Is knowing X helpful in explaining Y?

   (c) Using the maps library in S-plus, draw choropleth maps similar to Figure 3.1 of both the fitted verbal SAT scores and the spatial residuals from this fit. Is there evidence of spatial correlation in the response Y once the covariate X is accounted for?

   (d) Repeat your SAR model analysis above, again using slm but now assuming the CAR model of Section 3.3. Compare your estimates with those from the SAR model and interpret any changes.

   (e) One might imagine that the percentage of eligible students taking the exam should perhaps affect the variance of our model, not just the mean structure. To check this, refit the SAR model replacing your row-normalized weights with weights equal to the reciprocal of the percentage of students taking the SAT. Is this model sensible?

10. Consider the data www.biostat.umn.edu/~brad/data/Columbus.dat, taken from Anselin (1988, p. 189). These data record crime information for 49 neighborhoods in Columbus, OH, during 1980. Variables measured include NEIG, the neighborhood id value (1–49); HOVAL, its mean

housing value (in $1,000); INC, its mean household income (in $1,000); CRIME, its number of residential burglaries and vehicle thefts per thousand households; OPEN, a measure of the neighborhood's open space; PLUMB, the percentage of housing units without plumbing; DISCBD, the neighborhood centroid's distance from the central business district; X, an x-coordinate for the neighborhood centroid (in arbitrary digitizing units, not polygon coordinates); Y, the same as X for the y-coordinate; AREA, the neighborhood's area; and PERIM, the perimeter of the polygon describing the neighborhood.

(a) Use S+SpatialStats to construct spatial.neighbor objects for the neighborhoods of Columbus based upon centroid distances less than

   i. 3.0 units,
   ii. 7.0 units,
   iii. 15 units.

(b) For each of the four spatial neighborhoods constructed above, use the slm function to fit SAR models with CRIME as the dependent variable, and HOVAL, INC, OPEN, PLUMB, and DISCBD as the covariates. Compare your results and interpret your parameter estimates in each case.

(c) Repeat your analysis using Euclidean distances in the B matrix itself. That is, in equation (3.26), set $B = \rho W$ with the W_{ij} the Euclidean distance between location i and location j.

(d) Repeat part (b) for CAR models. Compare your estimates with those from the SAR model and interpret them.

## CHAPTER 4

Basics of Bayesian inference

In this chapter we provide a brief review of hierarchical Bayesian modeling and computing for readers not already familiar with these topics. Of course, in one chapter we can only scratch the surface of this rapidly expanding field, and readers may well wish to consult one of the many recent textbooks on the subject, either as preliminary work or on an as-needed basis. It should come as little surprise that the book we most highly recommend for this purpose is the one by Carlin and Louis (2000); the Bayesian methodology and computing material below roughly follows Chapters 2 and 5, respectively, in that text.

However, a great many other good Bayesian books are available, and we list a few of them and their characteristics. First we must mention the texts stressing Bayesian theory, including DeGroot (1970), Berger (1985), Bernardo and Smith (1994), and Robert (1994). These books tend to focus on foundations and decision theory, rather than computation or data analysis. On the more methodological side, a nice introductory book is that of Lee (1997), with O'Hagan (1994) and Gelman, Carlin, Stern, and Rubin (2004) offering more general Bayesian modeling treatments.

4.1 Introduction to hierarchical modeling and Bayes' Theorem

By modeling both the observed data and any unknowns as random variables, the Bayesian approach to statistical analysis provides a cohesive framework for combining complex data models and external knowledge or expert opinion. In this approach, in addition to specifying the distributional model $f(\mathbf{y}|\boldsymbol{\theta})$ for the observed data $\mathbf{y} = (y_1, \ldots, y_n)$ given a vector of unknown parameters $\boldsymbol{\theta} = (\theta_1, \ldots, \theta_k)$, we suppose that $\boldsymbol{\theta}$ is a random quantity sampled from a *prior* distribution $\pi(\boldsymbol{\theta}|\boldsymbol{\lambda})$, where $\boldsymbol{\lambda}$ is a vector of hyperparameters. For instance, y_i might be the empirical mammography rate in a sample of women aged 40 and over from county i, θ_i the underlying true mammography rate for all such women in this county, and $\boldsymbol{\lambda}$ a parameter controlling how these true rates vary across counties. If $\boldsymbol{\lambda}$ is known, inference concerning $\boldsymbol{\theta}$ is based on its *posterior* distribution,

$$p(\boldsymbol{\theta}|\mathbf{y},\boldsymbol{\lambda}) = \frac{p(\mathbf{y},\boldsymbol{\theta}|\boldsymbol{\lambda})}{p(\mathbf{y}|\boldsymbol{\lambda})} = \frac{p(\mathbf{y},\boldsymbol{\theta}|\boldsymbol{\lambda})}{\int p(\mathbf{y},\boldsymbol{\theta}|\boldsymbol{\lambda})\,d\boldsymbol{\theta}} = \frac{f(\mathbf{y}|\boldsymbol{\theta})\pi(\boldsymbol{\theta}|\boldsymbol{\lambda})}{\int f(\mathbf{y}|\boldsymbol{\theta})\pi(\boldsymbol{\theta}|\boldsymbol{\lambda})\,d\boldsymbol{\theta}} \ . \qquad (4.1)$$

Notice the contribution of both the data (in the form of the likelihood f) and the external knowledge or opinion (in the form of the prior π) to the posterior. Since, in practice, $\boldsymbol{\lambda}$ will not be known, a second stage (or *hyperprior*) distribution $h(\boldsymbol{\lambda})$ will often be required, and (4.1) will be replaced with

$$p(\boldsymbol{\theta}|\mathbf{y}) = \frac{p(\mathbf{y},\boldsymbol{\theta})}{p(\mathbf{y})} = \frac{\int f(\mathbf{y}|\boldsymbol{\theta})\pi(\boldsymbol{\theta}|\boldsymbol{\lambda})h(\boldsymbol{\lambda})\,d\boldsymbol{\lambda}}{\int f(\mathbf{y}|\boldsymbol{\theta})\pi(\boldsymbol{\theta}|\boldsymbol{\lambda})h(\boldsymbol{\lambda})\,d\boldsymbol{\theta}d\boldsymbol{\lambda}} \ .$$

Alternatively, we might replace $\boldsymbol{\lambda}$ by an estimate $\hat{\boldsymbol{\lambda}}$ obtained as the maximizer of the marginal distribution $p(\mathbf{y}|\boldsymbol{\lambda}) = \int f(\mathbf{y}|\boldsymbol{\theta})\pi(\boldsymbol{\theta}|\boldsymbol{\lambda})d\boldsymbol{\theta}$, viewed as a function of $\boldsymbol{\lambda}$. Inference could then proceed based on the *estimated* posterior distribution $p(\boldsymbol{\theta}|\mathbf{y},\hat{\boldsymbol{\lambda}})$, obtained by plugging $\hat{\boldsymbol{\lambda}}$ into equation (4.1). This approach is referred to as *empirical Bayes* analysis; see Berger (1985), Maritz and Lwin (1989), and Carlin and Louis (2000) for details regarding empirical Bayes methodology and applications.

The Bayesian inferential paradigm offers potentially attractive advantages over the classical, frequentist statistical approach through its more philosophically sound foundation, its unified approach to data analysis, and its ability to formally incorporate prior opinion or external empirical evidence into the results via the prior distribution π. Data analysts, formerly reluctant to adopt the Bayesian approach due to general skepticism concerning its philosophy and a lack of necessary computational tools, are now turning to it with increasing regularity as classical methods emerge as both theoretically and practically inadequate. Modeling the θ_i as random (instead of fixed) effects allows us to induce specific (e.g., spatial) correlation structures among them, hence among the observed data y_i as well. Hierarchical Bayesian methods now enjoy broad application in the analysis of spatial data, as the remainder of this book reveals.

A computational challenge in applying Bayesian methods is that for most realistic problems, the integrations required to do inference under (4.1) are generally not tractable in closed form, and thus must be approximated numerically. Forms for π and h (called *conjugate* priors) that enable at least partial analytic evaluation of these integrals may often be found, but in the presense of nuisance parameters (typically unknown variances), some intractable integrations remain. Here the emergence of inexpensive, high-speed computing equipment and software comes to the rescue, enabling the application of recently developed Markov chain Monte Carlo (MCMC) integration methods, such as the Metropolis-Hastings algorithm (Metropolis et al., 1953; Hastings, 1970) and the Gibbs sampler (Geman and Geman, 1984; Gelfand and Smith, 1990). This is the subject of Section 4.3.

Illustrations of Bayes' Theorem

Equation (4.1) is a generic version of what is referred to as *Bayes' Theorem* or *Bayes' Rule*. It is attributed to Reverend Thomas Bayes, an 18th-century nonconformist minister and part-time mathematician; a version of the result was published (posthumously) in Bayes (1763). In this subsection we consider a few basic examples of its use.

Example 4.1 Suppose we have observed a single normal (Gaussian) observation $Y \sim N\left(\theta, \sigma^2\right)$ with σ^2 known, so that the likelihood $f\left(y|\theta\right) = N\left(y|\theta, \sigma^2\right) \equiv \frac{1}{\sigma\sqrt{2\pi}} \exp(-\frac{(y-\theta)^2}{2\sigma^2})$, $y \in \Re$, $\theta \in \Re$, and $\sigma > 0$. If we specify the prior distribution as $\pi\left(\theta\right) = N\left(y \,\middle|\, \mu, \tau^2\right)$ with $\boldsymbol{\lambda} = (\mu, \tau^2)'$ fixed, then from (4.1) we can compute the posterior as

$$
\begin{aligned}
p\left(\theta|y\right) &= \frac{N\left(\theta|\mu, \tau^2\right) N\left(y|\theta, \sigma^2\right)}{p\left(y\right)} \\
&\propto N\left(\theta|\mu, \tau^2\right) N\left(y|\theta, \sigma^2\right) \\
&= N\left(\theta \,\middle|\, \frac{\sigma^2}{\sigma^2 + \tau^2}\mu + \frac{\tau^2}{\sigma^2 + \tau^2}y\,,\ \frac{\sigma^2\tau^2}{\sigma^2 + \tau^2}\right). \quad (4.2)
\end{aligned}
$$

That is, the posterior distribution of θ given y is also normal with mean and variance as given. The proportionality in the second row arises since the marginal distribution $p(y)$ does not depend on θ, and is thus constant with respect to the Bayes' Theorem calculation. The final equality in the third row results from collecting like (θ^2 and θ) terms in the exponential and then completing the square.

Note that the posterior mean $E(\theta|y)$ is a weighted average of the prior mean μ and the data value y, with the weights depending on our relative uncertainty with respect to the prior and the likelihood. Also, the posterior *precision* (reciprocal of the variance) is equal to $1/\sigma^2 + 1/\tau^2$, which is the sum of the likelihood and prior precisions. Thus, thinking of precision as "information," we see that in the normal/normal model, the information in the posterior is the total of the information in the prior and the likelihood.

Suppose next that instead of a single datum we have a set of n observations $\mathbf{y} = (y_1, y_2, \ldots, y_n)'$. From basic normal theory we know that $f(\bar{y}|\theta) = N(\theta, \sigma^2/n)$. Since $\bar{y}$ is sufficient for θ, from (4.2) we have

$$
\begin{aligned}
p(\theta|\mathbf{y}) = p\left(\theta|\bar{y}\right) &= N\left(\theta \,\middle|\, \frac{(\sigma^2/n)}{(\sigma^2/n) + \tau^2}\mu + \frac{\tau^2}{(\sigma^2/n) + \tau^2}\bar{y}\,,\ \frac{(\sigma^2/n)\tau^2}{(\sigma^2/n) + \tau^2}\right) \\
&= N\left(\theta \,\middle|\, \frac{\sigma^2}{\sigma^2 + n\tau^2}\mu + \frac{n\tau^2}{\sigma^2 + n\tau^2}\bar{y}\,,\ \frac{\sigma^2\tau^2}{\sigma^2 + n\tau^2}\right).
\end{aligned}
$$

Again we obtain a posterior mean that is a weighted average of the prior (μ) and data-supported ($\bar{y}$) values. ∎

In these two examples, the prior chosen leads to a posterior distribution for θ that is available in closed form, and is a member of the same distributional family as the prior. Such a prior is referred to as a *conjugate* prior. We will often use such priors in our work, since, when they are available, conjugate families are convenient and still allow a variety of shapes wide enough to capture our prior beliefs.

Note that setting $\tau^2 = \infty$ in the previous example corresponds to a prior that is arbitrarily vague, or *noninformative*. This then leads to a posterior of $p(\theta|y) = N(\theta|\bar{y}, \sigma^2/n)$, exactly the same as the likelihood for this problem. This arises since the limit of the conjugate (normal) prior here is actually a uniform, or "flat" prior, and thus the posterior is nothing but the likelihood (possibly renormalized to integrate to 1 as a function of θ). Of course, the flat prior is *improper* here, since the uniform does not integrate to anything finite over the entire real line; however, the posterior is still well defined since the likelihood can be integrated with respect to θ. Bayesians often use flat or otherwise improper noninformative priors, since prior feelings are often rather vague relative to the information in the likelihood, and in any case we typically want the data (and not the prior) to dominate the determination of the posterior.

Example 4.2 *(the general linear model).* Let $\mathbf{Y}$ be an $n \times 1$ data vector, X an $n \times p$ matrix of covariates, and adopt the likelihood and prior structure,

$$\mathbf{Y}|\boldsymbol{\beta} \sim N_n(X\boldsymbol{\beta}, \Sigma), \text{ i.e. } f(\mathbf{Y}|\boldsymbol{\beta}) \equiv N_n(\mathbf{Y}|X\boldsymbol{\beta}, \Sigma),$$
$$\boldsymbol{\beta} \sim N_p(A\boldsymbol{\alpha}, V), \text{ i.e. } \pi(\boldsymbol{\beta}) \equiv N(\boldsymbol{\beta}|A\boldsymbol{\alpha}, V).$$

Here $\boldsymbol{\beta}$ is a $p \times 1$ vector of regression coefficients and Σ is a $p \times p$ covariance matrix. Then it can be shown (now a classic result, first published by Lindley and Smith, 1972), that the marginal distribution of $\mathbf{Y}$ is

$$\mathbf{Y} \sim N\left(XA\boldsymbol{\alpha}, \Sigma + XVX^T\right),$$

and the posterior distribution of $\boldsymbol{\beta}|\mathbf{Y}$ is

$$\boldsymbol{\beta}|Y \sim N(D\mathbf{d}, D),$$
$$\text{where } D^{-1} = X^T\Sigma^{-1}X + V^{-1}$$
$$\text{and } \mathbf{d} = X^T\Sigma^{-1}\mathbf{Y} + V^{-1}A\boldsymbol{\alpha}.$$

Thus $E(\boldsymbol{\beta}|\mathbf{Y}) = D\mathbf{d}$ provides a point estimate for $\boldsymbol{\beta}$, with variability captured by the associated variance matrix D.

In particular, note that for a vague prior we may set $V^{-1} = 0$, so that $D^{-1} = X\Sigma^{-1}X$ and $\mathbf{d} = X^T\Sigma^{-1}\mathbf{Y}$. In the simple case where $\Sigma = \sigma^2 I_p$, the posterior becomes

$$\boldsymbol{\beta}|Y \sim N\left(\hat{\boldsymbol{\beta}}, \sigma^2(X'X)^{-1}\right),$$

where $\hat{\boldsymbol{\beta}} = (X'X)^{-1}X'\mathbf{y}$. Since the usual likelihood approach produces

$$\hat{\boldsymbol{\beta}} \sim N\left(\boldsymbol{\beta}, \sigma^2(X'X)^{-1}\right) ,$$

we once again we see "flat prior" Bayesian results that are formally equivalent to the usual likelihood approach. ∎

4.2 Bayesian inference

While the computing associated with Bayesian methods can be daunting, the subsequent inference is relatively straightforward, especially in the case of estimation. This is because once we have computed (or obtained an estimate of) the posterior, inference comes down merely to summarizing this distribution, since by Bayes' Rule the posterior summarizes everything we know about the model parameters in the light of the data. In the remainder of this section, we shall assume for simplicity that the posterior $p(\boldsymbol{\theta}|\mathbf{y})$ itself (and not merely an estimate of it) is available for summarization.

Bayesian methods for estimation are also reminiscent of corresponding maximum likelihood methods. This should not be surprising, since likelihoods form an important part of the Bayesian calculation; we have even seen that a normalized (i.e., standardized) likelihood can be thought of a posterior when this is possible. However, when we turn to hypothesis testing, the approaches have little in common. Bayesians (and many likelihoodists) have a deep and abiding antipathy toward p-values, for a long list of reasons we shall not go into here; the interested reader may consult Berger (1985, Sec. 4.3.3), Kass and Raftery (1995, Sec. 8.2), or Carlin and Louis (2000, Sec. 2.3.3).

4.2.1 Point estimation

To keep things simple, suppose for the moment that θ is univariate. Given the posterior $p(\theta|\mathbf{y})$, a sensible Bayesian point estimate of θ would be some measure of centrality. Three familiar choices are the posterior mean,

$$\hat{\theta} = E(\theta|\mathbf{y}) ,$$

the posterior median,

$$\hat{\theta} : \int_{-\infty}^{\hat{\theta}} p(\theta|\mathbf{y})d\theta = 0.5 ,$$

and the posterior mode,

$$\hat{\theta} : p(\hat{\theta}|\mathbf{y}) = \sup_{\theta} p(\theta|\mathbf{y}) .$$

Notice that the lattermost estimate is typically easiest to compute, since it does not require any integration: we can replace $p(\theta|\mathbf{y})$ by its unstandardized form, $f(\mathbf{y}|\theta)p(\theta)$, and get the same answer (since these two differ only by a multiplicative factor of $m(\mathbf{y})$, which does not depend on θ). Indeed, if the posterior exists under a flat prior $p(\theta) = 1$, then the posterior mode is nothing but the maximum likelihood estimate (MLE).

Note that for symmetric unimodal posteriors (e.g., a normal distribution), the posterior mean, median, and mode will all be equal. However, for multimodal or otherwise nonnormal posteriors, the mode will often be the poorest choice of centrality measure (consider for example the case of a steadily decreasing, one-tailed posterior; the mode will be the very first value in the support of the distribution — hardly central!). By contrast, the posterior mean will sometimes be overly influenced by heavy tails (just as the sample mean $\bar{y}$ is often nonrobust against outlying observations). As a result, the posterior median will often be the best and safest point estimate. It is also the most difficult to compute (since it requires both an integration and a rootfinder), but this difficulty is somewhat mitigated for posterior estimates computed via MCMC; see Section 4.3 below.

4.2.2 Interval estimation

The posterior allows us to make direct probability statements about not just its median, but any quantile. For example, suppose we can find the $\alpha/2$- and $(1 - \alpha/2)$-quantiles of $p(\theta|\mathbf{y})$, that is, the points q_L and q_U such that

$$\int_{-\infty}^{q_L} p(\theta|\mathbf{y})d\theta = \alpha/2 \text{ and } \int_{q_U}^{\infty} p(\theta|\mathbf{y})d\theta = 1 - \alpha/2 .$$

Then clearly $P(q_L < \theta < q_U|\mathbf{y}) = 1 - \alpha$; our confidence that θ lies in (q_L, q_U) is $100 \times (1 - \alpha)\%$. Thus this interval is a $100 \times (1 - \alpha)\%$ *credible set* (or simply *Bayesian confidence interval*) for θ. This interval is relatively easy to compute, and enjoys a direct interpretation ("the probability that θ lies in (q_L, q_U) is $(1 - \alpha)$") that the usual frequentist interval does not.

The interval just described is often called the *equal tail* credible set, for the obvious reason that is obtained by chopping an equal amount of support $(\alpha/2)$ off the top and bottom of $p(\theta|\mathbf{y})$. Note that for symmetric unimodal posteriors, this equal tail interval will be symmetric about this mode (which we recall equals the mean and median in this case). It will also be optimal in the sense that it will have shortest length among sets C satisfying

$$1 - \alpha \leq P(C|\mathbf{y}) = \int_C p(\theta|\mathbf{y})d\theta . \qquad (4.3)$$

Note that any such set C could be thought of as a $100 \times (1 - \alpha)\%$ credible set for θ. For posteriors that are not symmetric and unimodal, a better (shorter) credible set can be obtained by taking only those values of θ having posterior density greater than some cutoff $k(\alpha)$, where this cutoff is chosen to be as large as possible while C still satisfies equation (4.3). This *highest posterior density* (HPD) confidence set will always be of optimal length, but will typically be significantly more difficult to compute. The equal tail interval emerges as HPD in the symmetric unimodal case since there too it captures the "most likely" values of θ. Fortunately, many of the posteriors we will be interested in will be (at least approximately) symmetric unimodal, so the much simpler equal tail interval will often suffice.

4.2.3 Hypothesis testing and model choice

We have seen that Bayesian inference (point or interval) is quite straightforward given the posterior distribution, or an estimate thereof. By contrast, hypothesis testing is less straightforward, for two reasons. First, there is less agreement among Bayesians as to the proper approach to the problem. For years, posterior probabilities and Bayes factors were considered the only appropriate method. But these methods are only suitable with fully proper priors, and for relatively low-dimensional models. With the recent proliferation of very complex models with at least partly improper priors, other methods have come to the fore. Second, solutions to hypothesis testing questions often involve not just the posterior $p(\boldsymbol{\theta}|\mathbf{y})$, but also the *marginal* distribution, $m(\mathbf{y})$. Unlike the case of posterior and the predictive distributions, samples from the marginal distribution do not naturally emerge from most MCMC algorithms. Thus, the sampler must often be "tricked" into producing the necessary samples.

Recently, an approximate yet very easy-to-use model choice tool known as the Deviance Information Criterion (DIC) has gained popularity, as well as implementation in the WinBUGS software package. We will limit our attention in this subsection to Bayes factors, the DIC, and a related posterior predictive criterion due to Gelfand and Ghosh (1998). The reader is referred to Carlin and Louis (2000, Sections 2.3.3, 6.3, 6.4, and 6.5) for further techniques and information.

Bayes factors

We begin by setting up the hypothesis testing problem as a model choice problem, replacing the customary two hypotheses H_0 and H_A by two candidate parametric models M_1 and M_2 having respective parameter vectors $\boldsymbol{\theta}_1$ and $\boldsymbol{\theta}_2$. Under prior densities $\pi_i(\boldsymbol{\theta}_i)$, $i = 1, 2$, the marginal distributions of $\mathbf{Y}$ are found by integrating out the parameters,

$$p(\mathbf{y}|M_i) = \int f(\mathbf{y}|\boldsymbol{\theta}_i, M_i)\pi_i(\boldsymbol{\theta}_i)d\boldsymbol{\theta}_i \; , \; i = 1, 2 \; . \qquad (4.4)$$

Bayes' Theorem (4.1) may then be applied to obtain the posterior probabilities $P(M_1|\mathbf{y})$ and $P(M_2|\mathbf{y}) = 1 - P(M_1|\mathbf{y})$ for the two models. The quantity commonly used to summarize these results is the *Bayes factor*, BF, which is the ratio of the posterior odds of M_1 to the prior odds of M_1, given by Bayes' Theorem as

$$BF = \frac{P(M_1|\mathbf{y})/P(M_2|\mathbf{y})}{P(M_1)/P(M_2)} \tag{4.5}$$

$$= \frac{\left[\frac{p(\mathbf{y}|M_1)P(M_1)}{p(\mathbf{y})}\right] / \left[\frac{p(\mathbf{y}|M_2)P(M_2)}{p(\mathbf{y})}\right]}{P(M_1)/P(M_2)}$$

$$= \frac{p(\mathbf{y} \mid M_1)}{p(\mathbf{y} \mid M_2)} , \tag{4.6}$$

the ratio of the observed marginal densities for the two models. Assuming the two models are *a priori* equally probable (i.e., $P(M_1) = P(M_2) = 0.5$), we have that $BF = P(M_1|\mathbf{y})/P(M_2|\mathbf{y})$, the posterior odds of M_1.

Consider the case where both models share the same parametrization (i.e., $\boldsymbol{\theta}_1 = \boldsymbol{\theta}_2 = \boldsymbol{\theta}$), and both hypotheses are simple (i.e., $M_1 : \boldsymbol{\theta} = \boldsymbol{\theta}^{(1)}$ and $M_2 : \boldsymbol{\theta} = \boldsymbol{\theta}^{(2)}$). Then $\pi_i(\boldsymbol{\theta})$ consists of a point mass at $\boldsymbol{\theta}^{(i)}$ for $i = 1, 2$, and so from (4.4) and (4.6) we have

$$BF = \frac{f(\mathbf{y}|\boldsymbol{\theta}^{(1)})}{f(\mathbf{y}|\boldsymbol{\theta}^{(2)})} ,$$

which is nothing but the likelihood ratio between the two models. Hence, in the simple-versus-simple setting, the Bayes factor is precisely the odds in favor of M_1 over M_2 *given solely by the data.*

A popular "shortcut" method is the *Bayesian Information Criterion* (BIC) (also known as the *Schwarz Criterion*), the change in which across the two models is given by

$$\Delta BIC = W - (p_2 - p_1) \log n , \tag{4.7}$$

where p_i is the number of parameters in model M_i, $i = 1, 2$, and

$$W = -2 \log \left[\frac{\sup_{M_1} f(\mathbf{y}|\boldsymbol{\theta})}{\sup_{M_2} f(\mathbf{y}|\boldsymbol{\theta})}\right] ,$$

the usual likelihood ratio test statistic. Schwarz (1978) showed that for nonhierarchical (two-stage) models and large sample sizes n, BIC approximates $-2 \log BF$. An alternative to BIC is the *Akaike Information Criterion* (AIC), which alters (4.7) slightly to

$$\Delta AIC = W - 2(p_2 - p_1) . \tag{4.8}$$

Both AIC and BIC are *penalized likelihood ratio* model choice criteria, since both have second terms that act as a penalty, correcting for differences in size between the models (to see this, think of M_2 as the "full" model and M_1 as the "reduced" model).

The more serious (and aforementioned) limitation in using Bayes factors or their approximations is that they are not appropriate under noninformative priors. To see this, note that if $\pi_i(\boldsymbol{\theta}_i)$ is improper, then $p(\mathbf{y}|M_i) =$

$\int f(\mathbf{y}|\boldsymbol{\theta}_i, M_i)\pi_i(\boldsymbol{\theta}_i)d\boldsymbol{\theta}_i$ necessarily is as well, and so BF as given in (4.6) is not well defined. While several authors (see, e.g., Berger and Pericchi, 1996; O'Hagan, 1995) have attempted to modify the definition of BF to repair this deficiency, we suggest more informal yet general approaches described below.

The DIC criterion

Spiegelhalter et al. (2002) propose a generalization of the AIC, whose asymptotic justification is not appropriate for hierarchical (3 or more level) models. The generalization is based on the posterior distribution of the *deviance* statistic,

$$D(\boldsymbol{\theta}) = -2\log f(\mathbf{y}|\boldsymbol{\theta}) + 2\log h(\mathbf{y}) , \qquad (4.9)$$

where $f(\mathbf{y}|\boldsymbol{\theta})$ is the likelihood function and $h(\mathbf{y})$ is some standardizing function of the data alone. These authors suggest summarizing the *fit* of a model by the posterior expectation of the deviance, $\overline{D} = E_{\theta|y}[D]$, and the *complexity* of a model by the effective number of parameters p_D (which may well be less than the total number of model parameters, due to the borrowing of strength across random effects). In the case of Gaussian models, one can show that a reasonable definition of p_D is the expected deviance minus the deviance evaluated at the posterior expectations,

$$p_D = E_{\theta|y}[D] - D(E_{\theta|y}[\boldsymbol{\theta}]) = \overline{D} - D(\bar{\boldsymbol{\theta}}) . \qquad (4.10)$$

The *Deviance Information Criterion* (DIC) is then defined as

$$DIC = \overline{D} + p_D = 2\overline{D} - D(\bar{\boldsymbol{\theta}}) , \qquad (4.11)$$

with smaller values of DIC indicating a better-fitting model. Both building blocks of DIC and p_D, $E_{\theta|y}[D]$ and $D(E_{\theta|y}[\boldsymbol{\theta}])$), are easily estimated via MCMC methods (see below), enhancing the approach's appeal. Indeed, DIC may be computed automatically for any model in `WinBUGS`.

While the p_D portion of this expression does have meaning in its own right as an effective model size, DIC itself does not, since it has no absolute scale (due to the arbitrariness of the scaling constant $h(\mathbf{y})$, which is often simply set equal to zero). Thus only *differences* in DIC across models are meaningful. Relatedly, when DIC is used to compare nested models in standard exponential family settings, the unnormalized likelihood $L(\boldsymbol{\theta}; \mathbf{y})$ is often used in place of the normalized form $f(\mathbf{y}|\boldsymbol{\theta})$ in (4.9), since in this case the normalizing function $m(\boldsymbol{\theta}) = \int L(\boldsymbol{\theta}; \mathbf{y})d\mathbf{y}$ will be free of $\boldsymbol{\theta}$ and constant across models, hence contribute equally to the DIC scores of each (and thus have no impact on model selection). However, in settings where we require comparisons across different likelihood distributional forms, generally one must be careful to use the properly scaled joint density $f(\mathbf{y}|\boldsymbol{\theta})$ for each model.

Identification of what constitutes a *significant* difference is also a bit awkward; delta method approximations to $Var(DIC)$ have to date met with little success (Zhu and Carlin, 2000). In practice one typically adopts the informal approach of simply recomputing DIC a few times using different random number seeds, to get a rough idea of the variability in the estimates. With a large number of independent DIC replicates $\{DIC_l, \ l = 1, \ldots, N\}$, one could of course estimate $Var(DIC)$ by its sample variance,

$$\widehat{Var}(DIC) = \frac{1}{N-1}\sum_{l=1}^{N}(DIC_l - \overline{DIC})^2 \ .$$

But in any case, DIC is not intended for formal identification of the "correct" model, but rather merely as a method of comparing a collection of alternative formulations (all of which may be incorrect). This informal outlook (and DIC's approximate nature in markedly nonnormal models) suggests informal measures of its variability will often be sufficient. The p_D statistic is also helpful in its own right, since how close it is to the actual parameter count provides information about how many parameters are actually "needed" to adequately explain the data. For instance, a relatively low p_D may indicate collinear fixed effects or overshrunk random effects; see Exercise 1.

DIC is remarkably general, and trivially computed as part of an MCMC run without any need for extra sampling, reprogramming, or complicated loss function determination. Moreover, experience with DIC to date suggests it works remarkably well, despite the fact that no formal justification for it is yet available outside of posteriors that can be well approximated by a Gaussian distribution (a condition that typically occurs asymptotically, but perhaps not without a moderate to large sample size for many models). Still, DIC is by no means universally accepted by Bayesians as a suitable all-purpose model choice tool, as the discussion to Spiegelhalter et al. (2002) almost immediately indicates. Model comparison using DIC is not invariant to parametrization, so (as with prior elicitation) the most sensible parametrization must be carefully chosen beforehand. Unknown scale parameters and other innocuous restructuring of the model can also lead to subtle changes in the computed DIC value.

Finally, DIC will obviously depend on what part of the model specification is considered to be part of the likelihood, and what is not. Spiegelhalter et al. (2002) refer to this as the *focus* issue, i.e., determining which parameters are of primary interest, and which should "count" in p_D. For instance, in a hierarchical model with data distribution $f(\mathbf{y}|\boldsymbol{\theta})$, prior $p(\boldsymbol{\theta}|\eta)$ and hyperprior $p(\eta)$, one might choose as the likelihood either the obvious conditional expression $f(\mathbf{y}|\boldsymbol{\theta})$, or the *marginal* expression,

$$p(\mathbf{y}|\eta) = \int f(\mathbf{y}|\boldsymbol{\theta})p(\boldsymbol{\theta}|\eta)d\boldsymbol{\theta} \ . \tag{4.12}$$

We refer to the former case as "focused on $\boldsymbol{\theta}$," and the latter case as "focused on η." Spiegelhalter et al. (2002) defend the dependence of p_D and DIC on the choice of focus as perfectly natural, since while the two foci give rise to the same marginal density $m(y)$, the integration in (4.12) clearly suggests a different model complexity than the unintegrated version (having been integrated out, the θ parameters no longer "count" in the total). They thus argue that it is up to the user to think carefully about which parameters ought to be in focus before using DIC. Perhaps the one difficulty with this advice is that, in cases where the integration in (4.12) is not possible in closed form, the unintegrated version is really the only feasible choice. Indeed, the DIC tool in WinBUGS always focuses on the lowest level parameters in a model (in order to sidestep the integration issue), even when the user intends otherwise.

Posterior predictive loss criteria

An alternative to DIC that is also easily implemented using output from posterior simulation is the *posterior predictive loss* (performance) approach of Gelfand and Ghosh (1998). Using prediction with regard to replicates of the observed data, $Y_{\ell,rep}$, $\ell = 1, \ldots, n$, the selected models are those that perform well under a so-called *balanced* loss function. Roughly speaking, this loss function penalizes actions both for departure from the corresponding observed value ("fit") as well as for departure from what we expect the replicate to be ("smoothness"). The loss puts weights k and 1 on these two components, respectively, to allow for adjustment of relative regret for the two types of departure.

We avoid details here, but note that for squared error loss, the resulting criterion becomes

$$D_k = \frac{k}{k+1} \, G + P \,, \qquad (4.13)$$

where $G = \sum_{\ell=1}^{n} (\mu_\ell - y_{\ell,obs})^2$ and $P = \sum_{\ell=1}^{n} \sigma_\ell^2 \,.$

In (4.13), $\mu_\ell = E(Y_{\ell,rep}|\mathbf{y})$ and $\sigma_\ell^2 = Var(Y_{\ell,rep}|\mathbf{y})$, i.e., the mean and variance of the predictive distribution of $Y_{\ell,rep}$ given the observed data $\mathbf{y}$.

The components of D_k have natural interpretations. G is a goodness-of-fit term, while P is a penalty term. To clarify, we are seeking to penalize complexity and reward parsimony, just as DIC and other penalized likelihood criteria do. For a poor model we expect large predictive variance and poor fit. As the model improves, we expect to do better on both terms. But as we start to overfit, we will continue to do better with regard to goodness of fit, but also begin to inflate the variance (as we introduce multicollinearity). Eventually the resulting increased predictive variance penalty will exceed the gains in goodness of fit. So as with DIC, as we sort through a collection of models, the one with the smallest D_k is preferred. When $k = \infty$ (so that $D_k = D_\infty = G + P$), we will sometimes write D_∞ simply as D for brevity.

Two remarks are appropriate. First, we may report the first and second terms (excluding $k/(k+1)$) on the right side of (4.13), rather than reducing to the single number D_k. Second, in practice, ordering of models is typically insensitive to the particular choice of k.

The quantities μ_ℓ and σ_ℓ^2 can be readily computed from posterior samples. If under model m we have parameters $\boldsymbol{\theta}^{(m)}$, then

$$p(y_{\ell,rep}|\mathbf{y}) = \int p(y_{\ell,rep}|\boldsymbol{\theta}^{(m)})\, p(\boldsymbol{\theta}^{(m)}|\mathbf{y})\, d\boldsymbol{\theta}^{(m)} \; . \tag{4.14}$$

Hence each posterior realization (say, $\boldsymbol{\theta}^*$) can be used to draw a corresponding $y_{\ell,rep}$ from $p(y_{\ell,rep}|\boldsymbol{\theta}^{(m)} = \boldsymbol{\theta}^*)$. The resulting $y_{\ell,rep}^*$ has marginal distribution $p(y_{\ell,rep}|\mathbf{y})$. With samples from this distribution we can obtain μ_ℓ and σ_ℓ^2. Hence development of D_k requires an extra level of simulation, one for one with the posterior samples.

More general loss functions can be used, including the so-called deviance loss (based upon $p(y_\ell|\boldsymbol{\theta}^{(m)})$), again yielding two terms for D_k with corresponding interpretation and predictive calculation. This enables application to, say, binomial or Poisson likelihoods. We omit details here since in this book, only (4.13) is used for examples that employ this criterion rather than DIC.

We do not recommend a choice between the posterior predictive approach of this subsection and the DIC criterion of the previous subsection. Both involve summing a goodness-of-fit term and a complexity penalty. The fundamental difference is that the latter works in the parameter space with the likelihood, while the former works in predictive space with posterior predictive distributions. The latter addresses comparative explanatory performance, while the former addresses comparative predictive performance. So, if the objective is to use the model for explanation, we may prefer DIC; if instead the objective is prediction, we may prefer D_k.

4.3 Bayesian computation

As mentioned above, in this section we provide a brief introduction to Bayesian computing, following the development in Chapter 5 of Carlin and Louis (2000). The explosion in Bayesian activity and computing power of the last decade or so has caused a similar explosion in the number of books in this area. The earliest comprehensive treatment was by Tanner (1996), with books by Gilks et al. (1996), Gamerman (1997), and Chen et al. (2000) offering updated and expanded discussions that are primarily Bayesian in focus. Also significant are the computing books by Robert and Casella (1999) and Liu (2001), which, while not specifically Bayesian, still emphasize Markov chain Monte Carlo methods typically used in modern Bayesian analysis.

Without doubt, the most popular computing tools in Bayesian practice today are Markov chain Monte Carlo (MCMC) methods. This is due to their ability (in principle) to enable inference from posterior distributions of arbitrarily large dimension, essentially by reducing the problem to one of recursively solving a series of lower-dimensional (often unidimensional) problems. Like traditional Monte Carlo methods, MCMC methods work by producing not a closed form for the posterior in (4.1), but a *sample* of values $\{\boldsymbol{\theta}^{(g)},\ g = 1,\ldots,G\}$ from this distribution. While this obviously does not carry as much information as the closed form itself, a histogram or kernel density estimate based on such a sample is typically sufficient for reliable inference; moreover such an estimate can be made arbitrarily accurate merely by increasing the Monte Carlo sample size G. However, unlike traditional Monte Carlo methods, MCMC algorithms produce *correlated* samples from this posterior, since they arise from recursive draws from a particular Markov chain, the stationary distribution of which is the same as the posterior.

The convergence of the Markov chain to the correct stationary distribution can be guaranteed for an enormously broad class of posteriors, explaining MCMC's popularity. But this convergence is also the source of most of the difficulty in actually implementing MCMC procedures, for two reasons. First, it forces us to make a decision about when it is safe to stop the sampling algorithm and summarize its output, an area known in the business as *convergence diagnosis*. Second, it clouds the determination of the quality of the estimates produced (since they are based not on i.i.d. draws from the posterior, but on correlated samples. This is sometimes called the *variance estimation* problem, since a common goal here is to estimate the Monte Carlo variances (equivalently standard errors) associated with our MCMC-based posterior estimates.

In the remainder of this section, we introduce the two most popular MCMC algorithms, the Gibbs sampler and the Metropolis-Hastings algorithm. We then return to the convergence diagnosis and variance estimation problems.

4.3.1 The Gibbs sampler

Suppose our model features k parameters, $\boldsymbol{\theta} = (\theta_1,\ldots,\theta_k)'$. To implement the Gibbs sampler, we must assume that samples can be generated from each of the *full* or *complete* conditional distributions $\{p(\theta_i \mid \boldsymbol{\theta}_{j\neq i}, \mathbf{y}),\ i = 1,\ldots,k\}$ in the model. Such samples might be available directly (say, if the full conditionals were familiar forms, like normals and gammas) or indirectly (say, via a rejection sampling approach). In this latter case two popular alternatives are the adaptive rejection sampling (ARS) algorithm of

Gilks and Wild (1992), and the Metropolis algorithm described in the next subsection. In either case, under mild conditions, the collection of full conditional distributions uniquely determine the joint posterior distribution, $p(\boldsymbol{\theta}|\mathbf{y})$, and hence all marginal posterior distributions $p(\theta_i|\mathbf{y})$, $i = 1, \ldots, k$.

Given an arbitrary set of starting values $\{\theta_2^{(0)}, \ldots, \theta_k^{(0)}\}$, the algorithm proceeds as follows:

Gibbs Sampler: For $(t \in 1 : T)$, repeat:

Step 1: Draw $\theta_1^{(t)}$ from $p\left(\theta_1 \mid \theta_2^{(t-1)}, \theta_3^{(t-1)}, \ldots, \theta_k^{(t-1)}, \mathbf{y}\right)$

Step 2: Draw $\theta_2^{(t)}$ from $p\left(\theta_2 \mid \theta_1^{(t)}, \theta_3^{(t-1)}, \ldots, \theta_k^{(t-1)}, \mathbf{y}\right)$

$$\vdots$$

Step k: Draw $\theta_k^{(t)}$ from $p\left(\theta_k \mid \theta_1^{(t)}, \theta_2^{(t)}, \ldots, \theta_{k-1}^{(t)}, \mathbf{y}\right)$

Under mild regularity conditions that are generally satisified for most statistical models (see, e.g., Geman and Geman, 1984, or Roberts and Smith, 1993), one can show that the k-tuple obtained at iteration t, $(\theta_1^{(t)}, \ldots, \theta_k^{(t)})$, converges in distribution to a draw from the true joint posterior distribution $p(\theta_1, \ldots, \theta_k|\mathbf{y})$. This means that for t sufficiently large (say, bigger than t_0), $\{\boldsymbol{\theta}^{(t)}, t = t_0 + 1, \ldots, T\}$ is a (correlated) sample from the true posterior, from which any posterior quantities of interest may be estimated. For example, a histogram of the $\{\theta_i^{(t)}, t = t_0 + 1, \ldots, T\}$ themselves provides a simulation-consistent estimator of the marginal posterior distribution for θ_i, $p(\theta_i \mid \mathbf{y})$. We might also use a sample mean to estimate the posterior mean, i.e.,

$$\widehat{E}(\theta_i|\mathbf{y}) = \frac{1}{T - t_0} \sum_{t=t_0+1}^{T} \theta_i^{(t)} . \tag{4.15}$$

The time from $t = 0$ to $t = t_0$ is commonly known as the *burn-in* period; popular methods for selection of an appropriate t_0 are discussed below.

In practice, we may actually run m *parallel* Gibbs sampling chains, instead of only 1, for some modest m (say, $m = 5$). We will see below that such parallel chains may be useful in assessing sampler convergence, and anyway can be produced with no extra time on a multiprocessor computer. In this case, we would again discard all samples from the burn-in period, obtaining the posterior mean estimate,

$$\widehat{E}(\theta_i|\mathbf{y}) = \frac{1}{m(T - t_0)} \sum_{j=1}^{m} \sum_{t=t_0+1}^{T} \theta_{i,j}^{(t)} , \tag{4.16}$$

where now the second subscript on $\theta_{i,j}$ indicates chain number. Again we defer comment on how the issues how to choose t_0 and how to assess the quality of (4.16) and related estimators for the moment.

As a historical footnote, we add that Geman and Geman (1984) apparently chose the name "Gibbs sampler" because the distributions used in their context (image restoration, where the parameters were actually the colors of pixels on a screen) were Gibbs distributions (as previously seen in equation (3.9)). These were in turn named after J.W. Gibbs, a 19th-century American physicist and mathematician generally regarded as one of the founders of modern thermodynamics and statistical mechanics. While Gibbs distributions form an exponential family on potentials that includes most standard statistical models as special cases, most Bayesian applications do not require anywhere near this level of generality, typically dealing solely with standard statistical distributions (normal, gamma, etc.). Yet, despite a few attempts by some Bayesians to choose a more descriptive name (e.g., the "successive substitution sampling" (SSS) moniker due to Schervish and Carlin, 1992), the Gibbs sampler name has stuck. As such the Gibbs sampler is yet another example of Stigler's Law of Eponymy, which states that no scientific discovery is named for the person(s) who actually thought of it. (Interestingly, Stigler's Law of Eponymy is not due to Stigler (1999), meaning that it is an example of itself!)

4.3.2 The Metropolis-Hastings algorithm

The Gibbs sampler is easy to understand and implement, but requires the ability to readily sample from each of the full conditional distributions, $p(\theta_i \,|\, \boldsymbol{\theta}_{j \neq i}, \mathbf{y})$. Unfortunately, when the prior distribution $p(\boldsymbol{\theta})$ and the likelihood $f(\mathbf{y}|\boldsymbol{\theta})$ are not a conjugate pair, one or more of these full conditionals may not be available in closed form. Even in this setting, however, $p(\theta_i \,|\, \boldsymbol{\theta}_{j \neq i}, \mathbf{y})$ *will* be available up to a proportionality constant, since it is proportional to the portion of $f(\mathbf{y}|\boldsymbol{\theta}) \times p(\boldsymbol{\theta})$ that involves θ_i.

The *Metropolis algorithm* (or *Metropolis-Hastings algorithm*) is a rejection algorithm that attacks precisely this problem, since it requires only a function proportional to the distribution to be sampled, at the cost of requiring a rejection step from a particular *candidate* density. Like the Gibbs sampler, this algorithm was not developed by statistical data analysts for this purpose, but by statistical physicists working on the Manhattan Project in the 1940s seeking to understand the particle movement theory underlying the first atomic bomb (one of the coauthors on the original Metropolis et al. (1953) paper was Edward Teller, who is often referred to as "the father of the hydrogen bomb").

While as mentioned above our main interest in the algorithm is for generation from (typically univariate) full conditionals, it is most easily described (and theoretically supported) for the full multivariate θ vector. Thus, suppose for now that we wish to generate from a joint posterior distribution distribution $p(\boldsymbol{\theta}|\mathbf{y}) \propto h(\boldsymbol{\theta}) \equiv f(\mathbf{y}|\boldsymbol{\theta})p(\boldsymbol{\theta})$. We begin by specifying a candidate density $q(\boldsymbol{\theta}^*|\boldsymbol{\theta}^{(t-1)})$ that is a valid density function for every possible value of the conditioning variable $\boldsymbol{\theta}^{(t-1)}$, and satisfies $q(\boldsymbol{\theta}^*|\boldsymbol{\theta}^{(t-1)}) = q(\boldsymbol{\theta}^{(t-1)}|\boldsymbol{\theta}^*)$, i.e., q is *symmetric* in its arguments. Given a starting value $\boldsymbol{\theta}^{(0)}$ at iteration $t = 0$, the algorithm proceeds as follows:

Metropolis Algorithm: For $(t \in 1 : T)$, repeat:

1. Draw $\boldsymbol{\theta}^*$ from $q(\cdot|\boldsymbol{\theta}^{(t-1)})$

2. Compute the ratio $r = h(\boldsymbol{\theta}^*)/h(\boldsymbol{\theta}^{(t-1)}) = \exp[\log h(\boldsymbol{\theta}^*) - \log h(\boldsymbol{\theta}^{(t-1)})]$

3. If $r \geq 1$, set $\boldsymbol{\theta}^{(t)} = \boldsymbol{\theta}^*$;

   If $r < 1$, set $\boldsymbol{\theta}^{(t)} = \begin{cases} \boldsymbol{\theta}^* & \text{with probability } r \\ \boldsymbol{\theta}^{(t-1)} & \text{with probability } 1 - r \end{cases}$.

Then under generally the same mild conditions as those supporting the Gibbs sampler, a draw $\boldsymbol{\theta}^{(t)}$ converges in distribution to a draw from the true posterior density $p(\boldsymbol{\theta}|\mathbf{y})$. Note however that when the Metropolis algorithm (or the Metropolis-Hastings algorithm below) is used to update within a Gibbs sampler, it never samples from the full conditional distribution. Convergence using Metropolis steps, then, would be expected to be slower than that for a regular Gibbs sampler.

Recall that the steps of the Gibbs sampler were fully determined by the statistical model under consideration (since full conditional distributions for well-defined models are unique). By contrast, the Metropolis algorithm affords substantial flexibility through the selection of the candidate density q. This flexibility can be a blessing and a curse: while theoretically we are free to pick almost anything, in practice only a "good" choice will result in sufficiently many candidate acceptances. The usual approach (after $\boldsymbol{\theta}$ has been transformed to have support $\Re^k$, if necessary) is to set

$$q(\boldsymbol{\theta}^*|\boldsymbol{\theta}^{(t-1)}) = N(\boldsymbol{\theta}^*|\boldsymbol{\theta}^{(t-1)}, \widetilde{\Sigma}) , \tag{4.17}$$

since this distribution obviously satisfies the symmetry property, and is "self correcting" (candidates are always centered around the current value of the chain). Specification of q then comes down to specification of $\widetilde{\Sigma}$. Here we might try to mimic the posterior variance by setting $\widetilde{\Sigma}$ equal to an empirical estimate of the true posterior variance, derived from a preliminary sampling run.

The reader might well imagine an optimal choice of q would produce an empirical acceptance ratio of 1, the same as the Gibbs sampler (and with no apparent "waste" of candidates). However, the issue is rather more subtle than this: accepting all or nearly all of the candidates is often the result of an overly narrow candidate density. Such a density will "baby-step" around the parameter space, leading to high acceptance but also high autocorrelation in the sampled chain. An overly wide candidate density will also struggle, proposing leaps to places far from the bulk of the posterior's support, leading to high rejection and, again, high autocorrelation. Thus the "folklore" here is to choose $\widetilde{\Sigma}$ so that roughly 50% of the candidates are accepted. Subsequent theoretical work (e.g., Gelman et al., 1996) indicates even lower acceptance rates (25 to 40%) are optimal, but this result varies with the dimension and true posterior correlation structure of $\boldsymbol{\theta}$.

As a result, choice of $\widetilde{\Sigma}$ is often done *adaptively*. For instance, in one dimension (setting $\widetilde{\Sigma} = \widetilde{\sigma}$, and thus avoiding the issue of correlations among the elements of $\boldsymbol{\theta}$), a common trick is to simply pick some initial value of $\widetilde{\sigma}$, and then keep track of the empirical proportion of candidates that are accepted. If this fraction is too high (75 to 100%), we simply increase $\widetilde{\sigma}$; if it is too low (0 to 20%), we decrease it. Since certain kinds of adaptation can actually disturb the chain's convergence to its stationary distribution, the simplest approach is to allow this adaptation only during the burn-in period, a practice sometimes referred to as *pilot adaptation*. This is in fact the approach currently used by *WinBUGS*, where the default pilot period is 4000 iterations. A more involved alternative is to allow adaptation at *regeneration points* which, once defined and identified, break the Markov chain into independent sections. See, e.g., Mykland, Tierney and Yu (1995), Mira and Sargent (2000), and Hobert et al. (2002) for discussions of the use of regeneration in practical MCMC settings.

As mentioned above, in practice the Metropolis algorithm is often found as a substep in a larger Gibbs sampling algorithm, used to generate from awkward full conditionals. Such hybrid Gibbs-Metropolis applications were once known as "Metropolis within Gibbs" or "Metropolis substeps," and users would worry about how many such substeps should be used. Fortunately, it was soon realized that a single substep was sufficient to ensure convergence of the overall algorithm, and so this is now standard practice: when we encounter an awkward full conditional (say, for θ_i), we simply draw one Metropolis candidate, accept or reject it, and move on to θ_{i+1}. Further discussion of convergence properties and implementation of hybrid MCMC algorithms can be found in Tierney (1994) and Carlin and Louis (2000, Sec. 5.4.4).

We end this subsection with the important generalization of the Metropolis algorithm devised by Hastings (1970). In this variant we drop the requirement that q be symmetric in its arguments, which is often useful for bounded parameter spaces (say, $\theta > 0$) where Gaussian proposals as in (4.17) are not natural.

Metropolis-Hastings Algorithm: In Step 2 of the Metropolis algorithm above, replace the acceptance ratio r by

$$r = \frac{h(\boldsymbol{\theta}^*)q(\boldsymbol{\theta}^{(t-1)} \mid \boldsymbol{\theta}^*)}{h(\boldsymbol{\theta}^{(t-1)})q(\boldsymbol{\theta}^* \mid \boldsymbol{\theta}^{(t-1)})} \; . \tag{4.18}$$

Then again under mild conditions, a draw $\boldsymbol{\theta}^{(t)}$ converges in distribution to a draw from the true posterior density $p(\boldsymbol{\theta}|\mathbf{y})$ as $t \to \infty$.

In practice we often set $q(\boldsymbol{\theta}^* \mid \boldsymbol{\theta}^{(t-1)}) = q(\boldsymbol{\theta}^*)$, i.e., we use a proposal density that ignores the current value of the variable. This algorithm is sometimes referred to as a *Hastings independence chain*, so named because the proposals (though not the final $\boldsymbol{\theta}^{(t)}$ values) form an independent sequence. While easy to implement, this algorithm can be difficult to tune since it will converge slowly unless the chosen q is rather close to the true posterior (which is of course unknown in advance).

4.3.3 Slice sampling

An alternative to the Metropolis-Hastings algorithm that is still quite general is *slice sampling* (Neal, 2003). In its most basic form, suppose we seek to sample a univariate $\theta \sim f(\theta) \equiv h(\theta) / \int h(\theta) d\theta$, where $h(\theta)$ is known. Suppose we add a so-called *auxiliary variable* U such that $U|\theta \sim Unif(0, h(\theta))$. Then the joint distribution of θ and U is $p(\theta, u) \propto 1 \cdot I(U < h(\theta))$, where I denotes the indicator function. If we run a Gibbs sampler drawing from $U|\theta$ followed by $\theta|U$ at each iteration, we can obtain samples from $p(\theta, u)$, and hence from the marginal distribution of θ, $f(\theta)$. Sampling from $\theta|u$ requires a draw from a uniform distribution for θ over the set $S_U = \{\theta : U < h(\theta)\}$.

Figure 4.1 reveals why this approach is referred to as slice sampling. U "slices" the nonnormalized density, and the resulting "footprint" on the axis provides S_U. If we can enclose S_U in an interval, we can draw θ uniformly on this interval and simply retain it only if $U < h(\theta)$ (i.e., if $\theta \in S_U$). If $\boldsymbol{\theta}$ is instead multivariate, S_U is more complicated and now we would need a bounding rectangle.

Note that if $h(\theta) = h_1(\theta)h_2(\theta)$ where, say, h_1 is a standard density that is easy to sample, while h_2 is nonstandard and difficult to sample, then we can introduce an auxiliary variable U such that $U|\theta \sim U(0, h_2(\theta))$. Now $p(\theta, u) = h_1(\theta)I(U < h_2(\theta))$. Again $U|\theta$ is routine to sample, while to sample $\theta|U$ we would now draw θ from $h_1(\theta)$ and retain it only if θ is such that $U < h_2(\theta)$.

Slice sampling incurs problems similar to rejection sampling in that we may have to draw many θ's from h_1 before we are able to retain one. On the other hand, it has an advantage over the Metropolis-Hastings algorithm in that it always samples from the exact full conditional $p(\theta|u)$. As noted above, Metropolis-Hastings does not, and thus slice sampling would be expected to converge more rapidly. Nonetheless, overall comparison of computation time may make one method a winner for some cases, and the other a winner in other cases. We do remark that slice sampling is attractive for fitting a large range of point-referenced spatial data models, as we detail in Appendix Section A.6.

4.3.4 Convergence diagnosis

As mentioned above, the most problematic part of MCMC computation is deciding when it is safe to stop the algorithm and summarize the output.

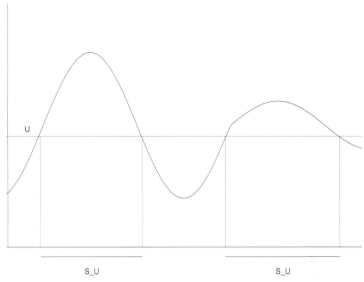

Figure 4.1 *Illustration of slice sampling. For this bimodal distribution, S_U is the union of two disjoint intervals.*

This means we must make a guess as to the iteration t_0 after which all output may be thought of as coming from the true stationary distribution of the Markov chain (i.e., the true posterior distribution). The most common approach here is to run a few (say, $m = 3$ or 5) *parallel* sampling chains, initialized at widely disparate starting locations that are overdispersed with respect to the true posterior. These chains are then plotted on a common set of axes, and these *trace plots* are then viewed to see if there is an identifiable point t_0 after which all m chains seem to be "overlapping" (traversing the same part of $\boldsymbol{\theta}$-space).

Sadly, there are obvious problems with this approach. First, since the posterior is unknown at the outset, there is no reliable way to ensure that the m chains are "initially overdispersed," as required for a convincing diagnostic. We might use extreme quantiles of the prior $p(\boldsymbol{\theta})$ and rely on the fact that the support of the posterior is typically a subset of that of the prior, but this requires a proper prior and in any event is perhaps doubtful in high-dimensional or otherwise difficult problems. Second, it is hard to see how to automate such a diagnosis procedure, since it requires a subjective judgment call by a human viewer. A great many papers have been written on various convergence diagnostic statistics that summarize MCMC output from one or many chains that may be useful when associated with various stopping rules; see Cowles and Carlin (1996) and Mengersen et al. (1999) for reviews of many such diagnostics.

Among the most popular diagnostic is that of Gelman and Rubin (1992). Here, we run a small number (m) of parallel chains with different starting points that are "initially overdispersed" with respect to the true posterior. (Of course, since we don't know the true posterior before beginning there is technically no way to ensure this; still, the rough location of the bulk of the posterior may be discernible from known ranges, the support of the (proper) prior, or perhaps a preliminary posterior mode-finding algorithm.) Running the m chains for $2N$ iterations each, we then try to see whether the variation within the chains for a given parameter of interest λ approximately equals the total variation across the chains during the latter N iterations. Specifically, we monitor convergence by the estimated *scale reduction factor*,

$$\sqrt{\hat{R}} = \sqrt{\left(\frac{N-1}{N} + \frac{m+1}{mN} \frac{B}{W} \right) \frac{df}{df - 2}} \, , \tag{4.19}$$

where B/N is the variance between the means from the m parallel chains, W is the average of the m within-chain variances, and df is the degrees of freedom of an approximating t density to the posterior distribution. Equation (4.19) is the factor by which the scale parameter of the t density might shrink if sampling were continued indefinitely; the authors show it must approach 1 as $N \to \infty$.

The approach is fairly intuitive and is applicable to output from any MCMC algorithm. However, it focuses only on detecting bias in the MCMC estimator; no information about the *accuracy* of the resulting posterior estimate is produced. It is also an inherently univariate quantity, meaning it must be applied to each parameter (or parametric function) of interest in turn, although Brooks and Gelman (1998) extend the Gelman and Rubin approach in three important ways, one of which is a multivariate generalization for simultaneous convergence diagnosis of every parameter in a model.

While the Gelman-Rubin-Brooks and other formal diagnostic approaches remain popular, in practice very simple checks often work just as well and may even be more robust against "pathologies" (e.g., multiple modes) in the posterior surface that may easily fool some diagnostics. For instance, sample autocorrelations in any of the observed chains can inform about whether slow traversing of the posterior surface is likely to impede convergence. Sample cross-correlations (i.e., correlations between two different parameters in the model) may identify ridges in the surface (say, due to collinearity between two predictors) that will again slow convergence; such parameters may need to be updated in multivariate blocks, or one of the parameters dropped from the model altogether. Combined with a visual inspection of a few sample trace plots, the user can at least get a good

feeling for whether posterior estimates produced by the sampler are likely
to be reliable.

4.3.5 Variance estimation

An obvious criticism of Monte Carlo methods generally is that no two analysts will obtain the same answer, since the components of the estimator
are random. This makes assessment of the variance of these estimators crucial. Combined with a central limit theorem, the result would be an ability
to test whether two Monte Carlo estimates were significantly different. For
example, suppose we have a single chain of N post-burn-in samples of a
parameter of interest λ, so that our basic posterior mean estimator (4.15)
becomes $\hat{E}(\lambda|\mathbf{y}) = \hat{\lambda}_N = \frac{1}{N} \sum_{t=1}^{N} \lambda^{(t)}$. Assuming the samples comprising
this estimator are independent, a variance estimate for it would be given
by

$$\widehat{Var}_{iid}(\hat{\lambda}_N) = s_\lambda^2/N = \frac{1}{N(N-1)} \sum_{t=1}^{N} (\lambda^{(t)} - \hat{\lambda}_N)^2 \,, \qquad (4.20)$$

i.e., the sample variance, $s_\lambda^2 = \frac{1}{N-1} \sum_{t=1}^{N} (\lambda^{(t)} - \hat{\lambda}_N)^2$, divided by N. But
while this estimate is easy to compute, it would very likely be an *under-
estimate* due to positive autocorrelation in the MCMC samples. One can
resort to *thinning*, which is simply retaining only every kth sampled value,
where k is the approximate lag at which the autocorrelations in the chain
become insignificant. However, MacEachern and Berliner (1994) show that
such thinning from a stationary Markov chain always increases the variance of sample mean estimators, and is thus suboptimal. This is intuitively
reminiscent of Fisher's view of sufficiency: it is never a good idea to throw
away information (in this case, $(k-1)/k$ of our MCMC samples) just to
achieve approximate independence among those that remain.

A better alternative is to use all the samples, but in a more sophisticated
way. One such alternative uses the notion of *effective sample size*, or *ESS*
(Kass et al. 1998, p. 99). *ESS* is defined as

$$ESS = N/\kappa(\lambda) \,,$$

where $\kappa(\lambda)$ is the *autocorrelation time* for λ, given by

$$\kappa(\lambda) = 1 + 2 \sum_{k=1}^{\infty} \rho_k(\lambda) \,, \qquad (4.21)$$

where $\rho_k(\lambda)$ is the autocorrelation at lag k for the parameter of interest λ.
We may estimate $\kappa(\lambda)$ using sample autocorrelations estimated from the
MCMC chain. The variance estimate for $\hat{\lambda}_N$ is then

$$\widehat{Var}_{ESS}(\hat{\lambda}_N) = s_\lambda^2/ESS(\lambda) = \frac{\kappa(\lambda)}{N(N-1)} \sum_{t=1}^{N} (\lambda^{(t)} - \hat{\lambda}_N)^2 \,.$$

Note that unless the $\lambda^{(t)}$ are uncorrelated, $\kappa(\lambda) > 1$ and $ESS(\lambda) < N$, so that $\widehat{Var}_{ESS}(\hat{\lambda}_N) > \widehat{Var}_{iid}(\hat{\lambda}_N)$, in concert with intuition. That is, since we have fewer than N effective samples, we expect some inflation in the variance of our estimate.

In practice, the autocorrelation time $\kappa(\lambda)$ in (4.21) is often estimated simply by cutting off the summation when the magnitude of the terms first drops below some "small" value (say, 0.1). This procedure is simple but may lead to a biased estimate of $\kappa(\lambda)$. Gilks et al. (1996, pp. 50–51) recommend an *initial convex sequence estimator* mentioned by Geyer (1992) which, while while still output-dependent and slightly more complicated, actually yields a consistent (asymptotically unbiased) estimate here.

A final and somewhat simpler (though also more naive) method of estimating $Var(\hat{\lambda}_N)$ is through *batching*. Here we divide our single long run of length N into m successive batches of length k (i.e., $N = mk$), with batch means $B_1, \ldots, B_m$. Clearly $\hat{\lambda}_N = \bar{B} = \frac{1}{m}\sum_{i=1}^{m} B_i$. We then have the variance estimate

$$\widehat{Var}_{batch}(\hat{\lambda}_N) = \frac{1}{m(m-1)} \sum_{i=1}^{m} (B_i - \hat{\lambda}_N)^2 , \qquad (4.22)$$

provided that k is large enough so that the correlation between batches is negligible, and m is large enough to reliably estimate $Var(B_i)$. It is important to verify that the batch means are indeed roughly independent, say, by checking whether the lag 1 autocorrelation of the B_i is less than 0.1. If this is not the case, we must increase k (hence N, unless the current m is already quite large), and repeat the procedure.

Regardless of which of the above estimates $\hat{V}$ is used to approximate $Var(\hat{\lambda}_N)$, a 95% confidence interval for $E(\lambda|\mathbf{y})$ is then given by

$$\hat{\lambda}_N \pm z_{.025}\sqrt{\hat{V}} ,$$

where $z_{.025} = 1.96$, the upper .025 point of a standard normal distribution. If the batching method is used with fewer than 30 batches, it is a good idea to replace $z_{.025}$ by $t_{m-1,.025}$, the upper .025 point of a t distribution with $m - 1$ degrees of freedom. WinBUGS offers both naive (4.20) and batched (4.22) variance estimates; this software is (at last!) the subject of the next section.

4.4 Computer tutorials

4.4.1 Basic Bayesian modeling in R or S-plus

In this subsection we merely point out that for simple (typically low-dimensional) Bayesian calculations employing standard likelihoods paired with conjugate priors, the built-in density, quantile, and plotting functions in standard statistical packages may well offer sufficient power; there is no need to use a "Bayesian" package per se. In such cases, statisticians might naturally turn to S-plus or R (the increasingly popular freeware package that is "not unlike S") due to their broad array of special functions (especially those offering summaries of standard distributions), graphics, interactive environments, and easy extendability.

As a concrete example, suppose we are observing a data value Y from a $Bin(n, \theta)$ distribution, with density proportional to

$$p(y|\theta) \propto \theta^y (1 - \theta)^{n-y} \ . \tag{4.23}$$

The $Beta(\alpha, \beta)$ distribution offers a conjugate prior for this likelihood, since its density is proportional to (4.23) as a function of θ, namely

$$p(\theta) \propto \theta^{\alpha-1}(1 - \theta)^{\beta-1} \ . \tag{4.24}$$

Using Bayes' Rule (4.1), it is clear that

$$
\begin{aligned}
p(\theta|y) \quad &\propto \quad \theta^{y+\alpha-1}(1 - \theta)^{n-y+\beta-1} \\
&\propto \quad Beta(y + \alpha, \ n - y + \beta) \ ,
\end{aligned}
\tag{4.25}
$$

another Beta distribution.

Now consider a setting where $n = 10$ and we observe $Y = y_{obs} = 7$. Choosing $\alpha = \beta = 1$ (i.e., a uniform prior for θ), the posterior is a $Beta(y_{obs} + 1, n - y_{obs} + 1) = Beta(8, 4)$ distribution. In either R or S-plus we can obtain a plot of this distribution by typing

```
> theta <- seq(from=0, to=1, length=101)
> yobs <- 7; n <- 10
> plot(theta, dbeta(theta, yobs+1, n-yobs+1), type="l",
    ylab="posterior density",xlab="")
```

The posterior median may be obtained as

```
> qbeta(.5, yobs+1, n-yobs+1)
```

while the endpoints of a 95% equal-tail credible interval are

```
> qbeta(c(.025, .975), yobs+1, n-yobs+1)
```

In fact, these points may be easily added to our posterior plot (see Figure 4.2) by typing

```
> abline(v=qbeta(.5, yobs+1, n-yobs+1))
> abline(v=qbeta(c(.025, .975), yobs+1, n-yobs+1), lty=2)
```

The pbeta and rbeta functions may be used similarly to obtain prespecified posterior probabilities (say, $Pr(\theta < 0.8|y_{obs})$) and random draws from the posterior, respectively.

Indeed, similar density, quantile, cumulative probability, and random generation routines are available in R or S-plus for a wide array of standard distributional families that often emerge as posteriors (gamma, normal, multivariate normal, Dirichlet, etc.). Thus in settings where MCMC techniques are unnecessary, these languages may offer the most sensible

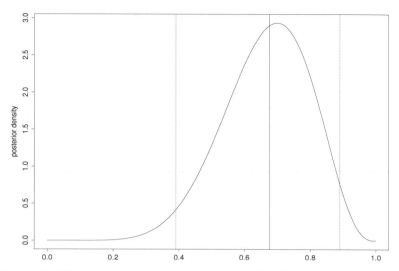

Figure 4.2 *Illustrative beta posterior, with vertical reference lines added at the .025, .5, and .975 quantiles.*

approach. They are especially useful in situations requiring code to be wrapped around statements like those above so that repeated posterior calculations may be performed. For example, when designing an experiment to be analyzed at some later date using a Bayesian procedure, we would likely want to simulate the procedure's performance in repeated sampling (the Bayesian analog of a power or "sample size" calculation). Such repeated sampling might be of the data for fixed parameters, or over both the data and the parameters. (We hasten to add that WinBUGS can be called from R, albeit in a special way; see www.stat.columbia.edu/~gelman/bugsR/. Future releases of WinBUGS may be available directly within R itself.)

4.4.2 Advanced Bayesian modeling in WinBUGS

In this subsection we provide a introduction to Bayesian data analysis in WinBUGS, the most well-developed and general Bayesian software package available to date. WinBUGS is the Windows successor to BUGS, a UNIX package whose name originally arose as a humorous acronym for Bayesian inference Using Gibbs Sampling. The package is freely available from the website http://www.mrc-bsu.cam.ac.uk/bugs/welcome.shtml. The sofware comes with a user manual, as well as two examples manuals that are enormously helpful for learning the language and various strategies for Bayesian data analysis.

We remark that for further examples of good applied Bayesian work, in addition to the fine book by Gilks et al. (1996), there are the series of "Bayesian case studies" books by Gatsonis et al. (1993, 1995, 1997, 1999, 2002, 2003), and the very recent Bayesian modeling book by Congdon (2001). While this lattermost text assumes a walking familiarity with the Bayesian approach, it also includes a great many examples and corresponding computer code for their implementation in WinBUGS.

WinBUGS has an interactive environment that enables the user to specify models (hierarchical) and it actually performs Gibbs sampling to generate posterior samples. Convergence diagnostics, model checks and comparisons, and other helpful plots and displays are also available. We will now look at some WinBUGS code for greater insight into its modeling language.

Example 4.3 The line example from the main WinBUGS manual will be considered in stages, in order to both check the installation and to illustrate the use of WinBUGS.

Consider a set of 5 (obviously artificial) (X, Y) pairs: (1, 1), (2, 3), (3, 3), (4, 3), (5, 5). We shall fit a simple linear regression of Y on X using the notation,

$$Y_i \sim N\left(\mu_i, \sigma^2\right),$$
$$\text{where } \mu_i = \alpha + \beta x_i.$$

As the WinBUGS code below illustrates, the language allows a concise expression of the model, where dnorm(a,b) denotes a normal distribution with mean a and *precision* (reciprocal of the variance) b, and dgamma(c,d) denotes a gamma distribution with mean c/d and variance c/d^2. The data means mu[i] are specified using a *logical* link (denoted by <-), instead of a *stochastic* one (denoted by $\sim$). The second logical expression allows the standard deviation σ to be estimated.

```
model
{
for(i in 1:N){
    Y[i] ~ dnorm(mu[i], tau)
    mu[i] <- alpha + beta * x[i]
}
sigma <- 1/sqrt(tau)
alpha ~ dnorm(0, 1.0E-6)
beta ~ dnorm(0, 1.0E-6)
tau ~ dgamma(1.0E-3, 1.0E-3)
}
```

The parameters in the Gibbs sampling order here will be α, β, and τ; note all are given proper but minimally informative prior distributions.

We next need to load in the data. The data can be represented using S-plus or R object notation as: list(x = c(1, 2, 3, 4, 5), Y = c(1,

3, 3, 3, 5), N = 5), or as a combination of an S-plus object and a rectangular array with labels at the head of the columns:

list(N=5)
x[ ]   Y[ ]
 1   1
 2   3
 3   3
 4   3
 5   5

Implementation of this code in WinBUGS is most easily accomplished by pointing and clicking through the menu on the Model/Specification, Inference/Samples, and Inference/Update tools; the reader may refer to www.statslab.cam.ac.uk/~krice/winbugsthemovie.html for an easy-to-follow Flash introduction to these steps. WinBUGS may also be called by R; see the functions written by Prof. Andrew Gelman for this purpose at www.stat.columbia.edu/~gelman/bugsR/. ∎

Example 4.4 Consider a basic kriging model of the form

$$\mathbf{Y} \ \sim \ MVN\left(\boldsymbol{\mu}, w^2 H(\phi) + v^2 I\right),$$

$$\text{where } \boldsymbol{\mu} \ = \ X\boldsymbol{\beta}.$$

Here I is an $N \times N$ identity matrix, while $\Sigma = w^2 H(\phi)$, an $N \times N$ correlation matrix of the form $H(\phi)_{ij} = \exp(-\phi d_{ij})$ where as usual d_{ij} is the distance between locations i and j.

What follows is some WinBUGS code to do this problem directly, i.e., using the multivariate normal distribution dnorm and constructing the H matrix directly using the exponential (exp) and power (pow) functions.

```
model
{
for(i in 1:N) {
  Y[i] ~ dnorm(mu[i], tauv)
  mu[i] <- inprod(X[i,],beta[]) + W[i]
  muW[i] <- 0
  }
for(i in 1:p) {beta[i] ~ dnorm(0.0, 0.0001)}
W[1:N] ~ dmnorm(muW[], Omega[,])
tauv ~ dgamma(0.001,0.001)
v <- 1/sqrt(tauv)
tauw ~ dgamma(0.001,0.001)
w <- 1/sqrt(tauw)
phi~ dgamma(0.01,0.01)

for (i in 1:N) {
```

```
  for(j in 1:N) {
    H[i,j] <- (1/tauw)*exp(-phi*pow(d[i,j],2)) } }
Omega[1:N,1:N] <- inverse(H[1:N,1:N])
}
```

We can also fit this model using the `spatial.exp` function now available in `WinBUGS` releases 1.4 and later:

```
model
{
for(i in 1:N) {
  Y[i] ~ dnorm(mu[i], tauv
  mu[i] <- inprod(X[i,],beta[]) + W[i]
  muW[i] <- 0
  }
for(i in 1:p) {beta[i] ~ dnorm(0.0, 0.0001)}
W[1:N] ~ spatial.exp(muW[], x[], y[], tauw, phi, 1)
tauv ~ dgamma(0.001,0.001)
v <- 1/sqrt(tauv)
tauw ~ dgamma(0.001,0.001)
w <- 1/sqrt(tauw)
phi ~ dgamma(0.01,0.01)
}
```

You are asked to compare the results of these two approaches using a "toy" ($N = 10$) data set in Exercise 4. ∎

4.5 Exercises

1. During her senior year in high school, Minnesota basketball sensation Carolyn Kieger scored at least 30 points in 9 consecutive games, helping her team win 7 of those games. The data for this remarkable streak are shown in Table 4.1. Notice that the rest of the team *combined* managed to outscore Kieger on only 2 of the 9 occasions.

   A local press report on the streak concluded (apparently quite sensibly) that Kieger was primarily responsible for the team's relatively good win-loss record during this period. A natural statistical model for testing this statement would be the *logistic regression* model,

   $$Y_i \overset{ind}{\sim} Bernoulli(p_i),$$
   $$\text{where} \quad \text{logit}(p_i) = \beta_0 + \beta_1 x_{1i} + \beta_2 x_{2i}.$$

   Here, Y_i is 1 if the team won game i and 0 if not, x_{1i} and x_{2i} are the corresponding points scored by Kieger and the rest of the team, respectively, and the logit transformation is defined as $\text{logit}(p_i) \equiv \log(p_i/(1 - p_i))$, so

| | Points scored by | | |
Game	Kieger	Rest of team	Game outcome
1	31	31	W, 62–49
2	31	16	W, 47–39
3	36	35	W, 71–64
4	30	42	W, 72–48
5	32	19	L, 64–51
6	33	37	W, 70–49
7	31	29	W, 60–37
8	33	23	W, 56–45
9	32	15	L, 57–47

Table 4.1 *Carolyn Kieger prep basketball data.*

that

$$p_i = \frac{\exp(\beta_0 + \beta_1 x_{1i} + \beta_2 x_{2i})}{1 + \exp(\beta_0 + \beta_1 x_{1i} + \beta_2 x_{2i})} \,.$$

(a) Using vague (or even flat) priors for the β_j, $j = 0, 1, 2$, fit this model to the data using the WinBUGS package. After downloading the program from http://www.mrc-bsu.cam.ac.uk/bugs/ you may wish to follow the models provided by the similar Surgical or Beetles examples (click on "Help" and pull down to "Examples Vol I" or "Examples Vol II"). Obtain posterior summaries for the β_j parameters, as well as a DIC score and effective number of parameters p_D. Also investigate MCMC convergence using trace plots, autocorrelations, and crosscorrelations (the latter from the "Correlations" tool under the "Inference" menu). Is this model acceptable, numerically or statistically?

(b) Fit an appropriate two-parameter reduction of the model in part (a). Center the remaining covariate(s) around their own mean to reduce crosscorrelations in the parameter space, and thus speed MCMC convergence. Is this model an improvement?

(c) Fit one additional two-parameter model, namely,

$$\text{logit}(p_i) = \beta_0 + \beta_1 z_i \,,$$

where $z_i = x_{1i}/(x_{1i} + x_{2i})$, the *proportion* of points scored by Kieger in game i. Again investigate convergence behavior, the β_j posteriors, and model fit relative to those in parts (a) and (b).

(d) For this final model, look at the estimated posteriors for the p_i themselves, and interpret the striking differences among them. What does this suggest might still be missing from our model?

2. Show that (4.17) is indeed a symmetric proposal density, as required by the conditions of the Metropolis algorithm.

3. Suppose now that θ is univariate but confined to the range $(0, \infty)$, with density proportional to $h(\theta)$.

   (a) Find the Metropolis acceptance ratio r assuming a Gaussian proposal density (4.17). Is this an efficient generation method?

   (b) Find the Metropolis acceptance ratio r assuming a Gaussian proposal density for $\eta \equiv \log \theta$. (Hint: Don't forget the Jacobian of this transformation!)

   (c) Finally, find the Metropolis-Hastings acceptance ratio r assuming a $Gamma(a, b)$ proposal density for θ.

4. Using the `WinBUGS` code and corresponding data set available from the web at `www.biostat.umn.edu/~brad/data/direct.bug`, attempt to fit the Bayesian kriging model in Example 4.4.

   (a) Using the "direct" code (which builds the $H(\phi)$ matrix explicitly).

   (b) Using the intrinsic `spatial.exp` function in `WinBUGS 1.4`.

   (c) Do your results in (a) and (b) agree? How do the runtimes compare?

   (d) Check to see if `WinBUGS` can handle the $N = 100$ case using the simulated data set `www.biostat.umn.edu/~brad/data/direct.bigdat` with a suitably modified version of your code.

5. Guo and Carlin (2004) consider a joint analysis of the AIDS longitudinal and survival data originally analyzed separately by Goldman et al. (1996) and Carlin and Louis (2000, Sec. 8.1). These data compare the effectiveness of two drugs, didanosine (ddI) and zalcitabine (ddC), in both preventing death and improving the longitudinal CD4 count trajectories in patients with late-stage HIV infection. The joint model used is one due to Henderson, Diggle, and Dobson (2000), which links the two submodels using bivariate Gaussian random effects. Specifically,

   Longitudinal model: For data $y_{i1}, y_{i2}, \ldots, y_{in_i}$ from the ith subject at times $s_{i1}, s_{i2}, \ldots, s_{i,n_i}$, let

   $$y_{ij} = \mu_i(s_{ij}) + W_{1i}(s_{ij}) + \epsilon_{ij} , \qquad (4.26)$$

   where $\mu_i(s) = \mathbf{x}_{1i}^T(s)\boldsymbol{\beta}_1$ is the mean response, $W_{1i}(s) = \mathbf{d}_{1i}^T(s)\mathbf{U}_i$ incorporates subject-specific random effects (adjusting the main trajectory for any subject), and $\epsilon_{ij} \sim N(0, \sigma_\epsilon^2)$ is a sequence of mutually independent measurement errors. This is the classic longitudinal random effects setting of Laird and Ware (1982).

   Survival model: Letting t_i is time to death for subject i, we assume the parametric model,

   $$t_i \sim \text{Weibull}\,(p, \mu_i(t)) ,$$

where $p > 0$ and

$$\log(\mu_i(t)) = \mathbf{x}_{2i}^T(t)\boldsymbol{\beta}_2 + W_{2i}(t) \ .$$

Here, $\boldsymbol{\beta}_2$ is the vector of fixed effects corresponding to the (possibly time-dependent) explanatory variables $\mathbf{x}_{2i}(t)$ (which may have elements in common with $\mathbf{x}_{1i}$), and $W_{2i}(t)$ is similar to $W_{1i}(s)$, including subject-specific covariate effects and an intercept (often called a *frailty*).

The specific joint model studied by Guo and Carlin (2004) assumes

$$W_{1i}(s) = U_{1i} + U_{2i}\,s \ , \quad \text{and} \qquad (4.27)$$
$$W_{2i}(t) = \gamma_1 U_{1i} + \gamma_2 U_{2i} + \gamma_3(U_{1i} + U_{2i}\,t) + U_{3i} \ , \qquad (4.28)$$

where $(U_{1i}, U_{2i})^T \stackrel{iid}{\sim} N(\mathbf{0}, \boldsymbol{\Sigma})$ and $U_{3i} \stackrel{iid}{\sim} N(0, \sigma_3^2)$, independent of the $(U_{1i}, U_{2i})^T$. The γ_1, γ_2, and γ_3 parameters in model (4.28) measure the association between the two submodels induced by the random intercepts, slopes, and fitted longitudinal value at the event time $W_{1i}(t)$, respectively.

(a) Use the code at www.biostat.umn.edu/~brad/software.html to fit the version of this model with $U_{3i} = 0$ for all i ("Model XII") in WinBUGS, as well as the further simplied version that sets $\gamma_3 = 0$ ("Model XI"). Which models fits better according to the DIC criterion?

(b) For your chosen model, investigate and comment on the posterior distributions of γ_1, γ_2, $\beta_{1,3}$ (the relative effect of ddI on the overall CD4 slope), and $\beta_{2,2}$ (the relative effect of ddI on survival).

(c) For each drug group separately, estimate the posterior distribution of the median survival time of a hypothetical patient with covariate values corresponding to a male who is AIDS-negative and intolerant of AZT at study entry. Do your answers change if you fit only the survival portion of the model (i.e., ignoring the longitudinal information)?

(d) Use the code at www.biostat.umn.edu/~brad/software.html to fit the SAS Proc NLMIXED code (originally written by Dr. Oliver Scha-benberger) for Models XI and XII above. Are the answers consistent with those you obtained from WinBUGS above? How do the computer runtimes compare? What is your overall conclusion about Bayesian versus classical estimation in this setting?

## CHAPTER 5

Hierarchical modeling for univariate spatial data

Having reviewed the basics of inference and computing under the hierarchical Bayesian modeling paradigm, we now turn our attention to its application in the setting of spatially arranged data. Many of the models discussed in Chapter 2 and Chapter 3 will be of interest, but now they may be introduced in either the first-stage specification, to directly model the data in a spatial fashion, *or* in the second-stage specification, to model spatial structure in the random effects. Parallel to the presentation in those two chapters, we begin with models for point-level data, proceed on to areal data models.

There is a substantial body of literature focusing on spatial prediction from a Bayesian perspective. This includes Le and Zidek (1992), Handcock and Stein (1993), Brown, Le, and Zidek (1994), Handcock and Wallis (1994), DeOliveira, Kedem, and Short (1997), Ecker and Gelfand (1997), Diggle, Tawn, and Moyeed (1998), and Karson et al. (1999). The work of Woodbury (1989), Abrahamsen (1993), and Omre and colleagues (Omre, 1987; Omre, 1988; Omre and Halvorsen, 1989; Omre, Halvorsen, and Berteig, 1989; and Hjort and Omre, 1994) is partially Bayesian in the sense that prior specification of the mean parameters and covariance function are elicited; however, no distributional assumption is made for the $Y(\mathbf{s})$.

5.1 Stationary spatial process models

The basic model we will work with is

$$Y(\mathbf{s}) = \mu(\mathbf{s}) + w(\mathbf{s}) + \epsilon(\mathbf{s}) \,, \qquad (5.1)$$

where the mean structure $\mu(\mathbf{s}) = \mathbf{x}^T(\mathbf{s})\boldsymbol{\beta}$. The residual is partitioned into two pieces, one spatial and one nonspatial. That is, the $w(\mathbf{s})$ are assumed to be realizations from a zero-centered stationary Gaussian spatial process (see Section 2.2), capturing residual spatial association, while the $\epsilon(\mathbf{s})$ are uncorrelated pure error terms. Thus the $w(\mathbf{s})$ introduce the partial sill (σ^2) and range (ϕ) parameters, while the $\epsilon(\mathbf{s})$ add the nugget effect (τ^2).

Valid correlation functions were discussed in Section 2.1. Recall that specifying the correlation function to be a function of the separation between sites yields a *stationary* model. If we further specify this dependence only through the distance $||s_i - s_j||$, we obtain *isotropy*; the most common such forms (exponential, Matérn , etc.) were presented in Subsection 2.1.3 and Tables 2.1 and 2.2.

Several interpretations can be attached to $\epsilon(\mathbf{s})$ and its associated variance τ^2. For instance, $\epsilon(\mathbf{s})$ can be viewed as a pure error term, as opposed to the spatial error term $w(\mathbf{s})$. Correspondingly, the nugget τ^2 is a variance component of $Y(\mathbf{s})$, as in σ^2. In other words, while $w(\mathbf{s} + \mathbf{h}) - w(\mathbf{s}) \rightarrow 0$ as $\mathbf{h} \rightarrow 0$ (if process realizations are continuous; see Subsection 2.2.3 and Section 10.1), $[w(\mathbf{s} + \mathbf{h}) + \epsilon(\mathbf{s} + \mathbf{h})] - [w(\mathbf{s}) - \epsilon(\mathbf{s})]$ will not. We are proposing residuals that are not spatially continuous, but not because the spatial process is not smooth. Instead, it is because we envision additional variabilty associated with $Y(\mathbf{s})$. This could be viewed as measurement error (as might be the case with data from certain monitoring devices) or more generally as "noise" associated with replication of measurement at location $\mathbf{s}$ (as might be the case with the sale of a single-family home at $\mathbf{s}$, in which case $\epsilon(\mathbf{s})$ would capture the effect of the particular seller, buyer, realtors, and so on).

Another view of τ^2 is that it represents *microscale* variability, i.e., variability at distances smaller than the smallest interlocation distance in the data. In this sense $\epsilon(\mathbf{s})$ could also be viewed as a spatial process, but with very rapid decay in association and with very small range. The dependence between the $\epsilon(\mathbf{s})$ would only matter at very high resolution. In this regard, Cressie (1993, pp. 112–113) suggests that $\epsilon(\mathbf{s})$ and τ^2 may themselves be partitioned into two pieces, one reflecting pure error and the other reflecting microscale error. In practice, we rarely know much about the latter, so in this book we employ $\epsilon(\mathbf{s})$ to represent only the former.

5.1.1 Isotropic models

Suppose we have data $Y(\mathbf{s}_i), i = 1, \ldots, n$, and let $\mathbf{Y} = (Y(\mathbf{s}_1), \ldots, Y(\mathbf{s}_n))^T$. The basic Gaussian isotropic kriging models of Section 2.4 are a special case of the general linear model, and therefore their Bayesian analysis can be viewed as a special case of Example 4.2. The problem just boils down to the appropriate definition of the Σ matrix. For example, in the case with a nugget effect,

$$\Sigma = \sigma^2 H(\phi) + \tau^2 I \, ,$$

where H is a correlation matrix with $H_{ij} = \rho(\mathbf{s}_i - \mathbf{s}_j; \phi)$ and ρ is a valid isotropic correlation function on $\Re^2$ indexed by a parameter (or parameters) ϕ. Collecting the entire collection of model parameters into a vector $\boldsymbol{\theta} = (\boldsymbol{\beta}, \sigma^2, \tau^2, \phi)^T$, a Bayesian solution requires an appropriate prior distribution $p(\boldsymbol{\theta})$. Parameter estimates may then be obtained from the posterior distribution, which by (4.1) is

$$p(\boldsymbol{\theta}|\mathbf{y}) \propto f(\mathbf{y}|\boldsymbol{\theta}) p(\boldsymbol{\theta}) , \qquad (5.2)$$

where

$$\mathbf{Y} \mid \boldsymbol{\theta} \sim N\left(X\boldsymbol{\beta}, \ \sigma^2 H(\phi) + \tau^2 I\right) . \qquad (5.3)$$

Typically, independent priors are chosen for the different parameters, i.e.,

$$p(\boldsymbol{\theta}) = p(\boldsymbol{\beta})p(\sigma^2)p(\tau^2)p(\phi) ,$$

and useful candidates are multivariate normal for $\boldsymbol{\beta}$ and inverse gamma for σ^2 and τ^2. Specification for ϕ of course depends upon the choice of ρ function; in the simple exponential case where $\rho(\mathbf{s}_i - \mathbf{s}_j; \phi) = \exp(-\phi||\mathbf{s}_i - \mathbf{s}_j||)$ (and ϕ is thus univariate), a gamma prior is often selected. As a general rule, one may adopt relatively noninformative priors for the mean parameters, since a proper posterior results even with $p(\boldsymbol{\beta})$ flat (improper uniform). However, improper priors for the variance/covariance parameters can lead to improper posteriors; see, e.g., Exercise 2. Since proper but very vague priors will lead to essentially improper posteriors (i.e., posteriors that are computationally indistinguishable from improper ones, hence MCMC convergence failure), the safest strategy is to choose informative specifications for σ^2, τ^2, and ϕ.

Since we will often want to make inferential statements about the parameters separately, we will need to obtain *marginal* posterior distributions. For example, a point estimate or credible interval for $\boldsymbol{\beta}$ arises from

$$
\begin{aligned}
p(\boldsymbol{\beta}|\mathbf{y}) &= \int \int \int p(\boldsymbol{\beta}, \sigma^2, \tau^2, \phi|\mathbf{y}) \, d\sigma^2 d\tau^2 d\phi \\
&\propto p(\boldsymbol{\beta}) \int \int \int f(\mathbf{y}|\boldsymbol{\theta}) p(\sigma^2)p(\tau^2)p(\phi)d\sigma^2 d\tau^2 d\phi .
\end{aligned}
$$

In principle this is simple, but in practice there will be no closed form for the above integrations. As such, we will often resort to MCMC or other numerical integration techniques, as described in Section 4.3.

Expression (5.3) can be recast as a hierarchical model by writing the first-stage specification as $\mathbf{Y}$ conditional not only on $\boldsymbol{\theta}$, but also on the vector of spatial random effects $\mathbf{W} = (w(\mathbf{s}_1), \dots, w(\mathbf{s}_n))^T$. That is,

$$\mathbf{Y} \mid \boldsymbol{\theta}, \mathbf{W} \sim N(X\boldsymbol{\beta} + \mathbf{W}, \tau^2 I) . \qquad (5.4)$$

The $Y(\mathbf{s}_i)$ are conditionally independent given the $w(\mathbf{s}_i)$. The second-stage specification is for $\mathbf{W}$, namely, $\mathbf{W}|\sigma^2, \phi \sim N(\mathbf{0}, \sigma^2 H(\phi))$ where $H(\phi)$ is as above. The model specification is completed by adding priors for $\boldsymbol{\beta}$ and τ^2 as well as for σ^2 and ϕ, the latter two of which may be viewed as hyperparameters. The parameter space is now augmented from $\boldsymbol{\theta}$ to $(\boldsymbol{\theta}, \mathbf{W})$, and its dimension is increased by n.

Regardless, the resulting $p(\boldsymbol{\theta}|\mathbf{y})$ is the same, but we have the choice of using Gibbs sampling (or some other MCMC method) to fit the model either as $f(\mathbf{y}|\boldsymbol{\theta})p(\boldsymbol{\theta})$, or as $f(\mathbf{y}|\boldsymbol{\theta},\mathbf{W})p(\mathbf{W}|\boldsymbol{\theta})p(\boldsymbol{\theta})$. The former is the result of marginalizing the latter over $\mathbf{W}$. Generally, we would prefer to work with the former (see Appendix Section A.6). Apart from the conventional wisdom that we should do as much marginalization in closed form as possible before implementing an MCMC algorithm (i.e., in as low a dimension as possible), the matrix $\sigma^2 H(\phi)+\tau^2 I$ is typically better behaved than $\sigma^2 H(\phi)$. To see this, note that if, say, $\mathbf{s}_i$ and $\mathbf{s}_j$ are very close to each other, $\sigma^2 H(\phi)$ will be close to singular while $\sigma^2 H(\phi) + \tau^2 I$ will not. Determinant and inversion calculation will also tend to be better behaved for the marginal model form than the conditional model form.

Interest is often in the spatial surface that involves $\mathbf{W}|\mathbf{y}$, as well as prediction for $W(\mathbf{s}_0)|\mathbf{y}$ for various choices of $\mathbf{s}_0$. At first glance it would appear that fitting the conditional model here would have an advantage, since realizations essentially from $p(\mathbf{W}|\mathbf{y})$ are directly produced in the process of fitting the model. However, since $p(\mathbf{W}|\mathbf{y}) = \int p(\mathbf{W}|\boldsymbol{\theta},\mathbf{y})p(\boldsymbol{\theta}|\mathbf{y})d\boldsymbol{\theta}$, posterior realizations of $\mathbf{W}$ can be obtained one for one via *composition* sampling using posterior realizations of $\boldsymbol{\theta}$. Specifically, if the values $\boldsymbol{\theta}^{(g)}$ are draws from an MCMC algorithm with stationary distribution $p(\boldsymbol{\theta}|\mathbf{y})$, then corresponding draws $\mathbf{W}^{(g)}$ from $p(\mathbf{W}|\boldsymbol{\theta}^{(g)},\mathbf{y})$ will have marginal distribution $p(\mathbf{W}|\mathbf{y})$, as desired. Thus we need not generate the $\mathbf{W}^{(g)}$ within the Gibbs sampler itself, but instead obtain them immediately given the output of the smaller, marginal sampler. Note that marginalization over $\mathbf{W}$ is only possible if the hierarchical form has a first-stage Gaussian specification, as in (5.4). We return to this matter in Section 5.3.

Next we turn to prediction of the response Y at a new value s_0 with associated covariate vector $\mathbf{x}(\mathbf{s}_0)$; this predictive step is the Bayesian "kriging" operation. Denoting the unknown value at that point by $Y(\mathbf{s}_0)$ and using the notations $Y_0 \equiv Y(\mathbf{s}_0)$ and $\mathbf{x}_0 \equiv \mathbf{x}(\mathbf{s}_0)$ for convenience, the solution in the Bayesian framework simply amounts to finding the predictive distribution,

$$
\begin{aligned}
p(y_0|\mathbf{y}, X, \mathbf{x}_0) &= \int p(y_0, \boldsymbol{\theta}|\mathbf{y}, X, \mathbf{x}_0)\, d\boldsymbol{\theta} \\
&= \int p(y_0|\mathbf{y}, \boldsymbol{\theta}, \mathbf{x}_0)\, p(\boldsymbol{\theta}|\mathbf{y}, X)\, d\boldsymbol{\theta} ,
\end{aligned}
\tag{5.5}
$$

where $p(y_0|\mathbf{y},\boldsymbol{\theta},\mathbf{x}_0)$ has a conditional normal distribution arising from the joint multivariate normal distribution of Y_0 and the original data $\mathbf{Y}$; see (2.18) and (2.19).

In practice, MCMC methods may again be readily used to obtain estimates of (5.5). Suppose we draw (after burn-in, etc.) our posterior sample $\boldsymbol{\theta}^{(1)}, \boldsymbol{\theta}^{(2)}, \ldots, \boldsymbol{\theta}^{(G)}$ from the posterior distribution $p(\boldsymbol{\theta}|\mathbf{y},X)$. Then the above predictive integral may be computed as a Monte Carlo mixture of the form

$$\widehat{p}\left(y_0|\mathbf{y}, X, \mathbf{x}_0\right) = \frac{1}{G} \sum_{g=1}^{G} p\left(y_0|\mathbf{y}, \boldsymbol{\theta}^{(g)}, \mathbf{x}_0\right) . \tag{5.6}$$

In practice we typically use composition sampling to draw, one for one for each $\boldsymbol{\theta}^{(g)}$, a $y_0^{(g)} \sim p\left(y_0|\mathbf{y}, \boldsymbol{\theta}^{(g)}, \mathbf{x}_0\right)$. The collection $\left\{y_0^{(1)}, y_0^{(2)}, \ldots, y_0^{(G)}\right\}$ is a sample from the posterior predictive density, and so can be fed into a histogram or kernel density smoother to obtain an approximate plot of the density, bypassing the mixture calculation (5.6). A point estimate and credible interval for the predicted Y_0 may be computed in the same manner as in the estimation case above. This is all routinely done in S-plus, R, or WinBUGS; see Subsection 5.1.2 for more details using the latter package.

Next suppose that we want to predict at a *set* of m sites, denoted, say, by $S_0 = \{\mathbf{s}_{01}, \mathbf{s}_{02}, \ldots, \mathbf{s}_{0m}\}$. We could individually predict at each of these points "independently" using the above method. But *joint* prediction may also be of interest, since it enables realizations from the same random spatial surface. As a result it allows estimation of posterior associations among the m predictions. We may form an unobserved vector $\mathbf{Y}_0 = (Y(\mathbf{s}_{01}), \ldots, Y(\mathbf{s}_{0m}))^T$ with associated design matrix X_0 having rows $\mathbf{x}(\mathbf{s}_{0j})^T$, and compute its joint predictive density as

$$
\begin{aligned}
p\left(\mathbf{y}_0|\mathbf{y}, X, X_0\right) &= \int p\left(\mathbf{y}_0|\mathbf{y}, \boldsymbol{\theta}, X_0\right) p\left(\boldsymbol{\theta}|\mathbf{y}, X\right) d\boldsymbol{\theta} \\
&\approx \frac{1}{G} \sum_{g=1}^{G} p\left(\mathbf{y}_0|\mathbf{y}, \boldsymbol{\theta}^{(g)}, X_0\right),
\end{aligned}
$$

where again $p\left(\mathbf{y}_0|\mathbf{y}, \boldsymbol{\theta}^{(j)}, X_0\right)$ is available from standard conditional normal formulae. We could also use composition to obtain, one for one for each $\boldsymbol{\theta}^{(g)}$, a collection of $\mathbf{y}_0^{(g)}$ and make any inferences we like based on this sample, either jointly or componentwise.

Often we are interested in not only the variables $Y(\mathbf{s})$, but also in functions of them, e.g., $\log Y(\mathbf{s})$ (if $Y(\mathbf{s}) > 0$), $I(Y(\mathbf{s}) > c)$, and so on. These functions are random variables as well. More generally we might be interested in functions $g(\mathbf{Y}_D)$ where $\mathbf{Y}_D = \{Y(\mathbf{s}) : \mathbf{s} \in D\}$. These include, for example, $(Y(\mathbf{s}_i) - Y(\mathbf{s}_j))^2$, which enter into the variogram, linear transformations $\sum_i \ell_i Y(\mathbf{s}_i)$, which include filters for spatial prediction at some location, and finite differences in specified directions, $[Y(\mathbf{s}+h\mathbf{u}) - Y(\mathbf{s})]/\mathbf{h}$, where $\mathbf{u}$ is a particular unit vector (see Subsection 10.1.2).

Functions of the form $g(\mathbf{Y}_D)$ also include block averages, i.e. $Y(A) = \frac{1}{|A|} \int_A g(Y(\mathbf{s})) d\mathbf{s}$. Block averages are developed in much more detail in Chapter 6. The case where $g(\mathbf{Y}_D) = I(Y(\mathbf{s}) \leq c)$ leads to the definition of the spatial CDF (SCDF) as in Section 10.3. Integration of a process or of a function of a process yields a new random variable, i.e., the integral is random and is usually referred to as a stochastic integral. An obvious but important point is that $E_A g(Y(\mathbf{s})) \neq g(E_A Y(\mathbf{s}))$ if g is not linear. Hence modeling $g(Y(\mathbf{s}))$ is not the same as modeling $g(Y(A))$. See Wakefield and Salway (2001) for further discussion.

5.1.2 Bayesian kriging in WinBUGS

Both the prediction techniques mentioned in the previous section (univariate and joint) are automated in WinBUGS (versions 1.3.1 and later). As usual we illustrate in the context of an example.

Example 5.1 Here we revisit the basic kriging model first considered in Example 4.4. Recall in that example we showed how to specify the model in the WinBUGS language directly, without making use of any special functions. Unfortunately, WinBUGS' standard matrix inversion routines are too slow for this approach to work for any but the smallest geostatistical data sets. However, the language does offer several special functions for Bayesian kriging, which we now describe.

First consider the pure spatial (no nugget effect) model that is compatible with WinBUGS 1.4. In this model (with corresponding computer code available at http://www.biostat.umn.edu/~brad/data2.html), Y are the observed responses with covariate data X, N is the number of observed data sites with spatial coordinates (x[],y[]), M is the number of missing data sites with spatial coordinates (x0[],y0[]), and we seek to predict the response $Y0$ given the observed covariate data $X0$.

In the example given at the website, the data were actually simulated from a (purely spatial) Gaussian field with a univariate mean structure. Specifically, the true parameter values are $\beta = 5.0$, $\phi = 1.05$, and spatial variance $\sigma^2 = 2.0$.

```
model
{
for (i in 1:N) { mu[i] <- inprod(X[i,],beta[]) }

for (i in 1:p) {beta[i] ~ dnorm(0.0, 0.0001)}
Y[1:N] ~ spatial.exp(mu[], x[], y[], spat.prec, phi, 1)
phi~dgamma(0.1,0.1)
spat.prec ~ dgamma(0.10, 0.10)
sigmasq <- 1/spat.prec

# Predictions Joint
Y0[1:M] ~ spatial.pred(mu0[], x0[], y0[], Y[])
for(j in 1:M) { mu0[j] <- inprod(X0[j,], beta[]) }
}
```

In this code, the spatial.exp command fits the exponential kriging model

directly to the observed data **Y**, meaning that we are forgoing the nugget
effect here. The final argument "1" in this command indicates an ordinary
exponential model; another option is "2," corresponding to a powered exponential model where the power used is 2 (i.e., spatial dependence between
two observations varies as the *square* of the distance between them). The
spatial.pred command handles the joint prediction (kriging) at the new
sites $X0$.

The following modification handles the modification where we add the
nugget to the spatial model in WinBUGS1.4:

```
model
{
    for (i in 1:N) {
        Y[i] ~dnorm(mu[i], error.prec)
        mu[i] <- inprod(X[i,],beta[]) + W[i]
        muW[i] <- 0
    }

    for (i in 1:p) {beta[i] ~dnorm(0.0, 0.0001)}
    tausq <- 1/error.prec
    W[1:N] ~ spatial.exp(muW[], x[], y[], spat.prec, phi, 1)
    phi~dgamma(0.1,0.1)
    spat.prec ~dgamma(0.10, 0.10)
    sigmasq <- 1/spat.prec

  # Predictions Joint
    W0[1:M] ~spatial.pred(muW0[], x0[], y0[], W[])
    for(j in 1:M) {
        muW0[j] <- 0
        Y0[j] <- inprod(X0[j,], beta[]) + W0[j]
    }
}
```

Here, the spatial.exp command is used not with the observed data **Y**,
but with the random effects vector **W**. Adding **W** into the mean structure
and placing an ordinary normal error structure on **Y** conditional on **W**
produces the "spatial plus nugget" error total structure we desire (see (5.3)
above). ∎

5.1.3 More general isotropic correlation functions

From Subsection 2.2.2, a correlation function $\rho(d, \phi)$ is valid only if it is
positive definite in d, $\rho(0, \phi) = 1$, and $|\rho(d, \phi)| \leq 1$ for all d. From Bochner's
Theorem (2.10), the characteristic function of a symmetric distribution
in R^r satisfies these constraints. From Khinchin's Theorem (e.g., Yaglom,

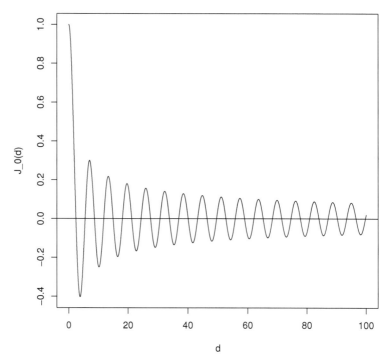

Figure 5.1 *A plot of $J_0(d)$ out to $d = 100$.*

1962, p. 106) as well as (2.12), the class of all valid functions $\rho(d, \phi)$ in $\Re^r$ can be expressed as

$$\rho(d, \phi) = \int_0^\infty \Omega_r(zd) dG_\phi(z) \,, \qquad (5.7)$$

where G_ϕ is nondecreasing integrable and $\Omega_r(x) = \left(\frac{2}{x}\right)^{\frac{r-2}{2}} \Gamma\left(\frac{r}{2}\right) J_{\left(\frac{r-2}{2}\right)}(x)$. Here again, $J_v(\cdot)$ is the Bessel function of the first kind of order v. For $r = 1, \Omega_1(x) = \cos(x)$; for $r = 2, \Omega_2(x) = J_0(x)$; for $r = 3, \Omega_3(x) = \sin(x)/x$; for $r = 4, \Omega_4(x) = \frac{2}{x} J_1(x)$; and for $r = \infty, \Omega_\infty(x) = \exp(-x^2)$. Specifically, $J_0(x) = \sum_{k=0}^\infty \frac{(-1)^k}{k!^2} \left(\frac{x}{2}\right)^{2k}$ and $\rho(d, \phi) = \int_0^\infty J_0(zd) dG_\phi(z)$ provides the class of all permissible correlation functions in $\Re^2$. Figure 5.1 provides a plot of $J_0(x)$ versus x, revealing that it is not monotonic. (This must be the case in order for $\rho(d, \phi)$ above to capture all correlation functions in $\Re^2$.)

In practice, a convenient simple choice for $G_\phi(z)$ is a step function that assigns positive mass (jumps or weights) w_ℓ at points (nodes) ϕ_ℓ, $\ell = 1, ..., p$

yielding, with $\mathbf{w} = (w_1, w_2, ..., w_p)$,

$$\rho(d, \phi, \mathbf{w}) = \sum_{\ell=1}^{p} w_\ell \Omega_n(\phi_\ell d) \ . \tag{5.8}$$

The forms in (5.8) are referred to as *nonparametric* variogram models in the literature to distinguish them from standard or parametric forms for $\rho(d, \phi)$, such as those given in Table 2.2. This is a separate issue from selecting a parametric or nonparametric methodology for parameter estimation. Sampson and Guttorp (1992), Shapiro and Botha (1991), and Cherry, Banfield, and Quimby (1996) use a step function for G_ϕ. Barry and Ver Hoef (1996) employ a mixture of piecewise linear variograms in R^1 and piecewise-planar models for sites in $\Re^2$. Hall, Fisher, and Hoffmann (1994) transform the problem from choosing ϕ_ℓ's and w_ℓ's in (5.8) to determining a kernel function and its associated bandwidth. Lele (1995) proposes iterative spline smoothing of the variogram yielding a ρ which is not obviously of the form (5.7). Most of these *nonparametric* models are fit to some version of the empirical semivariogram (2.9).

Sampson and Guttorp (1992) fit their model, using $\Omega_\infty(x)$ in (5.8), to the semivariogram cloud rather than to the smoothed Matheron semivariogram estimate. Their example involves a data set with 12 sites yielding only 66 points in the semivariogram cloud, making this feasible. Application of their method to a much larger (hence "noisier") data set would be expected to produce a variogram mixing hundreds and perhaps thousands of Gaussian forms. The resulting variogram will follow the semivariogram cloud too closely to be plausible.

Working in $\Re^2$, where again $\Omega_2(x) = J_0(x)$, under the Bayesian paradigm we can introduce (5.8) directly into the likelihood but keep p small (at most 5), allowing random w_ℓ or random ϕ_ℓ. This offers a compromise between the rather limiting standard parametric forms (Table 2.1) that specify two or three parameters for the covariance structure, and above nonparametric methods that are based upon a practically implausible (and potentially overfitting) mixture of hundreds of components. Moreover, by working with the likelihood, inference is conditioned upon the observed $\mathbf{y}$, rather than on a summary such as a smoothed version of the semivariogram cloud.

Returning to (5.7), when $n = 2$ we obtain

$$\rho(d, \phi) = \int_0^\infty \sum_{k=0}^\infty \frac{(-1)^k}{k!^2} \left(\frac{zd}{2}\right)^{2k} dG_\phi(z) \ . \tag{5.9}$$

Only if z is bounded, i.e., if G_ϕ places no mass on say $z > \phi_{max}$, can we interchange summation and integration to obtain

$$\rho(d, \phi) = \int_0^\infty \sum_{k=0}^\infty \frac{(-1)^k}{k!^2} \left(\frac{d}{2}\right)^{2k} \delta_{2k} \ , \tag{5.10}$$

where $\delta_{2k} = \int_0^{\phi_{\max}} z^{2k} dG_\phi(z)$. The simplest such choice for G_ϕ puts discrete mass w_ℓ at a finite set of values $\phi_\ell \in (0, \phi_{\max})$, $\ell = 1, ..., p$ resulting in a finite mixture of Bessels model for $\rho(d, \phi)$, which in turn yields

$$\gamma(d_{ij}) = \tau^2 + \sigma^2 \left(1 - \sum_{\ell=1}^{p} w_\ell J_0(\phi_\ell d_{ij}) \right) . \tag{5.11}$$

Under a Bayesian framework for a given p, if the w_ℓ's are each fixed to be $\frac{1}{p}$ with ϕ_ℓ's unknown (hence random), they are constrained by $0 < \phi_1 < \phi_2 < \cdots < \phi_p < \phi_{max}$ for identifiability. The result is an equally weighted mixture of random curves. If a random mixture of fixed curves is desired, then the w_ℓ's are random and the ϕ_ℓ's are systematically chosen to be $\phi_\ell = \left(\frac{\ell}{p+1} \right) \phi_{\max}$. We examine $p = 2, 3, 4, 5$ for fixed nodes and $p = 1, 2, 3, 4, 5$ for fixed weights. Mixture models using random w_ℓ's and random ϕ_ℓ's might be considered but, in our limited experience, the posteriors have exhibited weak identifiability in the parameters and thus are not recommended.

In choosing ϕ_{max}, we essentially determine the maximum number of sign changes we allow for the dampened sinusoidal Bessel correlation function over the range of d's of interest. For, say, $0 \le d \le d^{\max}$ where $d^{\max}$ is the maximum of the $d_{ij} = ||\mathbf{s}_i - \mathbf{s}_j||$, the larger ϕ is, the more sign changes $J_0(\phi d)$ will have over this range. This suggests making ϕ_{max} very large. However, as noted earlier in this section, we seek to avoid practically implausible ρ and γ, which would arise from an implausible $J_0(\phi d)$. For illustration, the plot in Figure 5.1 above allows several sign changes, to show the longer term stability of its oscillation. Letting κ be the value of x where $J_0(x) = 0$ attains its kth sign change (completes its $\frac{k-1}{2}$ period) we set $\kappa = \phi_{\max} d^{\max}$, thus determining $\phi_{\max}$. We reduce the choice of $\phi_{\max}$ to choosing the maximum number of Bessel periods allowable. For a given p, when the ϕ's are random, the posterior distribution for ϕ_p will reveal how close to $\phi_{\max}$ the data encourages ϕ_p to be.

Example 5.2 We return to the 1990 log-transformed scallop data, originally presented in Subsection 2.3.2. In 1990, 148 sites were sampled in the New York Bight region of the Atlantic Ocean, which encompasses the area from the tip of Long Island to the mouth of the Delaware River. These data have been analyzed by Ecker and Heltshe (1994), Ecker and Gelfand (1997, 1999), Kaluzny et al. (1998), and others. Figure 5.2 shows the semivariogram cloud (panel a) together with boxplots (panel b) formed from the cloud using the arbitrary lag $\delta = 0.05$. The 10,731 pairs of points that produce the semivariogram cloud do not reveal any distinct pattern. In a sense, this shows the folly of fitting a curve to this data: we have a weak signal, and a great deal of noise.

However, the boxplots and the Matheron empirical semivariograms each based on lag $\delta = 0.05$ (Figure 5.3) clearly exhibit spatial dependence, in the sense that when separation distances are small, the spatial variability

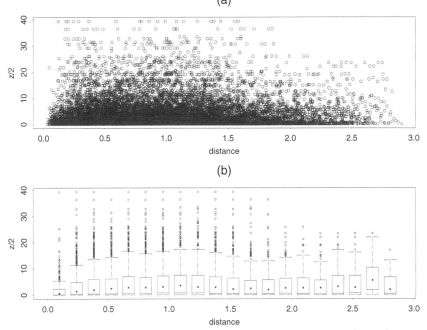

Figure 5.2 *Semivariogram cloud (a) and boxplot produced from 0.05 lag (b), 1993 scallop data.*

tends to be less. Here the attempt is to remove the noise to see whatever signal there may be. Of course, the severe skewness revealed by the boxplots (and expected from squared differences) raises the question of whether the bin averages are an appropriate summary (expression (2.9)); see Ecker and Gelfand (1997) in this regard. Clearly such displays and attempts to fit an empirical variogram must be viewed as part of the exploratory phase of our data analysis.

For the choice of ϕ_{max} in the nonparametric setup, we selected seven sign changes, or three Bessel periods. With $d_{ij}^{max} = 2.83$ degrees, ϕ_{max} becomes 7.5. A sensitivity analysis with two Bessel mixtures $(p = 2)$ having a fixed weight w_1 and random nodes was undertaken. Two, four, and five Bessel periods revealed little difference in results as compared with three. However, when one Bessel period was examined ($\phi_{max} = 3$), the model fit poorly and in fact ϕ_p was just smaller than 3. This is an indication that more flexibility (i.e., a larger value of ϕ_{max}) is required.

Several of the parametric models from Tables 2.1 and 2.2 and several nonparametric Bessel mixtures with different combinations of fixed and random parameters were fit to the 1990 scallop data. (Our analysis here parallels that of Ecker and Gelfand, 1997, although our results are not

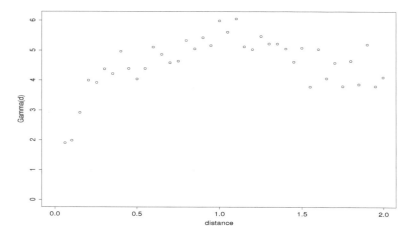

Figure 5.3 *Matheron empirical semivariograms for lag* $\delta = 0.05$.

identical to theirs since they worked with the 1993 version of the data set.)
Figure 5.4 shows the posterior mean of each respective semivariogram, while
Table 5.1 provides the value of model choice criteria for each model along
with the independence model, $\Sigma_{\mathbf{Y}} = (\tau^2 + \sigma^2)I$. Here we use the Gelfand
and Ghosh (1998) model selection criterion (4.13), as described in Subsec-
tion 4.2.3. However, since we are fitting variograms, we work somewhat less
formally using $Z_{ij,obs} = (Y(\mathbf{s}_i) - Y(\mathbf{s}_j))^2/2$. Since Z_{ij} is distributed as a
multiple of a χ_1^2 random variable, we use a loss associated with a gamma
family of distributions, obtaining a $D_{k,m}$ value of

$$
(k+1)\sum_{i,j}\left\{\log\left(\frac{\lambda_{ij}^{(m)} + kz_{ij,obs}}{k+1}\right) - \frac{\log(\lambda_{ij}^{(m)}) + k\log(z_{ij,obs})}{k+1}\right\}
$$
$$
+ \sum_{i,j}\left(\log(\lambda_{ij}^{(m)}) - E(\log(z_{ij,rep}) \mid \mathbf{y},m)\right) \qquad (5.12)
$$

for model m, where $\lambda_{ij}^{(m)} = E(z_{ij,rep} \mid \mathbf{y},m)$. The concavity of the log
function ensures that both summations on the right-hand side of (5.12) are
positive. (As an aside, in theory $z_{ij,obs} > 0$ almost surely, but in practice
we may observe some $z_{ij} = 0$ as, for example, with the log counts in the
scallop data example. A correction is needed and can be achieved by adding
ϵ to $z_{ij,obs}$ where ϵ is, say, one half of the smallest possible positive $z_{ij,obs}$.)
   Setting $k = 1$ in (5.12), we note that of the Bessel mixtures, the five-
component model with fixed ϕ's and random weights is best according
to the $D_{1,m}$ statistic. Here, given $\phi_{max} = 7.5$, the nodes are fixed to be
$\phi_1 = 1.25, \phi_2 = 2.5, \phi_3 = 3.75, \phi_4 = 5.0$, and $\phi_5 = 6.25$. One would expect
that the fit measured by the $G_{1,m}$ criterion should improve with increasing

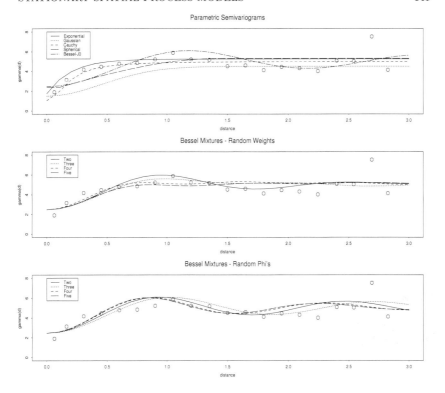

Figure 5.4 *Posterior means for various semivariogram models.*

p. However, the models do not form a nested sequence in p, except in some instances (e.g., the $p = 2$ model is a special case of the $p = 5$ model). Thus, the apparent poorer fit of the four-component fixed ϕ model relative to the three-component model is indeed possible. The random ϕ Bessel mixture models were all very close and, as a class, these models fit as well or better than the best parametric model. Hence, modeling mixtures of Bessel functions appears more sensitive to the choice of fixed ϕ's than to fixed weights. ∎

5.1.4 Modeling geometric anisotropy

As mentioned in Section 2.2.5, *anisotropy* refers to the situation where the spatial correlation between two observations depends upon the separation vector between their locations, rather than merely its length (i.e., the distance between the points). Thus here we have $Cov\left(Y\left(\mathbf{s}+\mathbf{h}\right),Y\left(\mathbf{s}\right)\right) = \rho\left(\mathbf{h};\phi\right).$

Ansiotropy is generally difficult to deal with, but there are special cases

Model	$G_{1,m}$	P_m	$D_{1,m}$
Parametric			
exponential	10959	13898	24857
Gaussian	10861	13843	24704
Cauchy	10683	13811	24494
spherical	11447	13959	25406
Bessel	11044	14037	25081
independent	11578	16159	27737
Semiparametric			
fixed ϕ_ℓ, random w_ℓ:			
two	11071	13968	25039
three	10588	13818	24406
four	10934	13872	24806
five	10567	13818	24385
random ϕ_ℓ, fixed w_ℓ:			
two	10673	13907	24580
three	10677	13959	24636
four	10636	13913	24549
five	10601	13891	24492

Table 5.1 *Model choice for fitted variogram models, 1993 scallop data.*

that are tractable yet still interesting. Among these, the most prominent in applications is *geometric anisotropy*. This refers to the situation where the coordinate space can be linearly transformed to an isotropic space. A linear transformation may correspond to rotation or stretching of the coordinate axes. Thus in general,

$$\rho\left(\mathbf{h}; \phi\right) = \rho_0\left(\|L\mathbf{h}\|; \phi\right) ,$$

where l is a $d \times d$ matrix describing the linear transformation. Of course, if L is the identity matrix, this reduces to the isotropic case.

We assume a second-order stationary normal model for $\mathbf{Y}$, arising from the customary model, $Y(\mathbf{s}) = \mu + w(\mathbf{s}) + \epsilon(\mathbf{s})$ as in (5.1). This yields $\mathbf{Y} \sim N(\mu \mathbf{1}, \Sigma(\boldsymbol{\alpha}))$, where $\boldsymbol{\alpha} = (\tau^2, \sigma^2, B)^T$, $B = L^T L$, and

$$\Sigma(\boldsymbol{\alpha}) = \tau^2 I + \sigma^2 H((\mathbf{h}'B\mathbf{h})^{\frac{1}{2}}) . \tag{5.13}$$

In (5.13), the matrix H has (i, j)th entry $\rho((\mathbf{h}'_{ij}B\mathbf{h}_{ij})^{\frac{1}{2}})$ where ρ is a valid correlation function and $\mathbf{h}_{ij} = \mathbf{s}_i - \mathbf{s}_j$. Common forms for ρ would be those in Table 2.2. In (5.13), τ^2 is the semiovariogram nugget and $\tau^2 + \sigma^2$ is the sill. The variogram is $2\gamma(\tau^2, \sigma^2, (\mathbf{h}'B\mathbf{h})^{\frac{1}{2}}) = 2(\tau^2 + \sigma^2(1 - \rho((\mathbf{h}'B\mathbf{h})^{\frac{1}{2}})))$.

Turning to $\Re^2$, B is 2×2 and the orientation of the associated ellipse, ω, is related to B by (see, e.g., Anton, 1984, p. 691)

$$\cot(2\omega) = \frac{b_{11} - b_{22}}{2b_{12}} . \tag{5.14}$$

The range in the direction η, where η is the angle $\mathbf{h}$ makes with the x-axis and which we denote as r_η, is determined by the relationship

$$\rho(r_\eta(\widetilde{\mathbf{h}}'_\eta B \widetilde{\mathbf{h}}_\eta)^{\frac{1}{2}}) = 0.05 , \tag{5.15}$$

where $\widetilde{\mathbf{h}}_\eta = (\cos\eta, \sin\eta)$ is a unit vector in direction η.

The *ratio of anisotropy* (Journel and Huijbregts, 1978, pp. 178–181), also called the *ratio of affinity* (Journel and Froidevaux, 1982, p. 228), which here we denote as λ, is the ratio of the major axis of the ellipse to the minor axis, and is related to B by

$$\lambda = \frac{r_\omega}{r_{(\pi-\omega)}} = \left(\frac{\widetilde{\mathbf{h}}'_{(\pi-\omega)} B \widetilde{\mathbf{h}}_{(\pi-\omega)}}{\widetilde{\mathbf{h}}'_\omega B \widetilde{\mathbf{h}}_\omega} \right)^{\frac{1}{2}} , \tag{5.16}$$

where again $\widetilde{\mathbf{h}}_\eta$ is the unit vector in direction η. Since (5.14), (5.15), and (5.16) are functions of B, posterior samples (hence inference) for them is straightforward given posterior samples of $\boldsymbol{\alpha}$.

A customary prior distribution for a positive definite matrix such as B is Wishart(R, p), where

$$\pi(b) \propto |B|^{\frac{p-n-1}{2}} \exp\left(-\frac{1}{2} tr(pBR^{-1}) \right) , \tag{5.17}$$

so that $E(B) = R$ and $p \geq n$ is a precision parameter in the sense that $Var(B)$ increases as p decreases. In $\Re^2$, the matrix $R = \begin{bmatrix} R_{11} & R_{12} \\ R_{12} & R_{22} \end{bmatrix}$. Prior knowledge is used to choose R, but we choose the prior precision parameter, p, to be as small as possible, i.e., $p = 2$.

A priori, it is perhaps easiest to assume that the process is isotropic, so we set $R = \delta I$ and then treat δ as fixed or random. For δ random, we model $p(B, \delta) = p(B|\delta)p(\delta)$, where $p(B|\delta)$ is the Wishart density given by (5.17) and $p(\delta)$ is an inverse gamma distribution with mean obtained from a rough estimate of the range and infinite variance (i.e., shape paramater equal to 2).

However, if we have prior evidence suggesting geometric anisotropy, we could attempt to capture it using (5.14), (5.15), or (5.16) with $\widetilde{\mathbf{h}}'_\eta R \widetilde{\mathbf{h}}_\eta$ replacing $\widetilde{\mathbf{h}}'_\eta B \widetilde{\mathbf{h}}_\eta$. For example, with a prior guess for ω, the angle of orientation of the major axis of the ellipse, a prior guess for λ, the ratio of major to minor axis (say, from a rose diagram), and a guess for the range in a specific direction (say, from a directional semivariogram), then (5.14), (5.15), and (5.16) provides a system of three linear equations in three unknowns to solve for R_{11}, R_{12}, and R_{22}. Alternatively, from three previous directional semivariograms, we might guess the range in three given directions, say, r_{η_1}, r_{η_2}, and r_{η_3}. Now, using (5.15), we again arrive at three linear equations with three unknowns in R_{11}, R_{12}, and R_{22}. One can also use an empirical semivariogram in $\Re^2$ constructed from prior data to provide guesses for R_{11}, R_{12}, and R_{22}. By computing a $0°$ and $90°$ directional semivariogram based on the ESC plot with rows where $h_y \approx 0$ for the former and columns where $h_x \approx 0$ in the latter, we obtain guesses for R_{11} and R_{22}, respectively. Finally, R_{12} can be estimated by examining a bin where neither $h_x \approx 0$ nor $h_y \approx 0$. Equating the empirical semivariogram to the theoretical semivariogram at the associated (x_i, y_j), with R_{11} and R_{22} already determined, yields a single equation to solve for R_{12}.

Example 5.3 Here we return again to the log-transformed sea scallop data of Subsection 2.3.2, and reexamine it for geometric anisotropy. Previous analyses (e.g., Ecker and Heltshe, 1994) have detected geometric anisotropy with the major axes of the ellipse oriented parallel to the coastline ($\approx 50°$ referenced counterclockwise from the x-axis). Kaluzny et al. (1998, p. 90) suggest that λ, the ratio of major axis to minor axis, is approximately 3. The 1993 scallop catches with 147 sites were analyzed in Ecker and Gelfand (1997) under isotropy. Referring back to the ESC plot in Figure 2.10, a geometrically anisotropic model seems reasonable. Here we follow Ecker and Gelfand (1999) and illustrate with a Gaussian correlation form, $\rho((\mathbf{h}'B\mathbf{h})^{\frac{1}{2}}) = \exp(-\mathbf{h}'B\mathbf{h})$.

We can use the 1990 scallop data to formulate isotropic and geometrically anisotropic prior specifications for R, the prior mean for B. The first has $R = \delta I$ with fixed $\widehat{\delta} = 0.0003$, i.e., a prior isotropic range of 100 km. Another has $\widehat{\delta} = 0.000192$, corresponding to a 125-km isotropic prior range to assess the sensitivity of choice of $\widehat{\delta}$, and a third has δ random. Under prior geometric anisotropy, we can use $\omega = 50°$, $\lambda = 3$, and $r_{50°} = 125$ km to obtain a guess for R. Solving (5.14), (5.15), and (5.16) gives $R_{11} = 0.00047$, $R_{12} = -0.00023$, and $R_{22} = 0.00039$. Using the customary directional semivariograms with the 1990 data, another prior guess for R can be built from the three prior ranges $r_{0°} = 50$ km, $r_{45°} = 125$ km, and $r_{135°} = 30$ km. Via (5.15), we obtain $R_{11} = 0.012$, $R_{12} = -0.00157$, and $R_{22} = 0.00233$. Using the ESC plot for the 1990 data, we use all bins where $h_x = h_{long} \approx 0$ ($90°$ semivariogram) to provide $R_{22} = 0.0012$, and bins where $h_y = h_{lat} \approx 0$ ($0°$ semivariogram) to provide $R_{11} = 0.00053$. Finally, we pick three bins with large bin counts (328, 285, 262) and along with the results of the $0°$ and $90°$ ESC plot directional semivariograms, we average the estimated R_{12} for each of these three bins to arrive at $R_{12} = -0.00076$.

The mean and 95% interval estimates for the isotropic prior specification with $\widehat{\delta} = 0.0003$, and the three geometrically anisotropic specifications are presented in Table 5.2. Little sensitivity to the prior specifications is ob-

	Isotropic prior fixed $\widehat{\psi} = 0.0003$	Geometrically anisotropic prior		
		ω, λ and $r_{50°}$	three ranges	ESC plot
τ^2	1.29 (1.00, 1.64)	1.43 (1.03, 1.70)	1.20 (1.01, 1.61)	1.33 (0.97, 1.73)
σ^2	2.43 (1.05, 5.94)	2.35 (1.27, 5.47)	2.67 (1.41, 5.37)	2.58 (1.26, 5.67)
sill	3.72 (2.32, 7.17)	3.80 (2.62, 6.69)	3.87 (2.66, 6.76)	3.91 (2.39, 7.09)
μ	2.87 (2.16, 3.94)	2.55 (1.73, 3.91)	3.14 (2.24, 3.99)	2.90 (2.14, 4.02)
ω	55.3 (26.7, 80.7)	64.4 (31.9, 77.6)	57.2 (24.5, 70.7)	60.7 (46.7, 75.2)
λ	2.92 (1.59, 4.31)	3.09 (1.77, 4.69)	3.47 (1.92, 4.73)	3.85 (2.37, 4.93)

Table 5.2 *Posterior means and 95% interval estimates for a stationary Gaussian model with Gaussian correlation structure under various prior specifications.*

served as expected, given that we use the smallest allowable prior precision. The posterior mean for the angle of orientation, ω, is about 60° and the ratio of the major ellipse axis to minor axis, λ, has a posterior mean of about 3 to 3.5. Furthermore, the value 1 is not in any of the three 95% interval estimates for λ, indicating that isotropy is inappropriate.

We next present posterior inference associated with the ESC plot-based prior specification. Figure 5.5 shows the posteriors for the nugget in panel (a), sill in panel (b), angle of orientation in panel (c), and the ratio of major axis to minor axis in panel (d). Figure 5.6 shows the mean posterior range plotted as a function of angle with associated individual 95% intervals. This plot is much more informative in revealing departure from isotropy than merely examining whether the 95% interval for λ contains 1. Finally, Figure 5.7 is a plot of the contours of the posterior mean surface of the semivariogram. Note that it agrees with the contours of the ESC plot given in Figure 2.10 reasonably well. ∎

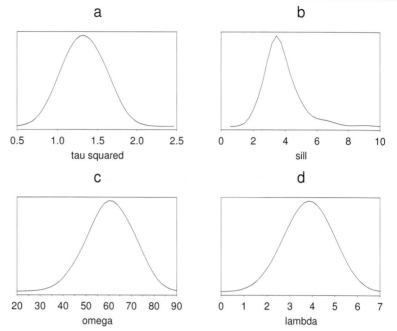

Figure 5.5 *Posterior distributions under the geometrically anisotropic prior formed from the ESC plot.*

5.2 Generalized linear spatial process modeling

In some point-referenced data sets we obtain measurements $Y(\mathbf{s})$ that would not naturally be modeled using a normal distribution; indeed, they need not even be continuous. For instance, $Y(\mathbf{s})$ might be a binary variable indicating whether or not measurable rain fell at location $\mathbf{s}$ in the past 24 hours, or a count variable indicating the number of insurance claims over the past five years by the residents of a single-family home at location $\mathbf{s}$. In an aggregate data context examining species range and richness, $Y(\mathbf{s})$ might indicate presence or absence of a particular species at $\mathbf{s}$ (although here, strictly speaking $\mathbf{s}$ is not a point, but really an area that is sufficiently small to be thought of as a point within the overall study area).

Following Diggle, Tawn, and Moyeed (1998), we formulate a hierarchical model analogous to those in Section 5.1, but with the Gaussian model for $Y(\mathbf{s})$ replaced by another suitable member of the class of exponential family models. Assume the observations $Y(\mathbf{s}_i)$ are conditionally independent given $\boldsymbol{\beta}$ and $w(\mathbf{s}_i)$ with distribution,

$$f(y(\mathbf{s}_i)|\boldsymbol{\beta}, w(\mathbf{s}_i), \gamma) = h(y(\mathbf{s}_i), \gamma)\exp\{\gamma[y(\mathbf{s}_i)\eta(\mathbf{s}_i) - \psi(\eta(\mathbf{s}_i))]\}\,,\quad(5.18)$$

where $g(\eta(\mathbf{s}_i)) = \mathbf{x}^T(\mathbf{s}_i)\boldsymbol{\beta} + w(\mathbf{s}_i)$ for some link function g, and γ is a disper-

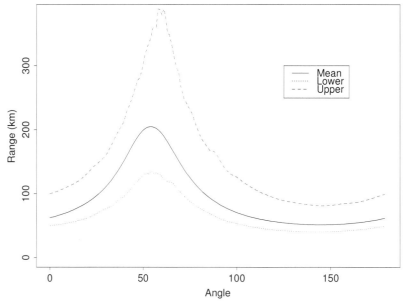

Figure 5.6 *Posterior range as a function of angle for the geometrically anisotropic prior formed from the ESC plot.*

sion parameter. We presume the $w(\mathbf{s}_i)$ to be spatial random effects coming from a Gaussian process, as in Section 5.1. The second-stage specification is $\mathbf{W} \sim N(\mathbf{0}, \sigma^2 H(\boldsymbol{\phi}))$ as before. Were the $w(\mathbf{s}_i)$ i.i.d., we would have a customary generalized linear mixed effects model (Breslow and Clayton, 1993). Hence (5.18) is still a generalized linear mixed model, but now with spatial structure in the random effects.

Two remarks are appropriate here. First, although we have defined a process for $w(\mathbf{s})$ we have not created a process for $Y(\mathbf{s})$. That is, all we have done is to create a joint distribution $f(y(\mathbf{s}_1), \dots, y(\mathbf{s}_n) | \boldsymbol{\beta}, \sigma^2, \boldsymbol{\phi}, \gamma)$, namely,

$$\int \left(\prod_{i=1}^{n} f(y(\mathbf{s}_i) | \boldsymbol{\beta}, w(\mathbf{s}_i), \gamma) \right) p(\mathbf{W} | \sigma^2, \boldsymbol{\phi}) d\mathbf{W} . \qquad (5.19)$$

Second, why not add a pure error term $\epsilon(\mathbf{s}_i)$ in the definition of $g(\eta(\mathbf{s}_i))$? This seems attractive in trying to separate a spatial effect from a pure error effect. But, upon reflection, this is not necessarily sensible, since $w(\mathbf{s}_i)$ is not a residual nor would $w(\mathbf{s}_i) + \epsilon(\mathbf{s}_i)$ be. In fact, the stochastic mechanism that is defined by f in (5.19) replaces the white noise term that arises from the Gaussian first-stage specification in (5.4).

We also note an important consequence of modeling with spatial ran-

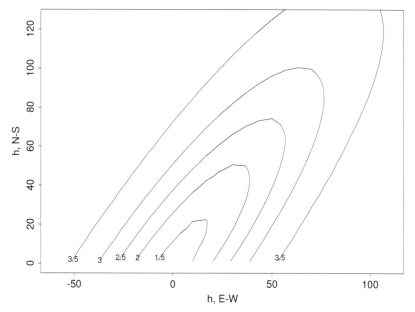

Figure 5.7 *Contours of the posterior mean semivariogram surface for the geometrically anisotropic prior formed from the ESC plot.*

dom effects (which incidentally is relevant for Sections 5.4 and 5.5 as well). Introducing these effects in the (transformed) mean, as below (5.18), encourages the means of the spatial variables at proximate locations to be close to each other (adjusted for covariates). Though marginal spatial dependence is induced between, say, $Y(\mathbf{s})$ and $Y(\mathbf{s}')$, the observed $Y(\mathbf{s})$ and $Y(\mathbf{s}')$ need *not* be close to each other. This would be the case even if $Y(\mathbf{s})$ and $Y(\mathbf{s}')$ had the same mean. As a result, second-stage spatial modeling is attractive when spatial explanation in the *mean* is of interest. Direct (first-stage) spatial modeling is appropriate to encourage proximate *observations* to be close.

Turning to computational issues, note that (5.19) cannot be integrated in closed form; we cannot marginalize over $\mathbf{W}$. Unlike the Gaussian case, a MCMC algorithm will have to update $\mathbf{W}$ as well as β, σ^2, ϕ, and γ. This same difficulty occurs with simulation-based model fitting of standard generalized linear mixed models (again see, e.g., Breslow and Clayton, 1993). In fact, the $w(\mathbf{s}_i)$ would likely be updated using a Metropolis step with a Gaussian proposal, or through adaptive rejection sampling (since their full conditional distributions will typically be log-concave); see Exercise 4.

Example 5.4 Non-Gaussian point-referenced spatial model. Here we consider a real estate data set, with observations at 50 locations in

Parameter	50%	(2.5%, 97.5%)
intercept	−1.096	(−4.198, 0.4305)
living area	0.659	(−0.091, 2.254)
age	0.009615	(−0.8653, 0.7235)
ϕ	5.79	(1.236, 9.765)
σ^2	1.38	(0.1821, 6.889)

Table 5.3 *Parameter estimates (posterior medians and upper and lower .025 points) for the binary spatial model.*

Baton Rouge, LA. The response $Y(\mathbf{s})$ is a binary variable, with $Y(\mathbf{s}) = 1$ indicating that the price of the property at location $\mathbf{s}$ is "high" (above the median price for the region), and $Y(\mathbf{s}) = 0$ indicating that the price is "low". Observed covariates include the house's age, total living area, and other area in the property. We fit the model given in (5.18) where $Y(\mathbf{s}) \sim Bernoulli(p(\mathbf{s}))$ and g is the logit link. The WinBUGS code and data for this example are at www.biostat.umn.edu/~brad/data2.html.

Table 5.3 provides the parameter estimates and Figure 5.8 shows the image plot with overlaid contour lines for the posterior mean surface of the latent $w(\mathbf{s})$ process. These are obtained by assuming vague priors for $\boldsymbol{\beta}$, a Uniform$(0, 10)$ prior for ϕ, and an Inverse Gamma$(0.1, 0.1)$ prior for σ^2. The image plot reveals negative residuals (i.e., lower prices) in the northern region, and generally positive residuals (higher prices) in the south-central region, although the southeast shows some lower price zones. The distribution of the contour lines indicate smooth flat stretches across the central parts, with downward slopes toward the north and southeast. The covariate effects are generally uninteresting, though living area seems to have a marginally significant effect on price class. ∎

5.3 Nonstationary spatial process models ⋆

Recognizing that isotropy is an assumption regarding spatial association that will rarely hold in practice, Subsection 5.1.4 proposed classes of covariance functions that were still stationary but anisotropic. However, we may wish to shed the stationarity assumption entirely and merely assume that $cov(Y(\mathbf{s}), Y(\mathbf{s}')) = C(\mathbf{s}, \mathbf{s}')$ where $C(\cdot, \cdot)$ is symmetric in its arguments. The choice of C must still be valid. Theoretical classes of valid nonstationary covariance functions can be developed (Rehman and Shapiro, 1996), but they are typically described through existence theorems, perhaps as functions in the complex plane.

We seek classes that are flexible but also offer attractive interpretation

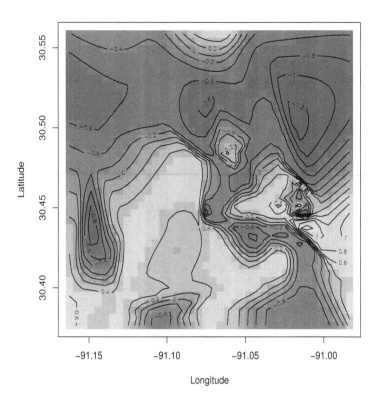

Figure 5.8 *Image plot of the posterior median surface of the latent spatial process* $w(\mathbf{s})$, *binary spatial model.*

and are computationally tractable. To this end, we prefer constructive approaches. We first observe that nonstationarity can be immediately introduced through scaling and through marginalization of stationary processes.

For the former, suppose $w(\mathbf{s})$ is a mean 0, variance 1 stationary process with correlation function ρ. Then $v(\mathbf{s}) = \sigma(\mathbf{s})w(\mathbf{s})$ is a nonstationary process. In fact,

$$var\ v(\mathbf{s}) = \sigma^2(\mathbf{s})$$
$$\text{and } cov(v(\mathbf{s}), v(\mathbf{s}')) = \sigma(\mathbf{s})\sigma(\mathbf{s}')\rho(\mathbf{s} - \mathbf{s}') , \quad (5.20)$$

so $v(\mathbf{s})$ could be used as a spatial error process, replacing $w(\mathbf{s})$ in (5.1). Where would $\sigma(\mathbf{s})$ come from? Since the use of $v(\mathbf{s})$ implies heterogeneous variance for $Y(\mathbf{s})$ we could follow the familiar course in regression modeling of setting $\sigma(\mathbf{s}) = g(x(\mathbf{s}))\sigma$ where $x(\mathbf{s})$ is a suitable positive covariate and g is a strictly increasing positive function. Hence, $var\ Y(\mathbf{s})$ increases in $x(\mathbf{s})$. Customary choices for $g(\cdot)$ are $(\cdot)$ or $(\cdot)^{\frac{1}{2}}$.

Instead, suppose we set $v(\mathbf{s}) = w(\mathbf{s}) + \delta z(\mathbf{s})$ with $z(\mathbf{s}) > 0$ and with δ being random with mean 0 and variance σ_δ^2. Then $v(\mathbf{s})$ is still a mean 0 process but now unconditionally, i.e., marginalizing over δ,

$$\begin{aligned} var\ v(\mathbf{s}) &= \sigma_w^2 + z^2(\mathbf{s})\sigma_\delta^2 \\ \text{and } cov(v(\mathbf{s}), v(\mathbf{s}')) &= \sigma_w^2 \rho(\mathbf{s} - \mathbf{s}') + z(\mathbf{s})z(\mathbf{s}')\sigma_\delta^2 \ . \end{aligned} \tag{5.21}$$

(There is no reason to impose $\sigma_w^2 = 1$ here.) Again, this model for $v(\mathbf{s})$ can replace that for $w(\mathbf{s})$ as above. Now where would $z(\mathbf{s})$ come from? One possibility is that $z(\mathbf{s})$ might be a function of the distance from s to some externality in the study region. (For instance, in modeling land prices, we might consider distance from the central business district.) Another possibility is that $z(\mathbf{s})$ is an explicit function of the location, e.g., of latitude or longitude, of eastings or northings (after some projection). Of course, we could introduce a vector $\boldsymbol{\delta}$ and a vector $\mathbf{z}(\mathbf{s})$ such that $\boldsymbol{\delta}^T\mathbf{z}(\mathbf{s})$ is a trend surface and then do a trend surface marginalization. In this fashion the spatial structure in the mean is converted to the association structure. And since $\mathbf{z}(\mathbf{s})$ varies with $\mathbf{s}$, the resultant association must be nonstationary.

In (5.20) the departure from stationarity is introduced in a multiplicative way, while through (5.21) it arises in an additive way. Evidently, we could create $v(\mathbf{s}) = \sigma(\mathbf{s})w(\mathbf{s}) + \delta z(\mathbf{s})$ yielding both types of departures from stationarity. But it is also evident that (5.20) and (5.21) are limited.

5.3.1 Deformation

In what is regarded as a landmark paper in spatial data analysis, Sampson and Guttorp (1992) introduced an approach to nonstationarity through *deformation*. The basic idea is to transform the geographic region D to a new region G, a region such that stationarity and, in fact, isotropy holds on G. The mapping $\mathbf{g}$ from D to G is bivariate, i.e., if $\mathbf{s} = (\ell_1, \ell_2)$, $\mathbf{g}(\ell_1, \ell_2) = (g_1(\ell_1, \ell_2), g_2(\ell_1, \ell_2))$. If C denotes the isotropic covariance function on G we have

$$cov(Y(\mathbf{s}), Y(\mathbf{s}')) = C(\|\mathbf{g}(\mathbf{s}) - \mathbf{g}(\mathbf{s}')\|) \ . \tag{5.22}$$

Thus, from (5.22) there are two unknown functions to estimate, $\mathbf{g}$ and C. The latter is assumed to be a parametric choice from a standard class of covariance function (as in Table 2.1). To determine the former is a challenging "fitting" problem. To what class of transformations shall we restrict ourselves? How shall we obtain the "best" member of this class? Sampson and Guttorp (1992) employ the class of thin plate splines and optimize a version of a two-dimensional nonmetric multidimensional scaling criterion (see, e.g., Mardia et al., 1979), providing an algorithmic solution. The solution is generally not well behaved, in the sense that $\mathbf{g}$ will be bijective, often folding over itself. Smith (1996) embedded this approach within a likelihood setting but worked instead with the class of radial basis functions.

Damian, Sampson, and Guttorp (2001) and Schmidt and O'Hagan (2002) have formulated fully Bayesian approaches to implement (5.22). The former still work with thin plate splines, but place priors over an identifiable parametrization (which depends upon the number of points, n being transformed). The latter elect not to model $\mathbf{g}$ directly but instead model the transformed locations. The set of n transformed locations are modeled as n realizations from a bivariate Gaussian spatial process (see Chapter 7) and a prior is placed on the process parameters. That is, $\mathbf{g}(\mathbf{s})$ arises as a random realization of a bivariate process at $\mathbf{s}$ rather than the value at $\mathbf{s}$ of a random bivariate transformation.

A fundamental limitation of the deformation approach is that implementation requires independent replications of the process in order to obtain an estimated sample covariance matrix for the set of $(Y(\mathbf{s}), ..., Y(\mathbf{s}_n))$. In practice, we rarely obtain i.i.d. replications of a spatial process. If we obtain repeated measurements at a particular location, they are typically collected across time. We would prefer to incorporate a temporal aspect in the modeling rather than attempting repairs (e.g., differencing and detrending) to achieve approximately i.i.d. observations. This is the focus of Chapter 8.

5.3.2 Kernel mixing of process variables

Kernel mixing provides an attractive way of introducing nonstationarity while retaining clear interpretation and permitting analytic calculation. Here we look at two distinct approaches, one due to Higdon (e.g., Higdon, 1998b, 2002; Higdon et al., 1999) and the other to Fuentes (e.g., Fuentes 2002a,b; Fuentes and Smith, 2001, 2003).

Kernel mixing has a long tradition in the statistical literature, especially in density estimation and regression modeling (Silverman, 1986). Kernel mixing is often done with distributions and we will look at this idea in a later subsection. Here, we focus on kernel mixing of random variables.

In fact, we work with bivariate kernels starting with stationary choices of the form $k(\mathbf{s} - \mathbf{s}')$, e.g., $k(\mathbf{s} - \mathbf{s}') = \exp\{-\frac{1}{2}(\mathbf{s} - \mathbf{s}')^T V(\mathbf{s} - \mathbf{s}')\}$. A natural choice for V would be diagonal with V_{11} and V_{22} providing componentwise scaling to the separation vector $\mathbf{s} - \mathbf{s}'$. Other choices of kernel function are available; specialization to versions based on Euclidean distance is immediate; again see, e.g., Silverman (1986). First we note the following. Let $z(\mathbf{s})$ be a white noise process, i.e., $E(z(\mathbf{s}) = 0)$, $var(z(\mathbf{s})) = \sigma^2$ and $cov(z(\mathbf{s}), z(\mathbf{s}')) = 0$. Let

$$w(\mathbf{s}) = \int_{\Re^2} k(\mathbf{s} - \mathbf{t})z(\mathbf{t})dt . \tag{5.23}$$

Rigorously speaking, (5.23) is not defined. More formally, the convolution should be written as $w(\mathbf{s}) = \int k(\mathbf{s} - \mathbf{t})\mathcal{X}(dt)$ where $\mathcal{X}(\mathbf{t})$ is two-dimensional Brownian motion. That is, $\int_A z(\mathbf{t})dt \equiv \int_A \mathcal{X}(dt) = \mathcal{X}(A) \sim N(0, \sigma^2|A|)$ and $cov(\mathcal{X}(A), \mathcal{X}(B)) = \sigma^2|A \cap B|$ where $|\cdot|$ denotes area.

The process $w(\mathbf{s})$ is said to arise through *kernel convolution*. By change of variable, (5.23) can be written as

$$w(\mathbf{s}) = \int_{\Re^2} k(\mathbf{u})z(\mathbf{s}+\mathbf{u})d\mathbf{u} \,, \qquad (5.24)$$

emphasizing that $w(\mathbf{s})$ arises as a kernel-weighted average of z's centered around $\mathbf{s}$. It is straightforward to show that $E[w(\mathbf{s})] = 0$, but also that

$$var\ w(\mathbf{s}) = \sigma^2 \int_{\Re^2} k^2(\mathbf{s}-\mathbf{t})d\mathbf{t} \,,$$
$$\text{and}\quad cov(w(\mathbf{s}), w(\mathbf{s}')) = \sigma^2 \int_{\Re^2} k(\mathbf{s}-\mathbf{t})k(\mathbf{s}'-\mathbf{t})d\mathbf{t} \,. \qquad (5.25)$$

A simple change of variables $(\mathbf{t} \to \mathbf{u} = \mathbf{s}' - \mathbf{t})$, shows that

$$cov(w(\mathbf{s}), w(\mathbf{s}')) = \sigma^2 \int_{\Re^2} k(\mathbf{s}-\mathbf{s}'+\mathbf{u})k(\mathbf{u})d\mathbf{u} \,, \qquad (5.26)$$

i.e., $w(\mathbf{s})$ is stationary. In fact, (5.23) is an established way of generating classes of stationary processes (see, e.g., Yaglom, 1962, Ch. 26).

We can extend (5.23) so that $z(\mathbf{s})$ is a mean 0 stationary spatial process with covariance function $\sigma^2 \rho(\cdot)$. Again $E[w(\mathbf{s})] = 0$ but now

$$var\ w(\mathbf{s}) = \sigma^2 \int_{\Re^2} \int_{\Re^2} k(\mathbf{s}-\mathbf{t})k(\mathbf{s}'-\mathbf{t})\rho(\mathbf{t}-\mathbf{t}')d\mathbf{t}d\mathbf{t}'$$
$$\text{and}\quad cov(w(\mathbf{s}), w(\mathbf{s}')) = \sigma^2 \int_{\Re^2} \int_{\Re^2} k(\mathbf{s}-\mathbf{t})k(\mathbf{s}'-\mathbf{t}')\rho(\mathbf{t}-\mathbf{t}')d\mathbf{t}d\mathbf{t}' \,. \qquad (5.27)$$

Interestingly, $w(\mathbf{s})$ is still stationary. We now use the change of variables $(\mathbf{t} \to \mathbf{u} = \mathbf{s}' - \mathbf{t},\ \mathbf{t}' \to \mathbf{u}' = \mathbf{s}' - \mathbf{t}')$ to obtain

$$cov(w(\mathbf{s}), w(\mathbf{s}')) = \sigma^2 \int_{\Re^2} \int_{\Re^2} k(\mathbf{s}-\mathbf{s}'+\mathbf{u})k(\mathbf{u}')\rho(\mathbf{u}-\mathbf{u}')d\mathbf{u}d\mathbf{u}' \,. \qquad (5.28)$$

Note that $w(\mathbf{s})$ can be proposed as a process having a covariance function as in (5.25) or in (5.27). We need not conceptualize or observe any $z(\mathbf{s})$'s. The integrations in (5.26) and (5.28) will not be possible to do explicitly except in certain special cases (see, e.g., Ver Hoef and Barry, 1998). Numerical integration across $\Re^2$ for (5.26) is straightforward. Numerical integration across $\Re^2 \times \Re^2$ for (5.28) may be a bit more difficult. Monte Carlo integration is not so attractive here: we would have to sample from the standardized density associated with k. But since $\mathbf{s}-\mathbf{s}'$ enters into the argument, we would have to do a separate Monte Carlo integration for each pair of locations $(\mathbf{s}_i, \mathbf{s}_j)$.

An alternative is to replace (5.23) with a finite sum approximation, i.e., to define

$$w(\mathbf{s}) = \sum_{j=1}^{L} k(\mathbf{s}-\mathbf{t}_j)z(\mathbf{t}_j) \qquad (5.29)$$

for locations $\mathbf{t}_j,\ j = 1,\ldots,L$. In the case of a white noise assumption for the z's,

$$var \ w(\mathbf{s}) = \sigma^2 \sum_{j=1}^{L} k^2(\mathbf{s} - \mathbf{t}_j)$$

$$\text{and } cov(w(\mathbf{s}), w(\mathbf{s}')) = \sigma^2 var \ w(\mathbf{s}) = \sigma^2 \sum_{j=1}^{L} k(\mathbf{s} - \mathbf{t}_j) k(\mathbf{s}' - \mathbf{t}_j) \ .$$

(5.30)

In the case of spatially correlated z's,

$$var \ w(\mathbf{s}) = \sigma^2 \sum_{j=1}^{L} \sum_{j'=1}^{L} k(\mathbf{s} - \mathbf{t}_j) k(\mathbf{s} - \mathbf{t}_{j'}) \rho(\mathbf{t}_j - \mathbf{t}_{j'})$$

$$\text{and } cov(w(\mathbf{s}), w(\mathbf{s}')) = \sigma^2 \sum_{j=1}^{L} \sum_{j'=1}^{L} k(\mathbf{s} - \mathbf{t}_j) k(\mathbf{s}' - \mathbf{t}_{j'}) \rho(\mathbf{t}_j - \mathbf{t}_{j'}) \ .$$

(5.31)

Expressions (5.30) and (5.31) can be calculated directly from (5.29) and, in fact, can be used to provide a limiting argument for (5.25) and (5.27); see Exercise 5.

Note that, while (5.30) and (5.31) are available explicitly, these forms reveal that the finite sum process in (5.29) is no longer stationary. While nonstationary specifications are the objective of this section, their creation through (5.29) is rather artificial as it arises from the arbitrary $\{\mathbf{t}_j\}$. We would prefer to modify (5.23) to achieve a class of nonstationary processes.

So, instead, suppose we allow the kernel in (5.23) to vary spatially. Notationally, we can write such an object as $k(\mathbf{s} - \mathbf{s}'; \mathbf{s})$. Illustratively, we might take $k(\mathbf{s} - \mathbf{s}'; \mathbf{s}) = \exp\{-\frac{1}{2}(\mathbf{s} - \mathbf{s}')^T V_{\mathbf{s}}(\mathbf{s} - \mathbf{s}')\}$. As above, we might take $V_{\mathbf{s}}$ to be diagonal with, if $\mathbf{s} = (\ell_1, \ell_2)$, $(V_{\mathbf{s}})_{11} = V(\ell_1)$ and $(V_{\mathbf{s}})_{22} = V(\ell_2)$. Higdon, Swall, and Kern (1999) adopt such a form with V taken to be a slowly varying function. We can insert $k(\mathbf{s} - \mathbf{s}'; \mathbf{s})$ into (5.23) in place of $k(\mathbf{s} - \mathbf{s}')$ with obvious changes to (5.25), (5.26), (5.27), and (5.28). Evidently, the process is now nonstationary. In fact, the variation in V provides insight into the departure from stationarity. For computational reasons Higdon et al. (1999) implement this modified version of (5.23) through a finite sum analogous to (5.29). A particularly attractive feature of employing a finite sum approximation is dimension reduction. If $z(\mathbf{s})$ is white noise we have an approach for handling large data sets (see Appendix Subsection A.5). That is, regardless of n, $\{w(\mathbf{s}_i)\}$ depends only on L latent variables z_j, $j = 1, \ldots, L$, and these variables are independent. Rather than fitting the model in the space of the $\{w(\mathbf{s}_i)\}$ we can work in the space of the z_ℓ.

Fuentes (2002a,b) offers a kernel mixing form that initially appears similar to (5.23) but is fundamentally different. Let

$$w(\mathbf{s}) = \int k(\mathbf{s} - \mathbf{t}) z_{\boldsymbol{\theta}(\mathbf{t})}(\mathbf{s}) d\mathbf{t} \ .$$

(5.32)

In (5.32), $k(\cdot)$ is as in (5.23) but $z_{\boldsymbol{\theta}}(\mathbf{s})$ denotes a mean 0 stationary spatial process with covariance function that is parametrized by $\boldsymbol{\theta}$. For instance $C(\cdot; \boldsymbol{\theta})$ might be $\sigma^2 \exp(-\phi \|\cdot\|^\alpha)$, a power exponential family with $\boldsymbol{\theta} = (\sigma^2, \phi, \alpha)$. In (5.32) $\boldsymbol{\theta}(\mathbf{t})$ indexes an uncountable number of processes. These processes are assumed independent across $\mathbf{t}$. Note that (5.32) is mixing an uncountable number of stationary spatial processes each at $\mathbf{s}$ while (5.23) is mixing a single process across all locations.

Formally, $w(\mathbf{s})$ has mean 0 and

$$
\begin{aligned}
var(w(\mathbf{s})) &= \int_{\Re^2} k^2(\mathbf{s} - \mathbf{t})C(0; \boldsymbol{\theta}(\mathbf{t}))dt \\
\text{and } cov(w(\mathbf{s}), w(\mathbf{s}')) &= \int_{\Re^2} k(\mathbf{s} - \mathbf{t})k(\mathbf{s}' - \mathbf{t})C(\mathbf{s} - \mathbf{s}'; \boldsymbol{\theta}(\mathbf{t}))dt \ .
\end{aligned}
\tag{5.33}
$$

Expression (5.33) reveals that (5.32) defines a nonstationary process. Suppose k is very rapidly decreasing and $\boldsymbol{\theta}(\mathbf{t})$ varies slowly. Then $w(\mathbf{s}) \approx k(0)z_{\boldsymbol{\theta}(\mathbf{t})}(\mathbf{s})$. But also, if $\mathbf{s} - \mathbf{s}'$ is small, $w(\mathbf{s})$ and $w(\mathbf{s}')$ will behave like observations from a stationary process with parameter $\boldsymbol{\theta}(\mathbf{s})$. Hence, Fuentes refers to the class of models in (5.32) as a nonstationary class that exhibits *local* stationarity.

In practice, one cannot work with (5.32) directly. Again, finite sum approximation is employed. Again, a finite set of locations $\mathbf{t}_1, \ldots, \mathbf{t}_L$ is selected and we set

$$
w(\mathbf{s}) = \sum_j k(\mathbf{s} - \mathbf{t}_j)z_j(\mathbf{s}) \ ,
\tag{5.34}
$$

writing $\boldsymbol{\theta}(\mathbf{t}_j)$ as j. Straightforwardly,

$$
\begin{aligned}
var(w(\mathbf{s}) &= \sum_{j=1}^L k^2(\mathbf{s} - \mathbf{t}_j)C_j(0) \\
\text{and } cov(w(\mathbf{s}), w(\mathbf{s}') &= \sum_{j=1}^L k(\mathbf{s} - \mathbf{t}_j)k(\mathbf{s}' - \mathbf{t}_j)C_j(\mathbf{s} - \mathbf{s}') \ .
\end{aligned}
\tag{5.35}
$$

In (5.34) it can happen that some $\mathbf{s}$'s may be far enough from each of the $\mathbf{t}_j$'s so that each $k(\mathbf{s} - \mathbf{t}_j) \approx 0$, whence $w(\mathbf{s}) \approx 0$. Of course, this cannot happen in (5.32) but we cannot work with this expression. A possible remedy was proposed in Banerjee et al. (2004). Replace (5.34) with

$$
w(\mathbf{s}) = \sum_{j=1}^L \alpha(\mathbf{s}, \mathbf{t}_j)z_j(\mathbf{s}) \ .
\tag{5.36}
$$

In (5.36), the $z_j(\mathbf{s})$ are as above, but $\alpha(\mathbf{s}, \mathbf{t}_j) = \gamma(\mathbf{s}, \mathbf{t}_j)/\sqrt{\sum_{j=1}^L \gamma^2(\mathbf{s}, \mathbf{t}_j)}$, where $\gamma(\mathbf{s}, \mathbf{t})$ is a decreasing function of the distance between $\mathbf{s}$ and $\mathbf{t}$, which may change with $\mathbf{s}$, i.e., $\gamma(\mathbf{s}, \mathbf{t}) = k_{\mathbf{s}}(\|\mathbf{s} - \mathbf{t}\|)$. (In the terminology of Higdon et al., 1999, $k_{\mathbf{s}}$ would be a spatially varying kernel function.) As a result, $\sum_{j=1}^L \alpha^2(\mathbf{s}, \mathbf{t}_j) = 1$, so regardless of where $\mathbf{s}$ is, not all of the weights in (5.36) can be approximately 0. Other standardizations for γ are possible; we have proposed this one because if all σ_j^2 are equal, then $var(w(\mathbf{s})) = \sigma^2$. That is, if each local process has the same variance, then this variance should be attached to $w(\mathbf{s})$. Furthermore, suppose $\mathbf{s}$ and $\mathbf{s}'$ are near to each other, whence $\gamma(\mathbf{s}, \mathbf{t}_j) \approx \gamma(\mathbf{s}', \mathbf{t}_j)$ and thus $\alpha(\mathbf{s}, \mathbf{t}_j) \approx \alpha(\mathbf{s}', \mathbf{t}_j)$. So, if in addition all $\phi_\ell = \phi$, then $cov(w(\mathbf{s}), w(\mathbf{s}')) \approx \sigma^2\rho(\mathbf{s} - \mathbf{s}'; \phi)$. So, if the process is in fact stationary over the entire region, we obtain essentially the second-order behavior of this process.

The alternative scaling $\widetilde{\alpha}(\mathbf{s}, \mathbf{t}_j) = \gamma(\mathbf{s}, \mathbf{t}_j)/\sum_{j'} \gamma(\mathbf{s}, \mathbf{t}_{j'})$ gives a weighted average of the component processes. Such weights would preserve an arbitrary constant mean. However, since, in our context, we are modeling a mean 0 process, such preservation is not a relevant feature.

Useful properties of the process in (5.36) are

$$E\left(w\left(\mathbf{s}\right)\right) = 0\,,$$

$$Var\left(w\left(\mathbf{s}\right)\right) = \sum_{j=1}^{L} \alpha^2(\mathbf{s},\mathbf{t}_j)\sigma_j^2\,,$$

$$\text{and}\quad cov(w(\mathbf{s}),w(\mathbf{s}')) = \sum_{j=1}^{L} \alpha(\mathbf{s},\mathbf{t}_j)\alpha(\mathbf{s}',\mathbf{t}_j)\sigma_j^2\rho(\mathbf{s}-\mathbf{s}';\phi_j)\,.$$

We have clearly defined a proper spatial process through (5.36). In fact, for arbitrary locations $\mathbf{s}_1,\ldots,\mathbf{s}_n$, let $\mathbf{w}_\ell^T = (w_\ell(\mathbf{s}_1)),\ldots,w_\ell(\mathbf{s}_n))$, $\mathbf{w}^T = (w(\mathbf{s}_1),\ldots,w(\mathbf{s}_n))$, and let A_ℓ be diagonal with $(A_\ell)_{ii} = \alpha(\mathbf{s}_i,\mathbf{t}_\ell)$. Then $\mathbf{w} \sim N(\mathbf{0}, \sum_{\ell=1}^{L} \sigma_\ell^2 A_\ell R(\phi_\ell)A_\ell)$ where $(\Sigma(\phi_\ell))_{ii'} = \rho(\mathbf{s}_i - \mathbf{s}_{i'}, \phi_\ell)$. Note that $L = 1$ is permissible in (5.36); $w(\mathbf{s})$ is still a nonstationary process. Finally, Fuentes and Smith (2003) and Banerjee et al. (2004) offer some discussion regarding precise number of and locations for the $\mathbf{t}_j$.

We conclude this subsection by noting that for a general nonstationary spatial process there is no sensible notion of a range. However, for the class of processes in (5.36) we can define a meaningful range. Under (5.36),

$$corr(w(\mathbf{s}),w(\mathbf{s}')) = \frac{\sum_{j=1}^{L} \alpha(\mathbf{s},\mathbf{t}_j)\alpha(\mathbf{s}',\mathbf{t}_j)\sigma_j^2\rho(\mathbf{s}-\mathbf{s}';\phi_j)}{\sqrt{\left(\sum_{j=1}^{L} \alpha^2(\mathbf{s},\mathbf{t}_j)\sigma_j^2\right)\left(\sum_{j=1}^{L} \alpha^2(\mathbf{s}',\mathbf{t}_j)\sigma_j^2\right)}}\,. \qquad (5.37)$$

Suppose ρ is positive and strictly decreasing asymptotically to 0 as distance tends to ∞, as is usually assumed. If ρ is, in fact, isotropic, let d_ℓ be the range for the ℓth component process, i.e., $\rho(d_\ell, \phi_\ell) = .05$, and let $\widetilde{d} = \max_\ell d_\ell$. Then (5.37) immediately shows that, at distance $\widetilde{d}$ between $\mathbf{s}$ and $\mathbf{s}'$, we have $corr(w(\mathbf{s}),w(\mathbf{s}')) \leq .05$. So $\widetilde{d}$ can be interpreted as a conservative range for $w(\mathbf{s})$. Normalized weights are not required in this definition. If ρ is only assumed stationary, we can similarly define the range in an arbitrary direction $\boldsymbol{\mu}$. Specifically, if $\boldsymbol{\mu}/\|\boldsymbol{\mu}\|$ denotes a unit vector in $\boldsymbol{\mu}$'s direction and if $d_{\boldsymbol{\mu},\ell}$ satisfies $\rho(d_{\boldsymbol{\mu},\ell}\boldsymbol{\mu}/\|\boldsymbol{\mu}\|;\phi_\ell) = .05$, we can take $\widetilde{d}_{\boldsymbol{\mu}} = \max_\ell \widetilde{d}_{\boldsymbol{\mu},\ell}$.

5.3.3 Mixing of process distributions

If $k(\mathbf{s}-\mathbf{t})$ is integrable and standardized to a density function and if f is also a density function, then

$$f_k(y) = \int k(y-x)f(x)dx \qquad (5.38)$$

is a density function. (It is, of course, the distribution of $X + Y - X$ where $X \sim f$, $Y - X \sim k$, and X and $Y - X$ are independent). In (5.38), $\mathbf{Y}$ can obviously be a vector of dimension n. But recall that we have specified the distribution for a spatial process through arbitrary finite dimensional distributions (see Section 2.2). This suggests that we can use (5.38) to build a process distribution.

Operating formally, let V_D be the set of all $V(\mathbf{s})$, $\mathbf{s} \in D$. Write $V_D = V_{0,D} + V_0 - V_{0,D}$ where $V_{0,D}$ is a realization of a mean 0 stationary Gaussian process over D, and $V_D - V_{0,D}$ is a realization of a white noise process with variance σ^2 over D. Write

$$f_k(V_D \mid \sigma) = \int \frac{1}{\sigma} k \left(\frac{1}{\sigma} (V_D - V_{0,D}) \right) f(V_{0,D}) dV_{0,D} . \qquad (5.39)$$

Formally, f_k is the distribution of the spatial process $v(\mathbf{s})$. In fact, $v(\mathbf{s})$ is just the customary model for the residuals in a spatial regression, i.e., $v(\mathbf{s}) = w(\mathbf{s}) + \epsilon(\mathbf{s})$ where $w(\mathbf{s})$ is a spatial process and $\epsilon(\mathbf{s})$ is a noise or nugget process.

Of course, in this familiar case there is no reason to employ the form (5.39). However it does reveal how, more generally, a spatial process can be developed through "kernel mixing" of a process distribution. More importantly, it suggests that we might introduce an alternative specification for $V_{0,D}$. For example, suppose $f(V_{0,D})$ is a discrete distribution, say, of the form $\sum_\ell p_\ell \delta(v_{\ell,D}^*)$ where $p_\ell \geq 0$, $\sum p_\ell = 1$, $\delta(\cdot)$ is the Dirac delta function, and $V_{\ell,D}^*$ is a surface over D. The sum may be finite or infinite. An illustration of the latter arises when $f(V_{0,D})$ is a realization from a finite discrete mixture (Duan and Gelfand, 2003) or from a Dirichlet process; see Gelfand, Kottas, and MacEachern (2003) for further details in this regard.

But then given $\{p_\ell\}$ and $\{v_{\ell,D}^*\}$, for any set of locations $\mathbf{s}_1, \ldots, \mathbf{s}_n$, if $\mathbf{V} = (v(\mathbf{s}_1), \ldots, v(\mathbf{s}_n))$,

$$f(\mathbf{V}) = \sum_\ell p_\ell N(\mathbf{v}_\ell^*, \sigma^2 I) , \qquad (5.40)$$

where $\mathbf{v}_\ell^* = (v_\ell^*(\mathbf{s}_1), \ldots, v_\ell^*(\mathbf{s}_n))^T$. So $v(\mathbf{s})$ is a continuous process that is non-Gaussian. But also, $E[v(\mathbf{s}_i)] = \sum_\ell p_\ell v_\ell^*(\mathbf{s}_i)$ and

$$var\ v(\mathbf{s}_i) = \sum_\ell p_\ell v_\ell^{2*}(\mathbf{s}_i) - (\sum_\ell p_\ell v_\ell^*(\mathbf{s}_i))^2$$
$$\text{and } cov(v(\mathbf{s}_i), v(\mathbf{s}_j)) = \sum_\ell p_\ell v_\ell^*(\mathbf{s}_i) v_\ell^*(\mathbf{s}_j) - (\sum_\ell p_\ell v_\ell^*(\mathbf{s}_i))(\sum_\ell p_\ell v_\ell^*(\mathbf{s}_j)) . \qquad (5.41)$$

This last expression shows that $v(\mathbf{s})$ is *not* a stationary process. However, a routine calculation shows that if the $v_{\ell,D}^*$ are continuous surfaces, the $v(\mathbf{s})$ process is mean square continuous and almost surely continuous (see Sections 2.2 and 10.1).

5.4 Areal data models

5.4.1 Disease mapping

A very common area of biostatistical and epidemiological interest is that of *disease mapping*. Here we typically have count data of the following sort:

$$Y_i = \text{observed number of cases of disease in county } i, \; i = 1, \ldots, I$$
$$E_i = \text{expected number of cases of disease in county } i, \; i = 1, \ldots, I$$

The Y_i are thought of as random variables, while the E_i are thought of as fixed and known functions of n_i, the number of persons at risk for the disease in county i. As a simple starting point, we might assume that

$$E_i = n_i \bar{r} \equiv n_i \left(\frac{\sum_i y_i}{\sum_i n_i} \right) ,$$

i.e., $\bar{r}$ is the overall disease rate in the entire study region. These E_i thus correspond to a kind of "null hypothesis," where we expect a constant disease rate in every county. This process is called *internal standardization*, since it centers the data (some counties will have observed rates higher than expected, and some less) but uses only the observed data to do so.

Internal standardization is "cheating" (or at least "empirical Bayes") in some sense, since we are "losing a degree of freedom" by estimating the grand rate r from our current data. An even better approach might be to make reference to an existing standard table of age-adjusted rates for the disease (as might be available for many types of cancer). Then after stratifying the population by age group, the E_i emerge as

$$E_i = \sum_j n_{ij} r_j ,$$

where n_{ij} is the person-years at risk in area i for age group j (i.e., the number of persons in age group j who live in area i times the number of years in the study), and r_j is the disease rate in age group j (taken from the standard table). This process is called *external standardization*. In either case, in its simplest form a disease map is just a display (in color or greyscale) of the raw disease rates overlaid on the areal units.

5.4.2 Traditional models and frequentist methods

If E_i is not too large (i.e, the disease is rare or the regions i are sufficiently small), the usual model for the Y_i is the Poisson model,

$$Y_i | \eta_i \sim Po(E_i \eta_i) ,$$

where η_i is the true *relative risk* of disease in region i. The maximum likelihood estimate (MLE) of η_i is readily shown to be

$$\hat{\eta}_i \equiv SMR_i = \frac{Y_i}{E_i} \, ,$$

the *standardized morbidity (or mortality) ratio* (SMR), i.e., the ratio of observed to expected disease cases (or deaths). Note that $Var(SMR_i) = Var(Y_i)/E_i^2 = \eta_i/E_i$, and so we might take $\widehat{Var}(SMR_i) = \hat{\eta}_i/E_i = Y_i/E_i^2$. This in turn permits calculation of traditional confidence intervals for η_i (although this is a bit awkward since the data are discrete), as well as hypothesis tests.

Example 5.5 To find a confidence interval for η_i, one might first assume that $\log SMR_i$ is roughly *normally* distributed. Using the delta method (Taylor series expansion), one can find that

$$Var[\log(SMR_i)] \approx \frac{1}{SMR_i^2} Var(SMR_i) = \frac{E_i^2}{Y_i^2} \times \frac{Y_i}{E_i^2} = \frac{1}{Y_i} \, .$$

An approximate 95% CI for $\log \eta_i$ is thus $\log SMR_i \pm 1.96/\sqrt{Y_i}$, and so (transforming back) an approximate 95% CI for η_i is

$$\left(SMR_i \exp(-1.96/\sqrt{Y_i}) \, , \, SMR_i \exp(1.96/\sqrt{Y_i}) \right) \, .$$

■

Example 5.6 Suppose we wish to test whether the true relative risk in county i is elevated or not, i.e.,

$$H_0 : \eta_i = 1 \ \text{ versus } \ H_A : \eta_i > 1 \, .$$

Under the null hypothesis, $Y_i \sim Po(E_i)$, so the p-value for this test is

$$p = Pr(X \geq Y_i|E_i) = 1 - Pr(X < Y_i|E_i) = 1 - \sum_{x=0}^{Y_i-1} \frac{\exp(-E_i)E_i^x}{x!} \, .$$

This is the (one-sided) p-value; if it is less than 0.05 we would typically reject H_0 and conclude that there is a statistically significant excess risk in county i. ■

5.4.3 Hierarchical Bayesian methods

The methods of the previous section are fine for detecting extra-Poisson variability (overdispersion) in the observed rates, but what if we seek to *estimate* and *map* the underlying relative risk surface $\{\eta_i, i = 1, \ldots, I\}$? In this case we might naturally think of a *random effects* model for the η_i, since we would likely want to assume that all the true risks come from a common underlying distribution. Random effects models also allow the

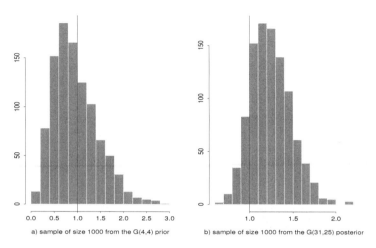

a) sample of size 1000 from the G(4,4) prior   b) sample of size 1000 from the G(31,25) posterior

Figure 5.9 *Samples of size 1000 from a Gamma(4,4) prior (a) and a Gamma(27+4, 21+4) posterior (b) for η_i.*

procedure to "borrow strength" across the various counties in order to come up with an improved estimate for the relative risk in each.

The random effects here, however, can be high dimensional, and are couched in a nonnormal (Poisson) likelihood. Thus, as in the previous sections of this chapter, the most natural way of handling this rather complex model is through hierarchical Bayesian modeling, as we now describe.

Poisson-gamma model

As a simple initial model, consider

$$Y_i \mid \eta_i \overset{ind}{\sim} Po(E_i\eta_i), i = 1, \ldots, I,$$

$$\text{and } \eta_i \overset{iid}{\sim} G(a, b),$$

where $G(a, b)$ denotes the *gamma* distribution with mean $\mu = a/b$ and variance $\sigma^2 = a/b^2$; note that this is the gamma parametrization used by the WinBUGS package. Solving these two equations for a and b we get $a = \mu^2/\sigma^2$ and $b = \mu/\sigma^2$. Suppose we set $\mu = 1$ (the "null" value) and $\sigma^2 = (0.5)^2$ (a fairly large variance for this scale). Figure 5.9(a) shows a sample of size 1000 from the resulting $G(4, 4)$ prior; note the vertical reference line at $\eta_i = \mu = 1$.

Inference about $\boldsymbol{\eta} = (\eta_1, \ldots, \eta_I)'$ is now based on the resulting posterior distribution, which in the Poisson-gamma emerges in closed form (thanks to the conjugacy of the gamma prior with the Poisson likelihood) as $\prod_i p(\eta_i|y_i)$, where $p(\eta_i|y_i)$ is $G(y_i + a, E_i + b)$. Thus a suitable point estimate of η_i might be the posterior mean,

$$E(\eta_i|\mathbf{y}) \;=\; E(\eta_i|y_i) \;=\; \frac{y_i + a}{E_i + b} \;=\; \frac{y_i + \frac{\mu^2}{\sigma^2}}{E_i + \frac{\mu}{\sigma^2}} \tag{5.42}$$

$$=\; \frac{E_i\left(\frac{y_i}{E_i}\right)}{E_i + \frac{\mu}{\sigma^2}} + \frac{\left(\frac{\mu}{\sigma^2}\right)\mu}{E_i + \frac{\mu}{\sigma^2}}$$

$$=\; w_i\,SMR_i + (1 - w_i)\mu\,, \tag{5.43}$$

where $w_i = E_i/[E_i + (\mu/\sigma^2)]$, so that $0 \le w_i \le 1$. Thus the Bayesian point estimate (5.43) is a *weighted average* of the the the data-based SMR for region i, and the prior mean μ. This estimate is approximately equal to SMR_i when w_i is close to 1 (i.e., when E_i is big, so the data are strongly informative, or when σ^2 is big, so the prior is weakly informative). On the other hand, (5.43) will be approximately equal to μ when w_i is close to 0 (i.e., when E_i is small, so the data are sparse, or when σ^2 is small, so that the prior is highly informative).

As an example, suppose in county i we observe $y_i = 27$ disease cases, when only $E_i = 21$ were expected. Under our $G(4,4)$ prior we obtain a $G(27+4, 21+4) = G(31, 25)$ posterior distribution; Figure 5.9(b) shows a sample of size 1000 drawn from this distribution. From (5.42) this distribution has mean $31/25 = 1.24$ (consistent with the figure), indicating slightly elevated risk (24%). However, the posterior probability that the true risk is bigger than 1 is $P(\eta_i > 1 \mid y_i) = .863$, which we can derive exactly (say, using 1 - pgamma(25,31) in S-plus), or estimate empirically as the proportion of samples in Figure 5.9(b) that are greater than 1. In either case, we see substantial but not overwhelming evidence of risk elevation in this county.

If we desired a $100 \times (1 - \alpha)\%$ confidence interval for η_i, the easiest approach would be to simply take the upper and lower $\alpha/2$-points of the $G(31, 25)$ posterior, since the resulting interval $\left(\eta_i^{(L)}, \eta_i^{(U)}\right)$, would be such that $P\left[\eta_i \in \left(\eta_i^{(L)}, \eta_i^{(U)}\right) \mid y_i\right] = 1 - \alpha$, by definition of the posterior distribution. This is the so-called *equal-tail credible interval* mentioned in Subsection 4.2.2. In our case, taking $\alpha = .05$ we obtain $(\eta_i^{(L)}, \eta_i^{(U)}) = (.842, 1.713)$, again indicating no "significant" elevation in risk for this county. (In Splus or R the appropriate commands here are qgamma(.025, 31)/25 and qgamma(.975, 31)/25.)

Finally, in a "real" data setting we would obtain not 1 but I point estimates, interval estimates, and posterior distributions, one for each county. Such estimates would often be summarized in a choropleth map, say, in Splus or ArcView. Full posteriors are obviously difficult to summarize spatially, but posterior means, variances, or confidence limits are easily mapped in this way. We shall explore this issue in the next subsection.

Poisson-lognormal models

The gamma prior of the preceding section is very convenient computationally, but suffers from a serious defect: it fails to allow for spatial correlation among the η_i. To do this we would need a *multivariate* version of the gamma distribution; such structures exist but are awkward both conceptually and computationally. Instead, the usual approach is to place some sort of multivariate *normal* distribution on the $\psi_i \equiv \log \eta_i$, the *log*-relative risks.

Thus, consider the following augmentation of our basic Poisson model:

$$
\begin{aligned}
Y_i \mid \psi_i &\overset{ind}{\sim} Po\left(E_i\, e^{\psi_i}\right), \\
\text{where } \psi_i &= \mathbf{x}_i'\boldsymbol{\beta} + \theta_i + \phi_i\,.
\end{aligned} \tag{5.44}
$$

The $\mathbf{x}_i$ are explanatory spatial covariates, having parameter coefficients $\boldsymbol{\beta}$. The covariates are *ecological*, or county (not individual) level, which may lead to problems of ecological bias (to be discussed later). However, the hope is that they will explain some (perhaps all) of the spatial patterns in the Y_i.

Next, the θ_i capture region-wide *heterogeneity* via an ordinary, exchangeable normal prior,

$$
\theta_i \overset{iid}{\sim} N(0\,,\; 1/\tau_h)\,, \tag{5.45}
$$

where τ_h is a precision (reciprocal of the variance) term that controls the magnitude of the θ_i. These random effects capture extra-Poisson variability in the log-relative risks that varies "globally," i.e., over the entire study region.

Finally, the ϕ_i are the parameters that make this a truly spatial model by capturing regional *clustering*. That is, they model extra-Poisson variability in the log-relative risks that varies "locally," so that nearby regions will have more similar rates. A plausible way to attempt this might be to try a point-referenced model on the parameters ϕ_i. For instance, writing $\boldsymbol{\phi} = (\phi_1, \ldots, \phi_I)'$, we might assume that

$$
\boldsymbol{\phi} \mid \mu, \boldsymbol{\lambda} \sim N_I(\mu, H(\boldsymbol{\lambda}))\,,
$$

where N_I denotes the I-dimensional normal distribution, μ is the (stationary) mean level, and $(H(\boldsymbol{\lambda}))_{ii'}$ gives the covariance between ϕ_i and $\phi_{i'}$ as a function of some hyperparameters $\boldsymbol{\lambda}$. The standard forms given in Table 2.1 remain natural candidates for this purpose.

While such models for the ϕ_i are very sensible, they turn out to be arduous to fit even in the isotropic case, due to the large amount of matrix inversion required. Moreover, in the areal data context, we would require a multivariate normal distribution for $\boldsymbol{\phi}$, which directly models association between ϕ_i and ϕ_j. But then we would also require some notion of "distance" between areal units. With units of roughly equal size laid out on a fairly regular grid, intercentroidal distance may be appropriate here. But with very irregular spatial units, such distance may make little sense. As a result, it is customary in hierarchical analyses of areal data to return to neighbor-based notions of proximity, and ultimately, to return to CAR specifications for ϕ (Section 3.3). In the present context (with the CAR model placed on the elements of ϕ rather than the elements of $\mathbf{Y}$), we will write

$$\phi \sim CAR(\tau_c) , \qquad (5.46)$$

where by this notation we mean the improper CAR (IAR) model in (3.16) with y_i replaced by ϕ_i, τ^2 replaced by $1/\tau_c$, and using the 0-1 (adjacency) weights w_{ij}. Thus τ_c is a *precision* (not variance) parameter in the CAR prior (5.46), just as τ_h is a precision parameter in the heterogeneity prior (5.45).

CAR models and their difficulties

Compared to point-level (geostatistical) models, CAR models are very convenient computationally, since our method of finding the posterior of θ and ϕ is itself a conditional algorithm, the Gibbs sampler. Recall from Subsection 4.3.1 that this algorithm operates by successively sampling from the *full conditional* distribution of each parameter (i.e., the distribution of each parameter given the data and every other parameter in the model). So for example, the full conditional of ϕ_i is

$$p(\phi_i|\phi_{j\neq i}, \boldsymbol{\theta}, \boldsymbol{\beta}, \mathbf{y}) \propto Po(y_i \mid E_i e^{\mathbf{x}_i'\boldsymbol{\beta}+\theta_i+\phi_i}) \times N(\phi_i \mid \bar{\phi}_i , 1/(\tau_c m_i)) , \quad (5.47)$$

meaning that we do not need to work with the joint distribution of ϕ at all. The conditional approach also eliminates the need for any matrix inversion.

While computationally convenient, CAR models have numerous theoretical and computational difficulties, some of which have already been noted in Section 3.3, and others of which only become acute now that we are using the CAR as a distribution for the random effects ϕ, rather than the data $\mathbf{Y}$ itself. We consider two of these issues.

1. Impropriety: Recall from the discussion surrounding (3.15) that the IAR prior we selected in (5.46) above is *improper*, meaning that it does not determine a legitimate probability distribution (one that integrates to 1). That is, the matrix $\Sigma_{\mathbf{y}}^{-1} = (D_w - W)$ is singular, and thus its inverse does not exist.

As mentioned in that earlier discussion, one possible fix for this situation is to include a "propriety parameter" ρ in the precision matrix, i.e., $\Sigma_{\mathbf{y}}^{-1} = (D_w - \rho W)$. Taking $\rho \in \left(1/\lambda_{(1)}, 1/\lambda_{(n)}\right)$, where $\lambda_{(1)}$ and $\lambda_{(n)}$ are the smallest and largest eigenvalues of $D_w^{-1/2} W D_w^{-1/2}$, respectively, ensures the existence of $\Sigma_{\mathbf{y}}$. Alternatively, using the scaled adjacency matrix $\widetilde{W} \equiv Diag(1/w_{i+})W$, $\Sigma_{\mathbf{y}}^{-1}$ can be written as $M^{-1}(I - \alpha\widetilde{W})$ where M is diagonal. One can then show (Carlin and Banerjee, 2003; Gelfand and

Vounatsou, 2002) that if $|\alpha| < 1$, then $(I - \alpha \widetilde{W})$ will be positive definite, resolving the impropriety problem without eigenvalue calculation.

Unfortunately, a difficulty with this fix (also already mentioned near the end of Subsection 3.3.1) is that this new prior typically does not deliver enough spatial similarity unless α is quite close to 1, thus getting us very close to the same problem again! Some authors (e.g., Carlin and Banerjee, 2003) recommend an informative prior that insists on larger α's (say, a $Beta(18, 2)$), but this is controversial since there will typically be little true prior information available regarding the magnitude of α.

A second possible fix (more common in practice) is simply to ignore the impropriety of the standard CAR model (5.46) and continue! After all, we are only using the CAR model as a *prior*; the *posterior* will typically still emerge as proper, so Bayesian inference may still proceed. This is the usual approach, but it also requires some care, as follows: this improper CAR prior is a *pairwise difference prior* (Besag et al., 1995) that is identified only up to an additive constant. Thus to identify an intercept term β_0 in the log-relative risk, we must add the constraint $\sum_{i=1}^{I} \phi_i = 0$. Note that in implementing a Gibbs sampler to fit (5.44), this constraint can be imposed *numerically* by recentering each sampled ϕ vector around its own mean following each Gibbs iteration.

2. *Selection of τ_c and τ_h:* Clearly the values of these two prior precision parameters will control the amount of extra-Poisson variability allocated to "heterogeneity" (the θ_i) and "clustering" (the ϕ_i). But they cannot simply be chosen to be arbitrarily large, since then the ϕ_i and θ_i would be *unidentifiable*: note that we see only a single Y_i in each county, yet we are attempting to estimate *two* random effects for each i! Eberly and Carlin (2000) investigate convergence and Bayesian learning for this data set and model, using fixed values for τ_h and τ_c.

Similarly, if we decide to place third-stage priors (*hyperpriors*) on τ_c and τ_h, they also cannot be arbitrarily vague for the same reason. Still, the gamma offers a conjugate family here, so we might simply take

$$\tau_h \sim G(a_h, b_h) \text{ and } \tau_c \sim G(a_c, b_c) \,.$$

To make this prior "fair" (i.e., equal prior emphasis on heterogeneity and clustering), it is tempting to simply set $a_h = a_c$ and $b_h = b_c$, but this would be incorrect for two reasons. First, the τ_h prior (5.45) uses the usual *marginal* specification, while the τ_c prior (5.46) is specified *conditionally*. Second, τ_c is multiplied by the number of neighbors m_i before playing the role of the (conditional) prior precision. Bernardinelli et al. (1995) note that the prior marginal standard deviation of ϕ_i is roughly equal to the prior conditional standard deviation divided by 0.7. Thus a scale that delivers

$$sd(\theta_i) = \frac{1}{\sqrt{\tau_h}} \approx \frac{1}{0.7\sqrt{\bar{m}\tau_c}} \approx sd(\phi_i) \qquad (5.48)$$

where $\bar{m}$ is the average number of neighbors may offer a reasonably "fair" specification. Of course, it is fundamentally unclear how to relate the marginal variance of a proper joint distribution with the conditional variance of an improper joint distribution.

Example 5.7 As an illustration of the Poisson-lognormal model (5.44), consider the data displayed in Figure 5.10. These data from Clayton and Kaldor (1987) are the observed (Y_i) and expected (E_i) cases of lip cancer for the $I = 56$ districts of Scotland during the period 1975–1980. One county-level covariate x_i, the percentage of the population engaged in agriculture, fishing or forestry (AFF), is also available (and also mapped in Figure 5.10). Modeling the log-relative risk as

$$\psi_i = \beta_0 + \beta_1 x_i + \theta_i + \phi_i \ , \tag{5.49}$$

we wish to investigate a variety of vague, proper, and arguably "fair" priors for τ_c and τ_h, find the estimated posterior of β_1 (the AFF effect), and find and map the fitted relative risks $E(\psi_i|\mathbf{y})$.

Recall that Y_i cannot inform about θ_i or ϕ_i, but only about their sum $\xi_i = \theta_i + \phi_i$. Making the reparameterization from $(\boldsymbol{\theta}, \boldsymbol{\phi})$ to $(\boldsymbol{\theta}, \boldsymbol{\xi})$, we have the joint posterior,

$$p(\boldsymbol{\theta}, \boldsymbol{\xi}|\mathbf{y}) \propto L(\boldsymbol{\xi}; \mathbf{y})p(\boldsymbol{\theta})p(\boldsymbol{\xi} - \boldsymbol{\theta}).$$

This means that

$$p(\theta_i \mid \theta_{j \neq i}, \boldsymbol{\xi}, \mathbf{y}) \propto p(\theta_i) \, p(\xi_i - \theta_i \mid \{\xi_j - \theta_j\}_{j \neq i}) \ .$$

Since this distribution is free of the data $\mathbf{y}$, the θ_i are *Bayesianly unidentified* (and so are the ϕ_i). But this does not preclude *Bayesian learning* (i.e., prior to posterior movement) about θ_i. No Bayesian learning would instead require

$$p(\theta_i|\mathbf{y}) = p(\theta_i) \ , \tag{5.50}$$

in the case where both sides are proper (a condition not satisfied by the CAR prior). Note that (5.50) is a much stronger condition than Bayesian unidentifiability, since the data have no impact on the *marginal* (not merely the conditional) posterior distribution.

Recall that, though they are unidentified, the θ_i and ϕ_i are interesting in their own right, as is

$$\alpha = \frac{sd(\boldsymbol{\phi})}{sd(\boldsymbol{\theta}) + sd(\boldsymbol{\phi})} \ ,$$

where $sd(\cdot)$ is the empirical marginal standard deviation function. That is, α is the proportion of the variability in the random effects that is due to clustering (hence $1 - \alpha$ is the proportion due to unstructured heterogeneity). Recall we need to specify vague but proper prior values τ_h and τ_c that lead to acceptable convergence behavior, yet still allow Bayesian learning. This prior should also be "fair," i.e., lead to $\alpha \approx 1/2$ *a priori*.

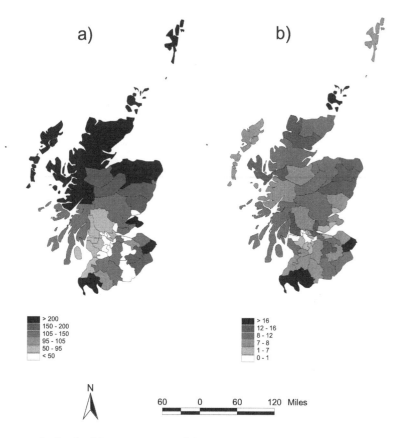

Figure 5.10 *Scotland lip cancer data: (a) crude standardized mortality ratios (observed / expected × 100); (b) AFF covariate values.*

Figure 5.11 contains the WinBUGS code for this problem, which is also available at http://www.biostat.umn.edu/~brad/data2.html. Note the use of vague priors for τ_c and τ_h as suggested by Best et al. (1999), and the use of the sd function in WinBUGS to greatly facilitate computation of α.

The basic posterior (mean, sd) and convergence (lag 1 autocorrelation) summaries for α, β_1, ξ_1, and ξ_{56} are given in Table 5.4. Besides the Best et al. (1999) prior, two priors inspired by equation (5.48) are also reported; see Carlin and Pérez (2000). The AFF covariate appears significantly different from 0 under all 3 priors, although convergence is *very* slow (very high values for llacf). The excess variability in the data seems mostly due to clustering $(E(\alpha|\mathbf{y}) > .50)$, but the posterior distribution for α does *not* seem robust to changes in the prior. Finally, convergence for the ξ_i (reason-

```
model
{
  for (i in 1 : regions) {
    O[i] ~ dpois(mu[i])
    log(mu[i]) <- log(E[i]) + beta0 + beta1*aff[i]/10
      + phi[i] + theta[i]
    theta[i] ~ dnorm(0.0,tau.h)
    xi[i] <- theta[i] + phi[i]
    SMRhat[i] <- 100 * mu[i]/E[i]
    SMRraw[i] <- 100 * O[i]/E[i]
  }
  phi[1:regions] ~ car.normal(adj[], weights[], num[], tau.c)

  beta0 ~ dnorm(0.0, 1.0E-5) # vague prior on grand intercept
  beta1 ~ dnorm(0.0, 1.0E-5) # vague prior on AFF effect

  tau.h ~ dgamma(1.0E-3,1.0E-3)  # ''fair'' prior from
  tau.c ~ dgamma(1.0E-1,1.0E-1)  #    Best et al. (1999)

  sd.h <- sd(theta[]) # marginal SD of heterogeneity effects
  sd.c <- sd(phi[])   # marginal SD of clustering effects
  alpha <- sd.c / (sd.h + sd.c)
}
```

Figure 5.11 WinBUGS *code to fit the Poisson-normal-CAR model to the Scottish lip cancer data.*

ably well identified) is rapid; convergence for the ψ_i (not shown) is virtually immediate.

Of course, a full analysis of these data would also involve a map of the posterior means of the raw and estimated SMR's, which we can do directly in GeoBUGS, the spatial statistics module supplied with WinBUGS Versions 1.4 and later. We investigate this in the context of the homework assignment in Exercise 12. ■

5.5 General linear areal data modeling

By analogy with Section 5.2, the areal unit measurements Y_i that we model need not be restricted to counts, as in our disease mapping setting. They may also be binary events (say, presence or absense of a particular facility in region i), or continuous measurements (say, population density, i.e., a region's total population divided by its area).

Again formulating a hierarchical model, Y_i may be described using a suitable first-stage member of the exponential family. Now given β and ϕ_i,

Priors for τ_c, τ_h	Posterior for α			Posterior for β		
	mean	sd	l1acf	mean	sd	l1acf
G(1.0, 1.0), G(3.2761, 1.81)	.57	.058	.80	.43	.17	.94
G(.1, .1), G(.32761, .181)	.65	.073	.89	.41	.14	.92
G(.1, .1), G(.001, .001)	.82	.10	.98	.38	.13	.91

Priors for τ_c, τ_h	Posterior for ξ_1			Posterior for ξ_{56}		
	mean	sd	l1acf	mean	sd	l1acf
G(1.0, 1.0), G(3.2761, 1.81)	.92	.40	.33	−.96	.52	.12
G(.1, .1), G(.32761, .181)	.89	.36	.28	−.79	.41	.17
G(.1, .1), G(.001, .001)	.90	.34	.31	−.70	.35	.21

Table 5.4 *Posterior summaries for the spatial model with Gamma hyperpriors for τ_c and τ_h, Scotland lip cancer data; "sd" denotes standard deviation while "l1acf" denotes lag 1 sample autocorrelation.*

analogous to (5.18) the Y_i are conditionally independent with density,

$$f(y_i|\boldsymbol{\beta}, \phi_i, \gamma) = h(y_i, \gamma) \exp\{\gamma[y_i\eta_i - \psi(\eta_i)]\}, \qquad (5.51)$$

where $g(\eta_i) = \mathbf{x}_i^T\boldsymbol{\beta} + \phi_i$ for some link function g with γ a dispersion parameter. The ϕ_i will be spatial random effects coming from a CAR model; the pairwise difference, intrinsic (IAR) form is most commonly used. As a result, we have a generalized linear mixed model with spatial structure in the random effects.

Note that in the previous section, independent homogeneity effects θ_i were also introduced into $g(\eta_i)$. If f were now normal this would clearly make no sense: we would have introduced independent normal errors twice! Even with a nonnormal first stage, as in Section 5.2, we are in a situation where the stochastic mechanism in f replaces these independent errors. This is why many practitioners prefer to fit models of the form (5.51) having only a spatial random effect. Computation is more stable and a "balanced" (or "fair") prior specification (as mentioned in connection with (5.48) above) is not an issue.

5.6 Comparison of point-referenced and areal data models

We conclude this chapter with a brief summary and comparison between point-referenced data models and areal unit data models. First, the former are defined with regard to an uncountable number of random variables. The process specification determines the n-dimensional joint distribution for the $Y(\mathbf{s}_i)$, $i = 1, \ldots, n$. For areal units, we envision only an n-dimensional distribution for the Y_i, $i = 1, \ldots, n$, which we write down to begin with.

Next, with point-referenced data, we model association directly. For example, if $\mathbf{Y} = (Y(\mathbf{s}_1), \ldots, Y(\mathbf{s}_n))'$ we specify $\Sigma_{\mathbf{Y}}$ using isotropy (or anisotropy), stationarity (or nonstationarity), and so on. With areal data $\mathbf{Y} = (Y_1, \ldots, Y_n)'$ and CAR (or SAR) specifications, we instead model $\Sigma_{\mathbf{Y}}^{-1}$ directly. For instance, with CAR models, Brook's Lemma enables us to reconstruct $\Sigma_{\mathbf{Y}}^{-1}$ from the conditional specifications; $\Sigma_{\mathbf{Y}}^{-1}$ provides conditional association structure (as in Section 3.3) but says nothing about *unconditional* association structure. When $\Sigma_{\mathbf{Y}}^{-1}$ is full rank, the transformation to $\Sigma_{\mathbf{Y}}$ is very complicated, and very nonlinear. Positive conditional association can become negative unconditional association. If the CAR is defined through distance-based w_{ij}'s there need not be any corresponding distance-based order to the unconditional associations. See Besag and Kooperberg (1995), Conlon and Waller (1999), Hrafnkelsson and Cressie (2003), and Wall (2004) for further discussion.

With regard to formal specification, in the most commonly employed point-level Gaussian case, the process is specified through a valid covariance function. With CAR modeling, the specification is instead done through Markov random fields (Section 3.2) employing the Hammersley-Clifford Theorem to ensure a unique joint distribution.

Explanation is a common goal of point-referenced data modeling, but often an even more important goal is spatial prediction or interpolation (i.e., kriging). This may be done at at new points, or for block averages (see Section 6.1). With areal units, again a goal is explanation, but now often supplemented by smoothing. Here the interpolation problem is to new areal units, the so-called modifiable areal unit problem (MAUP) as discussed in Sections 6.2 and 6.3.

Finally, with spatial processes, likelihood evaluation requires computation of a quadratic form involving $\Sigma_{\mathbf{Y}}^{-1}$ and the determinant of $\Sigma_{\mathbf{Y}}$. (With spatial random effects, this evaluation is deferred to the second stage of the model, but is still present.) With an increasing number of locations, such computation becomes very expensive (computing time is greater than order n^2), and may also become unstable, due to the enormous number of floating point operations required. We refer to this situation rather informally as a "big n" problem. Approaches for treating this problem are discussed in Appendix Subsection A.5. On the other hand, with CAR modeling the likelihood (or the second-stage model for the random effects, as the case may be) is written down immediately, since this model parametrizes $\Sigma_{\mathbf{Y}}^{-1}$ (rather than $\Sigma_{\mathbf{Y}}$). Full conditional distributions needed for MCMC sampling are immediate, and there is no big n problem. Also, for SAR the quadratic form is directly evaluated, while the determinant is usually evaluated efficiently (even for very large n) using sparse matrix methods; see, e.g., Pace and Barry (1997a,b).

5.7 Exercises

1. Derive the forms of the full conditionals for $\beta, \sigma^2, \tau^2, \phi$, and $\mathbf{W}$ in the exponential kriging model (5.1) and (5.3).

2. Assuming the likelihood in (5.3), suppose that ϕ is fixed and we adopt the prior $p(\beta, \sigma^2, \tau^2) \propto 1/(\sigma^2 \tau^2)$, a rather standard noninformative prior often chosen in nonspatial analysis settings. Show that the resulting posterior $p(\beta, \sigma^2, \tau^2 | \mathbf{y})$ is *improper*.

3. Derive the form of $p(y_0 | \mathbf{y}, \boldsymbol{\theta}, \mathbf{x}_0)$ in (5.5) via the usual conditional normal formulae; i.e., following Guttman (1982, pp. 69-72), if $\mathbf{Y} = (\mathbf{Y}_1^T, \mathbf{Y}_2^T)^T \sim N(\boldsymbol{\mu}, \boldsymbol{\Sigma})$ where

$$\boldsymbol{\mu} = \begin{pmatrix} \boldsymbol{\mu}_1 \\ \boldsymbol{\mu}_2 \end{pmatrix} \text{ and } \boldsymbol{\Sigma} = \begin{pmatrix} \Sigma_{11} & \Sigma_{12} \\ \Sigma_{21} & \Sigma_{22} \end{pmatrix},$$

then $\mathbf{Y}_2 | \mathbf{Y}_1 \sim N(\boldsymbol{\mu}_{2.1}, \Sigma_{2.1})$, where

$$\boldsymbol{\mu}_{2.1} = \boldsymbol{\mu}_2 + \Sigma_{21} \Sigma_{11}^{-1} (\mathbf{Y}_1 - \boldsymbol{\mu}_1) \text{ and } \Sigma_{2.1} = \Sigma_{22} - \Sigma_{21} \Sigma_{11}^{-1} \Sigma_{12}.$$

4. In expression (5.18), if $g(\theta) = \theta$ and the prior on β is a proper normal distribution,

   (a) Show that the full conditional distributions for the components of β are log-concave.

   (b) Show that the full conditional distributions for the $w(\mathbf{s}_i)$ are log-concave.

5.(a) Derive the variance and covariance relationships given in (5.25).

   (b) Derive the variance and covariance relationships given in (5.27).

6. The lithology data set (see www.biostat.umn.edu/~brad/data2.html) consists of measurements taken at 118 sample sites in the Radioactive Waste Management Complex region of the Idaho National Engineering and Environmental Laboratory. At each site, bore holes were drilled and measurements taken to determine the elevation and thickness of the various underground layers of soil and basalt. Understanding the spatial distribution of variables like these is critical to predicting fate and transport of groundwater and the (possibly harmful) constituents carried therein; see Leecaster (2002) for full details.

   For this problem, consider only the variables Northing, Easting, Surf Elevation, Thickness, and A-B Elevation, and only those records for which full information is available (i.e., extract only those data rows without an "NA" for any variable).

   (a) Produce image plots of the variables Thickness, Surf Elevation, and A-B Elevation. Add contour lines to each plot and comment on the descriptive topography of the region.

(b) Taking `log(Thickness)` as the response and `Surf Elevation` and `A-B Elevation` as covariates, fit a univariate Gaussian spatial model with a nugget effect, using the exponential and Matérn covariance functions. You may start with flat priors for the covariate slopes, Inverse Gamma(0.1, 0.1) priors for the spatial and nugget variances, and a Gamma(0.1, 0.1) prior for the spatial range parameter. Modify the priors and check for their sensitivity to the analysis. (*Hint:* You can use `WinBUGS` to fit the exponential model, but you must use the `krige.bayes()` function in `geoR` for the Matérn.)

(c) Perform Bayesian kriging on a suitable grid of values and create image plots of the posterior mean residual surfaces for the spatial effects. Overlay the plots with contour lines and comment on the consistency with the plots from the raw data in part (a).

(d) Repeat the above for a purely spatial model (without a nugget) and compare this model with the spatial+nugget model using a model choice criterion (say, DIC).

7. The real estate data set (`www.biostat.umn.edu/~brad/data2.html`) consists of information regarding 70 sales of single-family homes in Baton Rouge, LA, during the month of June 1989. It is customary to model log-selling price.

(a) Obtain the empirical variogram of the raw log-selling prices.

(b) Fit an ordinary least squares regression to log-selling price using living area, age, other area, and number of bathrooms as explanatory variables. Such a model is usually referred to as a *hedonic* model.

(c) Obtain the empirical variogram of the residuals to the least squares fit.

(d) Using an exponential spatial correlation function, attempt to fit the model $Y(\mathbf{s}) = \mathbf{x}^T(\mathbf{s})\boldsymbol{\beta} + W(\mathbf{s}) + \epsilon(\mathbf{s})$ as in equation (5.1) to the log-selling prices, obtaining estimates using `geoR` or `S+SpatialStats`.

(e) Predict the actual selling price for a home at location (longitude, latitude) = (−91.1174, 30.506) that has characteristics LivingArea = 938 sqft, OtherArea = 332sqft, Age = 25yrs, Bedrooms = 3, Baths = 1, and HalfBaths = 0. (*Reasonability check:* The actual log selling price for this location turned out to be 10.448.)

(f) Use `geoR` (grid-based integration routines) or `WinBUGS` (MCMC) to fit the above model in a Bayesian framework. Begin with the following prior specification: a flat prior for $\boldsymbol{\beta}$, IG(0.1,0.1) (`WinBUGS` parametrization) priors for $1/\sigma^2$ and $1/\tau^2$, and a Uniform(0,10) prior for ϕ. Also investigate prior robustness by experimenting with other choices.

(g) Obtain samples from the predictive distribution for log-selling price and selling price for the particular location mentioned above. Summarize this predictive distribution.

(h) Compare the classical and Bayesian inferences.

(i) *(advanced):* Hold out the first 20 observations in the data file, and fit the nonspatial (i.e., without the $W(\mathbf{s})$ term) and spatial models to the observations that remain. For both models, compute

- $\sum_{j=1}^{20}(Y(\mathbf{s}_{0j}) - \hat{Y}(\mathbf{s}_{0j}))^2$
- $\sum_{j=1}^{20} Var(Y(\mathbf{s}_{0j})|\mathbf{y})$
- the proportion of predictive intervals for $Y(\mathbf{s}_0)$ that are correct
- the proportion of predictions that are within 10% of the true value
- the proportion of predictions that are within 20% of the true value

Discuss the differences in predictive performance.

8. Suppose $Z_i = Y_i/n_i$ is the observed disease *rate* in each county, and we adopt the model $Z_i \stackrel{ind}{\sim} N(\eta_i, \sigma^2)$ and $\eta_i \stackrel{iid}{\sim} N(\mu, \tau^2)$, $i = 1, \dots, I$. Find $E(\eta_i|y_i)$, and express it as a weighted average of Z_i and μ. Interpret your result as the weights vary.

9. In fitting model (5.44) with priors for the θ_i and ϕ_i given in (5.45) and (5.46), suppose we adopt the hyperpriors $\tau_h \sim G(a_h, b_h)$ and $\tau_c \sim G(a_c, b_c)$. Find closed form expressions for the full conditional distributions for these two parameters.

10. The full conditional (5.47) does *not* emerge in closed form, since the CAR (normal) prior is not conjugate with the Poisson likelihood. However, prove that this full conditional *is* log-concave, meaning that the necessary samples can be generated using the adaptive rejection sampling (ARS) algorithm of Gilks and Wild (1992).

11. Confirm algebraically that, taken together, the expressions

$$\phi_i|\boldsymbol{\phi}_{j\neq i} \sim N(\phi_i \mid \bar{\phi}_i , \, 1/(\tau_c m_i)) , \; i = 1, \dots, I$$

are equivalent to the (improper) joint specification

$$p(\phi_1, \dots, \phi_I) \propto \exp\left\{ -\frac{\tau_c}{2} \sum_{i \, adj \, j} (\phi_i - \phi_j)^2 \right\} ,$$

i.e., the version of (3.16) corresponding to the usual, adjacency-based CAR model (5.46).

12. The Minnesota Department of Health is charged with investigating the possibility of geographical clustering of the rates for the late detection of colorectal cancer in the state's 87 counties. For each county, the late detection rate is simply the number of regional or distant case detections divided by the total cases observed in that county.

Information on several potentially helpful covariates is also available.

The most helpful is the county-level estimated proportions of persons who have been screened for colorectal cancer, as estimated from telephone interviews available biannually between 1993 and 1999 as part of the nationwide Behavioral Risk Factor Surveillance System (BRFSS).

(a) Use WinBUGS to model the log-relative risk using (5.49), fitting the heterogeneity plus clustering (CAR) model to these data. You will find the observed late detections Y_i, expected late detections E_i and screening covariates x_i in WinBUGS format in the file colorecbugs.dat on the webpage www.biostat.umn.edu/~brad/data2.html/. Find and summarize the posterior distributions of α (the proportion of excess variability due to clustering) and β_1 (the screening effect). Does MCMC convergence appear adequate in this problem?

(b) Use the poly.S function (see the webpage again) in S-plus to obtain a boundary file for the counties of the state of Minnesota. Do this in S-plus by typing

- source(''poly.S'')
- mkpoly(''minnesota'')

The result should be a file called minnesota.txt. Now open this file in WinBUGS, and pull down to Import Splus on the Map menu. Kill and restart WinBUGS, and then pull down to Adjacency Tool again from the Map menu; "minnesota" should now be one of the options! Click on adj map to see the adjacency map, adj matrix to print out the adjacency matrix, and show region to find any given region of interest.

(c) To use GeoBUGS to map the raw and fitted SMR's, pull down to Mapping Tool on the Map menu. Customize your maps by playing with cuts and colors. (Remember you will have to have saved those Gibbs samples during your WinBUGS run in order to summarize them!)

(d) Save your maps as .odc files, and then (if possible) as .ps files by selecting the "print to file" option within WinBUGS. (Your PC must have installed on it a printer driver that is specifically designated for a postscript printer.) Though the resulting file might not have a ".ps" extension, what ultimately gets saved to disk should indeed be a postscript file.

(e) Since the screening covariate was estimated from the BRFSS survey, we should really account for survey measurement error, since this may be substantial for rural counties having few respondents. To do this, replace the observed covariate x_i in the log-relative risk model (5.49) by T_i, the true (unobserved) rate of colorectal screening in county i. Following Xia and Carlin (1998), we then further augment our hierarchical model with

$$T_i \overset{iid}{\sim} N(\mu_0, 1/\lambda) \quad \text{and} \quad x_i | T_i \overset{ind}{\sim} N(T_i, 1/\delta) \, ,$$

That is, x_i is acknowledged as an imperfect (albeit unbiased) measure of the true screening rate T_i. Treat the measurement error precision λ and prior precision δ either as known "tuning parameters," or else assign them gamma hyperprior distributions, and recompute the posterior for β_1. Observe and interpret any changes.

(f) A more realistic errors-in-covariates model might assume that the precision of x_i given T_i should be proportional to the survey sample size r_i in each county. Write down (but do not fit) a hierarchical model that would address this problem.

## CHAPTER 6

Spatial misalignment

In this chapter we tackle the problem of spatial misalignment. By this we mean the summary or analysis of spatial data at a different level of spatial resolution than it was originally collected. For example, we might wish to obtain the spatial distribution of some variable at the county level, even though it was originally collected at the census tract level. We might have a very low-resolution global climate model for weather prediction, and seek to predict more locally (i.e., at higher resolution). For areal unit data, our purpose might be simply to understand the variable's distribution at a new level of spatial aggregation (the so-called *modifiable areal unit problem*, or MAUP), or perhaps so we can relate it to another variable that is already available at this level (say, a demographic census variable collected over the tracts). For data modeled through a spatial process we would envision block averaging at different spatial scales, (the so-called *change of support problem*, or COSP), again possibly for connection with another variable observed at a particular scale. For either type of data, our goal in the first case is typically one of spatial *interpolation*, while in the second it is one of spatial *regression*.

In addition to our presentation here, we also encourage the reader to look at the excellent review paper by Gotway and Young (2002). These authors give nice discussions of (as well as both traditional and Bayesian approaches for) the MAUP and COSP, spatial regression, and the *ecological fallacy*. This last term refers to the fact that relationships observed between variables measured at the ecological (aggregate) level may not accurately reflect (and will often overstate) the relationship between these same variables measured at the individual level. Discussion of this problem dates at least to Robinson (1950); see Wakefield (2001, 2003, 2004) for more modern treatments of this difficult subject.

As in previous sections, we group our discussion according to whether the data is suitably modeled using a spatial process as opposed to a CAR or SAR model. Here the former assumption leads to more general modeling, since point-level data may be naturally aggregated to block level, but the reverse procedure may or may not be possible; e.g., if the areal data are counts or proportions, what would the point-level variables be? However, since block-level summary data are quite frequent in practice (often due to confidentiality restrictions), methods associated with such data are also of great importance. We thus consider point-level and block-level modeling.

6.1 Point-level modeling

6.1.1 Gaussian process models

Consider a univariate variable that is modeled through a spatial process. In particular, assume that it is observed either at points in space, or over areal units (e.g., counties or zip codes), which we will refer to as *block* data. The *change of support problem* is concerned with inference about the values of the variable at points or blocks different from those at which it has been observed.

Motivating data set

A solution to the change of support problem is required in many health science applications, particularly spatial and environmental epidemiology. To illustrate, consider again the data set of ozone levels in the Atlanta, GA metropolitan area, originally reported by Tolbert et al. (2000). Ozone measures are available at between 8 and 10 fixed monitoring sites during the 92 summer days (June 1 through August 31) of 1995. Similar to Figure 1.3 (which shows 8-hour maximum ozone levels), Figure 6.1 shows the 1-hour daily maximum ozone measures at the 10 monitoring sites on July 15, 1995, along with the boundaries of the 162 zip codes in the Atlanta metropolitan area. Here we might be interested in predicting the ozone level at different points on the map (say, the two points marked **A** and **B**, which lie on opposite sides of a single city zip), or the average ozone level over a particular zip (say, one of the 36 zips falling within the city of Atlanta, the collection of which are encircled by the dark boundary on the map). The latter problem is of special interest, since in this case relevant health outcome data are available only at the zip level. In particular, for each day and zip, we have the number of pediatric ER visits for asthma, as well as the total number of pediatric ER visits. Thus an investigation of the relationship between ozone exposure and pediatric asthma cannot be undertaken until the mismatch in the support of the two variables is resolved. Situations like this are relatively common in health outcome settings, since personal privacy concerns often limit statisticians' access to data other than at the areal or block level.

In many earth science and population ecology contexts, presence/absence is typically recorded at essentially point-referenced sites while relevant climate layers are often downscaled to grid cells at some resolution. A previous study of the Atlanta ozone data by Carlin et al. (1999) realigned the point-level ozone measures to the zip level by using an ARC/INFO universal kriging

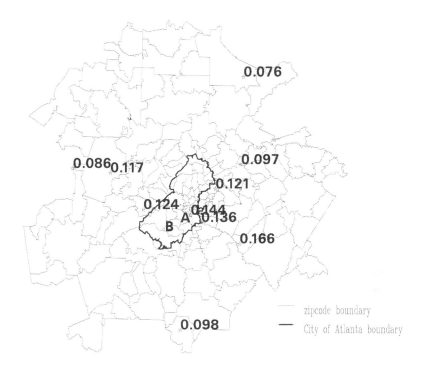

Figure 6.1 *Zip code boundaries in the Atlanta metropolitan area and 1-hour daily maximum ozone levels at the 10 monitoring sites for July 15, 1995.*

procedure to fit a smooth ozone exposure surface, and subsequently took the kriged value at each zip centroid as the ozone value for that zip. But this approach uses a single centroid value to represent the ozone level in the entire zip, and fails to properly capture variability and spatial association by treating these kriged estimates as observed values.

Model Assumptions and Analytic Goals

Let $Y(\mathbf{s})$ denote the spatial process (e.g., ozone level) measured at location $\mathbf{s}$, for $\mathbf{s}$ in some region of interest D. In our applications $D \subset \Re^2$ but our development works in arbitrary dimensions. A realization of the process is a surface over D. For point-referenced data the realization is observed at a finite set of sites, say, $\mathbf{s}_i, i = 1, 2, \ldots, I$. For block data we assume the observations arise as block averages. That is, for a block $B \subset D$,

$$Y(B) = |B|^{-1} \int_B Y(\mathbf{s}) d\mathbf{s} \,, \tag{6.1}$$

where $|B|$ denotes the area of B (see, e.g., Cressie, 1993). The integration in (6.1) is an average of random variables, hence a random or stochastic integral. Thus, the assumption of an underlying spatial process is only appropriate for block data that can be sensibly viewed as an averaging over point data; examples of this would include rainfall, pollutant level, temperature, and elevation. It would be inappropriate for, say, population, since there is no "population" at a particular point. It would also be inappropriate for most proportions. For instance, if $Y(B)$ is the proportion of college-educated persons in B, then $Y(B)$ is continuous but even were we to conceptualize an individual at every point, $Y(\mathbf{s})$ would be binary.

In general, we envision four possibilities. First, starting with point data $Y(\mathbf{s}_1), \ldots, Y(\mathbf{s}_I)$, we seek to predict at new locations, i.e., to infer about $Y(\mathbf{s}_1'), \ldots, Y(\mathbf{s}_K')$ (points to points). Second, starting with point data, we seek to predict at blocks, i.e., to infer about $Y(B_1), \ldots, Y(B_K)$ (points to blocks). Third, starting with block data $Y(B_1), ..., Y(B_I)$, we seek to predict at a set of locations, i.e., to infer about $Y(\mathbf{s}_1'), \ldots, Y(\mathbf{s}_K')$ (blocks to points). Finally, starting with block data, we seek to predict at new blocks, i.e., to infer about $Y(B_1'), \ldots, Y(B_K')$ (blocks to blocks).

All of this prediction may be collected under the umbrella of kriging, as in Sections 2.4 and 5.1. Our kriging here will be implemented within the Bayesian framework enabling full inference (a posterior predictive distribution for every prediction of interest, joint distributions for all pairs of predictions, etc.) and avoiding asymptotics. We will however use rather noninformative priors, so that our results will roughly resemble those of a likelihood analysis.

Inference about blocks through averages as in (6.1) is not only formally attractive but demonstrably preferable to *ad hoc* approaches. One such approach would be to average over the observed $Y(\mathbf{s}_i)$ in B. But this presumes there is at least one observation in any B, and ignores the information about the spatial process in the observations outside of B. Another *ad hoc* approach would be to simply predict the value at some central point of B. But this value has larger variability than (and may be biased for) the block average.

In the next section, we develop the methodology for spatial data at a single time point; the general spatiotemporal case is similar and described in Section 8.2. Example 6.1 then applies our approaches to the Atlanta ozone data pictured in Figure 6.1.

6.1.2 Methodology for the point-level realignment

We start with a stationary Gaussian process specification for $Y(\mathbf{s})$ having mean function $\mu(\mathbf{s}; \boldsymbol{\beta})$ and covariance function $c(\mathbf{s}-\mathbf{s}'; \boldsymbol{\theta}) = \sigma^2 \rho(\mathbf{s}-\mathbf{s}'; \boldsymbol{\phi})$, so that $\boldsymbol{\theta} = (\sigma^2, \boldsymbol{\phi})^T$. Here μ is a trend surface with coefficient vector $\boldsymbol{\beta}$, while σ^2 is the process variance and $\boldsymbol{\phi}$ denotes the parameters associated with the stationary correlation function ρ. Beginning with point data observed at sites $\mathbf{s}_1, ..., \mathbf{s}_I$, let $\mathbf{Y}_s^T = (Y(\mathbf{s}_1), \ldots, Y(\mathbf{s}_I))$. Then

$$\mathbf{Y}_s \mid \boldsymbol{\beta}, \boldsymbol{\theta} \sim N(\boldsymbol{\mu}_s(\boldsymbol{\beta}), \sigma^2 H_s(\boldsymbol{\phi})) \ , \tag{6.2}$$

where $\boldsymbol{\mu}_s(\boldsymbol{\beta})_i = \mu(\mathbf{s}_i; \boldsymbol{\beta})$ and $(H_s(\boldsymbol{\phi}))_{ii'} = \rho(\mathbf{s}_i - \mathbf{s}_{i'}; \boldsymbol{\phi})$.

Given a prior on $\boldsymbol{\beta}, \sigma^2$, and $\boldsymbol{\phi}$, models such as (6.2) are straightforwardly fit using simulation methods as described in Section 5.1, yielding posterior samples $(\boldsymbol{\beta}_g^*, \boldsymbol{\theta}_g^*)$, $g = 1, \ldots, G$ from $f(\boldsymbol{\beta}, \boldsymbol{\theta} \mid \mathbf{Y}_s)$.

Then for prediction at a set of new locations $\mathbf{Y}_{s'}^T = (Y(\mathbf{s}_1'), ..., Y(\mathbf{s}_K'))$, we require only the predictive distribution,

$$f(\mathbf{Y}_{s'} \mid \mathbf{Y}_s) = \int f(\mathbf{Y}_{s'} \mid \mathbf{Y}_s, \boldsymbol{\beta}, \boldsymbol{\theta}) f(\boldsymbol{\beta}, \boldsymbol{\theta} \mid \mathbf{Y}_s) d\boldsymbol{\beta} d\boldsymbol{\theta} \ . \tag{6.3}$$

By drawing $\mathbf{Y}_{s',g}^* \sim f(\mathbf{Y}_{s'} \mid \mathbf{Y}_s, \boldsymbol{\beta}_g^*, \boldsymbol{\theta}_g^*)$ we obtain a sample from (6.3) via composition which provides any desired inference about $\mathbf{Y}_{s'}$ and its components.

Under a Gaussian process,

$$f\left(\begin{pmatrix} \mathbf{Y}_s \\ \mathbf{Y}_{s'} \end{pmatrix} \Big| \boldsymbol{\beta}, \boldsymbol{\theta} \right) = N\left(\begin{pmatrix} \boldsymbol{\mu}_s(\boldsymbol{\beta}) \\ \boldsymbol{\mu}_{s'}(\boldsymbol{\beta}) \end{pmatrix}, \sigma^2 \begin{pmatrix} H_s(\boldsymbol{\phi}) & H_{s,s'}(\boldsymbol{\phi}) \\ H_{s,s'}^T(\boldsymbol{\phi}) & H_{s'}(\boldsymbol{\phi}) \end{pmatrix} \right) , \tag{6.4}$$

with entries defined as in (6.2). Hence, $\mathbf{Y}_{s'} \mid \mathbf{Y}_s, \boldsymbol{\beta}, \boldsymbol{\theta}$ is distributed as

$$N \left(\boldsymbol{\mu}_{s'}(\boldsymbol{\beta}) + H_{s,s'}^T(\boldsymbol{\phi}) H_s^{-1}(\boldsymbol{\phi})(\mathbf{Y}_s - \boldsymbol{\mu}_s(\boldsymbol{\beta})) , \right.$$
$$\left. \sigma^2 [H_{s'}(\boldsymbol{\phi}) - H_{s,s'}^T(\boldsymbol{\phi}) H_s^{-1}(\boldsymbol{\phi}) H_{s,s'}(\boldsymbol{\phi})] \right) \ . \tag{6.5}$$

Sampling from (6.5) requires the inversion of $H_s(\boldsymbol{\phi}_g^*)$, which will already have been done in sampling $\boldsymbol{\phi}_g^*$, and then the square root of the $K \times K$ covariance matrix in (6.5).

Turning next to prediction for $\mathbf{Y}_B^T = (Y(B_1), ..., Y(B_K))$, the vector of averages over blocks $B_1, ..., B_K$, we again require the predictive distribution, which is now

$$f(\mathbf{Y}_B \mid \mathbf{Y}_s) = \int f(\mathbf{Y}_B \mid \mathbf{Y}_s; \boldsymbol{\beta}, \boldsymbol{\theta}) f(\boldsymbol{\beta}, \boldsymbol{\theta} \mid \mathbf{Y}_s) d\boldsymbol{\beta} d\boldsymbol{\theta} \ . \tag{6.6}$$

Under a Gaussian process, we now have

$$f\left(\begin{pmatrix} \mathbf{Y}_s \\ \mathbf{Y}_B \end{pmatrix} \Big| \boldsymbol{\beta}, \boldsymbol{\theta} \right) = N\left(\begin{pmatrix} \boldsymbol{\mu}_s(\boldsymbol{\beta}) \\ \boldsymbol{\mu}_B(\boldsymbol{\beta}) \end{pmatrix}, \sigma^2 \begin{pmatrix} H_s(\boldsymbol{\phi}) & H_{s,B}(\boldsymbol{\phi}) \\ H_{s,B}^T(\boldsymbol{\phi}) & H_B(\boldsymbol{\phi}) \end{pmatrix} \right) , \tag{6.7}$$

where

$$(\boldsymbol{\mu}_B(\boldsymbol{\beta}))_k = E(Y(B_k) \mid \boldsymbol{\beta}) = |B_k|^{-1} \int_{B_k} \mu(\mathbf{s}; \boldsymbol{\beta}) d\mathbf{s} \ ,$$

$$(H_B(\boldsymbol{\phi}))_{kk'} = |B_k|^{-1} |B_{k'}|^{-1} \int_{B_k} \int_{B_{k'}} \rho(\mathbf{s} - \mathbf{s}'; \boldsymbol{\phi}) d\mathbf{s}' d\mathbf{s} \ ,$$

$$\text{and} \quad (H_{s,B}(\boldsymbol{\phi}))_{ik} = |B_k|^{-1} \int_{B_k} \rho(\mathbf{s}_i - \mathbf{s}'; \boldsymbol{\phi}) d\mathbf{s}' \ .$$

Analogously to (6.5), $\mathbf{Y}_B | \mathbf{Y}_s, \boldsymbol{\beta}, \boldsymbol{\theta}$ is distributed as

$$
N \left(\boldsymbol{\mu}_B(\boldsymbol{\beta}) + H_{s,B}^T(\boldsymbol{\phi}) H_s^{-1}(\boldsymbol{\phi}) (\mathbf{Y}_s - \boldsymbol{\mu}_s(\boldsymbol{\beta})) , \right.
$$
$$
\left. \sigma^2 \left[H_B(\boldsymbol{\phi}) - H_{s,B}^T(\boldsymbol{\phi}) H_s^{-1}(\boldsymbol{\phi}) H_{s,B}(\boldsymbol{\phi}) \right] \right) . \tag{6.8}
$$

The major difference between (6.5) and (6.8) is that in (6.5), given $(\boldsymbol{\beta}_g^*, \boldsymbol{\theta}_g^*)$, numerical values for all of the entries in $\boldsymbol{\mu}_{s'}(\boldsymbol{\beta})$, $H_{s'}(\boldsymbol{\phi})$, and $H_{s,s'}(\boldsymbol{\phi})$ are immediately obtained. In (6.8) every analogous entry requires an integration as above. Anticipating irregularly shaped B_k's, Riemann approximation to integrate over these regions may be awkward. Instead, noting that each such integration is an expectation with respect to a uniform distribution, we propose Monte Carlo integration. In particular, for each B_k we propose to draw a set of locations $\mathbf{s}_{k,\ell}$, $\ell = 1, 2, ..., L_k$, distributed independently and uniformly over B_k. Here L_k can vary with k to allow for very unequal $|B_k|$.

Hence, we replace $(\boldsymbol{\mu}_B(\boldsymbol{\beta}))_k$, $(H_B(\boldsymbol{\phi}))_{kk'}$, and $(H_{s,B}(\boldsymbol{\phi}))_{ik}$ with

$$
(\widehat{\boldsymbol{\mu}}_B(\boldsymbol{\beta}))_k = L_k^{-1} \sum_\ell \mu(\mathbf{s}_{k,\ell}; \boldsymbol{\beta}) ,
$$
$$
(\widehat{H}_B(\boldsymbol{\phi}))_{kk'} = L_k^{-1} L_{k'}^{-1} \sum_\ell \sum_{\ell'} \rho(\mathbf{s}_{k\ell} - \mathbf{s}_{k'\ell'}; \boldsymbol{\phi}) , \tag{6.9}
$$
$$
\text{and } (\widehat{H}_{s,B}(\boldsymbol{\phi}))_{ik} = L_k^{-1} \sum_\ell \rho(\mathbf{s}_i - \mathbf{s}_{k\ell}; \boldsymbol{\phi}) .
$$

In our notation, the "hat" denotes a Monte Carlo integration that can be made arbitrarily accurate and has nothing to do with the data $\mathbf{Y}_s$. Note also that the same set of $\mathbf{s}_{k\ell}$'s can be used for each integration and with each $(\boldsymbol{\beta}_g^*, \boldsymbol{\theta}_g^*)$; we need only obtain this set once. In obvious notation we replace (6.7) with the $(I+K)$-dimensional multivariate normal distribution $\widehat{f} \left((\mathbf{Y}_s, \mathbf{Y}_B)^T \, \big| \, \boldsymbol{\beta}, \boldsymbol{\theta} \right)$.

It is useful to note that if we define $\widehat{Y}(B_k) = L_k^{-1} \sum_\ell Y(\mathbf{s}_{k\ell})$, then $\widehat{Y}(B_k)$ is a Monte Carlo integration for $Y(B_k)$ as given in (6.1). With an obvious definition for $\widehat{\mathbf{Y}}_B$, it is apparent that

$$
\widehat{f} \left((\mathbf{Y}_s, \mathbf{Y}_B)^T \mid \boldsymbol{\beta}, \boldsymbol{\theta} \right) = f \left((\mathbf{Y}_s, \widehat{\mathbf{Y}}_B)^T \mid \boldsymbol{\beta}, \boldsymbol{\theta} \right) \tag{6.10}
$$

where (6.10) is interpreted to mean that the approximate joint distribution of $(\mathbf{Y}_s, \mathbf{Y}_B)$ is the exact joint distribution of $\mathbf{Y}_s, \widehat{\mathbf{Y}}_B$. In practice, we will work with $\widehat{f}$, converting to $\widehat{f}(\mathbf{Y}_B \mid \mathbf{Y}_s, \boldsymbol{\beta}, \boldsymbol{\theta})$ to sample $\mathbf{Y}_B$ rather than sampling the $\widehat{Y}(B_k)$'s through the $Y(\mathbf{s}_{k\ell})$'s. But, evidently, we are sampling $\widehat{\mathbf{Y}}_B$ rather than $\mathbf{Y}_B$.

As a technical point, we might ask when $\widehat{\mathbf{Y}}_B \xrightarrow{P} \mathbf{Y}_B$. An obvious sufficient condition is that realizations of the $Y(\mathbf{s})$ process are almost surely continuous. In the stationary case, Kent (1989) provides sufficient conditions on $c(\mathbf{s} - \mathbf{t}; \boldsymbol{\theta})$ to ensure this. Alternatively, Stein (1999a) defines $Y(\mathbf{s})$ to be *mean square continuous* if $\lim_{\mathbf{h} \to 0} E(Y(\mathbf{s} + \mathbf{h}) - Y(\mathbf{s}))^2 = 0$ for all $\mathbf{s}$. But then $Y(\mathbf{s} + \mathbf{h}) \xrightarrow{P} Y(\mathbf{s})$ as $\mathbf{h} \to 0$, which is sufficient to guarantee that $\widehat{\mathbf{Y}}_B \xrightarrow{P} \mathbf{Y}_B$. Stein notes that if $Y(\mathbf{s})$ is stationary, we only require $c(\cdot\,; \boldsymbol{\theta})$ continuous at $\mathbf{0}$ for mean square continuity. (See Subsection 2.2.3 and Section 10.1 for further discussion of smoothness of process realizations.)

Finally, starting with block data $\mathbf{Y}_B^T = (Y(B_1), \ldots, Y(B_I))$, analogous to (6.2) the likelihood is well defined as

$$f(\mathbf{Y}_B \mid \boldsymbol{\beta}, \boldsymbol{\theta}) = N(\boldsymbol{\mu}_B(\boldsymbol{\beta})\,,\ \sigma^2 H_B(\boldsymbol{\phi})). \tag{6.11}$$

Hence, given a prior on $\boldsymbol{\beta}$ and $\boldsymbol{\theta}$, the Bayesian model is completely specified. As above, evaluation of the likelihood requires integrations. So, we replace (6.11) with

$$\widehat{f}(\mathbf{Y}_B \mid \boldsymbol{\beta}, \boldsymbol{\theta}) = N(\widehat{\boldsymbol{\mu}}_B(\boldsymbol{\beta})\,,\ \sigma^2 \widehat{H}_B(\boldsymbol{\phi})). \tag{6.12}$$

Simulation-based fitting is now straightforward, as below (6.2), albeit somewhat more time consuming due to the need to calculate $\widehat{\boldsymbol{\mu}}_B(\boldsymbol{\beta})$ and $\widehat{H}_B(\boldsymbol{\phi})$.

To predict for $\mathbf{Y}_{s'}$ we require $f(\mathbf{Y}_{s'} \mid \mathbf{Y}_B)$. As above, we only require $f(\mathbf{Y}_B, \mathbf{Y}_{s'} \mid \boldsymbol{\beta}, \boldsymbol{\theta})$, which has been given in (6.7). Using (6.10) we now obtain $\widehat{f}(\mathbf{Y}_{s'} \mid \mathbf{Y}_B, \boldsymbol{\beta}, \boldsymbol{\theta})$ to sample $\mathbf{Y}_{s'}$. Note that $\widehat{f}$ is used in (6.12) to obtain the posterior samples and again to obtain the predictive samples. Equivalently, the foregoing discussion shows that we can replace $\mathbf{Y}_B$ with $\widehat{\mathbf{Y}}_B$ throughout. To predict for new blocks $B_1', ..., B_K'$, let $\mathbf{Y}_{B'}^T = (Y(B_1'), ..., Y(B_K'))$. Now we require $f(\mathbf{Y}_{B'} \mid \mathbf{Y}_B)$, which in turn requires $f(\mathbf{Y}_B, \mathbf{Y}_{B'} \mid \boldsymbol{\beta}, \boldsymbol{\theta})$. The approximate distribution $\widehat{f}(\mathbf{Y}_B, \mathbf{Y}_{B'} \mid \boldsymbol{\beta}, \boldsymbol{\theta})$ employs Monte Carlo integrations over the B_k''s as well as the B_i's, and yields $\widehat{f}(\mathbf{Y}_{B'} \mid \mathbf{Y}_B, \boldsymbol{\beta}, \boldsymbol{\theta})$ to sample $\mathbf{Y}_{B'}$. Again $\widehat{f}$ is used to obtain both the posterior and predictive samples.

Note that in all four prediction cases, we can confine ourselves to an $(I + K)$-dimensional multivariate normal. Moreover, we have only an $I \times I$ matrix to invert repeatedly in the model fitting, and a $K \times K$ matrix whose square root is required for the predictive sampling.

For the modifiable areal unit problem (i.e., prediction at new blocks using data for a given set of blocks), suppose we take as our point estimate for a generic new set B_0 the posterior mean,

$$E(Y(B_0) \mid \mathbf{Y}_B) = E\{\mu(B_0; \boldsymbol{\beta}) + \mathbf{H}_{B,B_0}^T(\boldsymbol{\phi}) H_B^{-1}(\boldsymbol{\phi})(\mathbf{Y}_B - \boldsymbol{\mu}_B(\boldsymbol{\beta})) \mid \mathbf{Y}_B\}\,,$$

where $\mathbf{H}_{B,B_0}(\boldsymbol{\phi})$ is $I \times 1$ with i^{th} entry equal to $cov(Y(B_i), Y(B_0) \mid \boldsymbol{\theta})/\sigma^2$. If $\mu(\mathbf{s}; \boldsymbol{\beta}) \equiv \mu_i$ for $\mathbf{s} \in B_i$, then $\mu(B_0; \boldsymbol{\beta}) = |B_0|^{-1} \sum_i |B_i \cap B_0| \mu_i$. But $E(\mu_i \mid \mathbf{Y}_B) \approx Y(B_i)$ to a first-order approximation, so in this case $E(Y(B_0) \mid \mathbf{Y}_B) \approx |B_0|^{-1} \sum_i |B_i \cap B_0| Y(B_i)$, the areally weighted estimate.

Example 6.1 We now use the foregoing approach to perform point-point and point-block inference for the Atlanta ozone data pictured in Figure 6.1.

Recall that the target points are those marked **A** and **B** on the map, while the target blocks are the 36 Atlanta city zips. The differing block sizes suggest use of a different L_k for each k in equation (6.9). Conveniently, our GIS (ARC/INFO) can generate random points over the whole study area, and then allocate them to each zip. Thus L_k is proportional to the area of the zip, $|B_k|$. Illustratively, our procedure produced 3743 randomly chosen locations distributed over the 36 city zips, an average L_k of nearly 104.

Suppose that log-ozone exposure $Y(\mathbf{s})$ follows a second-order stationary spatial Gaussian process, using the exponential covariance function $c(\mathbf{s}_i - \mathbf{s}_{i'}; \boldsymbol{\theta}) = \sigma^2 e^{-\phi\|\mathbf{s}_i - \mathbf{s}_{i'}\|}$. A preliminary exploratory analysis of our data set suggested that a constant mean function $\mu(\mathbf{s}_i; \boldsymbol{\beta}) = \mu$ is adequate for our data set. We place the customary flat prior on μ, and assume that $\sigma^2 \sim IG(a, b)$ and $\phi \sim G(c, d)$. We chose $a = 3$, $b = 0.5$, $c = 0.03$, and $d = 100$, corresponding to fairly vague priors. We then fit this three-parameter model using an MCMC implementation, which ran 3 parallel sampling chains for 1000 iterations each, sampling μ and σ^2 via Gibbs steps and ϕ through Metropolis-Hastings steps with a $G(3, 1)$ candidate density. Convergence of the sampling chains was virtually immediate. We obtained the following posterior medians and 95% equal-tail credible intervals for the three parameters: for μ, 0.111 and (0.072, 0.167); for σ^2, 1.37 and (1.18, 2.11); and for ϕ, 1.62 and (0.28, 4.13).

Figure 6.2 maps summaries of the posterior samples for the 36 target blocks (city zips) and the 2 target points (A and B); specifically, the posterior medians, $q_{.50}$, upper and lower .025 points, $q_{.975}$ and $q_{.025}$, and the lengths of the 95% equal-tail credible intervals, $q_{.975} - q_{.025}$. The zip-level medians show a clear spatial pattern, with the highest predicted block averages occurring in the southeastern part of the city near the two high observed readings (0.144 and 0.136), and the lower predictions in the north apparently the result of smoothing toward the low observed value in this direction (0.076). The interval lengths reflect spatial variability, with lower values occurring in larger areas (which require more averaging) or in areas nearer to observed monitoring stations (e.g., those near the southeastern, northeastern, and western city boundaries). Finally, note that our approach allows sensibly differing predicted medians for points A and B, with A being higher due to the slope of the fitted surface. Previous centroid-based analyses (like that of Carlin et al., 1999) would instead implausibly impute the same fitted value to both points, since both lie within the same zip. ■

6.2 Nested block-level modeling

We now turn to the case of variables available (and easily definable) only as block-level summaries. For example, it might be that disease data are known at the county level, but hypotheses of interest pertain to sociodemographically depressed census tracts. We refer to regions on which data

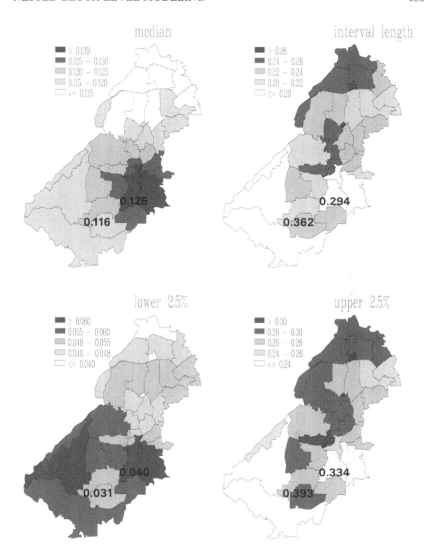

Figure 6.2 *Posterior point-point and point-block summaries, static spatial model, Atlanta ozone data for July 15, 1995.*

are available as "source" zones and regions for which data are needed as "target" zones.

As mentioned earlier, the block-block interpolation problem has a rich literature and is often referred to as the *modifiable areal unit problem* (see, e.g., Cressie, 1996). In the case of an *extensive* variable (i.e., one whose value for a block can be viewed as a sum of sub-block values, as in the case of population, disease counts, productivity, or wealth), areal weighting offers a simple imputation strategy. While rather naive, such allocation proportional to area has a long history and is routinely available in GIS software.

The validity of simple areal interpolation obviously depends on the spatial variable in question being more or less evenly distributed across each region. For instance, Tobler (1979) introduced the so-called *pycnophylactic* approach. He assumed population density to be a continuous function of location, and proposed a simple "volume preserving" (with regard to the observed areal data) estimator of that function. This method is appropriate for continuous outcome variables but is harder to justify for count data, especially counts of human populations, since people do not generally spread out continuously over an areal unit; they tend to cluster.

Flowerdew and Green (1989) presented an approach wherein the variable of interest is count data and which uses information about the distribution of a binary covariate in the target zone to help estimate the counts. Their approach applies Poisson regression iteratively, using the EM algorithm, to estimate target zone characteristics. While subsequent work (Flowerdew and Green, 1992) extended this EM approach to continuous (typically normally distributed) outcome variables, neither of these papers reflects a fully inferential approach to the population interpolation problem.

In this section we follow Mugglin and Carlin (1998), and focus on the setting where the target zonation of the spatial domain D is a refinement of the source zonation, a situation we term *nested* misalignment. In the data setting we describe below, the source zones are U.S. census tracts, while the target zones (and the zones on which covariate data are available) are U.S. census block groups.

Methodology for nested block-level realignment

Consider the diagram in Figure 6.3. Assume that a particular rectangular tract of land is divided into two regions (I and II), and spatial variables (say, disease counts) y_1 and y_2 are known for these regions (the source zones). But suppose that the quantity of interest is Y_3, the unknown corresponding count in Region III (the target zone), which is comprised of subsections (IIIa and IIIb) of Regions I and II.

As already mentioned, a crude way to approach the problem is to assume that disease counts are distributed evenly throughout Regions I and II, and so the number of affected individuals in Region III is just

$$y_1 \left[\frac{area(IIIa)}{area(I)} \right] + y_2 \left[\frac{area(IIIb)}{area(II)} \right] . \qquad (6.13)$$

This simple areal interpolation approach is available within many GIS's.

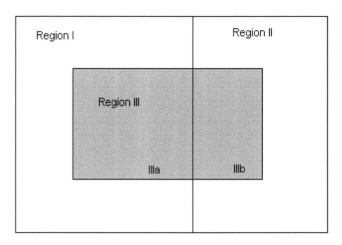

Figure 6.3 *Regional map for motivating example.*

However, (6.13) is based on an assumption that is likely to be unviable, and also offers no associated estimate of uncertainty.

Let us now assume that the entire tract can be partitioned into smaller subsections, where on each subsection we can measure some other variable that is correlated with the disease count for that region. For instance, if we are looking at a particular tract of land, in each subsection we might record whether the land is predominantly rural or urban in character. We do this in the belief that this variable affects the likelihood of disease. Continuous covariates could also be used (say, the median household income in the subsection). Note that the subsections could arise simply as a refinement of the original scale of aggregation (e.g., if disease counts were available only by census tract, but covariate information arose at the census block group level), or as the result of overlaying a completely new set of boundaries (say, a zip code map) onto our original map. The statistical model is easier to formulate in the former case, but the latter case is of course more general, and is the one motivated by modern GIS technology (and to which we return in Section 6.3.

To facilitate our discussion in the former case, we consider a data set on the incidence of leukemia in Tompkins County, New York, that was originally presented and analyzed by Waller et al. (1994), and available on the web at www.biostat.umn.edu/~brad/data/tompkins.dat. As seen in Figure 6.4, Tompkins County, located in west-central New York state, is roughly centered around the city of Ithaca, NY. The county is divided into 23 census tracts, with each tract further subdivided into between 1 and 5 block groups, for a total of 51 such subregions. We have leukemia counts available at the tract level, and we wish to predict them at the block

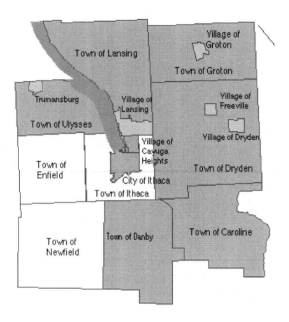

Figure 6.4 *Map of Tompkins County, NY.*

group level with the help of population counts and covariate information available on this more refined scale. In this illustration, the two covariates we consider are whether the block group is coded as "rural" or "urban," and whether or not the block group centroid is located within 2 kilometers of a hazardous chemical waste site. There are two waste sites in the county, one in the northeast corner and the other in downtown Ithaca, near the county's center. (For this data set, we in fact have leukemia counts at the block group level, but we use only the tract totals in the model-fitting process, reserving the refined information to assess the accuracy of our results.) In this example, the unequal population totals in the block groups will play the weighting role that unequal areas would have played in (6.13).

Figure 6.5 shows a census tract-level disease map produced by the GIS MapInfo. The data record the block group-level population counts n_{ij} and covariate values u_{ij} and w_{ij}, where u_{ij} is 1 if block group j of census tract i is classified as urban, 0 if rural, and w_{ij} is 1 if the block group centroid is within 2 km of a waste site, 0 if not. Typical of GIS software, MapInfo permits allocation of the census tract totals to the various block groups proportional to block group area or population. We use our hierarchical Bayesian method to incorporate the covariate information, as well as to obtain variance estimates to accompany the block group-level point estimates.

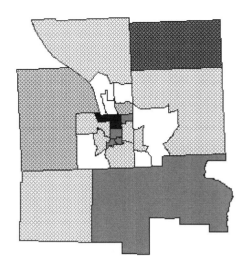

Figure 6.5 *GIS map of disease counts by census tract, Tompkins County, NY.*

As in our earlier disease mapping discussion (Subsection 5.4.1), we introduce a first-stage Poisson model for the disease counts,

$$Y_{ij}|m_{k(i,j)} \stackrel{ind}{\sim} Po(E_{ij}m_{k(i,j)}), \ i = 1, \dots, I, \ j = 1, \dots, J_i \ ,$$

where $I = 23$, J_i varies from 1 to 5, Y_{ij} is the disease count in block group j of census tract i, and E_{ij} is the corresponding "expected" disease count, computed as $E_{ij} = n_{ij}\lambda$ where n_{ij} is the population count in the cell and λ is the overall probability of contracting the disease. This "background" probability could be estimated from our data; here we take $\lambda = 5.597 \times 10^{-4}$, the crude leukemia rate for the 8-county region studied by Waller et al. (1994), an area that includes Tompkins County. Hence, $m_{k(i,j)}$ is the relative risk of contracting leukemia in block group (i,j), and $k = k(i,j) = 1, 2, 3,$ or 4 depending on the covariate status of the block group. Specifically, we let

$$k(i,j) = \begin{cases} 1 \ , & \text{if } (i,j) \text{ is rural, not near a waste site} \\ 2 \ , & \text{if } (i,j) \text{ is urban, not near a waste site} \\ 3 \ , & \text{if } (i,j) \text{ is rural, near a waste site} \\ 4 \ , & \text{if } (i,j) \text{ is urban, near a waste site} \end{cases} .$$

Defining $\mathbf{m} = (m_1, m_2, m_3, m_4)$ and again adopting independent and minimally informative gamma priors for these four parameters, we seek estimates of $p(m_k|\mathbf{y})$, where $\mathbf{y} = (y_{1.}, \dots, y_{I.})$, and $y_{i.} = \sum_{j=1}^{J_i} y_{ij}$, the census tract disease count totals. We also wish to obtain block group-specific mean and variance estimates $E[Y_{ij}|\mathbf{y}]$ and $Var[Y_{ij}|\mathbf{y}]$, to be plotted in a disease map at the block group (rather than census tract) level. Finally, we may also wish to estimate the distribution of the total disease count in some conglomeration of block groups (say, corresponding to some village or city).

By the conditional independence of the block group counts we have $Y_i.|\mathbf{m} \overset{ind}{\sim} Po(\sum_{k=1}^{4} s_k m_k)$, $i = 1, \ldots, I$, where $s_k = \sum_{j:k(i,j)=k} E_{ij}$, the sum of the expected cases in block groups j of region i corresponding to covariate pattern k, $k = 1, \ldots, 4$. The likelihood $L(\mathbf{m}; \mathbf{y})$ is then the product of the resulting $I = 23$ Poisson kernels. After multiplying this by the prior distribution term $\prod_{k=1}^{4} p(m_k)$, we can obtain forms proportional to the four full conditional distributions $p(m_k|m_{l\neq k}, \mathbf{y})$, and sample these sequentially via univariate Metropolis steps.

Once again it is helpful to reparameterize to $\delta_k = \log(m_k)$, $k = 1, \ldots, 4$, and perform the Metropolis sampling on the log scale. We specify reasonably vague $Gamma(a,b)$ priors for the m_k by taking $a = 2$ and $b = 10$ (similar results were obtained with even less informative Gamma priors unless a was quite close to 0, in which case convergence was unacceptably poor). For this "base prior," convergence obtains after 200 iterations, and the remaining 1800 iterations in 5 parallel MCMC chains are retained as posterior samples from $p(\mathbf{m}|\mathbf{y})$.

A second reparametrization aids in interpreting our results. Suppose we write

$$\delta_{k(i,j)} = \theta_0 + \theta_1 u_{ij} + \theta_2 w_{ij} + \theta_3 u_{ij} w_{ij} , \qquad (6.14)$$

so that θ_0 is an intercept, θ_1 is the effect of living in an urban area, θ_2 is the effect of living near a waste site, and θ_3 is the urban/waste site interaction. This reparametrization expresses the log-relative risk of disease as a linear model, a common approach in spatial disease mapping (Besag et al., 1991; Waller et al., 1997). A simple 1-1 transformation converts our $(m_1^{(g)}, m_2^{(g)}, m_3^{(g)}, m_4^{(g)})$ samples to $(\theta_0^{(g)}, \theta_1^{(g)}, \theta_2^{(g)}, \theta_3^{(g)})$ samples on the new scale, which in turn allows direct investigation of the main effects of urban area and waste site proximity, as well as the effect of interaction between these two. Figure 6.6 shows the histograms of the posterior samples for θ_i, $i = 0, 1, 2, 3$. We note that θ_0, θ_1, and θ_3 are not significantly different from 0 as judged by the 95% BCI, while θ_2 is "marginally significant" (in a Bayesian sense) at this level. This suggests a moderately harmful effect of residing within 2 km of a waste site, but no effect of merely residing in an urban area (in this case, the city of Ithaca). The preponderance of negative $\theta_3^{(g)}$ samples is somewhat surprising; we might have expected living near an urban waste site to be associated with an increased (rather than decreased) risk of leukemia. This is apparently the result of the high leukemia rate in a few rural block groups *not* near a waste site (block groups 1 and 2 of tract 7, and block group 2 of tract 20), forcing θ_3 to adjust for the relatively lower overall rate near the Ithaca waste site.

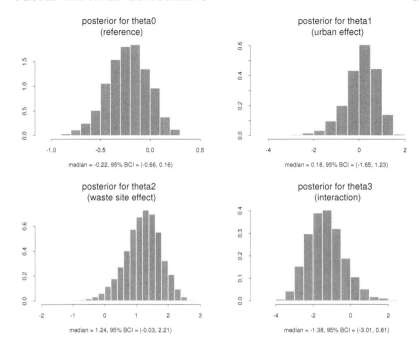

Figure 6.6 *Posterior histograms of sampled log-relative risk parameters, Tompkins County, NY, data set.*

Individual block group estimation

To create the block group-level estimated disease map, for those census tracts having $J_i > 1$, we obtain a conditional binomial distribution for Y_{ij} given the parameters $\mathbf{m}$ and the census tract totals $\mathbf{y}$, so that

$$E(Y_{ij}|\mathbf{y}) = E[E(Y_{ij}|\mathbf{m}, \mathbf{y})] \approx \frac{y_{i\cdot}}{G} \sum_{g=1}^{G} p_{ij}^{(g)} \,, \qquad (6.15)$$

where p_{ij} is the appropriate binomial probability arising from conditioning a Poisson random variable on the sum of itself and a second, independent Poisson variable. For example, for $p_{11}^{(g)}$ we have

$$p_{11}^{(g)} = \frac{1617 m_1^{(g)}}{(1617 + 702)m_1^{(g)} + (1526 + 1368)m_3^{(g)}} \,,$$

as determined by the covariate patterns in the first four rows of the data set. Note that when $J_i = 1$ the block group total equals the known census tract total, hence no estimation is necessary.

The resulting collection of estimated block group means $E(Y_{ij}|\mathbf{y})$ are included in the data set on our webpage, along with the actual case counts y_{ij}. (The occasional noninteger values of y_{ij} in the data are not errors, but arise from a few cases in which the precise block group of occurrence is unknown, resulting in fractional counts being allocated to several block groups.) Note that, like other interpolation methods, the sum of the estimated cases in each census tract is the same as the corresponding sum for the actual case counts. The GIS maps of the $E(Y_{ij}|\mathbf{y})$ and the actual y_{ij} shown in Figure 6.7 reveal two pockets of elevated disease counts (in the villages of Cayuga Heights and Groton).

To get an idea of the variability inherent in the posterior surface, we might consider mapping the estimated posterior variances of our interpolated counts. Since the block group-level variances do not involve aggregation across census tracts, these variances may be easily estimated as $Var(Y_{ij}|\mathbf{y}) = E(Y_{ij}^2|\mathbf{y}) - [E(Y_{ij}|\mathbf{y})]^2$, where the $E(Y_{ij}|\mathbf{y})$ are the estimated means (already calculated), and

$$
\begin{aligned}
E(Y_{ij}^2|\mathbf{y}) &= E[E(Y_{ij}^2|\mathbf{m},\mathbf{y})] = E[y_{i.}p_{ij}(1-p_{ij}) + y_{i.}^2(p_{ij})^2] \\
&\approx \frac{1}{G}\sum_{g=1}^{G}\left[y_{i.}p_{ij}^{(g)}(1-p_{ij}^{(g)}) + y_{i.}^2(p_{ij}^{(g)})^2\right],
\end{aligned}
\tag{6.16}
$$

where p_{ij} is again the appropriate binomial probability for block group (i,j); see Mugglin and Carlin (1998) for more details.

We remark that most of the census tracts are composed of homogeneous block groups (e.g., all rural with no waste site nearby); in these instances the resulting binomial probability for each block group is free of $\mathbf{m}$. In such cases, posterior means and variances are readily available without any need for mixing over the Metropolis samples, as in equations (6.15) and (6.16).

Aggregate estimation: Block groups near the Ithaca, NY, waste site

In order to assess the number of leukemia cases we expect in those block groups within 2 km of the Ithaca waste site, we can sample the predictive distributions for these blocks, sum the results, and draw a histogram of these sums. Twelve block groups in five census tracts fall within these 2-km radii: all of the block groups in census tracts 11, 12, and 13, plus two of the three (block groups 2 and 3) in tract 6 and three of the four (block groups 2, 3, and 4) in tract 10. Since the totals in census tracts 11, 12, and 13 are known to our analysis, we need only sample from two binomial distributions, one each for the conglomerations of near-waste site block groups within tracts 6 and 10. Defining the sum over the twelve block groups as Z, we have

$$
Z^{(g)} = Y_{6,(2,3)}^{(g)} + Y_{10,(2,3,4)}^{(g)} + y_{11,.} + y_{12,.} + y_{13,.} .
$$

a) interpolated mean estimates

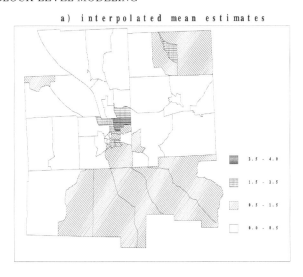

b) actual data values

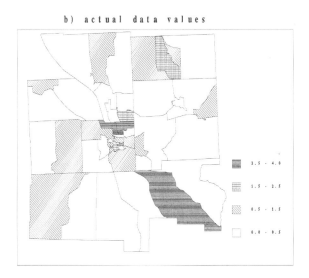

Figure 6.7 *GIS maps of interpolated (a) and actual (b) block group disease counts, Tompkins County, NY.*

A histogram of these values is shown in Figure 6.8. The estimated median value of 10 happens to be exactly equal to the true value of 10 cases in this area. The sample mean, 9.43, is also an excellent estimate. Note that the minimum and maximum values in Figure 6.8, $Z = 7$ and $Z = 11$, are imposed by the data structure: there must be at least as many cases as the

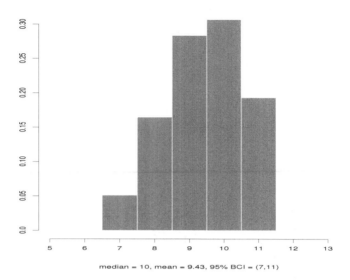

median = 10, mean = 9.43, 95% BCI = (7,11)

Figure 6.8 *Histogram of sampled disease counts, total of all block groups having centroid within 2 km of the Ithaca, NY, waste site.*

total known to have occurred in census tracts 11, 12, and 13 (which is 7), and there can be no more than the total number known to have occurred in tracts 6, 10, 11, 12, and 13 (which is 11).

Finally, we may again compare our results to those produced by a GIS under either area-based or population-based interpolation. The former produces a mean estimate of 9.28, while the latter gives 9.59. These are close to the Bayesian mean 9.43, but neither approach produces an associated confidence interval, much less a full graphical display of the sort given in Figure 6.8.

6.3 Nonnested block-level modeling

The approach of the previous section (see also Mugglin and Carlin, 1998, and Mugglin et al., 1999) offered a hierarchical Bayesian method for interpolation and smoothing of Poisson responses with covariates in the nested case. In the remainder of this section we develop a framework for hierarchical Bayesian interpolation, estimation, and spatial smoothing over *nonnested* misaligned data grids. In Subsection 6.3.1 we summarize a data set collected in response to possible contamination resulting from the former Feed Materials Production Center (FMPC) in southwestern Ohio with the foregoing analytic goals. In Subsection 6.3.2 we develop the theory of our modeling approach in a general framework, as well as our MCMC approach and a particular challenge that arises in its implementation for the FMPC data. Finally in Example 6.2 we set forth the conclusions resulting from our analysis of the FMPC data.

6.3.1 Motivating data set

Risk-based decision making is often used for prioritizing cleanup efforts at U.S. Superfund sites. Often these decisions will be based on estimates of the past, present, and future potential health impacts. These impact assessments usually rely on estimation of the number of outcomes, and the accuracy of these estimates will depend heavily on the ability to estimate the number of individuals at risk. Our motivating data set is connected with just this sort of risk assessment.

In the years 1951–1988 near the town of Ross in southwestern Ohio, the former Feed Materials Production Center (FMPC) processed uranium for weapons production. Draft results of the Fernald Dosimetry Reconstruction Project, sponsored by the Centers for Disease Control and Prevention (CDC), indicated that during production years the FMPC released radioactive materials (primarily radon and its decay products and, to a lesser extent, uranium and thorium) from the site. Although radioactive liquid wastes were released, the primary exposure to residents of the surrounding community resulted from breathing radon decay products. The potential for increased risk of lung cancer is thus the focus of intense local public interest and ongoing public health studies (see Devine et al., 1998).

Estimating the number of adverse health outcomes in the population (or in subsets thereof) requires estimation of the number of individuals at risk. Population counts, broken down by age and sex, are available from the U.S. Census Bureau according to federal census block groups, while the areas of exposure interest are dictated by both direction and distance from the plant. Rogers and Killough (1997) construct an exposure "windrose," which consists of 10 concentric circular bands at 1-kilometer radial increments divided into 16 compass sectors (N, NNW, NW, WNW, W, etc.). Through the overlay of such a windrose onto U.S. Geological Survey (USGS) maps, they provide counts of the numbers of "structures" (residential buildings, office buildings, industrial building complexes, warehouses, barns, and garages) within each subdivision (*cell*) of the windrose.

Figure 6.9 shows the windrose centered at the FMPC. We assign numbers to the windrose cells, with 1 to 10 indexing the cells starting at the plant and running due north, then 11 to 20 running from the plant to the north-northwest, and so on. Structure counts are known for each cell; the hatching pattern in the figure indicates the areal density (structures per square kilometer) in each cell.

Also shown in Figure 6.9 are the boundaries of 39 Census Bureau block groups, for which 1990 population counts are known. These are the source

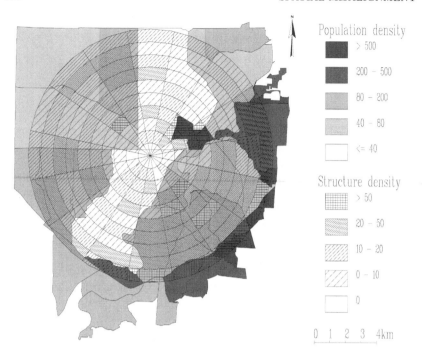

Figure 6.9 *Census block groups and 10-km windrose near the FMPC site, with 1990 population density by block group and 1980 structure density by cell (both in counts per km²).*

zones for our interpolation problem. Shading intensity indicates the population density (persons per square kilometer) for each block group. The intersection of the two (nonnested) zonation systems results in 389 regions we call *atoms,* which can be aggregated appropriately to form either cells or block groups.

The plant was in operation for 38 years, raising concern about the potential health risks it has caused, a question that has been under active investigation by the CDC for some time. Present efforts to assess the impact of the FMPC on cancer morbidity and mortality require the analysis of this misaligned data set; in particular, it is necessary to interpolate gender- and age group-specific population counts to the windrose exposure cells. These numbers of persons at risk could then be combined with cell-specific dose estimates obtained by Killough et al. (1996) and estimates of the cancer risk per unit dose to obtain expected numbers of excess cancer cases by cell.

In fact, such an expected death calculation was made by Devine et al. (1998), using traditional life table methods operating on the Rogers and

Killough (1997) cell-level population estimates (which were in turn derived simply as proportional to the structure counts). However, these estimates were only for the total population in each cell; sex- and age group-specific counts were obtained by "breaking out" the totals into subcategories using a standard table (i.e., the *same* table in each cell, regardless of its true demographic makeup). In addition, the uncertainty associated with the cell-specific population estimates was quantified in a rather ad hoc way.

6.3.2 Methodology for nonnested block-level realignment

We confine our model development to the case of two misaligned spatial grids. Given this development, the extension to more than two grids will be conceptually apparent. The additional computational complexity and bookkeeping detail will also be evident.

Let the first grid have regions indexed by $i = 1, ..., I$, denoted by B_i, and let $S_B = \bigcup_i B_i$. Similarly, for the second grid we have regions C_j, $j = 1, ..., J$ with $S_C = \bigcup_j C_j$. In some applications $S_B = S_C$, i.e., the B-cells and the C-cells offer different partitions of a common region. Nested misalignment (e.g., where each C_j is contained entirely in one and only one B_i) is evidently a special case. Another possibility is that one data grid contains the other; say, $S_B \subset S_C$. In this case, there will exist some C cells for which a portion lies outside of S_B. In the most general case, there is no containment and there will exist B-cells for which a portion lies outside of S_C and C-cells for which a portion lies outside of S_B. Figure 6.10 illustrates this most general situation.

Atoms are created by intersecting the two grids. For a given B_i, each C-cell which intersects B_i creates an atom (which possibly could be a union of disjoint regions). There may also be a portion of B_i which does not intersect with any C_j. We refer to this portion as the *edge* atom associated with B_i, i.e., a B-edge atom. In Figure 6.10, atoms B_{11} and B_{21} are B-edge atoms. Similarly, for a given C_j, each B-cell which intersects with C_j creates an atom, and we analogously determine C-edge atoms (atoms C_{11} and C_{22} in Figure 6.10). It is crucial to note that each nonedge atom can be referenced relative to an appropriate B-cell, say B_i, and denoted as B_{ik}. It also can be referenced relative to an appropriate C cell, say C_j, and denoted by $C_{j\ell}$. Hence, there is a one-to-one mapping within $S_B \cap S_C$ between the set of ik's and the set of $j\ell$'s, as shown in Figure 6.10 (which also illustrates our convention of indexing atoms by area, in descending order). Formally we can define the function c on nonedge B-atoms such that $c(B_{ik}) = C_{j\ell}$, and the *inverse* function b on C-atoms such that $b(C_{j\ell}) = B_{ik}$. For computational purposes we suggest creation of "look-up" tables to specify these functions. (Note that the possible presence of both types of edge cell precludes a single "ij" atom numbering system, since such a system could index cells on either S_B or S_C, but not their union.)

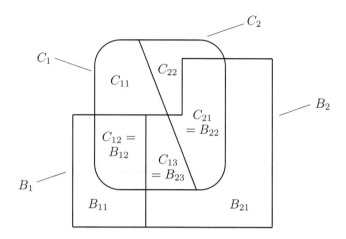

Figure 6.10 *Illustrative representation of areal data misalignment.*

Without loss of generality we refer to the first grid as the *response* grid, that is, at each B_i we observe a response Y_i. We seek to explain Y_i using a variety of covariates. Some of these covariates may, in fact, be observed on the response grid; we denote the value of this vector for B_i by $\mathbf{W}_i$. But also, some covariates are observed on the second or *explanatory* grid. We denote the value of this vector for C_j by $\mathbf{X}_j$.

We seek to explain the observed Y's through both $\mathbf{X}$ and $\mathbf{W}$. The misalignment between the $\mathbf{X}$'s and Y's is the obstacle to standard regression methods. What levels of $\mathbf{X}$ should be assigned to Y_i? We propose a fully model-based approach in the case where the Y's and X's are aggregated measurements. The advantage of a model-based approach implemented within a Bayesian framework is full inference both with regard to estimation of model parameters and prediction using the model.

The assumption that the Y's are aggregated measurements means Y_i can be envisioned as $\sum_k Y_{ik}$, where the Y_{ik} are unobserved or latent and the summation is over all atoms (including perhaps an edge atom) associated with B_i. To simplify, we assume that the X's are also scalar aggregated measurements, i.e., $X_j = \sum_\ell X_{j\ell}$ where the summation is over all atoms associated with C_j. As for the $\mathbf{W}$'s, we assume that each component is either an aggregated measurement or an *inheritable* measurement. For component r, in the former case $W_i^{(r)} = \sum_k W_{ik}^{(r)}$ as with Y_i; in the latter case $W_{ik}^{(r)} = W_i^{(r)}$.

In addition to (or perhaps in place of) the $\mathbf{W}_i$ we will introduce B-cell random effects μ_i, $i = 1, ..., I$. These effects are employed to capture spatial association among the Y_i's. The μ_i can be given a spatial prior specification. A Markov random field form (Besag, 1974; Bernardinelli and Montomoli, 1992), as described below, is convenient. Similarly we will introduce C-cell random effects ω_j, $j = 1, ..., J$ to capture spatial association among the X_j's. It is assumed that the latent Y_{ik} inherit the effect μ_i and that the latent $X_{j\ell}$ inherit the effect ω_j.

For aggregated measurements that are counts, we assume the latent variables are conditionally independent Poissons. As a result, the observed measurements are Poissons as well and the conditional distribution of the latent variables given the observed is a product multinomial. We note that it is not required that the Y's be count data. For instance, with aggregated measurements that are continuous, a convenient distributional assumption is conditionally independent gammas, in which case the latent variables would be rescaled to product Dirichlet. An alternative choice is the normal, whereupon the latent variables would have a distribution that is a product of conditional multivariate normals. In this section we detail the Poisson case.

As mentioned above, area naturally plays an important role in allocation of spatial measurements. Letting $|A|$ denote the area of region A, if we apply the standard assumption of allocation proportional to area to the $X_{j\ell}$ in a stochastic fashion, we would obtain

$$X_{j\ell} \mid \omega_j \sim Po(e^{\omega_j}|C_{j\ell}|) , \qquad (6.17)$$

assumed independent for $\ell = 1, 2, ..., L_j$. Then $X_j \mid \omega_j \sim Po(e^{\omega_j}|C_j|)$ and $(X_{j1}, X_{j2}, ..., X_{j,L_j} \mid X_j, \omega_j) \sim Mult(X_j; q_{j1}, ..., q_{j,L_j})$ where $q_{j\ell} = |C_{j\ell}|/|C_j|$.

Such strictly area-based modeling cannot be applied to the Y_{ik}'s since it fails to connect the Y's with the X's (as well as the $\mathbf{W}$'s). To do so we again begin at the atom level. For nonedge atoms we use the previously mentioned look-up table to find the $X_{j\ell}$ to associate with a given Y_{ik}. It is convenient to denote this $X_{j\ell}$ as X'_{ik}. Ignoring the $\mathbf{W}_i$ for the moment, we assume

$$Y_{ik} \mid \mu_i, \theta_{ik} \sim Po\left(e^{\mu_i}|B_{ik}| \, h(X'_{ik}/|B_{ik}| \, ; \theta_{ik})\right) , \qquad (6.18)$$

independent for $k = 1, \ldots, K_i$. Here h is a preselected parametric function, the part of the model specification that adjusts an expected proportional-to-area allocation according to X'_{ik}. Since (6.17) models expectation for $X_{j\ell}$ proportional to $|C_{j\ell}|$, it is natural to use the *standardized* form $X'_{ik}/|B_{ik}|$ in (6.18). Particular choices of h include $h(z \, ; \theta_{ik}) = z$ yielding $Y_{ik} \mid \mu_i \sim Po(e^{\mu_i}X'_{ik})$, which would be appropriate if we choose not to use $|B_{ik}|$ explicitly in modeling $E(Y_{ik})$. In our FMPC implementation, we actually select $h(z \, ; \theta_{ik}) = z + \theta_{ik}$ where $\theta_{ik} = \theta/(K_i|B_{ik}|)$ and $\theta > 0$; see equation (6.23) below and the associated discussion.

If B_i has no associated edge atom, then

$$Y_i \mid \mu_i, \boldsymbol{\theta}, \{X_{j\ell}\} \sim Po\left(e^{\mu_i} \sum_k |B_{ik}| \, h(X'_{ik}/|B_{ik}| \, ; \theta_{ik})\right) . \qquad (6.19)$$

If B_i has an edge atom, say B_{iE}, since there is no corresponding $C_{j\ell}$, there is no corresponding X'_{iE}. Hence, we introduce a latent X'_{iE} whose distribution is determined by the nonedge atoms that are neighbors of B_{iE}. Paralleling equation (6.17), we model X'_{iE} as

$$X'_{iE} \mid \omega^*_i \sim Po(e^{\omega^*_i}|B_{iE}|) , \qquad (6.20)$$

thus adding a new set of random effects $\{\omega^*_i\}$ to the existing set $\{\omega_j\}$. These two sets together are assumed to have a single CAR specification. An alternative is to model $X'_{iE} \sim Po\left(|B_{iE}| \left(\sum_{N(B_{iE})} X'_t / \sum_{N(B_{iE})} |B_t|\right)\right)$, where $N(B_{iE})$ is the set of neighbors of B_{iE} and t indexes this set. Effectively, we multiply $|B_{iE}|$ by the overall count per unit area in the neighboring nonedge atoms. While this model is somewhat more data-dependent than the (more model-dependent) one given in (6.20), we remark that it can actually lead to better MCMC convergence due to the improved identifiability in its parameter space: the spatial similarity of the structures in the edge zones is being modeled directly, rather than indirectly via the similarity of the ω^*_i and the ω_j.

Now, with an X'_{ik} for all ik, (6.18) is extended to all B-atoms and the conditional distribution of Y_i is determined for all i as in (6.19). But also $Y_{i1}, ..., Y_{ik_i} \mid Y_i, \mu_i, \theta_{ik}$ is distributed Multinomial$(Y_i; q_{i1}, ..., q_{ik_i})$, where $q_{ik} = |B_{ik}| \, h(X'_{ik}/|B_{ik}| \, ; \theta_{ik})/\sum_k |B_{ik}| \, h(X'_{ik}/|B_{ik}| \, ; \theta_{ik})$.

To capture the spatial nature of the B_i we may adopt an IAR model for the μ_i, i.e.,

$$p(\mu_i \mid \mu_{i'}, i' \neq i) = N\left(\sum_{i'} w_{ii'} \mu_{i'}/w_{i.} \, , \, 1/(\lambda_\mu w_{i.})\right) \qquad (6.21)$$

where $w_{ii} = 0$, $w_{ii'} = w_{i'i}$ and $w_{i.} = \sum_{i'} w_{ii'}$. Below, we set $w_{ii'} = 1$ for $B_{i'}$ a neighbor of B_i and $w_{ii'} = 0$ otherwise, the standard "0-1 adjacency" form.

Similarly we assume that

$$f(\omega_j \mid \omega_{j'}, j' \neq j) = N\left(\sum_{j'} v_{jj'} \omega_{j'}/v_{j.} \, , \, 1/(\lambda_\omega v_{j.})\right) .$$

We adopt a proper Gamma prior for λ_μ and also for λ_ω. When $\boldsymbol{\theta}$ is present we require a prior that we denote by $f(\boldsymbol{\theta})$. The choice of $f(\boldsymbol{\theta})$ will likely be vague but its form depends upon the adopted parametric form of h.

The entire specification can be given a representation as a graphical model, as in Figure 6.11. In this model the arrow from $\{X_{j\ell}\} \rightarrow \{X'_{ik}\}$

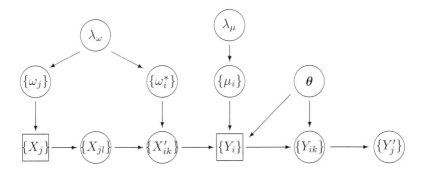

Figure 6.11 *Graphical version of the model, with variables as described in the text. Boxes indicate data nodes, while circles indicate unknowns.*

indicates the inversion of the $\{X_{jl}\}$ to $\{X'_{ik}\}$, augmented by any required edge atom values X'_{iE}. The $\{\omega^*_i\}$ would be generated if the X'_{iE} are modeled using (6.20). Since the $\{Y_{ik}\}$ are not observed, but are distributed as multinomial given the fixed block group totals $\{Y_i\}$, this is a predictive step in our model, as indicated by the arrow from $\{Y_i\}$ to $\{Y_{ik}\}$ in the figure. In fact, as mentioned above the further predictive step to impute Y'_j, the Y total associated with X_j in the jth target zone, is of key interest. If there are edge atoms C_{jE}, this will require a model for the associated Y'_{jE}. Since there is no corresponding B-atom for C_{jE} a specification such as (6.18) is not appropriate. Rather, we can imitate the above modeling for X'_{iE} using (6.20) by introducing $\{\mu^*_j\}$, which along with the μ_i follow the prior in (6.21). The $\{\mu^*_j\}$ and $\{Y'_{jE}\}$ would add two consecutive nodes to the right side of Figure 6.11, connecting from λ_μ to $\{Y'_j\}$.

The entire distributional specification overlaid on this graphical model has been supplied in the foregoing discussion and (in the absence of C_{jE} edge atoms, as in Figure 6.9) takes the form

$$\prod_i f(Y_{i1}, ..., Y_{ik_i} \mid Y_i, \boldsymbol{\theta}) \prod_i f(Y_i \mid \mu_i, \boldsymbol{\theta}, \{X'_{ik}\}) f(\{X'_{ik}\} \mid \omega^*_i, \{X_{j\ell}\})$$
$$\times \prod_j f(X_{j1}, ..., X_{jL_j} \mid X_j) \prod_j f(X_j \mid \omega_j)$$
$$\times f(\{\mu_i\} \mid \lambda_\mu) f(\lambda_\mu) f(\{\omega_j\}, \{\omega^*_i\} \mid \lambda_\omega) f(\lambda_\omega) f(\boldsymbol{\theta}) .$$
$$(6.22)$$

Bringing in the $\mathbf{W}_i$ merely revises the exponential term in (6.18) from $\exp(\mu_i)$ to $\exp(\mu_i + \mathbf{W}_{ik}^T \boldsymbol{\beta})$. Again, for an inherited component of $\mathbf{W}_i$, say, $W_i^{(r)}$, the resulting $W_{ik}^{(r)} = W_i^{(r)}$. For an aggregated component of $\mathbf{W}_i$, again, say, $W_i^{(r)}$, we imitate (6.17) assuming $W_{ik}^{(r)} \mid \mu_i^{(r)} \sim Po(e^{\mu_i^{(r)}} |B_{ik}|)$, independent for $k = 1, ..., K_i$. A spatial prior on the $\mu_i^{(r)}$ and a Gaussian (or perhaps flat) prior on $\boldsymbol{\beta}$ completes the model specification.

Finally, on the response grid, for each B_i rather than observing a single Y_i we may observe Y_{im}, where $m = 1, 2, ..., M$ indexes levels of factors such as sex, race, or age group. Here we seek to use these factors, in an ANOVA fashion, along with the X_j (and $\mathbf{W}_i$) to explain the Y_{im}. Ignoring $\mathbf{W}_i$, the resultant change in (6.18) is that Y_{ikm} will be Poisson with μ_i replaced by μ_{im}, where μ_{im} has an appropriate ANOVA form. For example, in the case of sex and age classes, we might have a sex main effect, an age main effect, and a sex-age interaction effect. In our application these effects are not nested within i; we include only a spatial overall mean effect indexed by i.

Regarding the MCMC implementation of our model, besides the usual concerns about appropriate choice of Metropolis-Hastings candidate densities and acceptability of the resulting convergence rate, one issue deserves special attention. Adopting the identity function for h in (6.18) produces the model $Y_{ik} \sim Po\,(e^{\mu_i}\,(X'_{ik}))$, which in turn implies $Y_{i.} \sim Po\,(e^{\mu_i}(X'_{i.}))$. Suppose however that $Y_{i.} > 0$ for a particular block group i, but in some MCMC iteration no structures are allocated to any of the atoms of the block group. The result is a flawed probabilistic specification. To ensure $h > 0$ even when $z = 0$, we revised our model to $h(z\,;\,\theta_{ik}) = z + \theta_{ik}$ where $\theta_{ik} = \theta/(K_i|B_{ik}|)$ with $\theta > 0$, resulting in

$$Y_{ik} \sim Po\left(e^{\mu_i}\left(X'_{ik} + \frac{\theta}{K_i}\right)\right) . \qquad (6.23)$$

This adjustment eliminates the possibility of a zero-valued Poisson parameter, but does allow for the possibility of a nonzero population count in a region where there are no structures observed. When conditioned on $Y_{i.}$, we find $(Y_{i1}, \ldots, Y_{iK_i} \mid Y_{i.}) \sim \text{Mult}(Y_{i.}\,;\,p_{i1}, \ldots, p_{iK_i})$, where

$$p_{ik} = \frac{X'_{ik} + \theta/K_i}{X'_{i.} + \theta} \quad \text{and} \quad Y_{i.} \sim Po\,(e^{\mu_i}(X'_{i.} + \theta)) . \qquad (6.24)$$

Our basic model then consists of (6.23) to (6.24) together with

$$\mu_i \overset{iid}{\sim} N\left(\eta_\mu, 1/\tau_\mu\right), \quad X_{jl} \sim Po\,(e^{\omega_j}|C_{jl}|) \Rightarrow X_{j.} \sim Po\,(e^{\omega_j}|C_j|) ,$$
$$(X_{j1}, \ldots, X_{jL_j} \mid X_{j.}) \sim \text{Mult}(X_{j.}\,;\,q_{j1}, \ldots, q_{jL_j}), \text{ where } q_{jl} = |C_{jl}|/|C_j|,$$
$$X'_{iE} \sim Po\left(e^{\omega_i^*}|B_{iE}|\right) , \quad \text{and} \quad (\omega_j, \omega_i^*) \sim \text{CAR}(\lambda_\omega) ,$$
$$(6.25)$$

where X'_{iE} and ω_i^* refer to edge atom structure counts and log relative risk parameters, respectively. While θ could be estimated from the data, in our implementation we simply set $\theta = 1$; Mugglin et al. (2000, Sec. 6) discuss the impact of alternate selections.

Example 6.2 *(FMPC data analysis).* We turn now to the particulars of the FMPC data analysis, examining two different models in the context of the misaligned data as described in Section 6.3.1. In the first case we take up the problem of total population interpolation, while in the second we consider age- and sex-specific population interpolation.

Total population interpolation model

We begin by taking $\eta_\mu = 1.1$ and $\tau_\mu = 0.5$ in (6.25). The choice of mean value reflects the work of Rogers and Killough (1997), who found population per household (PPH) estimates for four of the seven townships in which the windrose lies. Their estimates ranged in value from 2.9 to 3.2, hence our choice of $\eta_\mu = 1.1 \approx \log(3)$. The value $\tau_\mu = 0.5$ is sufficiently small to make the prior for μ_i large enough to support all feasible values of μ_i (two prior standard deviations in either direction would enable PPH values of 0.18 to 50.8).

For $\boldsymbol{\omega} = \{\omega_j, \omega_i^*\}$ we adopted a CAR prior and fixed $\lambda_\omega = 10$. We did not impose any centering of the elements of $\boldsymbol{\omega}$ around 0, allowing them to determine their own mean level in the MCMC algorithm. Since most cells have four neighbors, the value $\lambda_\omega = 10$ translates into a conditional prior standard deviation for the ω's of $\sqrt{1/(10 \cdot 4)} = .158$, hence a marginal prior standard deviation of roughly $.158/.7 \approx .23$ (Bernardinelli et al., 1995). In any case, we found $\lambda_\omega < 10$ too vague to allow MCMC convergence. Typical posterior medians for the ω's ranged from 2.2 to 3.3 for the windrose ω_j's and from 3.3 to 4.5 for the edge ω_i^*s.

Running 5 parallel sampling chains, acceptable convergence obtains for all parameters within 1,500 iterations. We discarded this initial sample and then continued the chains for an additional 5,000 iterations each, obtaining a final posterior sample of size 25,000. From the resulting samples, we can examine the posterior distributions of any parameters we wish. It is instructive first to examine the distributions of the imputed structure counts X_{jl}. For example, consider Figure 6.12, which shows the posterior distributions of the structure counts in cell 106 (the sixth one from the windrose center in the SE direction), for which $L_j = 4$. The known cell total $X_{106,\cdot}$ is 55. Note that the structure values indicated in the histograms are integers. The vertical bars in each histogram indicate how the 55 structures would be allocated if imputed proportionally to area. In this cell we observe good general agreement between these naively imputed values and our histograms, but the advantage of assessing variability from the full distributional estimates is immediately apparent.

Population estimates per cell for cells 105 through 110 (again in the SE direction, from the middle to outer edge of the windrose) are indicated in Figure 6.13. Vertical bars here represent estimates calculated by multiplying the number of structures in the cell by a fixed (map-wide) constant representing population per household (PPH), a method roughly equivalent to that employed by Rogers and Killough (1997), who as mentioned

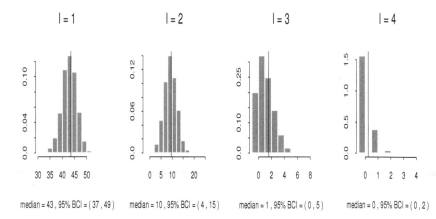

Figure 6.12 *Posterior distributions of structure estimates for the four atoms of cell 106 (SE6). Vertical bars represent structure values if imputed proportionally to area. Here and in the next figure, "median" denotes posterior median, and "95% BCI" denotes the equal-tail Bayesian confidence interval.*

above actually used four different PPH values. Our reference lines use a constant value of 3 (the analogue of our prior mean). While cells 105 and 106 indicate good general agreement in these estimates, cells 107 through 110 display markedly different population estimates, where our estimates are substantially higher than the constant-PPH estimates. This is typical of cells toward the outer edge of the southeast portion of the windrose, since the suburbs of Cincinnati encroach on this region. We have population data only (no structures) in the southeastern edge atoms, so our model must estimate both the structures and the population in these regions. The resulting PPH is higher than a mapwide value of 3 (one would expect suburban PPH to be greater than rural PPH) and so the CAR model placed on the $\{\omega_j, \omega_i^*\}$ parameters induces a spatial similarity that can be observed in Figure 6.13.

We next implement the $\{Y_i.\} \rightarrow \{Y_{ik}\}$ step. From the resulting $\{Y_{ik}\}$ come the $\{Y_{j.}'\}$ cell totals by appropriate reaggregation. Figure 6.14 shows the population densities by atom $(Y_{ik}/|B_{ik}|)$, calculated by taking the posterior medians of the population distributions for each atom and dividing by atom area in square kilometers. This figure clearly shows the encroachment by suburban Cincinnati on the southeast side of our map, with some spatial smoothing between the edge cells and the outer windrose cells. Finally, Figure 6.15 shows population densities by cell $(Y_{j.}'/|C_j|)$, where the atom-level populations have been aggregated to cells before calculating densities. Posterior standard deviations, though not shown, are also available for each cell. While this figure, by definition, provides less detail than Fig-

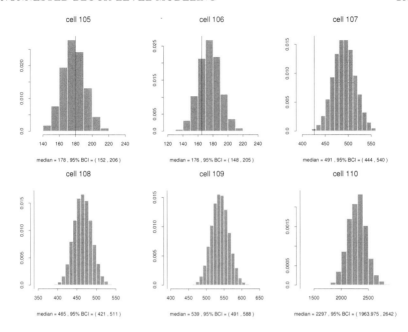

Figure 6.13 *Posterior distributions of populations in cells 105 to 110. Vertical bars represent estimates formed by multiplying structures per cell by a constant population per household (PPH) of 3.0.*

ure 6.14, it provides information at the scale appropriate for combination with the exposure values of Killough et al. (1996). Moreover, the scale of aggregation is still fine enough to permit identification of the locations of Cincinnati suburban sprawl, as well as the communities of Ross (contained in cells ENE 4-5 and NE 4), Shandon (NW 4-5), New Haven (WSW 5-6), and New Baltimore (SSE 4-5).

Age and sex effects

Recall from Section 6.3.1 that we seek population counts not only by cell but also by sex and age group. This is because the dose resulting from a given exposure will likely differ depending on gender and age, and because the risk resulting from that dose can also be affected by these factors. Again we provide results only for the year 1990; the extension to other timepoints would of course be similar. Population counts at the block group level by sex and age group are provided by the U.S. Census Bureau. Specifically, age is recorded as counts in 18 quinquennial (5-year) intervals: 0–4, 5–9, ..., 80–84, and 85+. We consider an additive extension of our basic model (6.23)–(6.25) to the sex- and age group-specific case; see Mugglin et al.

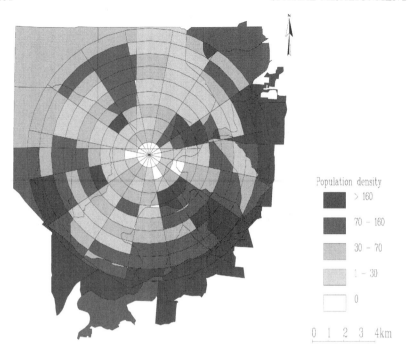

Figure 6.14 *Imputed population densities* (persons/km^2) *by atom for the FMPC region.*

(2000) for results from a slightly more complex additive-plus-interaction model.

We start with the assumption that the population counts in atom k of block group i for gender g at age group a is Poisson-distributed as

$$Y_{ikga} \sim Po\left(e^{\delta_{iga}}\left(X'_{ik} + \frac{\theta}{K_i}\right)\right), \quad \text{where} \quad \delta_{iga} = \mu_i + g\alpha + \sum_{a=1}^{17}\beta_a I_a ,$$

$g{=}0$ for males and 1 for females, and I_a is a 0–1 indicator for age group a ($a = 1$ for ages 5-9, $a = 2$ for 10-14, etc.). The μ_i are block group-specific baselines (in our parametrization, they are the logs of the fitted numbers of males in the 0–4 age bracket), and α and the $\{\beta_a\}$ function as main effects for sex and age group, respectively. Note the α and $\{\beta_a\}$ parameters are not specific to any one block group, but rather apply to all 39 block groups in the map.

With each μ_i now corresponding only to the number of baby boys (not the whole population) in block group i, we expect its value to be decreased accordingly. Because there are 36 age-sex divisions, we modified the prior mean η_μ to $-2.5 \approx \log(3/36)$. We placed vague independent $N(0,10^2)$

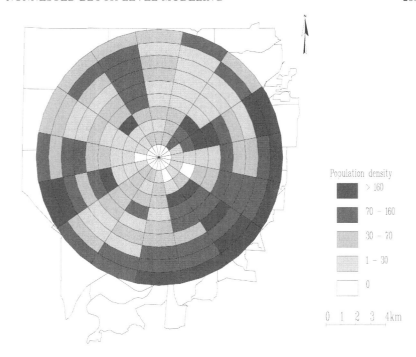

Figure 6.15 *Imputed population densities* (persons/km²) *by cell for the FMPC windrose.*

priors on α and the βs, and kept all other prior values the same as in Section 6.2. Convergence of the MCMC algorithm obtains in about 1,500 iterations. (The slowest parameters to converge are those pertaining to the edge atoms, where we have no structure data. Some parameters converge much faster: the α and β_a parameters, for example, converge by about 500 iterations.) We then ran 5,000 iterations for each of 5 chains, resulting in a final sample of 25,000.

Population interpolation results are quite similar to those outlined in Section 6.2, except that population distributions are available for each cell at any combination of age and sex. While we do not show these results here, we do include a summary of the main effects for age and sex. Table 6.1 shows the posterior medians and 2.5% and 97.5% quantiles for the α and β_a parameters. Among the β_a parameters, we see a significant negative value of β_4 (ages 20–24), reflecting a relatively small group of college-aged residents in this area. After a slight increase in the age distribution for ages 30–44, we observe increasingly negative values as a increases, indicating the expected decrease in population with advancing age. ∎

Effect	Parameter	Median	2.5%	97.5%
Gender	α	0.005	−0.012	0.021
Ages 5–9	β_1	0.073	0.033	0.116
Ages 10–14	β_2	0.062	0.021	0.106
Ages 15–19	β_3	−0.003	−0.043	0.041
Ages 20–24	β_4	−0.223	−0.268	−0.177
Ages 25–29	β_5	−0.021	−0.063	0.024
Ages 30–34	β_6	0.137	0.095	0.178
Ages 35–39	β_7	0.118	0.077	0.160
Ages 40–44	β_8	0.044	0.001	0.088
Ages 45–49	β_9	−0.224	−0.270	−0.180
Ages 50–54	β_{10}	−0.404	−0.448	−0.357
Ages 55–59	β_{11}	−0.558	−0.609	−0.507
Ages 60–64	β_{12}	−0.627	−0.677	−0.576
Ages 65–69	β_{13}	−0.896	−0.951	−0.839
Ages 70–74	β_{14}	−1.320	−1.386	−1.255
Ages 75–79	β_{15}	−1.720	−1.797	−1.643
Ages 80–84	β_{16}	−2.320	−2.424	−2.224
Ages 85+	β_{17}	−2.836	−2.969	−2.714

Table 6.1 *Quantiles and significance of gender and age effects for the age-sex additive model.*

6.4 Misaligned regression modeling

The methods of the preceding sections allow us to realign spatially misaligned data. The results of such methods may be interesting in and of themselves, but in many cases our real interest in data realignment will be as a precursor to fitting *regression* models relating the (newly realigned) variables.

For instance, Agarwal, Gelfand, and Silander (2002) apply the ideas of Section 6.2 in a rasterized data setting. Such data are common in remote sensing, where satellites can collect data (say, land use) over a pixelized surface, which is often fine enough so that town or other geopolitical boundaries can be (approximately) taken as the union of a collection of pixels.

The focal area for the Agarwal et al. (2002) study is the tropical rainforest biome within Toamasina (or Tamatave) Province of Madagascar. This province is located along the east coast of Madagascar, and includes the greatest extent of tropical rainforest in the island nation. The aerial extent of Toamasina Province is roughly 75,000 square km. Four georeferenced GIS coverages were constructed for the province: town boundaries with associated 1993 population census data, elevation, slope, and land cover. Ultimately, the total number of towns was 159, and the total number of

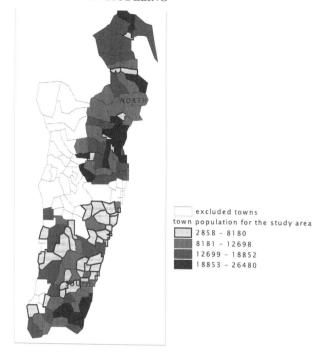

Figure 6.16 *Northern and southern regions within the Madagascar study region, with population overlaid (see also color insert).*

pixels was 74,607. For analysis at a lower resolution, the above 1-km raster layers are aggregated into 4-km pixels.

Figure 6.16 (see also color insert Figure C.4) shows the town-level map for the 159 towns in the Madagascar study region. In fact, there is an escarpment in the western portion where the climate differs from the rest of the region. It is a seasonally dry grassland/savanna mosaic. Also, the northern part is expected to differ from the southern part, since the north has fewer population areas with large forest patches, while the south has more villages with many smaller forest patches and more extensive road development, including commercial routes to the national capital west of the study region. The north and south regions with a transition zone were created as shown in Figure 6.16.

The joint distribution of land use and population count is modeled at the pixel level. Let L_{ij} denote the land use value for the jth pixel in the ith town and let P_{ij} denote the population count for the jth pixel in the ith town. Again, the L_{ij} are observed but only $P_{i.} = \sum_j P_{ij}$ are observed at the town level. Collect the L_{ij} and P_{ij} into town-level vectors $\mathbf{L}_i$ and $\mathbf{P}_i$, and overall vectors $\mathbf{L}$ and $\mathbf{P}$.

Covariates observed at each pixel include an elevation, E_{ij}, and a slope, S_{ij}. To capture spatial association between the L_{ij}, pixel-level spatial effects φ_{ij} are introduced; to capture spatial association between the $P_{i.}$, town-level spatial effects δ_i are introduced. That is, the spatial process governing land use may differ from that for population.

The joint distribution, $p(\mathbf{L}, \mathbf{P} \mid \{E_{ij}\}, \{S_{ij}\}, \{\varphi_{ij}\}, \{\delta_i\})$ is specified by factoring it as

$$p(\mathbf{P} \mid \{E_{ij}\}, \{S_{ij}\}, \{\delta_i\}) \, p(\mathbf{L} \mid \mathbf{P}, \{E_{ij}\}, \{S_{ij}\}, \{\varphi_{ij}\}) \, . \qquad (6.26)$$

Conditioning is done in this fashion in order to explain the effect of population on land use. Causality is *not* asserted; the conditioning could be reversed. (Also, implicit in (6.26) is a marginal specification for $\mathbf{L}$ and a conditional specification for $\mathbf{P} \mid \mathbf{L}$.)

Turning to the first term in (6.26), the P_{ij} are assumed conditionally independent given the E's, S's, and δ's. In fact, we assume $P_{ij} \sim \text{Poisson}(\lambda_{ij})$, where

$$\log \lambda_{ij} = \beta_0 + \beta_1 E_{ij} + \beta_2 S_{ij} + \delta_i \, . \qquad (6.27)$$

Thus $P_{i.} \sim \text{Poisson}(\lambda_{i.})$, where $\log \lambda_{i.} = \log \sum_j \lambda_{ij} = \log \sum_j \exp(\beta_0 + \beta_1 E_{ij} + \beta_2 S_{ij} + \delta_i)$. In other words, the P_{ij} inherit the spatial effect associated with $P_{i.}$. Also, $\{P_{ij}\} \mid P_{i.} \sim \text{Multinomial}(P_{i.}; \{\gamma_{ij}\})$, where $\gamma_{ij} = \lambda_{ij}/\lambda_{i.}$.

In the second term in (6.26), conditional independence of the L_{ij} given the P's, E's, S's, and φ's is assumed. To facilitate computation, we aggregate to 4 km × 4 km resolution. (The discussion regarding Figure 3.2 in Subsection 3.1.1 supports this.) Since L_{ij} lies between 0 and 16, it is assumed that $L_{ij} \sim \text{Binomial}(16, q_{ij})$, i.e., that the sixteen 1 km × 1 km pixels that comprise a given 4 km × 4 km pixel are i.i.d. Bernoulli random variables with q_{ij} such that

$$\log \left(\frac{q_{ij}}{1 - q_{ij}} \right) = \alpha_0 + \alpha_1 E_{ij} + \alpha_2 S_{ij} + \alpha_3 P_{ij} + \varphi_{ij} \, . \qquad (6.28)$$

For the town-level spatial effects, a conditionally autoregressive (CAR) prior is assumed using only the adjacent towns for the mean structure, with variance τ_δ^2, and similarly for the pixel effects using only adjacent pixels, with variance τ_φ^2.

To complete the hierarchical model specification, priors for $\boldsymbol{\alpha}, \boldsymbol{\beta}, \tau_\delta^2$, and τ_φ^2 (when the φ_{ij} are included) are required. Under a binomial, with proper priors for τ_δ^2 and τ_φ^2, a flat prior for $\boldsymbol{\alpha}$ and $\boldsymbol{\beta}$ will yield a proper posterior. For τ_δ^2 and τ_φ^2, inverse Gamma priors may be adopted. Figure 6.17 offers a graphical representation of the full model.

We now present a brief summary of the data analysis. At the 4 km x 4 km pixel scale, two versions of the model in (6.28) were fit, one with

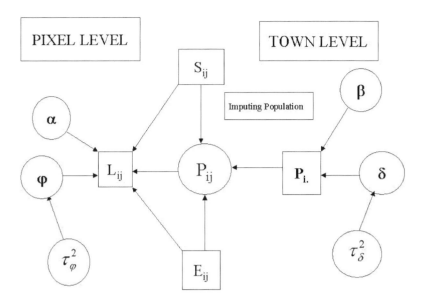

Figure 6.17 *Graphical representation of the land use-population model.*

the φ_{ij} (Model 2) and one without them (Model 1). Models 1 and 2 were fitted separately for the northern and southern regions. The results are summarized in Table 6.2, point (posterior median) and interval (95% equal tail) estimate. The population-count model results are little affected by the inclusion of the φ_{ij}. For the land-use model this is not the case. Interval estimates for the fixed effects coefficients are much wider when the φ_{ij} are included. This is not surprising from the form in (6.28). Though the P_{ij} are modeled and are constrained by summation over j and though the ϕ_{ij} are modeled dependently through the CAR specification, since neither is observed, strong collinearity between the P_{ij} and ϕ_{ij} is expected, inflating the variability of the α's.

Specifically, for the population count model in (6.27), in all cases the elevation coefficient is significantly negative; higher elevation yields smaller expected population. Interestingly, the elevation coefficient is more nega-tive in the north. The slope variable is intended to provide a measure of the differential in elevation between a pixel and its neighbors. However, a crude algorithm is used within the ARC/INFO software for its calculation, diminishing its value as a covariate. Indeed, higher slope would typically encourage lower expected population. While this is roughly true for the south under either model, the opposite emerges for the north. The infer-ence for the town-level spatial variance component τ_δ^2 is consistent across

Model:	M_1		M_2	
Region:	North	South	North	South
Population model parameters:				
β_1	−.577	−.245	−.592	−.176
(elev)	(−.663,−.498)	(−.419,−.061)	(−.679,−.500)	(−.341,.019)
β_2	.125	−.061	.127	−.096
(slope)	(.027,.209)	(−.212,.095)	(.014,.220)	(−.270,.050)
τ_{δ^2}	1.32	1.67	1.33	1.71
	(.910,2.04)	(1.23,2.36)	(.906,1.94)	(1.22,2.41)
Land use model parameters:				
α_1	.406	−.081	.490	.130
(elev)	(.373,.440)	(−.109,−.053)	(.160,.857)	(−.327,.610)
α_2	.015	.157	.040	−.011
(slope)	(−.013,.047)	(.129,.187)	(−.085,.178)	(−.152,.117)
α_3	−5.10	−3.60	−4.12	−8.11
($\times 10^{-4}$)	(−5.76,−4.43)	(−4.27,−2.80)	(−7.90,−.329)	(−14.2,−3.69)
τ_{φ^2}	—	—	6.84	5.85
			(6.15,7.65)	(5.23,6.54)

Table 6.2 *Parameter estimation (point and interval estimates) for Models 1 and 2 for the northern and southern regions.*

all models. Homogeneity of spatial variance for the population model is acceptable.

Turning to (6.28), in all cases the coefficient for population is significantly negative. There is a strong relationship between land use and population size; increased population increases the chance of deforestation, in support of the primary hypothesis for this analysis. The elevation coefficients are mixed with regard to significance. However, for both Models 1 and 2, the coefficient is always at least .46 larger in the north. Elevation more strongly encourages forest cover in the north than in the south. This is consistent with the discussion of the preceding paragraph but, apparently, the effect is weaker in the presence of the population effect. Again, the slope covariate provides inconsistent results; but is insignificant in the presence of spatial effects. Inference for the pixel-level spatial variance component

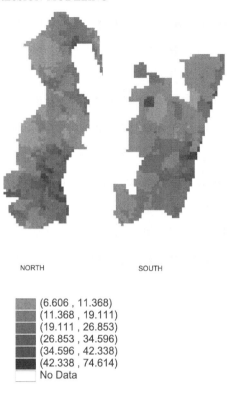

NORTH   SOUTH

	(6.606 , 11.368)
	(11.368 , 19.111)
	(19.111 , 26.853)
	(26.853 , 34.596)
	(34.596 , 42.338)
	(42.338 , 74.614)
	No Data

Figure 6.18 *Imputed population (on the square root scale) at the pixel level for north and south regions.*

does not criticize homogeneity across regions. Note that τ_φ^2 is significantly larger than τ_δ^2. Again, this is expected. With a model having four population parameters to explain 3186 $q'_{ij}s$ as opposed to a model having three population parameters to explain 115 $\lambda'_i s$, we would expect much more variability in the $\varphi'_{ij}s$ than in the $\delta'_i s$. Finally, Figure 6.18 shows the imputed population at the 4 km × 4 km pixel level.

The approach of Section 6.3 will be difficult to implement with more than two mutually misaligned areal data layers, due mostly to the multiple labeling of atoms and the needed higher-way look-up table. However, the approach of this section suggests a simpler strategy for handling this situation. First, rasterize all data layers to a common scale of resolution. Then, build a suitable latent regression model at that scale, with conditional distributions for the response and explanatory variables constrained by the observed aggregated measurements for the respective layers.

Zhu, Carlin, and Gelfand (2003) consider regression in the point-block misalignment setting, illustrating with the Atlanta ozone data pictured in

Figure 6.1. Recall that in this setting the problem is to relate several air quality indicators (ozone, particulate matter, nitrogen oxides, etc.) and a range of sociodemographic variables (age, gender, race, and a socioeconomic status surrogate) to the response, pediatric emergency room (ER) visit counts for asthma in Atlanta, GA. Here the air quality data is collected at fixed monitoring stations (point locations) while the sociodemographic covariates and response variable is collected by zip code (areal summaries). In fact, the air quality data is available as daily averages at each monitoring station, and the response is available as daily counts of visits in each zip code. Zhu et al. (2003) use the methods of Section 6.1 to realign the data, and then fit a Poisson regression model on this scale. Since the data also involves a temporal component, we defer further details until Subsection 8.5.4.

6.5 Exercises

1. Suppose we estimate the average value of some areal variable $Y(B)$ over a block B by the predicted value $Y(\mathbf{s}^*)$, where $\mathbf{s}^*$ is some central point of B (say, the population-weighted centroid). Prove that $Var(Y(\mathbf{s}^*)) \geq Var(Y(B))$ for any $\mathbf{s}^*$ in B. Is this result still true if $Y(\mathbf{s})$ is nonstationary?

2. Derive the form for $H_B(\boldsymbol{\phi})$ given below (6.7). (*Hint:* This may be easiest to do by gridding the B_k's, or through a limiting Monte Carlo integration argument.)

3. Suppose g is a differentiable function on $\Re^+$, and suppose $Y(\mathbf{s})$ is a mean-zero stationary process. Let $Z(\mathbf{s}) = g(Y(\mathbf{s}))$ and $Z(B) = \frac{1}{|B|} \int_B Z(\mathbf{s}) d\mathbf{s}$. Approximate $Var(Z(B))$ and $Cov(Z(B), Z(B'))$. (*Hint:* Try the delta method here.)

4. Define a process (for convenience, on $\Re^1$) such that $\widehat{Y}(B)$ defined as above (6.10) does *not* converge almost surely to $Y(B)$.

5. Consider the subset of the 1993 scallop data sites formed by the rectangle having opposite vertices (73.0W, 39.5N) and (72.5W, 40.0N) (refer to Figure 6.19 as well as Figure 2.16). This rectangle includes 20 locations; the full scallop data are provided both in S+SpatialStats and also at www.biostat.umn.edu/~brad/data/myscallops.dat (the latter has the advantage of including our transformed variable, log(tcatch+1)).

   (a) Krige the block average of log(tcatch+1) for this region by simulating from the posterior predictive distribution given all of the 1993 data. Adopt the model and prior structure in Example 6.1, and use equation (6.6) implemented through (6.12) to carry out the generation.

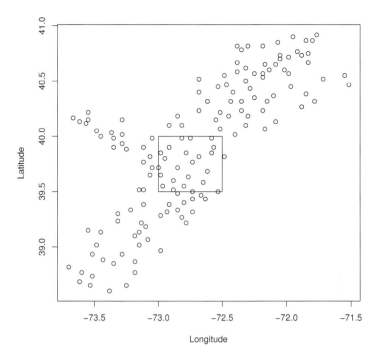

Figure 6.19 *1993 scallop data, with rectangle over which a block average is desired.*

(b) Noting the caveats regarding vague priors mentioned just below equation (5.3), change to a more informative prior specification on the spatial variance components. Are your findings robust to this change?

6. Suppose that Figure 6.20 gives a (nested) subdivision of the region in Figure 6.3, where we assume the disease count in each subsection is Poisson-distributed with parameter m_1 or m_2, depending on which value (1 or 2) a subregional binary measurement assumes. Suppose further that these Poisson variables are independent given the covariate. Let the observed disease counts in Region I and Region II be $y_1 = 632$ and $y_2 = 311$, respectively, and adopt independent $Gamma(a, b)$ priors for m_1 and m_2 with $a = 0.5$ and $b = 100$, so that the priors have mean 50 (roughly the average observed count per subregion) and variance 5000.

(a) Derive the full conditional distributions for m_1 and m_2, and obtain estimates of their marginal posterior densities using MCMC or some other approach. (*Hint:* To improve the numerical stability of your algorithm, you may wish to transform to the log scale. That is, reparametrize to $\delta_1 = \log(m_1)$ and $\delta_2 = \log(m_2)$, remembering to multiply by the Jacobian ($\exp(\delta_i), i = 1, 2$) for each transformation.)

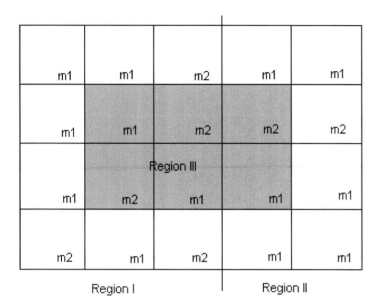

Figure 6.20 *Subregional map for motivating example.*

(b) Find an estimate of $E(Y_3|\mathbf{y})$, the predictive mean of Y_3, the total disease count in the shaded region. (*Hint:* First estimate $E(Y_{3a}|\mathbf{y})$ and $E(Y_{3b}|\mathbf{y})$, where Y_{3a} and Y_{3b} are the subtotals in the left (Region I) and right (Region II) portions of Region III.)

(c) Obtain a sample from the posterior predictive distribution of Y_3, $p(y_3|\mathbf{y})$. Is your answer consistent with the naive one obtained from equation (6.13)?

7. For the Tompkins County data, available on our website at address www.biostat.umn.edu/~brad/data/tompkins.dat and with supporting information on StatLib at lib.stat.cmu.edu/datasets/csb/, obtain smoothed estimates of the underlying block group-level relative risks of disease by modifying the log-relative risk model (6.14) to

$$\delta_{k(i,j)} = \theta_0 + \theta_1 u_{ij} + \theta_2 w_{ij} + \theta_3 u_{ij} w_{ij} + \phi_k ,$$

where we assume

(a) $\phi_k \overset{iid}{\sim} N(0, 1/\tau)$ (*global* smoothing), and

(b) $\boldsymbol{\phi} \sim CAR(\lambda)$, i.e., $\phi_k \mid \phi_{k' \neq k} \sim N\left(\bar{\phi}_k , \frac{1}{\lambda n_k}\right)$ (*local* smoothing).

Do your estimates significantly differ? How do they change as you change λ?

8. For the FMPC data and model in Section 6.3,

   (a) Write an explicit expression for the full Bayesian model, given in shorthand notation in equation (6.22).

   (b) For the full conditionals for μ_i, X_{ji}, and X'_{iE}, show that the Gaussian, multinomial, and Poisson (respectively) are sensible choices as Metropolis-Hastings proposal densities, and give the rejection ratio (4.18) in each case.

CHAPTER 7

Multivariate spatial modeling

In this chapter we take up the problem of multivariate spatial modeling. A conditioning approach, along the lines of the way misalignment was treated in Chapter 6 (e.g., X followed by $Y|X$) offers one possibility (see Subsection 7.2.2). However, a broader objective is the provision of *joint* spatial modeling for multivariate measurements at point-referenced locations or over areal units.

Spatial data collected at point locations is often multivariate. For example, at a particular environmental monitoring station, levels of several pollutants would typically be measured (e.g., ozone, nitric oxide, carbon monoxide, $PM_{2.5}$, etc.). In atmospheric modeling, at a given site we may observe surface temperature, precipitation, and wind speed. In examining commercial real estate markets, for an individual property we may observe both selling price and total rental income. In each of these illustrations, we anticipate both dependence between measurements at a particular location, and association between measurements across locations.

To add generality, one could envision a latent multivariate spatial process defined over locations in a region. For instance, for a given tree in a stand, we might be interested in the amount of energy devoted to flowering, to total seed production, and to amount of photosynthate produced. None of the foregoing quantities can be directly observed, though we anticipate the foregoing sort of dependence structure. In fact, as we illustrate in Section 10.2, in certain modeling settings we can attach to each location a vector of random effects, e.g., a vector of spatially varying regression coefficients. Just as one could expect dependence among the components of an estimated vector of regression coefficients $\widehat{\boldsymbol{\beta}}$, one could expect a spatially varying coefficient vector $\boldsymbol{\beta}(s)$ to exhibit both within and between location dependence.

Using the generic notation $\mathbf{Y}(s)$ to denote a $p \times 1$ vector of random variables at location $\mathbf{s}$, we seek flexible, interpretable, and computationally tractable models to describe the process $\{\mathbf{Y}(\mathbf{s}) : \mathbf{s} \in D\}$. As earlier, such processes are described through finite dimensional distributions, i.e., by providing $p(\mathbf{Y})$ where $\mathbf{Y} = (\mathbf{Y}(\mathbf{s}_i), ..., \mathbf{Y}(\mathbf{s}_n))$.

The crucial object is the cross-covariance $C(\mathbf{s}, \mathbf{s}') \equiv cov(\mathbf{Y}(\mathbf{s}), \mathbf{Y}(\mathbf{s}'))$, a

$p \times p$ matrix that need not be symmetric (i.e., $cov(Y_j(\mathbf{s}), Y_{j'}(\mathbf{s}'))$ need not equal $cov(Y_{j'}(\mathbf{s}), Y_j(\mathbf{s}'))$. Hence, there is no notion of positive definitiveness associated with $C(\mathbf{s}, \mathbf{s}')$ except in a limiting sense. That is, as $\|\mathbf{s} - \mathbf{s}'\| \to 0$, $C(\mathbf{s}, \mathbf{s})$ is the covariance matrix associated with the vector $\mathbf{Y}(\mathbf{s})$. We have (weak) stationarity if C depends upon $\mathbf{s}$ and $\mathbf{s}'$ only through the separation vector $\mathbf{s} - \mathbf{s}'$; we have isotropy if C depends upon $\mathbf{s}$ and $\mathbf{s}'$ through the distance $\|\mathbf{s} - \mathbf{s}'\|$. Again, since our primary focus in this text is Gaussian process models (or mixtures of such processes), specification of $C(\mathbf{s}, \mathbf{s}')$ is all we need to provide all finite dimensional distributions.

As in the univariate case, it is evident that not every matrix $C(\mathbf{s}, \mathbf{s}')$ which we might propose will be *valid*. Indeed, validity for a cross-covariance matrix is clearly more demanding than for a covariance function (see Section 2.2.2). We require that for an arbitrary number of and choice of locations, the resulting $np \times np$ covariance matrix for $\mathbf{Y}$ must be positive definite. Formal mathematical investigation of this problem (e.g., existence theorems, sufficient conditions, results for general Hermitian forms) have some history in the literature; see, e.g., Rehman and Shapiro (1996). Consistent with our objective of using multivariate spatial process models in an applied context, we prefer constructive approaches for such cross-covariance functions. The next three sections describe approaches based upon separability, coregionalization, moving averages, and convolution. Nonstationarity can be introduced following the univariate approaches in Section 5.3; however, no details are presented here (see, e.g., Gelfand, Schmidt, Banerjee, and Sirmans, 2004, Sec. 4). Finally, Section 7.4 describes multivariate CAR models for areal data. As usual, our analytic approaches for these high-dimensional models will be predominantly hierarchical Bayesian.

7.1 Separable models

Perhaps the most obvious specification of a valid cross-covariance function for a p-dimensional $\mathbf{Y}(\mathbf{s})$ is to let ρ be a valid correlation function for a univariate spatial process, let T be a $p \times p$ positive definite matrix, and let

$$C(\mathbf{s}, \mathbf{s}') = \rho(\mathbf{s}, \mathbf{s}') \cdot T . \qquad (7.1)$$

In (7.1), $T \equiv (T_{ij})$ is interpreted as the covariance matrix associated with $\mathbf{Y}(\mathbf{s})$, and ρ attenuates association as $\mathbf{s}$ and $\mathbf{s}'$ become farther apart. The covariance matrix for $\mathbf{Y}$ resulting from (7.1) is easily shown to be

$$\Sigma_{\mathbf{Y}} = H \otimes T , \qquad (7.2)$$

where $(H)_{ij} = \rho(\mathbf{s}_i, \mathbf{s}_j)$ and $\otimes$ denotes the Kronecker product. $\Sigma_{\mathbf{Y}}$ is evidently positive definite since H and T are. In fact, $\Sigma_{\mathbf{Y}}$ is convenient to work with since $|\Sigma_{\mathbf{Y}}| = |H|^p |T|^n$ and $\Sigma_{\mathbf{Y}}^{-1} = H^{-1} \otimes T^{-1}$. This means that updating $\Sigma_{\mathbf{Y}}$ requires working with a $p \times p$ and an $n \times n$ matrix, rather than an $np \times np$ one. Moreover, if we permute the rows of $\mathbf{Y}$ to $\tilde{\mathbf{Y}}$ where

$\tilde{\mathbf{Y}}^T = (Y_1(\mathbf{s}_1), \dots, Y_1(\mathbf{s}_n), Y_2(\mathbf{s}_1), \dots, Y_2(\mathbf{s}_n), \dots, Y_p(\mathbf{s}_1), \dots, Y_p(\mathbf{s}_n))$, then $\Sigma_{\tilde{\mathbf{Y}}} = T \otimes H$.

In fact, working in the fully Bayesian setting, additional advantages accrue to (7.1). With ϕ and T a priori independent and an inverse Wishart prior for T, the full conditional distribution for T, that is, $p(T|\mathbf{W}, \phi)$, is again an inverse Wishart (e.g., Banerjee, Gelfand, and Polasek, 2000). If the Bayesian model is to be fitted using a Gibbs sampler, updating T requires a draw of a $p \times p$ matrix from a Wishart distribution, substantially faster than updating the $np \times np$ matrix $\Sigma_{\mathbf{Y}}$.

What limitations are associated with (7.1)? Clearly $C(\mathbf{s}, \mathbf{s}')$ is symmetric, i.e., $cov(Y_\ell(\mathbf{s}_i), Y_{\ell'}(\mathbf{s}_{i'})) = cov(Y_{\ell'}(\mathbf{s}_i), Y_\ell(\mathbf{s}_{i'}))$ for all i, i', ℓ, and ℓ'. Moreover, it is easy to check that if ρ is stationary, the generalized correlation, also referred to as the coherence in the time series literature (see, e.g., Wei, 1990), is such that

$$\frac{cov(Y_\ell(s), Y_{\ell'}(s+h))}{\sqrt{cov(Y_\ell(s), Y_\ell(s+h))cov(Y_{\ell'}(s), Y_{\ell'}(s+h))}} = \frac{T_{\ell\ell'}}{\sqrt{T_{\ell\ell}T_{\ell'\ell'}}}, \qquad (7.3)$$

regardless of $\mathbf{s}$ and h. Also, if ρ is isotropic and strictly decreasing, then the spatial range (see Section 2.1.3) is identical for each component of $\mathbf{Y}(\mathbf{s})$. This must be the case since only one correlation function is introduced in (7.1). This seems the most unsatisfying restriction, since if, e.g., $\mathbf{Y}(\mathbf{s})$ is a vector of levels of different pollutants at $\mathbf{s}$, then why should the range for all pollutants be the same? In any event, some preliminary marginal examination of the $Y_\ell(\mathbf{s}_i)$ for each ℓ, $\ell = 1, \dots, p$, might help to clarify the feasibility of a common range.

Additionally, (7.1) implies that, for each component of $\mathbf{Y}(\mathbf{s})$, correlation between measurements tends to 1 as distance between measurements tends to 0. For some variables, including those in our illustration, such an assumption is appropriate. For others it may not be, in which case microscale variability (captured through a nugget) is a possible solution. Formally, suppose independent $\epsilon(\mathbf{s}) \sim N\left(\mathbf{0}, Diag(\boldsymbol{\tau}^2)\right)$, where $Diag(\boldsymbol{\tau}^2)$ is a $p \times p$ diagonal matrix with (i, i) entry τ_i^2, are included in the modeling. That is, we write $\mathbf{Y}(\mathbf{s}) = \mathbf{V}(\mathbf{s}) + \epsilon(\mathbf{s})$ where $\mathbf{V}(\mathbf{s})$ has the covariance structure in (7.1). An increased computational burden results, since the full conditional distribution for T is no longer an inverse Wishart, and likelihood evaluation requires working with an $np \times np$ matrix.

In a sequence of papers by Le and Zidek and colleagues (mentioned in the next subsection), it was proposed that $\Sigma_{\mathbf{Y}}$ be taken as a random covariance matrix drawn from an inverse Wishart distribution centered around (7.2). In other words, an extra hierarchical level is added to the modeling for $\mathbf{Y}$. In this fashion, we are not specifying a spatial process for $\mathbf{Y}(\mathbf{s})$; rather, we are creating a joint distribution for $\mathbf{Y}$ with a flexible covariance matrix. Indeed, the resulting $\Sigma_{\mathbf{Y}}$ will be nonstationary. In fact, the entries will have no connection to the respective $\mathbf{s}_i$ and $\mathbf{s}_j$. This may be unsatisfactory since we expect to obtain many inconsistencies with regard to distance between points and corresponding association across components. We may be able to obtain the posterior distribution of, say, $Corr(Y_\ell(\mathbf{s}_i), Y_\ell(\mathbf{s}_j))$, but there will be no notion of a range.

The form in (7.1) was presented in Mardia and Goodall (1993) who used it in conjunction with maximum likelihood estimation. Banerjee and Gelfand (2002) discuss its implementation in a fully Bayesian context, as we outline in the next subsection.

7.1.1 Spatial prediction, interpolation, and regression

Multivariate spatial process modeling is required when we are analyzing several point-referenced data layers, when we seek to explain or predict for one layer given the others, or when the layers are not all collected at the same locations. The last of these is a type of spatial misalignment that can also be viewed as a missing data problem, in the sense that we are missing observations to completely align all of the data layers. For instance, in monitoring pollution levels, we may observe some pollutants at one set of monitoring sites, and other pollutants at a diffferent set of sites. Alternatively, we might have data on temperature, elevation, and wind speed, but all at different locations.

More formally, suppose we have a conceptual response $Z(\mathbf{s})$ along with a conceptual vector of covariates $\mathbf{x}(\mathbf{s})$ at each location $\mathbf{s}$. However, in the sampling, the response and the covariates are observed at possibly different locations. To set some notation, let us partition our set of sites into three mutually disjoint groups: let S_Z be the sites where only the response $Z(\mathbf{s})$ has been observed, S_X the set of sites where only the covariates have been observed, S_{ZX} the set where both $Z(\mathbf{s})$ and the covariates have been observed, and finally S_U the set of sites where no observations have been taken.

In this context we can formalize three types of inference questions. One concerns $Y(\mathbf{s})$ when $\mathbf{s} \in S_X$, which we call *interpolation*. The second concerns $Y(\mathbf{s})$ for $\mathbf{s}$ belonging to S_U, which we call *prediction*. Evidently, prediction and interpolation are similar but interval estimates will be at least as tight for the latter compared with the former. The last concerns the functional relationship between $X(\mathbf{s})$ and $Y(\mathbf{s})$ at an arbitrary site $\mathbf{s}$, along with other covariate information at $\mathbf{s}$, say $\mathbf{U}(\mathbf{s})$. We capture this through $E[Y(\mathbf{s})|X(\mathbf{s}), \mathbf{U}(\mathbf{s})]$, and refer to it as *spatial regression*. Figure 7.1 offers a graphical clarification of the foregoing definitions.

In the usual stochastic regressors setting one is interested in the relationship between $Y(\mathbf{s})$ and $X(\mathbf{s})$ where the pairs $(X(\mathbf{s}_i), Y(\mathbf{s}_i))$, $i = 1, \ldots, n$ (suppressing $\mathbf{U}(\mathbf{s}_i)$) are independent. For us, they are dependent with the dependence captured through a spatial characterization. Still, one may be interested in the regression of $Y(\mathbf{s})$ on $X(\mathbf{s})$ at an arbitrary $\mathbf{s}$. Note that

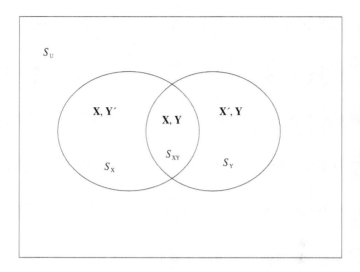

Figure 7.1 *A graphical representation of the S sets. Interpolation applies to lo-cations in S_X, prediction applies to locations in S_U, and regression applies to all locations.* $\mathbf{X}_{aug} = (\mathbf{X}, \mathbf{X}')$, $\mathbf{Y}_{aug} = (\mathbf{Y}, \mathbf{Y}')$.

there is no conditional spatial process, $Y(\mathbf{s}) \mid X(\mathbf{s})$, associated with the bivariate spatial process $(X(\mathbf{s}), Y(\mathbf{s}))$; how would one define the joint distribution of $Y(\mathbf{s}_i) \mid X(\mathbf{s}_i)$ and $Y(\mathbf{s}_{i'}) \mid X(\mathbf{s}_{i'})$?

We also note that our modeling structure here differs considerably from that of Diggle, Tawn, and Moyeed (1998). These authors specify a univariate spatial process in order to introduce unobserved spatial effects (say, $V(\mathbf{s})$) into the modeling, after which the $Y(\mathbf{s})$'s are conditionally independent given the $V(\mathbf{s})$'s. In other words, the $V(\mathbf{s})$'s are intended to capture spatial association in the means of the $Y(\mathbf{s})$'s. For us, the $X(\mathbf{s})$'s are also modeled through a spatial process, but they are observed and introduced as an explanatory variable with a regression coefficient. Hence, along with the $Y(\mathbf{s})$'s, we require a bivariate spatial process.

Here we provide a fully Bayesian examination of the foregoing questions. In Subsection 7.1.2 we study the case where $Y(\mathbf{s})$ is Gaussian, but in Subsection 7.1.4 we allow the response to be binary.

The Gaussian interpolation problem is addressed from an empirical Bayes perspective in a series of papers by Zidek and coworkers. For instance, Le and Zidek (1992) and Brown, Le, and Zidek (1994) develop a Bayesian interpolation theory (both spatial and temporal) for multivariate random

spatial data. Le, Sun, and Zidek (1997) extend this methodology to account
for misalignment, i.e., where possibly not all monitored sites measured the
same set of pollutants (data missing by design). Their method produces
the joint predictive distribution for several locations and different time
points using all available data, thus allowing for simultaneous temporal and
spatial interpolation without assuming the random field to be stationary.
Their approach provides a first-stage multivariate normal distribution for
the observed data. However, this distribution does not arise from a spatial
Gaussian process.

Framing multivariate spatial prediction (often referred to as *cokriging*)
in the context of linear regression dates at least to Corsten (1989) and
Stein and Corsten (1991). In this work, the objective is to carry out predictions for a possible future observation. Stein and Corsten (1991) advocate
looking at the prediction problem under a regression setup. They propose
trend surface modeling of the point source response using polynomials in
the coordinates. Typically in trend surface analysis (Cressie, 1993), spatial structure is modeled through the mean but observations are assumed
to be independent. Instead, Stein and Corsten (1991) retain familiar spatial dependence structure but assume the resultant covariances and cross-
covariances (and hence the dispersion matrix) are known. In this context,
Stein et al. (1991) use restricted maximum likelihood to estimate unknown
spatial dependence structure parameters.

7.1.2 Regression in the Gaussian case

Assume for the moment a single covariate with no misalignment, and let
$\mathbf{X} = (X(\mathbf{s}_1), \dots, X(\mathbf{s}_n))^T$ and $\mathbf{Y} = (Y(\mathbf{s}_1), \dots, Y(\mathbf{s}_n))^T$. be the measurements on the covariates and the response, respectively. Supposing that $X(\mathbf{s})$
is continuous and that is it meaningful to model it in a spatial fashion, our
approach is to envision (perhaps after a suitable transformation) a bivariate
Gaussian spatial process,

$$\mathbf{W}(\mathbf{s}) = \begin{pmatrix} X(\mathbf{s}) \\ Y(\mathbf{s}) \end{pmatrix} \sim \mathrm{N}(\boldsymbol{\mu}(\mathbf{s}), T), \qquad (7.4)$$

which generates the data. We assume $\mathbf{W}(\mathbf{s})$ is distributed as in (7.1).

With misalignment, let $\mathbf{X}$ be the vector of observed $X(\mathbf{s})$'s at the sites
in $S_{XY} \cup R_X$, while $\mathbf{Y}$ will be the vector of $Y(\mathbf{s})$'s the sites in $S_{XY} \cup S_Y$.
If we let $\mathbf{X}'$ denote the vector of missing X observations in S_Y and $\mathbf{Y}'$ the
vector of missing Y observations in S_X, then in the preceding discussion
we can replace $\mathbf{X}$ and $\mathbf{Y}$ by the augmented vectors $\mathbf{X}_{aug} = (\mathbf{X}, \mathbf{X}')$ and
$\mathbf{Y}_{aug} = (\mathbf{Y}, \mathbf{Y}')$; see Figure 7.1 for clarification. After permutation to line
up the X's and Y's, they can be collected into a vector $\mathbf{W}_{aug}$. In the
Bayesian model specification, $\mathbf{X}'$ and $\mathbf{Y}'$ are viewed as latent (unobserved)
vectors. In implementing a Gibbs sampler for model-fitting, we update the

model parameters given $\mathbf{X}'$ and $\mathbf{Y}'$ (i.e., given $\mathbf{W}_{aug}$), and then update $(\mathbf{X}', \mathbf{Y}')$ given $\mathbf{X}$, $\mathbf{Y}$, and the model parameters. The latter updating is routine since the associated full conditional distributions are normal. Such augmentation proves computationally easier with regard to bookkeeping since we retain the convenient Kronecker form for $\Sigma_{\mathbf{W}}$. That is, it is easier to marginalize over $\mathbf{X}'$ and $\mathbf{Y}'$ after simulation than before. For convenience of notation, we suppress the augmentation in the sequel.

In what we have called the prediction problem, it is desired to predict the outcome of the response variable at some unobserved site. Thus we are interested in the posterior predictive distribution $p(y(\mathbf{s}_0)|\mathbf{y}, \mathbf{x})$. We note that $x(\mathbf{s}_0)$ is also not observed here. On the other hand, the interpolation problem may be regarded as a method of imputing missing data. Here the covariate $x(\mathbf{s}_0)$ is observed but the response is "missing." Thus our attention shifts to the posterior predictive distribution For the regression problem, the distribution of interest is $p(E[Y(\mathbf{s}_0)|x(\mathbf{s}_0)] \mid x(\mathbf{s}_0), \mathbf{y}, \mathbf{x})$.

For simplicity, suppose $\boldsymbol{\mu}(\mathbf{s}) = (\mu_1, \mu_2)^T$, independent of the site coordinates. (With additional fixed site-level covariates for $Y(\mathbf{s})$, say $\mathbf{U}(\mathbf{s})$, we would replace μ_2 with $\mu_2(\mathbf{s}) = \boldsymbol{\alpha}^T \mathbf{U}(\mathbf{s})$.) Then, from (7.4), for the pair $(X(\mathbf{s}), Y(\mathbf{s}))$, $p(y(\mathbf{s})|x(\mathbf{s}), \beta_0, \beta_1, \sigma^2)$ is $N\left(\beta_0 + \beta_1 x(\mathbf{s}), \sigma^2\right)$. That is, $E[Y(\mathbf{s})|x(\mathbf{s})] = \beta_0 + \beta_1 x(\mathbf{s})$, where

$$\beta_0 = \mu_2 - \frac{T_{12}}{T_{11}}\mu_1, \ \beta_1 = \frac{T_{12}}{T_{11}}, \text{ and } \sigma^2 = T_{22} - \frac{T_{12}^2}{T_{11}} . \qquad (7.5)$$

So, given samples from the joint posterior distribution of (μ_1, μ_2, T, ϕ), we directly have samples from the posterior distributions for the parameters in (7.5), and thus from the posterior distribution of $E[Y(\mathbf{s})|x(\mathbf{s})]$.

Rearrangement of the components of $\mathbf{W}$ as below (7.2) yields

$$\left(\begin{array}{c} \mathbf{X} \\ \mathbf{Y} \end{array} \right) \sim N\left(\left(\begin{array}{c} \mu_1 \mathbf{1} \\ \mu_2 \mathbf{1} \end{array} \right) , \ T \otimes H\left(\phi\right) \right) , \qquad (7.6)$$

which simplifies calculation of the conditional distribution of $\mathbf{Y}$ given $\mathbf{X}$.

Assuming an inverse Wishart prior for T, completing the Bayesian specification requires a prior for μ_1, μ_2, and ϕ. For (μ_1, μ_2), for convenience we would take a vague but proper bivariate normal prior. A suitable prior for ϕ depends upon the choice of $\rho(h; \phi)$. Then we use a Gibbs sampler to simulate the necessary posterior distributions. The full conditionals for μ_1 and μ_2 are in fact Gaussian distributions, while that of the T matrix is inverted Wishart as already mentioned. The full conditional for the ϕ parameter finds ϕ arising in the entries in H, and so is not available in closed form. Metropolis or slice sampling can be employed for its updating.

Under the above framework, interpolation presents no new problems. Let $\mathbf{s}_0$ be a new site at which we would like to predict the variable of interest.

We first modify the $H(\phi)$ matrix forming the new matrix H^* as follows:

$$H^*(\phi) = \begin{pmatrix} H(\phi) & \mathbf{h}(\phi) \\ \mathbf{h}(\phi)^T & \rho(0; \phi) \end{pmatrix}, \qquad (7.7)$$

where $\mathbf{h}(\phi)$ is the vector with components $\rho(\mathbf{s}_0 - \mathbf{s}_j; \phi)$, $j = 1, 2, \ldots, n$. It then follows that

$$\mathbf{W}^* \equiv (\mathbf{W}(\mathbf{s}_0), \ldots, \mathbf{W}(s_n))^T \sim \mathrm{N}\left(\mathbf{1}_{n+1} \otimes \begin{pmatrix} \mu_1 \\ \mu_2 \end{pmatrix}, \; H^*(\phi) \otimes T\right).$$
$$(7.8)$$

Once again a simple rearrangement of the above vector enables us to arrive at the conditional distribution $p(\mathbf{y}(\mathbf{s}_0)|\mathbf{x}(\mathbf{s}_0), \mathbf{y}, \mathbf{x}, \boldsymbol{\mu}, T, \phi)$ as a Gaussian distribution. The predictive distribution for the interpolation problem, $p(\mathbf{y}(\mathbf{s}_0)|\mathbf{y}, \mathbf{x})$, can now be obtained by marginalizing over the parameters, i.e.,

$$p(y(\mathbf{s}_0)|\mathbf{y}, \mathbf{x}) = \int p(y(\mathbf{s}_0)|x(\mathbf{s}_0), \mathbf{y}, \mathbf{x}, \boldsymbol{\mu}, T, \phi) \, p(\boldsymbol{\mu}, T, \phi|x(\mathbf{s}_0), \mathbf{y}, \mathbf{x}) . \quad (7.9)$$

For prediction, we do not have $x(\mathbf{s}_0)$. But this does not create any new problems, as it may be treated as a latent variable and incorporated into $\mathbf{x}'$. This only results in an additional draw within each Gibbs iteration, and is a trivial addition to the computational task.

7.1.3 Avoiding the symmetry of the cross-covariance matrix

In the spirit of Le and Zidek (1992), we can avoid the symmetry in $\Sigma_{\mathbf{W}}$ noted above (7.3). Instead of directly modeling $\Sigma_{\mathbf{W}}$ as $H(\phi) \otimes T$, we can add a further hierarchical level, with $p(\Sigma_{\mathbf{W}}|\phi, T)$ following an inverted Wishart distribution with mean $H(\phi) \otimes T$. All other specifications remain as before. Note that the marginal model (i.e., marginalizing over $\Sigma_{\mathbf{W}}$) is no longer Gaussian. However, using standard calculations, the resulting cross-covariance matrix is a function of $\rho(\mathbf{s} - \mathbf{s}'; \phi)$, retaining desirable spatial interpretation. Once again we resort to the Gibbs sampler to arrive at the posteriors, although in this extended model the number of parameters has increased substantially, since the elements of $\Sigma_{\mathbf{W}}$ are being introduced as new parameters.

The full conditionals for the means μ_1 and μ_2 are still Gaussian and it is easily seen that the full conditional for $\Sigma_{\mathbf{W}}$ is inverted Wishart. The full conditional distribution for ϕ is now proportional to $p(\Sigma_{\mathbf{W}}|\phi, T)p(\phi)$; a Metropolis step may be employed for its updating. Also, the full conditional for T is no longer inverted Wishart and a Metropolis step with an inverted Wishart proposal is used to sample the T matrix. All told, this is indeed a much more computationally demanding proposition since we now have to deal with the $2n \times 2n$ matrix $\Sigma_{\mathbf{W}}$ with regard to sampling, inversion, determinants, etc.

7.1.4 Regression in a probit model

Now suppose we have binary response from a point-source spatial dataset. At each site, $Z(\mathbf{s})$ equals 0 or 1 according to whether we observed "failure" or "success" at that particular site. Thus, a realization of the process can be partitioned into two disjoint subregions, one for which $Z(\mathbf{s}) = 0$, the other $Z(\mathbf{s}) = 1$, and is called a *binary map* (DeOliveira, 2000). Again, the process is only observed at a finite number of locations. Along with this binary response we have a set of covariates observed at each site. We follow the latent variable approach for probit modeling as in, e.g., DeOliveira, (2000). Let $Y(\mathbf{s})$ be a latent spatial process associated with the sites and let $X(\mathbf{s})$ be a process that generates the values of a particular covariate, in particular, one that is misaligned with $Z(\mathbf{s})$ and is sensible to model in a spatial fashion. For the present we assume $X(\mathbf{s})$ is univariate but extension to the multivariate case is apparent. Let $Z(\mathbf{s}) = 1$ if and only if $Y(\mathbf{s}) > 0$. We envision our bivariate process $\mathbf{W}(\mathbf{s}) = (X(\mathbf{s}), Y(\mathbf{s}))^T$ distributed as in (7.4), but where now $\boldsymbol{\mu}(\mathbf{s}) = (\mu_1, \ \mu_2 + \boldsymbol{\alpha}^T \mathbf{U}(\mathbf{s}))^T$, with $\mathbf{U}(\mathbf{s})$ regarded as a $p \times 1$ vector of fixed covariates. Note that the conditional variance of $Y(\mathbf{s})$ given $X(\mathbf{s})$ is not identifiable. Thus, without loss of generality, we set $T_{22} = 1$, so that the T matrix has only two parameters.

Now, we formulate a probit regression model as follows:

$$P\left(Z(\mathbf{s}) = 1 \mid x(\mathbf{s}), \mathbf{U}(\mathbf{s}), \boldsymbol{\alpha}, \mu_1, \mu_2, T_{11}, T_{12}\right)$$
$$= \Phi\left(\left[\beta_0 + \beta_1 X(\mathbf{s}) + \boldsymbol{\alpha}^T \mathbf{U}(\mathbf{s})\right] \Big/ \sqrt{1 - \tfrac{T_{12}^2}{T_{11}}}\right). \qquad (7.10)$$

Here, as in (7.5), $\beta_0 = \mu_2 - (T_{12}/T_{11})\mu_1$, and $\beta_1 = T_{12}/T_{11}$.

The posterior of interest is $p(\mu_1, \mu_2, \boldsymbol{\alpha}, T_{11}, T_{12}, \boldsymbol{\phi}, \mathbf{y} \mid \mathbf{x}, \mathbf{z})$, where $\mathbf{z} = (z(\mathbf{s}_1), \ldots, z(\mathbf{s}_n))^T$ is a vector of 0's and 1's. The fitting again uses MCMC. Here, $\mathbf{X} = (X(\mathbf{s}_1), \ldots, X(\mathbf{s}_n))^T$ and $\mathbf{Y} = (Y(\mathbf{s}_1), \ldots, Y(\mathbf{s}_n))^T$ as in Subsection 7.1.2, except that $\mathbf{Y}$ is now unobserved, and introduced only for computational convenience. Analogous to (7.6),

$$\begin{pmatrix} \mathbf{X} \\ \mathbf{Y} \end{pmatrix} \sim N\left(\begin{pmatrix} \mu_1 \mathbf{1} \\ \mu_2 \mathbf{1} + \mathbf{U}\boldsymbol{\beta} \end{pmatrix}, \ T \otimes H(\boldsymbol{\phi})\right), \qquad (7.11)$$

where $\mathbf{U} = (U(\mathbf{s}_1), \ldots, U(\mathbf{s}_n))^T$.

From (7.11), the full conditional distribution for each latent $Y(\mathbf{s}_i)$ is a univariate normal truncated to a set of the form $\{Y(\mathbf{s}_i) > 0\}$ or $\{Y(\mathbf{s}_i) < 0\}$. The full conditionals for μ_1 and μ_2 are both univariate normal, while that of $\boldsymbol{\beta}$ is multivariate normal with the appropriate dimension. For the elements of the T matrix, we may simulate first from a Wishart distribution (as mentioned in Subsection 7.1.2) and then proceed to scale it by T_{22}, or we may proceed individually for T_{12} and T_{11} using Metropolis-Hastings over a restricted convex subset of a hypercube (Chib and Greenberg, 1998). Finally, $\boldsymbol{\phi}$ can be simulated using a Metropolis step, as in Subsection 7.1.2.

Misalignment is also treated as in Subsection 7.1.2, introducing appropriate latent $\mathbf{X'}$ and $\mathbf{Y'}$.

With posterior samples from $p(\mu_1, \mu_2, \boldsymbol{\alpha}, T_{11}, T_{12}, \boldsymbol{\phi} \mid \mathbf{x}, \mathbf{z})$, we immediately obtain samples from the posterior distributions for β_0 and β_1. Also, given $x(\mathbf{s}_0)$, (7.10) shows how to obtain samples from the posterior for a particular probability, such as $p(P(Z(\mathbf{s}_0) = 1 \mid x(\mathbf{s}_0), \mathbf{U}(\mathbf{s}_0), \boldsymbol{\alpha}, \mu_1, \mu_2, T_{11}, T_{12}) \mid x(\mathbf{s}_0), \mathbf{x}, \mathbf{z})$ at an unobserved site $\mathbf{s}_0$, clarifying the regression structure. Were $x(\mathbf{s}_0)$ not observed, we could still consider the chance that $Z(\mathbf{s}_0)$ equals 1. This probability, $P(Z(\mathbf{s}_0) = 1 \mid \mathbf{U}(\mathbf{s}_0), \boldsymbol{\alpha}, \mu_1, \mu_2, T_{11}, T_{12})$, arises by averaging over $X(\mathbf{s}_0)$, i.e.,

$$\int P(Z(\mathbf{s}_0) = 1 \mid x(\mathbf{s}_0), \mathbf{U}(\mathbf{s}_0), \boldsymbol{\alpha}, \mu_1, \mu_2, T_{11}, T_{12}) \ p(x(\mathbf{s}_0)|\mu_1, T_{11}) \ dx(\mathbf{s}_0) \ .$$
(7.12)

In practice, we would replace the integration in (7.12) by a Monte Carlo integration. Then, plugging into this Monte Carlo integration, the foregoing posterior samples would yield essentially posterior realizations of (7.12).

Both the prediction problem and the interpolation problem may be viewed as examples of *indicator kriging* (e.g., Solow, 1986; DeOliveira, 2000). For the prediction case we seek $p(z(\mathbf{s}_0)|\mathbf{x}, \mathbf{z})$; realizations from this distribution arise if we can obtain realizations from $p(y(\mathbf{s}_0)|\mathbf{x}, \mathbf{z})$. But

$$p(y(\mathbf{s}_0) \mid \mathbf{x}, \mathbf{z}) = \int p(y(\mathbf{s}_0) \mid \mathbf{x}, \mathbf{y}) \ p(\mathbf{y}|\mathbf{x}, \mathbf{z}) d\,\mathbf{y} \ .$$
(7.13)

Since the first distribution under the integral in (7.13) is a univariate normal, as in Subsection 7.1.2, the posterior samples of $\mathbf{Y}$ immediately provide samples of $Y(\mathbf{s}_0)$. For the interpolation case we seek $p(z(\mathbf{s}_0) \mid x(\mathbf{s}_0), \mathbf{x}, \mathbf{z})$. Again we only need realizations from $p(y(\mathbf{s}_0) \mid x(\mathbf{s}_0), \mathbf{x}, \mathbf{z})$, but

$$p(y(\mathbf{s}_0) \mid x(\mathbf{s}_0), \mathbf{x}, \mathbf{z}) = \int p(y(\mathbf{s}_0) \mid x(\mathbf{s}_0), \mathbf{x}, \mathbf{y}) \ p(\mathbf{y} \mid x(\mathbf{s}_0), \mathbf{x}, \mathbf{z}) d\,\mathbf{y} \ . \quad (7.14)$$

As with (7.13), the first distribution under the integral in (7.14) is a univariate normal.

Example 7.1 *(Gaussian model).* Our examples are based upon an ecological dataset collected over a west-facing watershed in the Negev Desert in Israel. The species under study is called an isopod, and builds its residence by making burrows. Some of these burrows thrive through the span of a generation while others do not. We study the following variables at each of 1129 sites. The variable "dew" measures time in minutes (from 8 a.m.) to evaporation of the morning dew. The variables "shrub" and "rock" density are percentages (the remainder is sand) characterizing the environment around the burrows. In our first example we try to explain shrub density (Y) through dew duration (X). In our second example we try to explain burrow survival (Z) through shrub density, rock density, and dew duration, treating only the last one as random and spatial. We illustrate the Gaussian

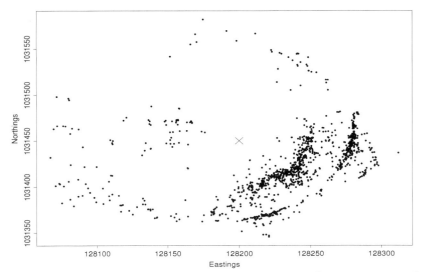

Figure 7.2 *Spatial locations of the isopod burrows data. The axes represent the eastings and the northings on a UTM projection.*

case for the first example with 694 of the sites offering both measurements, 204 sites providing only the shrub density, and 211 containing only the dew measurements.

The spatial locations are displayed in Figure 7.2 using rescaled planar coordinates after UTM projection. The rectangle in Figure 7.2 is roughly 300 km by 250 km. Hence the vector $\mathbf{X}$ consists of $694 + 211 = 905$ measurements, while the vector $\mathbf{Y}$ consists of $694 + 204 = 898$ measurements. For these examples we take the exponential correlation function, $\rho(h; \phi) = e^{-\phi h}$. We assign a vague inverse gamma specification for the parameter ϕ, namely an IG(2, 1/0.024). This prior has infinite variance and suggests a range $(3/\phi)$ of 125 km, which is roughly half the maximum pairwise distance in our region. We found little inference sensitivity to the mean of this prior. The remaining prior specifications are all rather noninformative, i.e., a $N\left(\mathbf{0}, Diag(10^5, 10^5)\right)$ prior for (μ_1, μ_2) and an $IW\left(2, Diag(0.001, 0.001)\right)$ for T. That is, $E(T_{11}) = E(T_{22}) = 0.001$, $E(T_{12}) = 0$, and the variances of the T_{ij}'s do not exist.

Table 7.1 provides the 95% credible intervals for the regression parameters and the decay parameter ϕ. The significant negative association between dew duration and shrub density is unexpected but is evident on a scatterplot of the 714 sites having both measurements. The intercept β_0 is significantly high, while the slope β_1 is negative. The maximum distance in the sample is approximately 248.1 km, so the spatial range, computed from the point estimate of $3/\phi$ from Table 7.1, is approximately 99.7 km, or about 40% of the maximum distance.

Parameter	Quantiles		
	2.5%	50%	97.5%
μ_1	73.118	73.885	74.665
μ_2	5.203	5.383	5.572
T_{11}	95.095	105.220	117.689
T_{12}	−4.459	−2.418	−0.528
T_{22}	5.564	6.193	6.914
$T_{12}/\sqrt{T_{11}T_{22}}$ (nonspatial corr. coef.)	−0.171	-0.095	−0.021
β_0 (intercept)	5.718	7.078	8.463
β_1 (slope)	−0.041	−0.023	−0.005
σ^2	5.582	6.215	6.931
ϕ	0.0091	0.0301	0.2072

Table 7.1 *Posterior quantiles for the shrub density/dew duration example.*

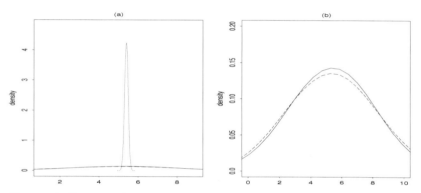

Figure 7.3 *Posterior distributions for inference at the location s_0, denoted by "×" in the previous figure. Line legend: dotted line denotes $p\left(E[Y(\mathbf{s}_0)|x(\mathbf{s}_0)] \mid \mathbf{x}, \mathbf{y}\right)$ (regression); solid line denotes $p\left(y(\mathbf{s}_0) \mid x(\mathbf{s}_0), \mathbf{x}, \mathbf{y}\right)$ (prediction); and dashed line denotes $p\left(y(\mathbf{s}_0) \mid \mathbf{x}, \mathbf{y}\right)$ (interpolation).*

In Figure 7.3(a) we show the relative performances (using posterior density estimates) of prediction, interpolation, and regression at a somewhat central location $\mathbf{s}_0$, indicated by an "×" in Figure 7.2. The associated $X(\mathbf{s}_0)$ has the value 73.10 minutes. Regression (dotted line), since it models the means rather than predicting a variable, has substantially smaller variability than prediction (solid line) or interpolation (dashed line). In Figure 7.3(b), we "zoom in" on the latter pair. As expected, interpolation has less variability due to the specification of $x(\mathbf{s}_0)$. It turns out that in all cases, the observed value falls within the associated intervals. Finally, in Figure 7.4 we present a three-dimensional surface plot of $E(Y(\mathbf{s})|\mathbf{x}, \mathbf{y})$ over

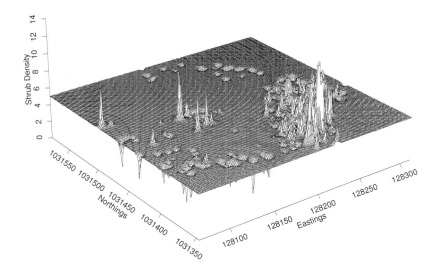

Figure 7.4 *For the Gaussian analysis case, a three-dimensional surface plot of* $E(Y(\mathbf{s})|\mathbf{x},\mathbf{y})$ *over the isopod burrows regional domain.*

the region. This plot reveals the spatial pattern in shrub density over the watershed. Higher measurements are expected in the eastern and particularly the southeastern part of the region, while relatively fewer shrubs are found in the northern and western parts. ■

Example 7.2 *(Probit model).* Our second example uses a smaller data set, from the same region as Figure 7.2, which has 246 burrows of which 43 do not provide the dew measurements. Here the response is binary, governed by the success $(Y = 1)$ or failure $(Y = 0)$ of a burrow at a particular site. The explanatory variables (dew duration, shrub density, and rock density) relate, in some fashion, to water retention. Dew measurements are taken as the X's in our modeling with shrub and rock density being U_1 and U_2, respectively. The prior specifications leading to the probit modeling again have vague bivariate normal priors for (μ_1, μ_2) and also for β, which is two-dimensional in this example. For ϕ we again assign a noninformative inverse gamma specification, the IG(2, 0.024). We generate T_{11} and T_{12} through scaling a Wishart distribution for T with prior $IW(2, Diag(0.001, 0.001))$.

In Table 7.2, we present the 95% credible intervals for the parameters in the model. The positive coefficient for dew is expected. It is interesting to note that shrub and rock density seem to have a negative impact on the success of the burrows. This leads us to believe that although high shrub and rock density may encourage the hydrology, it is perhaps not

| | | Quantiles | |
Parameter	2.5%	50%	97.5%
μ_1	75.415	76.095	76.772
μ_2	0.514	1.486	2.433
T_{11}	88.915	99.988	108.931
T_{12}	0.149	0.389	0.659
ϕ	0.0086	0.0302	0.2171
β_0 (intercept)	0.310	1.256	2.200
β_1 (dew slope)	0.032	0.089	0.145
α_1 (shrub)	−0.0059	−0.0036	−0.0012
α_2 (rock)	−0.00104	−0.00054	−0.00003

Table 7.2 *Posterior quantiles for the burrow survival example.*

conducive to the growth of food materials for the isopods, or encourages predation of the isopods. The spatial range parameter again explains about 40% of the maximum distance. Figure 7.5 presents the density estimates for the posteriors $p\left(P(Z(\mathbf{s}_0) = 1 \mid x(s_0), \mathbf{U}(\mathbf{s}_0), \boldsymbol{\alpha}, \beta_0, \beta_1) \mid x(s_0), \mathbf{x}, \mathbf{z}\right)$ and $p\left(P(Z(\mathbf{s}_0) = 1 \mid \mathbf{U}(\mathbf{s}_0), \boldsymbol{\alpha}, \mu_1, \mu_2, T_{11}, T_{12}) \mid \mathbf{x}, \mathbf{z}\right)$, with $\mathbf{s}_0$ being a central location and $x(\mathbf{s}_0) = 74.7$ minutes (after 8 a.m.), to compare performance of interpolation and prediction. As expected, interpolation provides a slightly tighter posterior distribution. ■

7.2 Coregionalization models ⋆

7.2.1 Coregionalization models and their properties

We now consider a constructive modeling strategy to add flexibility to (7.1) while retaining interpretability and computational tractability. Our approach is through the *linear model of coregionalization* (LMC), as for example in Grzebyk and Wackernagel (1994) and Wackernagel (1998). The term "coregionalization" is intended to denote a model for measurements that covary jointly over a region.

The most basic coregionalization model, the so-called *intrinsic specification*, dates at least to Matheron (1982). It arises as $\mathbf{Y}(\mathbf{s}) = \mathbf{A}\mathbf{w}(\mathbf{s})$ where the components of $\mathbf{w}(\mathbf{s})$ are i.i.d. spatial processes. If the $w_j(\mathbf{s})$ have mean 0 and are stationary with variance 1 and correlation function $\rho(h)$, then $E(\mathbf{Y}(\mathbf{s}))$ is $\mathbf{0}$ and the cross-covariance matrix, $\Sigma_{\mathbf{Y}(\mathbf{s}),\mathbf{Y}(\mathbf{s}')} \equiv C(\mathbf{s} - \mathbf{s}') = \rho(\mathbf{s} - \mathbf{s}')\mathbf{A}\mathbf{A}^T$. Letting $\mathbf{A}\mathbf{A}^T = \mathbf{T}$ this immediately reveals the equivalence between this simple intrinsic specification and the separable covariance specification as in Section 7.1 above. As in Subsection 2.1.2, the term

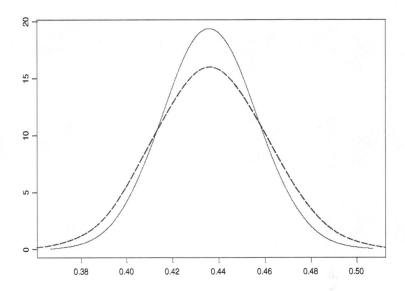

Figure 7.5 *Estimated posterior densities for the probit data analysis: solid line indicates* $P\left(Z(s_0) = 1 \mid X(s_0), \mathbf{U}(s_0), \alpha, \beta_0, \beta_1\right)$, *while dashed line indicates* $P\left(Z(s_0) = 1 \mid \mathbf{U}(s_0), \alpha, \mu_1, \mu_2, T_{11}, T_{12}\right)$.

"intrinsic" is taken to mean that the specification only requires the first and second moments of differences in measurement vectors and that the first moment difference is $\mathbf{0}$ and the second moments depend on the locations only through the separation vector $\mathbf{s} - \mathbf{s}'$. In fact here $E(\mathbf{Y}(\mathbf{s}) - \mathbf{Y}(\mathbf{s}')) = \mathbf{0}$ and $\frac{1}{2}\Sigma_{\mathbf{Y}(\mathbf{s})-\mathbf{Y}(\mathbf{s}')} = G(\mathbf{s}-\mathbf{s}')$ where $G(\mathbf{h}) = C(\mathbf{0})-C(\mathbf{h}) = \mathbf{T}-\rho(\mathbf{s}-\mathbf{s}')\mathbf{T} = \gamma(\mathbf{s} - \mathbf{s}')\mathbf{T}$ where γ is a valid variogram. Of course, as in the $p = 1$ case, we need not begin with a covariance function but rather just specify the process through γ and $\mathbf{T}$. A more insightful interpretation of "intrinsic" is that given in equation (7.3). We assume A is full rank and, for future reference, we note that $\mathbf{A}$ can be assumed to be lower triangular. No additional richness accrues to a more general $\mathbf{A}$.

A more general LMC arises if again $\mathbf{Y}(\mathbf{s}) = \mathbf{A}\mathbf{w}(\mathbf{s})$ but now the $w_j(\mathbf{s})$ are independent but no longer identically distributed. In fact, let the $w_j(\mathbf{s})$ process have mean μ_j, variance 1, and correlation function $\rho_j(h)$. Then $E(\mathbf{Y}(\mathbf{s})) = \mathbf{A}\boldsymbol{\mu}$ where $\boldsymbol{\mu}^T = (\mu_1, \cdots, \mu_p)$ and the cross-covariance matrix associated with $\mathbf{Y}(\mathbf{s})$ is now

$$\Sigma_{\mathbf{Y}(\mathbf{s}),\mathbf{Y}(\mathbf{s}')} \equiv C(\mathbf{s} - \mathbf{s}') = \sum_{j=1}^{p} \rho_j(\mathbf{s} - \mathbf{s}')\mathbf{T}_j \, , \qquad (7.15)$$

where $\mathbf{T}_j = \mathbf{a}_j \mathbf{a}_j^T$ with $\mathbf{a}_j$ the jth column of $\mathbf{A}$. Note that $\sum_j \mathbf{T}_j = \mathbf{T}$. More importantly, we note that such linear combination produces stationary spatial processes. We return to this point in Section 7.2.3.

The one-to-one relationship between $\mathbf{T}$ and lower triangular $\mathbf{A}$ is standard. For future use, when $p = 2$ we have $a_{11} = \sqrt{T_{11}}$, $a_{21} = \frac{T_{12}}{\sqrt{T_{11}}}$ and $a_{22} = \sqrt{T_{22} - \frac{T_{12}^2}{T_{11}}}$. When $p = 3$ we add $a_{31} = \frac{T_{13}}{\sqrt{T_{11}}}$, $a_{32} = \frac{T_{11}T_{23} - T_{12}T_{13}}{\sqrt{T_{11}T_{22} - T_{12}^2}\sqrt{T_{11}}}$ and $a_{33} = \sqrt{T_{33} - \frac{T_{13}^2}{T_{11}} - \frac{(T_{11}T_{23} - T_{12}T_{13})^2}{T_{11}(T_{11}T_{22} - T_{12}^2)}}$.

Lastly, if we introduce monotonic isotropic correlation functions, we will be interested in the range associated with $Y_j(\mathbf{s})$. An advantage to (7.15) is that each $Y_j(\mathbf{s})$ has its own range. In particular, for $p = 2$ the range for $Y_1(\mathbf{s})$ solves $\rho_1(d) = 0.05$, while the range for $Y_2(\mathbf{s})$ solves the weighted average correlation,

$$\frac{a_{21}^2\rho_1(d) + a_{22}^2\rho_2(d)}{a_{21}^2 + a_{22}^2} = 0.05 \,. \tag{7.16}$$

Since ρ_1 and ρ_2 are monotonic the left side of (7.16) is decreasing in d. Hence, solving (7.16) is routine. If we have $p = 3$, we need in addition the range for $Y_3(\mathbf{s})$. We require the solution of

$$\frac{a_{31}^2\rho_1(d) + a_{32}^2\rho_2(d) + a_{33}^2\rho_3(d)}{a_{31}^2 + a_{32}^2 + a_{33}^2} = 0.05 \,. \tag{7.17}$$

The left side of (7.17) is again decreasing in d. The form for general p is clear.

In practice, the ρ_j are parametric classes of functions. Hence the range d is a parametric function that is not available explicitly. However, within a Bayesian context, when models are fitted using simulation-based methods, we obtain posterior samples of the parameters in the ρ_j's, as well as $\mathbf{A}$. Each sample, when inserted into the left side of (7.16) or (7.17), enables solution for a corresponding d. In this way, we obtain posterior samples of each of the ranges, one-for-one with the posterior parameter samples.

Extending in a different fashion, we can define a process having a general *nested* covariance model (see, e.g., Wackernagel, 1998) as

$$\mathbf{Y}(\mathbf{s}) = \sum \mathbf{Y}^{(u)}(\mathbf{s}) = \sum_{u=1}^{r} \mathbf{A}^{(u)}\mathbf{w}^{(u)}(\mathbf{s}) \,, \tag{7.18}$$

where the $\mathbf{Y}^{(u)}$ are independent intrinsic LMC specifications with the components of $\mathbf{w}^{(u)}$ having correlation function ρ_u. The cross-covariance matrix associated with (7.18) takes the form

$$C(\mathbf{s} - \mathbf{s}') = \sum_{u=1}^{r} \rho_u(\mathbf{s} - \mathbf{s}')\mathbf{T}^{(u)} \,, \tag{7.19}$$

with $\mathbf{T}^{(u)} = \mathbf{A}^{(u)}(\mathbf{A}^{(u)})^T$. The $\mathbf{T}^{(u)}$ are full rank and are referred to as *coregionalization matrices*. Expression (7.19) can be compared to (7.15). Note that r need not be equal to p, but $\mathbf{\Sigma}_{\mathbf{Y}(\mathbf{s})} = \sum_u \mathbf{T}^{(u)}$. Also, recent work of Vargas-Guzmán et al. (2002) allows the $\mathbf{w}^{(u)}(\mathbf{s})$, hence the $\mathbf{Y}^{(u)}(\mathbf{s})$ in (7.18), to be dependent.

Returning to the more general LMC, in applications we introduce (7.15) as a spatial random effects component of a general multivariate spatial model for the data. That is, we assume

$$\mathbf{Y}(\mathbf{s}) = \boldsymbol{\mu}(\mathbf{s}) + \mathbf{v}(\mathbf{s}) + \boldsymbol{\epsilon}(\mathbf{s}) \,, \tag{7.20}$$

where $\boldsymbol{\epsilon}(\mathbf{s})$ is a white noise vector, i.e., $\boldsymbol{\epsilon}(\mathbf{s}) \sim N(\mathbf{0}, \mathbf{D})$ where $\mathbf{D}$ is a $p \times p$ diagonal matrix with $(D)_{jj} = \tau_j^2$. In (7.20), $\mathbf{v}(\mathbf{s}) = \mathbf{Aw}(\mathbf{s})$ following (7.15) as above, but further assuming that the $w_j(\mathbf{s})$ are mean-zero Gaussian processes. Lastly $\boldsymbol{\mu}(\mathbf{s})$ arises from $\mu_j(\mathbf{s}) = \mathbf{X}_j^T(\mathbf{s})\boldsymbol{\beta}_j$. Each component can have its own set of covariates with its own coefficient vector.

As in Section 5.1, (7.20) can be viewed as a hierarchical model. At the first stage, given $\{\boldsymbol{\beta}_j, j = 1, \cdots, p\}$ and $\{\mathbf{v}(\mathbf{s}_i)\}$, the $\mathbf{Y}(\mathbf{s}_i)$, $i = 1, \cdots, n$ are conditionally independent with $\mathbf{Y}(\mathbf{s}_i) \sim N(\boldsymbol{\mu}(\mathbf{s}_i) + \mathbf{v}(\mathbf{s}_i), \mathbf{D})$. At the second stage, the joint distribution of $\mathbf{v} \equiv (\mathbf{v}(\mathbf{s}_1), \cdots, \mathbf{v}(\mathbf{s}_n))^T$ is $N(\mathbf{0}, \sum_{j=1}^p \mathbf{H}_j \otimes \mathbf{T}_j)$, where $\mathbf{H}_j$ is $n \times n$ with $(\mathbf{H}_j)_{ii'} = \rho_j(\mathbf{s}_i - \mathbf{s}_{i'})$. Concatenating the $\mathbf{Y}(\mathbf{s}_i)$ into an $np \times 1$ vector $\mathbf{Y}$ (and similarly $\boldsymbol{\mu}(\mathbf{s}_i)$ into $\boldsymbol{\mu}$), we can marginalize over $\mathbf{v}$ to obtain

$$p(\mathbf{Y} \mid \{\boldsymbol{\beta}_j\}, \mathbf{D}, \{\rho_j\}, \mathbf{T}) = N \left(\boldsymbol{\mu} \,, \sum_{j=1}^p (\mathbf{H}_j \otimes \mathbf{T}_j) + \mathbf{I}_{n \times n} \otimes \mathbf{D} \right) \,. \tag{7.21}$$

Prior distributions on $\{\boldsymbol{\beta}_j\}$, $\{\tau_j^2\}$, $\mathbf{T}$, and the parameters of the ρ_j complete the Bayesian hierarchical model specification.

7.2.2 Unconditional and conditional Bayesian specifications

Equivalence of likelihoods

The LMC of the previous section can be developed through a conditional approach rather than a joint modeling approach. This idea has been elaborated in, e.g., Royle and Berliner (1999) and Berliner (2000), who refer to it as a hierarchical modeling approach to multivariate spatial modeling and prediction. In the context of say $\mathbf{v}(\mathbf{s}) = \mathbf{Aw}(\mathbf{s})$ where the $w_j(\mathbf{s})$ are mean-zero Gaussian processes, by taking $\mathbf{A}$ to be lower triangular the equivalence and associated reparametrization are easy to see. Upon permutation of the components of $\mathbf{v}(\mathbf{s})$ we can, without loss of generality, write $p(\mathbf{v}(\mathbf{s})) = p(v_1(\mathbf{s}))p(v_2(\mathbf{s})|v_1(\mathbf{s})) \cdots p(v_p(\mathbf{s})|v_1(\mathbf{s}), \cdots, v_{p-1}(\mathbf{s}))$. In the case of $p = 2$, $p(v_1(\mathbf{s}))$ is clearly $N(0, T_{11})$, i.e. $v_1(\mathbf{s}) = \sqrt{T_{11}}w_1(\mathbf{s}) = a_{11}w_1(\mathbf{s})$, $a_{11} > 0$. But $p(v_2(\mathbf{s})|v_1(\mathbf{s})) \sim N \left(\frac{T_{12}v_1(\mathbf{s})}{T_{11}}, T_{22} - \frac{T_{12}^2}{T_{11}} \right)$, i.e. $N \left(\frac{a_{21}}{a_{11}}v_1(\mathbf{s}), a_{22}^2 \right)$. In fact, from the previous section we have $\Sigma_{\mathbf{v}} = \sum_{j=1}^{p} \mathbf{H}_j \otimes \mathbf{T}_j$. If we permute the rows of $\mathbf{v}$ to $\tilde{\mathbf{v}} = \left(\mathbf{v}^{(1)}, \mathbf{v}^{(2)}\right)^T$, where $\mathbf{v}^{(l)} = (v_l(\mathbf{s}_1), \dots, v_l(\mathbf{s}_n))^T$ for $l = 1, 2$, then $\Sigma_{\mathbf{v}} = \sum_{j=1}^{p} \mathbf{T}_j \otimes \mathbf{H}_j$. Again with $p = 2$ we can calculate $E(\mathbf{v}^{(2)}|\mathbf{v}^{(1)}) = \frac{a_{21}}{a_{11}}\mathbf{v}^{(1)}$ and $\Sigma_{\mathbf{v}^{(2)}|\mathbf{v}^{(1)}} = a_{22}^2 \mathbf{H}_2$. But this is exactly the mean and covariance structure associated with variables $\{v_2(\mathbf{s}_i)\}$ given $\{v_1(\mathbf{s}_i)\}$, i.e., with $v_2(\mathbf{s}_i) = \frac{a_{21}}{a_{11}}v_1(\mathbf{s}_i) + a_{22}w_2(\mathbf{s}_i)$. Note that as in Subsection 7.1.1, there is no notion of a *conditional* process here. Again there is only a joint distribution for $\mathbf{v}^{(1)}, \mathbf{v}^{(2)}$ given any n and any $\mathbf{s}_1, \cdots, \mathbf{s}_n$, hence a conditional distribution for $\mathbf{v}^{(2)}$ given $\mathbf{v}^{(1)}$.

Suppose we write $v_1(\mathbf{s}) = \sigma_1 w_1(\mathbf{s})$ where $\sigma_1 > 0$ and $w_1(\mathbf{s})$ is a mean 0 spatial process with variance 1 and correlation function ρ_1 and we write $v_2(\mathbf{s})|v_1(\mathbf{s}) = \alpha v_1(\mathbf{s}) + \sigma_2 w_2(\mathbf{s})$ where $\sigma_2 > 0$ and $w_2(\mathbf{s})$ is a mean 0 spatial process with variance 1 and correlation function ρ_2. The parametrization $(\alpha, \sigma_1, \sigma_2)$ is obviously equivalent to (a_{11}, a_{12}, a_{22}), i.e., $a_{11} = \sigma_1$, $a_{21} = \alpha \sigma_1$, $a_{22} = \sigma_2$ and hence to $\mathbf{T}$, i.e., to (T_{11}, T_{12}, T_{22}), that is, $T_{11} = \sigma_1^2$, $T_{12} = \alpha \sigma_1^2$, $T_{22} = \alpha^2 \sigma_1^2 + \sigma_2^2$.

Extension to general p is straightforward but notationally messy. We record the transformations for $p = 3$ for future use. First, $v_1(\mathbf{s}) = \sigma_1 w_1(\mathbf{s})$, $v_2(\mathbf{s})|v_1(\mathbf{s}) = \alpha^{(2|1)}v_1(\mathbf{s}) + \sigma_2 w_2(\mathbf{s})$ and $v_3(\mathbf{s})|v_1(\mathbf{s}), v_2(\mathbf{s}) = \alpha^{(3|1)}v_1(\mathbf{s}) + \alpha^{(3|2)}v_2(\mathbf{s}) + \sigma_3 w_3(\mathbf{s})$. Then $a_{11} = \sigma_1$, $a_{21} = \alpha^{(2|1)}\sigma_1$, $a_{22} = \sigma_2$, $a_{31} = \alpha^{(3|1)}\sigma_1$, $a_{32} = \alpha^{(3|2)}\sigma_2$ and $a_{33} = \sigma_3$. But also $a_{11} = \sqrt{T_{11}}$, $a_{21} = \frac{T_{12}}{\sqrt{T_{11}}}$, $a_{22} = \sqrt{T_{22} - \frac{T_{12}^2}{T_{11}}}$, $a_{31} = \frac{T_{13}}{\sqrt{T_{11}}}$, $a_{32} = \sqrt{\frac{T_{11}T_{23} - T_{12}T_{13}}{T_{11}(T_{11}T_{22} - T_{12}^2)}}$, and $a_{33} = \sqrt{T_{33} - \frac{T_{13}^2}{T_{11}} - \frac{(T_{11}T_{23} - T_{12}T_{13})^2}{T_{11}(T_{11}T_{12} - T_{12}^2)}}$.

Advantages to working with the conditional form of the model are certainly computational and possibly mechanistic or interpretive. For the former, with the "σ, α" parametrization, the likelihood factors and thus, with a matching prior factorization, models can be fitted componentwise. Rather than the $pn \times pn$ covariance matrix involved in working with $\mathbf{v}$ we obtain p covariance matrices each of dimension $n \times n$, one for $\mathbf{v}^{(1)}$, one for $\mathbf{v}^{(2)}|\mathbf{v}^{(1)}$, etc. Since likelihood evaluation with spatial processes is more than an order n^2 calculation, there can be substantial computational savings in using the conditional model. If there is some natural chronology or perhaps causality in events, then this would determine a natural order for conditioning and hence suggest natural conditional specifications. For example, in the illustrative commercial real estate setting of Example 7.3, we have the income (I) generated by an apartment block and the selling price (P) for the block. A natural modeling order here is I, then P given I.

Equivalence of prior specifications

Working in a Bayesian context, it is appropriate to ask about choice of parametrization with regard to prior specification. Suppose we let ϕ_j be the parameters associated with the correlation function ρ_j. Let $\boldsymbol{\phi}^T = (\boldsymbol{\phi}_1, \cdots, \boldsymbol{\phi}_p)$. Then the distribution of $\mathbf{v}$ depends upon $\mathbf{T}$ and $\boldsymbol{\phi}$. Suppose we assume *a priori* that $p(\mathbf{T}, \boldsymbol{\phi}) = p(\mathbf{T})p(\boldsymbol{\phi}) = p(\mathbf{T}) \prod_j p(\boldsymbol{\phi}_j)$. Then reparametrization, using obvious notation, to the $(\boldsymbol{\sigma}, \boldsymbol{\alpha})$ space results on a prior $p(\boldsymbol{\sigma}, \boldsymbol{\alpha}, \boldsymbol{\phi}) = p(\boldsymbol{\sigma}, \boldsymbol{\alpha}) \prod_j p(\boldsymbol{\phi}_j)$.

Standard prior specification for $\mathbf{T}$ would of course be an inverse Wishart, while standard modeling for $(\sigma^2, \boldsymbol{\alpha})$ would be a product inverse gamma by normal form. In the present situation, when will they agree? We present the details for the $p = 2$ case. The Jacobian from $\mathbf{T} \to (\sigma_1, \sigma_2, \alpha)$ is $|\mathbf{J}| = \sigma_1^2$, hence in the reverse direction it is $1/T_{11}$. Also $|\mathbf{T}| = T_{11}T_{22} - T_{12}^2 = \sigma_1^2 \sigma_2^2$ and

$$
\mathbf{T}^{-1} = \frac{1}{T_{11}T_{22} - T_{12}^2} \begin{pmatrix} T_{22} & -T_{12} \\ -T_{12} & T_{11} \end{pmatrix} = \frac{1}{\sigma_1^2 \sigma_2^2} \begin{pmatrix} \alpha^2 \sigma_1^2 + \sigma_2^2 & -\alpha \sigma_1^2 \\ -\alpha \sigma_1^2 & \sigma_1^2 \end{pmatrix}.
$$

After some manipulation we have the following result:

Result 1: $\mathbf{T} \sim IW_2(\nu, (\nu'\mathbf{D})^{-1})$; that is,

$$
p(\mathbf{T}) \propto |\mathbf{T}|^{-\frac{\nu+3}{2}} \exp\left\{ -\frac{1}{2}tr(\nu'\mathbf{D}\mathbf{T}^{-1}) \right\},
$$

where $\mathbf{D} = Diag(d_1, d_2)$ and $\nu' = \nu - 3$ if and only if

$$
\sigma_1^2 \sim IG\left(\frac{\nu-1}{2}, \frac{d_1}{2}\right), \ \sigma_2^2 \sim IG\left(\frac{\nu+1}{2}, \frac{d_2}{2}\right), \ \text{and} \ \alpha|\sigma_2^2 \sim N\left(0, \frac{\sigma_2^2}{d_1}\right).
$$

Note also that the prior in $(\boldsymbol{\sigma}, \boldsymbol{\alpha})$ space factors into $p(\sigma_1^2)p(\sigma_2^2, \alpha)$ to match the likelihood factorization.

This result is obviously order dependent. If we condition in the reverse order, σ_1^2, σ_2^2, and α no longer have the same meanings. In fact, writing this parametrization as $(\tilde{\sigma}_1^2, \tilde{\sigma}_2^2, \tilde{\alpha})$, we obtain equivalence to the above inverse Wishart prior for $\mathbf{T}$ if and only if $\tilde{\sigma}_1^2 \sim IG\left(\frac{\nu+1}{2}, \frac{d_1}{2}\right)$, $\tilde{\sigma}_2^2 \sim IG\left(\frac{\nu-1}{2}, \frac{d_2}{2}\right)$, and $\tilde{\alpha}|\tilde{\sigma}_1^2 \sim N\left(0, \frac{\tilde{\sigma}_1^2}{d_1}\right)$.

The result can be extended to $p > 2$ but the expressions become messy. However, if $p = 3$ we have:

Result 2: $\mathbf{T} \sim IW_3(\nu, (\nu'\mathbf{D})^{-1})$, that is,

$$
p(\mathbf{T}) \propto |\mathbf{T}|^{-\frac{\nu+4}{2}} \exp\left\{ -\frac{1}{2}tr(\nu'\mathbf{D}\mathbf{T}^{-1}) \right\},
$$

where now where $\mathbf{D} = Diag(d_1, d_2, d_3)$ and $\nu' = \nu - 3 + 1$ if and only if $\sigma_1^2 \sim IG\left(\frac{\nu-2}{2}, \frac{d_1}{2}\right)$, $\sigma_2^2 \sim IG\left(\frac{\nu}{2}, \frac{d_2}{2}\right)$, $\sigma_3^2 \sim IG\left(\frac{\nu+2}{2}, \frac{d_3}{2}\right)$, $\alpha^{(2|1)}|\sigma_2^2 \sim N\left(0, \frac{\sigma_2^2}{d_1}\right)$, $\alpha^{(3|1)}|\sigma_3^2 \sim N\left(0, \frac{\sigma_3^2}{d_1}\right)$, and $\alpha^{(3|2)}|\sigma_3^2 \sim N\left(0, \frac{\sigma_3^2}{d_2}\right)$. Though there is a one-to-one transformation from $\mathbf{T}$-space to $(\boldsymbol{\sigma}, \boldsymbol{\alpha})$-space, a Wishart prior with nondiagonal $\mathbf{D}$ implies a nonstandard prior on $(\boldsymbol{\sigma}, \boldsymbol{\alpha})$-space. Moreover, it implies that the prior in $(\boldsymbol{\sigma}, \boldsymbol{\alpha})$-space will not factor to match the likelihood factorization.

Returning to the model in (7.20), the presence of white noise in (7.20) causes difficulties with the attractive factorization of the likelihood under conditioning. Consider again the $p = 2$ case. If

$$
\begin{aligned}
Y_1(\mathbf{s}) &= \mathbf{X}_1^T(\mathbf{s})\boldsymbol{\beta}_1 + v_1(\mathbf{s}) + \epsilon_1(\mathbf{s}) \\
\text{and } Y_2(\mathbf{s}) &= \mathbf{X}_2^T(\mathbf{s})\boldsymbol{\beta}_1 + v_2(\mathbf{s}) + \epsilon_2(\mathbf{s}) ,
\end{aligned} \tag{7.22}
$$

then the conditional form of the model writes

$$
\begin{aligned}
Y_1(\mathbf{s}) &= \mathbf{X}_1^T(\mathbf{s})\boldsymbol{\beta}_1 + \sigma_1 w_1(\mathbf{s}) + \tau_1 u_1(\mathbf{s}) \\
\text{and } Y_2(\mathbf{s})|Y_1(\mathbf{s}) &= \mathbf{X}_2^T(\mathbf{s})\boldsymbol{\beta}_2 + \alpha Y_1(\mathbf{s}) + \sigma_2 w_2(\mathbf{s}) + \tau_2 u_2(\mathbf{s}) .
\end{aligned} \tag{7.23}
$$

In (7.23), $w_1(\mathbf{s})$ and $w_2(\mathbf{s})$ are as above with $u_1(\mathbf{s})$, $u_2(\mathbf{s}) \sim N(0,1)$, independent of each other and the $w_l(\mathbf{s})$. But then, unconditionally, $Y_2(\mathbf{s})$ equals

$$
\begin{aligned}
&\mathbf{X}_2^T(\mathbf{s})\tilde{\boldsymbol{\beta}}_2 + \alpha \left(\mathbf{X}_1^T(\mathbf{s})\boldsymbol{\beta}_1 + \sigma_1 w_1(\mathbf{s}) + \tau_1 u_1(\mathbf{s}) \right) + \sigma_2 w_2(\mathbf{s}) + \tau_2 u_2(\mathbf{s}) \\
&= \mathbf{X}_2^T(\mathbf{s})\tilde{\boldsymbol{\beta}}_2 + \mathbf{X}_1^T(\mathbf{s})\alpha\boldsymbol{\beta}_1 + \alpha\sigma_1 w_1(\mathbf{s}) + \sigma_2 w_2(\mathbf{s}) + \alpha\tau_1 u_1(\mathbf{s})) + \tau_2 u_2(\mathbf{s}) .
\end{aligned} \tag{7.24}
$$

In attempting to align (7.24) with (7.33) we require $\mathbf{X}_2(\mathbf{s}) = \mathbf{X}_1(\mathbf{s})$, whence $\boldsymbol{\beta}_2 = \tilde{\boldsymbol{\beta}}_2 + \alpha\boldsymbol{\beta}_1$. We also see that $v_2(\mathbf{s}) = \alpha\sigma_1 w_1(\mathbf{s}) + \sigma_2 w_2(\mathbf{s})$. But, perhaps most importantly, $\epsilon_2(\mathbf{s}) = \alpha\tau_1 u_1(\mathbf{s}) + \tau_2 u_2(\mathbf{s})$. Hence $\epsilon_1(\mathbf{s})$ and $\epsilon_2(\mathbf{s})$ are not independent, violating the white noise modeling assumption associated with (7.33). If we have a white noise component in the model for $Y_1(\mathbf{s})$ and also in the conditional model for $Y_2(\mathbf{s})|Y_1(\mathbf{s})$ we do not have a white noise component in the unconditional model specification. Obviously, the converse is true as well.

If $u_1(\mathbf{s}) = 0$, i.e., the $Y_1(\mathbf{s})$ process is purely spatial, then, again with $\mathbf{X}_2(\mathbf{s}) = \mathbf{X}_1(\mathbf{s})$, the conditional and marginal specifications agree up to reparametrization. More precisely, the parameters for the unconditional model are $\boldsymbol{\beta}_1$, $\boldsymbol{\beta}_2$, τ_2^2 with T_{11}, T_{12}, T_{22}, ϕ_1, and ϕ_2. For the conditional model we have $\boldsymbol{\beta}_1$, $\boldsymbol{\beta}_2$, τ_2^2 with σ_1, σ_2, α, ϕ_1, and ϕ_2. We can appeal to the equivalence of (T_{11}, T_{12}, T_{22}) and $(\sigma_1, \sigma_2, \alpha)$ as above. Also note that if we extend (7.33) to $p > 2$, in order to enable conditional and marginal specifications to agree, we will require a common covariate vector and that $u_1(\mathbf{s}) = u_2(\mathbf{s}) = \cdots = u_{p-1}(\mathbf{s}) = 0$, i.e., that all but one of the processes is purely spatial.

7.2.3 Spatially varying coregionalization models

A possible extension of the LMC would replace $\mathbf{A}$ by $\mathbf{A}(\mathbf{s})$ and thus define

$$
\mathbf{Y}(\mathbf{s}) = \mathbf{A}(\mathbf{s})\mathbf{w}(\mathbf{s}) . \tag{7.25}
$$

We refer to the model in (7.25) as a *spatially varying LMC*. Following the notation in Section 7.2.1, let $\mathbf{T}(\mathbf{s}) = \mathbf{A}(\mathbf{s})\mathbf{A}(\mathbf{s})^T$. Again $\mathbf{A}(\mathbf{s})$ can be taken to be lower triangular for convenience. Now $C(\mathbf{s}, \mathbf{s}')$ is such that

$$C(\mathbf{s}, \mathbf{s}') = \sum \rho_j(\mathbf{s} - \mathbf{s}')\mathbf{a}_j(\mathbf{s})\mathbf{a}_j(\mathbf{s}') , \qquad (7.26)$$

with $\mathbf{a}_j(\mathbf{s})$ the jth column of $\mathbf{A}(\mathbf{s})$. Letting $\mathbf{T}_j(\mathbf{s}) = \mathbf{a}_j(\mathbf{s})\mathbf{a}_j^T(\mathbf{s})$, again, $\sum \mathbf{T}_j(\mathbf{s}) = \mathbf{T}(\mathbf{s})$. We see from (7.26) that $\mathbf{Y}(\mathbf{s})$ is no longer stationary. Extending the intrinsic specification for $\mathbf{Y}(\mathbf{s})$, $C(\mathbf{s}, \mathbf{s}') = \rho(\mathbf{s} - \mathbf{s}')\mathbf{T}(\mathbf{s})$, which is a multivariate version of the case of a spatial process with a spatially varying variance.

This motivates natural definition of $\mathbf{A}(\mathbf{s})$ through its one-to-one correspondence with $\mathbf{T}(\mathbf{s})$ (again from Section 7.2.1) since $\mathbf{T}(\mathbf{s})$ is the covariance matrix for $\mathbf{Y}(\mathbf{s})$. In the univariate case choices for $\sigma^2(\mathbf{s})$ include $\sigma^2(\mathbf{s}, \theta)$, i.e., a parametric function of location; $\sigma^2(x(\mathbf{s})) = g(x(\mathbf{s}))\sigma^2$ where $x(\mathbf{s})$ is some covariate used to explain $\mathbf{Y}(\mathbf{s})$ and $g(\cdot) > 0$ (then $g(x(\mathbf{s}))$ is typically $x(\mathbf{s})$ or $x^2(\mathbf{s})$); or $\sigma^2(\mathbf{s})$ is itself a spatial process (e.g., $\log \sigma^2(\mathbf{s})$ might be a Gaussian process). In practice, $\mathbf{T}(\mathbf{s}) = g(x(\mathbf{s}))\mathbf{T}$ will likely be easiest to work with.

Note that all of the discussion in Section 7.2.2 regarding the relationship between conditional and unconditional specifications is applicable here. Particularly, if $p = 2$ and $\mathbf{T}(\mathbf{s}) = g(x(\mathbf{s}))\mathbf{T}$ then (T_{11}, T_{12}, T_{22}) is equivalent to $(\sigma_1, \sigma_2, \alpha)$, and we have $a_{11}(\mathbf{s}) = \sqrt{g(x(\mathbf{s})}\sigma_1$, $a_{22}(\mathbf{s}) = \sqrt{g(x(\mathbf{s})}\sigma_2$, and $a_{21} = \sqrt{g(x(\mathbf{s})}\alpha\sigma_1$.

7.2.4 Model-fitting issues

This subsection starts by discussing the computational issues in fitting the joint multivariate model presented in Subsection 7.2.1. It will be shown that it is a challenging task to fit this joint model. On the other hand, making use of the equivalence of the joint and conditional models, as discussed in Section 7.2.2, we demonstrate that it is much simpler to fit the latter.

Fitting the joint model

Different from previous approaches that have employed the coregionalization model, our intent is to follow the Bayesian paradigm. For this purpose, the model specification is complete only after assigning prior distributions to all unknown quantities in the model. The posterior distribution of the set of parameters is obtained after combining the information about them in the likelihood (see equation (7.21)) with their prior distributions.

Observing equation (7.21), we see that the parameter vector defined as $\boldsymbol{\theta}$ consists of $\{\boldsymbol{\beta}_j\}, \mathbf{D}, \{\rho_j\}, \mathbf{T}, j = 1, \cdots, p$. Adopting a prior that assumes independence across j we take $p(\boldsymbol{\theta}) = \prod_j p(\boldsymbol{\beta}_j) p(\rho_j) p(\tau_j^2) p(\mathbf{T})$. Hence $p(\boldsymbol{\theta}|\mathbf{y})$ is given by

$$p(\boldsymbol{\theta}|\mathbf{y}) \propto p(\mathbf{y} \mid \{\boldsymbol{\beta}_j\}, \mathbf{D}, \{\rho_j\}, \mathbf{T}) \, p(\boldsymbol{\theta}) .$$

For the elements of $\boldsymbol{\beta}_j$, a normal mean-zero prior distribution with large variance can be assigned, resulting in a full conditional distribution that will also be normal. Inverse gamma distributions can be assigned to the elements of $\mathbf{D}$, the variances of the p white noise processes. If there is no information about such variances, the means of these inverse gammas could be based on the least squares estimates of the independent models with large variances. Assigning inverse gamma distributions to τ_j^2 will result in inverse gamma full conditionals. The parameters of concern are the elements of ρ_j and $\mathbf{T}$. Regardless of what prior distributions we assign, the full conditional distributions will not have a standard form. For example, if we assume that ρ_j is the exponential correlation function, $\rho_j(h) = \exp(-\phi_j h)$, a gamma prior distribution can be assigned to the ϕ_j's. In order to obtain samples of the ϕ_j's we can use the Metropolis-Hastings algorithm with, for instance, log-normal proposals centered at the current $\log \phi_j$.

We now consider how to sample $\mathbf{T}$, the covariance matrix among the responses at each location $\mathbf{s}$. Due to the one-to-one relationship between $\mathbf{T}$ and the lower triangular $\mathbf{A}$, one can either assign a prior to the elements of $\mathbf{A}$, or set a prior on the matrix $\mathbf{T}$. The latter seems to be more natural, since $\mathbf{T}$ is interpreted as the covariance matrix of the elements of $\mathbf{Y}(\mathbf{s})$. As $\mathbf{T}$ must be positive definite, we use an inverse Wishart prior distribution with ν degrees of freedom and mean $\mathbf{D}^*$, i.e., the scale matrix is $(\nu - p - 1)(\mathbf{D}^*)^{-1}$. If there is no information about the prior mean structure of $\mathbf{T}$, rough estimates of the elements of the diagonal of $\mathbf{D}^*$ can be obtained using ordinary least squares estimates based on the independent spatial models for each $Y_j(\mathbf{s})$, $j = 1, \cdots, p$. A small value of $\nu(> p + 1)$ would be assigned to provide high uncertainty in the resulting prior distribution.

To sample from the full conditional for $\mathbf{T}$, Metropolis-Hastings updates are a place to start. In our experience, random walk Wishart proposals do not work well, and importance sampled Wishart proposals have also proven problematic. Instead, we recommend updating the elements of $\mathbf{T}$ individually. In fact, it is easier to work in the unconstrained space of the components of $\mathbf{A}$, so we would reparametrize the full conditional from $\mathbf{T}$ to $\mathbf{A}$. Random walk normal proposals for the $\mathbf{a}$'s with suitably tuned variances will mix well, at least for $p = 2$ or 3. For larger p, repeated decomposition of $\mathbf{T}$ to $\mathbf{A}$ may prove too costly.

Fitting the conditional model

Section 7.2.2 showed the equivalence of conditional and unconditional specifications in terms of $\mathbf{v}(\mathbf{s})$. Here we write the multivariate model for $\mathbf{Y}(\mathbf{s})$ in its conditional parametrization and see that the inference procedure is simpler than for the multivariate parametrization. Following the discussion in Section 7.2.2, for a general p, the conditional parametrization is

$$Y_1(\mathbf{s}) \quad = \quad \mathbf{X}_1^T(\mathbf{s})\boldsymbol{\beta}_1 + \sigma_1 w_1(\mathbf{s})$$

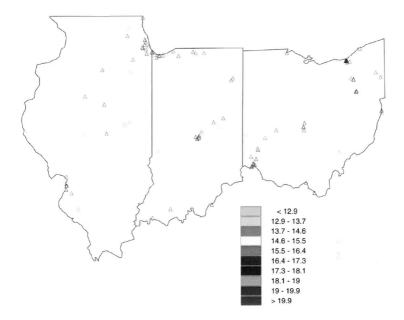

	< 12.9
	12.9 - 13.7
	13.7 - 14.6
	14.6 - 15.5
	15.5 - 16.4
	16.4 - 17.3
	17.3 - 18.1
	18.1 - 19
	19 - 19.9
	> 19.9

COLOR FIGURE C.1 Map of PM2.5 sampling sites over three midwestern U.S. states; color indicates range of average monitored PM2.5 level over the year 2001 (see figure legend).

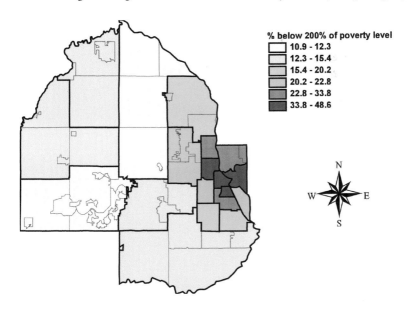

% below 200% of poverty level
	10.9 - 12.3
	12.3 - 15.4
	15.4 - 20.2
	20.2 - 22.8
	22.8 - 33.8
	33.8 - 48.6

COLOR FIGURE C.2 ArcView map of percent of surveyed population with household income below 200% of the federal poverty limit, regional survey units in Hennepin County, MN.

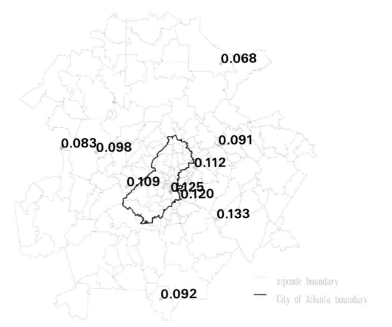

COLOR FIGURE C.3 Zip code boundaries in the Atlanta metropolitan area and 8-hour maximum ozone levels (ppm) at 10 monitoring sites for July 15, 1995.

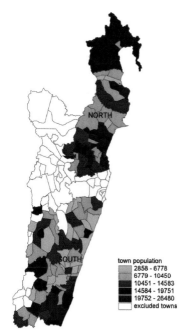

COLOR FIGURE C.4 Northern and southern regions within the Madagascar study region, with population overlaid.

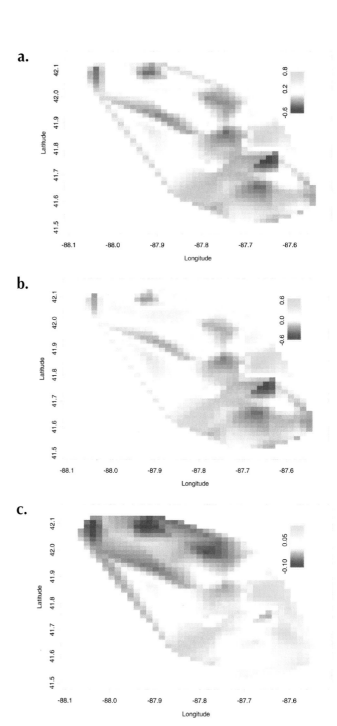

COLOR FIGURE C.5 Image plots of the spatial processes of (a) net income, (b) price, and (c) risk.

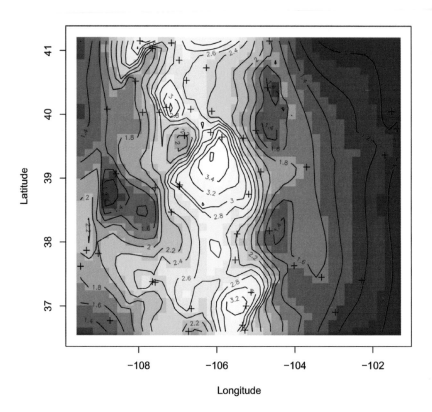

COLOR FIGURE C.6 Map of the region in Colorado that forms the spatial domain. The data for the illustrations come from 50 locations, marked by "+" signs in this region.

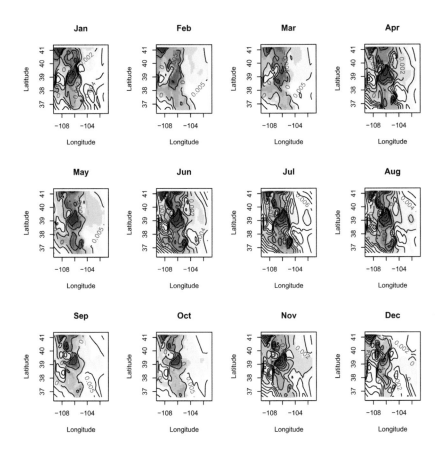

COLOR FIGURE C.7 Time-sliced image-contour plots displaying the posterior mean surface of the spatial residuals corresponding to the slope process in the temperature given precipitation model.

I.I.D. model (with covariates)

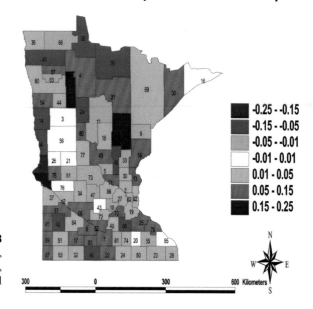

■	-0.25 - -0.15
■	-0.15 - -0.05
■	-0.05 - -0.01
□	-0.01 - 0.01
■	0.01 - 0.05
■	0.05 - 0.15
■	0.15 - 0.25

COLOR FIGURE C.8
Posterior median frailties,
i.i.d. model with covariates,
Minnesota county-level
infant mortality data.

CAR model (with covariates)

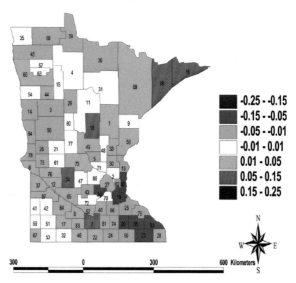

■	-0.25 - -0.15
■	-0.15 - -0.05
■	-0.05 - -0.01
□	-0.01 - 0.01
■	0.01 - 0.05
■	0.05 - 0.15
■	0.15 - 0.25

COLOR FIGURE C.9
Posterior median frailties,
CAR model with covariates,
Minnesota county-level
infant mortality data.

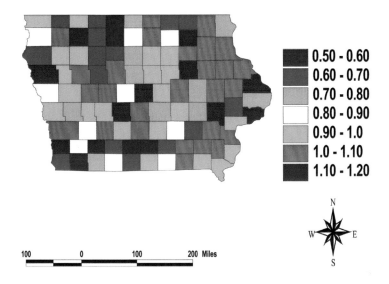

COLOR FIGURE C.10 Fitted spatiotemporal frailties, Iowa counties, 1986.

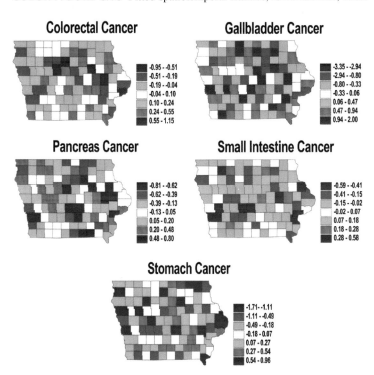

COLOR FIGURE C.11 Posterior mean spatial frailties, Iowa cancer data, static spatial MCAR model.

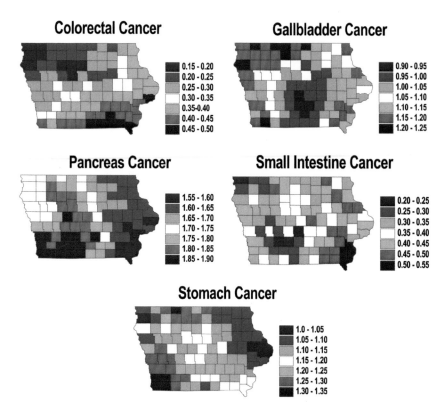

COLOR FIGURE C.12 Posterior mean spatially varying coefficients, Iowa cancer data, static spatial MCAR model.

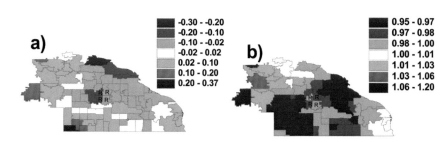

COLOR FIGURE C.13 Maps of posterior means for the ϕ_i (a) and the ρ_i (b) in the full spatial MCAR model, assuming the data to be interval-censored.

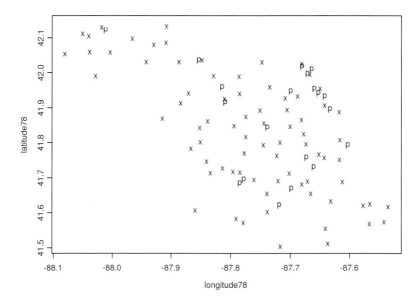

Figure 7.6 *Locations of the 78 sites* (x) *used to fit the (price, income) model, and the 20 sites used for prediction* (p).

$$Y_2(\mathbf{s}) \mid Y_1(\mathbf{s}) = \mathbf{X}_2^T(\mathbf{s})\boldsymbol{\beta}_2 + \alpha^{2|1}Y_1(\mathbf{s}) + \sigma_2 w_2(\mathbf{s})$$

$$\vdots \qquad (7.27)$$

$$Y_p(\mathbf{s}) \mid Y_1(\mathbf{s}), \cdots, Y_p(\mathbf{s}) = \mathbf{X}_p^T(\mathbf{s})\boldsymbol{\beta}_p + \alpha^{p|1}Y_1(\mathbf{s})$$

$$+ \cdots + \alpha^{p|p-1}Y_{p-1}(\mathbf{s}) + \sigma_p w_p(\mathbf{s}) .$$

In (7.27), the set of parameters to be estimated is $\boldsymbol{\theta}_c = \{\boldsymbol{\beta}, \boldsymbol{\alpha}, \boldsymbol{\sigma}^2, \boldsymbol{\phi}\}$, where $\boldsymbol{\alpha}^T = (\alpha^{2|1}, \alpha^{3|1}, \alpha^{3|2}, \cdots, \alpha^{p|p-1})$, $\boldsymbol{\beta}^T = (\boldsymbol{\beta}_1, \cdots, \boldsymbol{\beta}_p)$, $\boldsymbol{\sigma}^2 = (\sigma_1^2, \cdots, \sigma_p^2)$, and $\boldsymbol{\phi}$ is as defined in Subsection 7.2.2. The likelihood is given by

$$f_c(\mathbf{Y}|\boldsymbol{\theta}_c) = f(\mathbf{Y}_1|\boldsymbol{\theta}_{c_1})\, f(\mathbf{Y}_2|\mathbf{Y}_1, \boldsymbol{\theta}_{c_2}) \cdots f(\mathbf{Y}_p|\mathbf{Y}_1, \cdots, \mathbf{Y}_{p-1}, \boldsymbol{\theta}_{c_p}) .$$

If $\pi(\boldsymbol{\theta}_c)$ is taken to be $\prod_{j=1}^{p} \pi(\boldsymbol{\theta}_{c_j})$ then this equation implies that the conditioning yields a factorization into p models each of which can be fitted separately. Prior specification of the parameters was discussed in Subsection 7.2.2. With those forms, standard univariate spatial models that can be fit using the GeoBUGS package arise.

Example 7.3 *(Commercial real estate example).* The selling price of commercial real estate, for example an apartment property, is theoretically the expected income capitalized at some (risk-adjusted) discount rate. (See Kinnard, 1971, and Lusht, 1997, for general discussions of the basics of commercial property valuation theory and practice.) Here we consider a

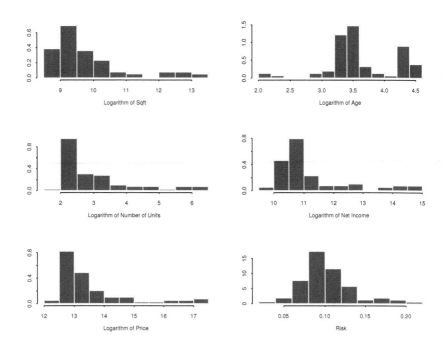

Figure 7.7 *Histograms of the logarithm of the variables.*

data set consisting of 78 apartment buildings, with 20 additional transactions held out for prediction of the selling price based on four different
models. The locations of these buildings are shown in Figure 7.6. The aim
here is to fit a joint model for selling price and net income and obtain a
spatial surface associated with the risk, which, for any transaction, is given
by net income/price. For this purpose we fit a model using the following covariates: average square feet of a unit within the building (sqft), the age of
the building (age), the number of units within the building (unit), the selling price of the transaction (P), and the net income (I). Figure 7.7 shows
the histograms of these variables on the log scale. Using the conditional
parametrization, the model is

$$
\begin{aligned}
I(\mathbf{s}) &= sqft(\mathbf{s})\beta_{I1} + age(\mathbf{s})\beta_{I2} + unit(\mathbf{s})\beta_{I3} + \sigma_1 w_1(\mathbf{s}) \\
P(\mathbf{s})|I(\mathbf{s}) &= sqft(\mathbf{s})\beta_{P1} + age(\mathbf{s})\beta_{P2} + unit(\mathbf{s})\beta_{P3} \qquad (7.28)\\
&\quad + I(\mathbf{s})\alpha^{(2|1)} + \sigma_2 w_2(\mathbf{s}) + \epsilon(\mathbf{s}) \ .
\end{aligned}
$$

Notice that $I(\mathbf{s})$ is considered to be purely spatial since, adjusted for building characteristics, we do not anticipate a microscale variability component.

Parameter	Mean	2.50%	Median	97.50%
β_{I1}	0.156	-0.071	0.156	0.385
β_{I2}	-0.088	-0.169	-0.088	-0.008
β_{I3}	0.806	0.589	0.804	1.014
β_{P1}	0.225	0.010	0.229	0.439
β_{P2}	-0.092	-0.154	-0.091	-0.026
β_{P3}	-0.150	-0.389	-0.150	0.093
$\alpha^{(2|1)}$	0.858	0.648	0.856	1.064
σ_1^2	0.508	0.190	0.431	1.363
σ_2^2	0.017	0.006	0.014	0.045
τ_2^2	0.051	0.036	0.051	0.071
ϕ_I	3.762	1.269	3.510	7.497
ϕ_P	1.207	0.161	1.072	3.201
range$_I$	0.969	0.429	0.834	2.291
range$_P$	1.2383	0.554	1.064	2.937
corr(I, P)	0.971	0.912	0.979	0.995
T_{II}	0.508	0.190	0.431	1.363
T_{IP}	0.435	0.158	0.369	1.136
T_{PP}	0.396	0.137	0.340	1.000

Table 7.3 *Posterior summaries, joint model of price and income.*

The need for white noise in the price component results from the fact that two identical properties at essentially the same location need not sell for the same price due to the motivation of the seller, the buyer, the brokerage process, etc. (If a white noise component for $I(\mathbf{s})$ were desired, we would fit the joint model as described near the beginning of Subsection 7.2.4.) The model in (7.28) is in accordance with the conditional parametrization in Subsection 7.2.2. The prior distributions were assigned as follows. For all the coefficients of the covariates, including $\alpha^{(2|1)}$, we assigned a normal 0 mean distribution with large variance. For σ_1^2 and σ_2^2 we used inverse gammas with infinite variance. We use exponential correlation functions and the decay parameters ϕ_j, $j = 1, 2$ have a gamma prior distribution arising from a mean range of one half the maximum interlocation distance, with infinite variance. Finally, τ_2^2, the variance of $\epsilon(\cdot)$, has an inverse gamma prior centered at the ordinary least squares variance estimate obtained from an independent model for log selling price given log net income.

Table 7.3 presents the posterior summaries of the parameters of the model. For the income model the age coefficient is significantly negative, the coefficient for number of units is significantly positive. Notice further that the correlation between net income and price is very close to 1. Nevertheless,

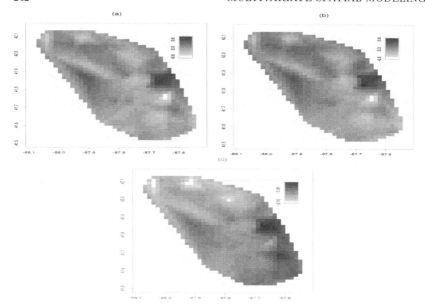

Figure 7.8 *Image plots of the spatial processes of (a) net income, (b) price, and (c) risk; see also color insert.*

for the conditional price model age is still significant. Also we see that price shows a bigger range than net income. Figure 7.8 (see also color insert Figure C.5) shows the spatial surfaces associated with the three processes: net income, price, and risk. It is straightforward to show that the logarithm of the spatial surface for risk is obtained through $(1-\alpha^{(2|1)})\sigma_1 w_1(\mathbf{s})-\sigma_2 w_2(\mathbf{s})$. Therefore, based on the posterior samples of $\alpha^{(2|1)}$, $w_1(\mathbf{s})$, σ_1, σ_2, and $w_2(\mathbf{s})$ we are able to obtain samples for the spatial surface for risk. From Figure 7.8(c), we note that the spatial risk surface tends to have smaller values than the other surfaces. Since $\log R(\mathbf{s}) = \log I(\mathbf{s}) - \log P(\mathbf{s})$ with $R(\mathbf{s})$ denoting the risk at location $\mathbf{s}$, the strong association between $I(\mathbf{s})$ and $P(\mathbf{s})$ appears to result in some cancellation of spatial effect for log risk. Actually, we can obtain the posterior distribution of the variance of the spatial process for $\log R(\mathbf{s})$. It is $(1-\alpha^{(2|1)})^2\sigma_1^2+\sigma_2^2$. The posterior mean of this variance is 0.036 and the 95% credible interval is given by $(0.0087, 0.1076)$ with median equal 0.028. The posterior variance of the noise term is given by τ_2^2, which is in Table 7.3. If we compare the medians of the posteriors of the variance of the spatial process of the risk and the variance of the white noise, we see that the spatial process presents a smaller variance; the variability of the risk process is being more explained by the residual component.

In order to examine the comparative performance of the model proposed above we decided to run four different models for the selling price using each

Model	$\sum_{j=1}^{20} e_j^2$	$\sum_{j=1}^{20} Var(P(\mathbf{s}_j)|\mathbf{y})$
Independent, nonspatial	2.279	3.277
Independent, spatial	1.808	2.963
Conditional, nonspatial	0.932	1.772
Conditional, spatial	0.772	1.731

Table 7.4 *Squared error and sum of the variances of the predictions for the 20 sites left out in the fitting of the model.*

one to predict at the locations marked with p in Figure 7.6. For all these models we used the same covariates as described before. Model 1 comprises an independent model for price, i.e., without a spatial component or net income. Model 2 has a spatial component and is not conditioned on net income. In Model 3 the selling price is conditioned on the net income but without a spatial component, and Model 4 has net income as a covariate and also a spatial component. Table 7.4 shows both $\sum_{j=1}^{20} e_j^2$, where $e_j = P(\mathbf{s}_j) - E(P(\mathbf{s}_j)|\mathbf{y}, \text{model})$ and $P(\mathbf{s}_j)$ is the observed log selling price for the jth transaction, and $\sum_{j=1}^{20} Var(P(\mathbf{s}_j)|\mathbf{y}, \text{model})$. Recall from equation (4.13) that the former is a measurement of predictive goodness of fit, while the latter is a measure of predictive variability. It is clear from the table that the model conditioned on net income and with a spatial component is best, both in terms of fit and predictive variability. ∎

7.3 Other constructive approaches ⋆

Here we consider two additional constructive strategies for building valid cross-covariance functions. The first is referred to as a moving average approach in Ver Hoef and Barry (1998). It is a multivariate version of the kernel convolution development of Subsection 5.3.2 that convolves process variables to produce a new process. The second approach convolves valid covariance functions to produce a valid cross-covariance function.

For the first approach, expressions (5.23) and (5.24) suggest several ways to achieve multivariate extension. Again with $\mathbf{Y}(\mathbf{s}) = (Y_1(\mathbf{s}), \ldots, Y_p(\mathbf{s}))^T$, define

$$Y_\ell(\mathbf{s}) = \int_{\Re^2} k_\ell(\mathbf{u}) Z(\mathbf{s} + \mathbf{u}) d\mathbf{u} \,, \quad \ell = 1, \ldots, p. \qquad (7.29)$$

In this expression, $Z(\cdot)$ is a mean 0 stationary process with correlation function $\rho(\cdot)$, and k_ℓ is a kernel associated with the ℓth component of $\mathbf{Y}(\mathbf{s})$. In practice, $k_\ell(\mathbf{u})$ would be parametric, i.e., $k_\ell(\mathbf{u}; \boldsymbol{\theta}_\ell)$. The resulting cross-covariance matrix for $\mathbf{Y}(\mathbf{s})$ has entries

$$C_{\ell,\ell'}(\mathbf{s},\mathbf{s}') = \sigma^2 \int_{\Re^2} \int_{\Re^2} k_\ell(\mathbf{s}-\mathbf{s}'+\mathbf{u})k_{\ell'}(\mathbf{u}')\rho(\mathbf{u}-\mathbf{u}')d\mathbf{u}d\mathbf{u}' . \qquad (7.30)$$

This cross-covariance matrix is necessarily valid. It is stationary and, as may be easily verified, is symmetric, i.e. $cov(Y_\ell(\mathbf{s}), Y_{\ell'}(\mathbf{s}')) = C_{\ell\ell'}(\mathbf{s}-\mathbf{s}') = C_{\ell'\ell}(\mathbf{s}-\mathbf{s}') = cov(Y_{\ell'}(\mathbf{s}), Y_\ell(\mathbf{s}'))$. Since the integration in (7.30) will not be possible to do explicitly except in certain special cases, finite sum approximation of (7.29), analogous to (5.29), is an alternative.

An alternative extension to (5.24) introduces *lags* $\mathbf{h}_\ell$, defining

$$Y_\ell(\mathbf{s}) = \int_{\Re^2} k(\mathbf{u})Z(\mathbf{s}+\mathbf{h}_\ell+\mathbf{u})d\mathbf{u} , \; \ell = 1,\ldots,p.$$

Now

$$C_{\ell,\ell'}(\mathbf{s},\mathbf{s}') = \sigma^2 \int_{\Re^2} \int_{\Re^2} k(\mathbf{s}-\mathbf{s}'+\mathbf{u})k(\mathbf{u}')\rho(\mathbf{h}_\ell-\mathbf{h}_{\ell'}+\mathbf{u}-\mathbf{u}')d\mathbf{u}d\mathbf{u}'$$

Again the resulting cross-covariance matrix is valid; again the process is stationary. However now it is easy to verify that the cross-covariance matrix is not symmetric. Whether a lagged relationship between the variables is appropriate in a purely spatial specification would depend upon the application. However, in practice the $\mathbf{h}_\ell$ would be unknown and would be considered as model parameters. A fully Bayesian treatment of such a model has not yet been discussed in the literature.

For the second approach, suppose $C_\ell(\mathbf{s})$, $\ell = 1,...,p$ are each squared integrable stationary covariance functions valid in two-dimensional space. We now show $C_{\ell\ell}(\mathbf{s}) = \int_{R^2} C_\ell(\mathbf{s}-\mathbf{t})C_\ell(\mathbf{t})dt$, the convolution of C_ℓ with itself, is again a valid covariance function. Writing $\widehat{C}_\ell(\mathbf{w}) = \int e^{-i\mathbf{w}^T\mathbf{h}}C_\ell(\mathbf{h})d\mathbf{h}$, by inversion, $C_\ell(\mathbf{s}) = \int e^{i\mathbf{w}^T\mathbf{s}}\frac{\widehat{C}_\ell(\mathbf{w})}{(2\pi)^2}d\mathbf{w}$. But also, from (2.11), $\widehat{C}_{\ell\ell}(\mathbf{w}) \equiv \int e^{-i\mathbf{w}^T\mathbf{s}}C_{\ell\ell}(\mathbf{s})d\mathbf{s} = \int e^{-i\mathbf{s}^T\mathbf{s}}\int C_\ell(\mathbf{s}-\mathbf{t})C_\ell(\mathbf{t})dtd\mathbf{s} = \int \int e^{i\mathbf{w}^T(\mathbf{s}-\mathbf{t})}C_\ell(\mathbf{s}-\mathbf{t})e^{i\mathbf{w}^T\mathbf{t}}C_\ell(\mathbf{t})dtd\mathbf{s} = (\widehat{C}_\ell(\mathbf{w}))^2$. Self-convolution of C_ℓ produces the square of the Fourier transform. However, since $C_\ell(\cdot)$ is valid, Bochner's Theorem (Subsection 2.2.2) tells us that $\widehat{C}_\ell(\mathbf{w})/(2\pi)^2 C(0)$ is a spectral density symmetric about 0. But then due to the squared integrability assumption, up to proportionality, so is $(\widehat{C}_\ell(w))^2$, and thus $C_{\ell\ell}(\cdot)$ is valid.

The same argument ensures that

$$C_{\ell\ell'}(\mathbf{s}) = \int_{R^2} C_\ell(\mathbf{s}-\mathbf{t})C_{\ell'}(\mathbf{t})dt \qquad (7.31)$$

is also a valid stationary covariance function; cross-convolution provides a valid covariance function. (Now $\widehat{C}_{\ell\ell'}(w) = \widehat{C}_\ell(w)\widehat{C}_{\ell'}(w)$.) Moreover, it can be shown that $C(\mathbf{s}-\mathbf{s}')$ defined by $(C(\mathbf{s}-\mathbf{s}'))_{\ell\ell'} = C_{\ell\ell'}(\mathbf{s}-\mathbf{s}')$ is a valid $p \times p$ cross-covariance function (see Majumdar and Gelfand, 2003). It is also the case that if each C_ℓ is isotropic, then so is $C(\mathbf{s}-\mathbf{s}')$. To see this, suppose $\|\mathbf{h}_1\| = \|\mathbf{h}_2\|$. We need only show that $C_{\ell\ell'}(\mathbf{h}_1) = C_{\ell\ell'}(\mathbf{h}_2)$. But $\mathbf{h}_1 = P\mathbf{h}_2$ where P is orthogonal. Hence, $C_{\ell\ell'}(\mathbf{h}_1) = \int C_\ell(\mathbf{h}_1 - \mathbf{t})C_{\ell'}(\mathbf{t})dt = \int C_\ell(P(\mathbf{h}_1 - \mathbf{t}))C_{\ell'}(P\mathbf{t})dt = \int C_\ell(\mathbf{h}_2 - \tilde{\mathbf{t}})C_{\ell'}(\tilde{\mathbf{t}})d\tilde{\mathbf{t}} = C_{\ell\ell'}(\mathbf{h}_2)$.

We note that the range associated with $C_{\ell\ell}$ is not the same as that for C_ℓ but that if the C_ℓ's have distinct ranges then so will the components, $Y_\ell(\mathbf{s})$. Computational issues associated with using $C(s-s')$ in model-fitting are also discussed in Majumdar and Gelfand (2003). We note that (7.31) can in most cases be conveniently computed by transformation to polar coordinates and then using Monte Carlo integration.

7.4 Multivariate models for areal data

In this section we explore the extension of univariate CAR methodology (Sections 3.3 and 5.4.3) to the multivariate setting. Such models can be employed to introduce multiple, dependent spatial random effects associated with areal units (as standard CAR models do for a single set of random effects). In this regard, Kim et al. (2001) presented a "twofold CAR" model to model counts for two different types of disease over each areal unit. Similarly, Knorr-Held and Best (2000) have developed a "shared component" model for the above purpose, but their methodology too seems specific to the bivariate situation. Knorr-Held and Rue (2002) illustrate sophisticated MCMC blocking approaches in a model placing three conditionally independent CAR priors on three sets of spatial random effects in a shared component model setting.

Multivariate CAR models can also provide coefficients in a multiple regression setting that are dependent and spatially varying at the areal unit level. For example, Gamerman et al. (2002) investigate a Gaussian Markov random field (GMRF) model (a multivariate generalization of the pairwise difference IAR model) and compare various MCMC blocking schemes for sampling from the posterior that results under a Gaussian multiple linear regression likelihood. They also investigate a "pinned down" version of this model that resolves the impropriety problem by centering the ϕ_i vectors around some mean location. These authors also place the spatial structure on the spatial regression coefficients themselves, instead of on extra intercept terms (that is, in (5.49) we would drop the ϕ_i, and replace β_1 by β_{1i}, which would now be assumed to have a CAR structure). Assunção et al. (2002) refer to these models as *space-varying coefficient* models, and illustrate in the case of estimating fertility schedules. Assunção (2003) offers a nice review of the work to date in this area. Also working with areal units, Sain and Cressie (2002) offer multivariate GMRF models, proposing a generalization that permits asymmetry in the spatial cross-correlation matrix. They use this approach to jointly model the counts of white and minority persons residing in the census block groups of St. James Parish, LA, a region containing several hazardous waste sites.

In the remainder of this section, we present a general, CAR-based approach for multivariate spatial random variables at areal unit level that is applicable to either problem.

7.4.1 Motivating data set

Child growth is usually monitored using anthropometric indicators such as height adjusted for age (HAZ), weight adjusted for height (WHZ), and weight adjusted for age (WAZ). Independent analysis of each of these indicators is normally carried out to identify factors influencing growth that may range from genetic and environmental factors (e.g., altitude, seasonality) to differences in nutrition and social deprivation. Substantial variation in growth is common within as well as between populations. Recently, geographical variation in child growth has been thoroughly investigated for the country of Papua New Guinea in Mueller et al. (2001). Independent spatial analyses for each of the anthropometric growth indicators identified complex geographical patterns of child growth finding areas where children are taller but skinnier than average, others where they are heavier but shorter, and areas where they are both short and light. These geographical patterns could be linked to differences in diet and subsistence agriculture, leading to the analysis presented here; see Gelfand and Vounatsou (2003) for further discussion.

The data for our illustration comes from the 1982–1983 Papua New Guinea National Nutrition Survey (NNS) (Heywood et al., 1988). The survey includes anthropometric measures (age, height, weight) of approximately 28,000 children under 5 years of age, as well as dietary, socioeconomic, and demographic data about those children and their families. Dietary data include the type of food that respondents had eaten the previous day. Subsequently, the data were coded to 14 important staples and sources of protein. Each child was assigned to a village and each village was assigned to one of 4566 environmental zones (resource mapping units, or RMUs) into which Papua New Guinea has been divided for agriculture planning purposes. A detailed description of the data is given in Mueller et al. (2001).

The nutritional scores, height adjusted for age (HAZ), and weight adjusted for age (WAZ) that describe the nutritional status of a child were obtained using the method of Cole and Green (1992), which yields age-adjusted standard normal deviate Z-scores. The data set was collected at 537 RMUs. To overcome sparseness and to facilitate computation, we collapsed to 250 spatial units. In the absence of digitized boundaries, Delaunay tessellations were used to create the neighboring structure in the spatial units.

Because of the complex, multidimensional nature of human growth, a bivariate model that considers differences in height and weight jointly might be more appropriate for analyzing child growth data in general and to identify geographical patterns of growth in particular. We propose the use of Bayesian hierarchical spatial models and with *multivariate CAR* (MCAR) specifications to analyze the bivariate pairs of indicators, HAZ and WAZ, of child growth. Our modeling reveals bivariate spatial random effects at RMU level, justifying the MCAR specification.

7.4.2 Multivariate CAR (MCAR) theory

In this section we broadly follow the notation and approach of Gelfand and Vounatsou (2003); the approach of Carlin and Banerjee (2003) is similar.

For a vector of univariate variables $\phi = (\phi_1, \phi_2, \ldots, \phi_n)$, zero-centered CAR specifications were detailed in Section 3.3. For the MCAR model we instead let $\phi^T = (\phi_1, \phi_2, \ldots, \phi_n)$ where each ϕ_i is a $p \times 1$. Following Mardia (1988), the zero-centered MCAR sets

$$\phi_i \mid \phi_{j \neq i}, \Sigma_i \sim N \left(\sum_j B_{ij} \phi_j, \, \Sigma_i \right), \, i = 1, \ldots, n, \qquad (7.32)$$

where each B_{ij} is $p \times p$, as is each Σ_i. As in the univariate case, Brook's lemma (3.7) yields a joint density for ϕ of the form

$$p(\phi \mid \{\Sigma_i\}) \propto \exp \left\{ -\frac{1}{2} \phi^T \Gamma^{-1} (I - \tilde{B}) \phi \right\},$$

where Γ is block diagonal with blocks Σ_i and $\tilde{B}$ is an $np \times np$ with (i, j)th block B_{ij}.

As in the univariate case, symmetry of $\Gamma^{-1}(I - \tilde{B})$ is required. A convenient special case sets $B_{ij} = b_{ij} I_{p \times p}$, yielding the symmetry condition $b_{ij} \Sigma_j = b_{ji} \Sigma_i$, analogous to (3.14). If as in Subsection 3.3.1 we take $b_{ij} = w_{ij}/w_{i+}$ and $\Sigma_i = w_{i+}^{-1} \Sigma$, then the symmetry condition is satisfied.

Kronecker product notation simplifies the form of $\Gamma^{-1}(I - \tilde{B})$. That is, setting $\tilde{B} = B \otimes I$ with B as in (3.13) and $\Gamma = D_W^{-1} \otimes \Sigma$ so

$$\Gamma^{-1}(I - \tilde{B}) = (D_W \otimes \Sigma^{-1})(I - B \otimes I) = (D_W - W) \otimes \Sigma^{-1}.$$

Again, the singularity of $D_W - W$ implies that $\Gamma^{-1}(I - B)$ is singular. We denote this distribution by $MCAR(1, \Sigma)$. As in Chapter 3 and Chapter 5, in practice we work with the proper full conditional distributions in (7.32), imposing p linear constraints.

To consider remedies to the impropriety, analogous to the univariate case, we rewrite (7.32) in the general form

$$E(\phi_i \mid \phi_{j \neq i}, \Gamma) = R_i \sum_j B_{ij} \phi_j.$$

Now $\Gamma^{-1}(I - \tilde{B})$ is revised to $\Gamma^{-1}(I - \tilde{B}_R)$ where $\tilde{B}_R$ has (i, j)th block

$R_i B_{ij}$. In general, then, the symmetry condition becomes $(\Sigma_i^{-1} R_i B_{ij})^T = \Sigma_j^{-1} R_j B_{ji}$, or $\Sigma_j B_{ij}^T R_i^T = R_j B_{ji} \Sigma_i$ (see Mardia, 1988, expression (2.4) in this regard).

If $B_{ij} = b_{ij} I_{p \times p}$ and $b_{ij} = w_{ij}/w_{i+}$, the symmetry condition simplifies to

$$w_{j+} \Sigma_j R_i^T = w_{i+} R_j \Sigma_i \ .$$

Finally, if in addition we take $\Sigma_i = w_{i+}^{-1} \Sigma$, we obtain $\Sigma R_i^T = R_j \Sigma$, which reveals that we must have $R_i = R_j = R$, and thus

$$\Sigma R^T = R\Sigma \ . \tag{7.33}$$

For any arbitrary positive definite Σ, a generic solution to (7.33) is $R = \rho \Sigma^t$. Hence, regardless of t, (7.33) introduces a total of $\binom{p+1}{2} + 1$ parameters. Thus, without loss of generality, we can set $t = 0$, hence $R = \rho I$. Calculation as above yields

$$\Sigma_\phi^{-1} = \Gamma^{-1}(I - \tilde{B}_R) = (D_W - \rho W) \otimes \Sigma^{-1}. \tag{7.34}$$

Hence, under the same restriction to ρ as in the univariate case, a nonsingular covariance matrix results. We denote this model by $MCAR(\rho, \Sigma)$.

If Σ is constrained to be diagonal with elements σ_l^2, R can be diagonal with elements ρ_l, yielding the case of p independent CAR specifications. Routine calculations show that each ρ_l must therefore satisfy the same restrictions as above to ensure a nonsingular covariance matrix. This modeling introduces $2p$ parameters and is conceptually less satisfying than allowing a more general Σ. However, only if all $\rho_l = \rho$ is this a special version of the general Σ case, since a diagonal R with a general Σ does not satisfy (7.33).

Lastly, following Gelfand and Vounatsou (2003), we provide a generalization of the $MCAR(\rho, \Sigma)$ model that permits the introduction of a spatial autoregression coefficient for each component of ϕ_i, i.e., a vector $\rho^T = (\rho_1, \rho_2, \ldots, \rho_p)$. First, suppose we rearrange the rows of the $np \times 1$ vector ϕ to block by components, rather than by units. That is, we write $\phi = (\phi_{11}, \phi_{21}, \ldots, \phi_{n1}, \phi_{12}, \ldots, \phi_{n2}, \ldots, \phi_{1p}, \ldots, \phi_{np})^T$, so that $\phi^T = P\phi$ where P is orthogonal. But also from (7.34),

$$\Sigma_{\phi'}^{-1} = \Sigma^{-1} \otimes (D_W - \rho W) \ . \tag{7.35}$$

Let $D_W^{-1/2} W D_W^{-1/2} = Q \Delta Q^T$ where Δ is diagonal with entries λ_i, which are the eigenvalues of $D_W^{-1/2} W D_W^{-1/2}$, and Q is orthogonal. Then, if $T_{\rho_j} = D_W - \rho_j W$, it is evident that $T_{\rho_j} = D_W^{1/2} Q \Omega_j Q^T D_W^{1/2}$, where Ω_j is diagonal with $(\Omega_j)_{ii} = 1 - \rho_j \lambda_i$. Also, $T_{\rho_j} = A_j A_j^T$ where $A_j = D_W^{1/2} Q \Omega_j^{1/2} Q^T$. Note that A_j^{-1} exists if $\Omega_j^{-1/2}$ exists. But if ρ_j satisfies our earlier restrictions i.e., $\rho_j \in (\lambda_{min}^{-1}, \lambda_{max}^{-1})$, then $1 - \rho_j \lambda_i > 0$ for each i so $\Omega_j^{-1/2}$ exists.

Next let $G_j = A_1 A_j^{-1}, j = 1, \ldots, p$, and let G be block diagonal with blocks $G_1, \ldots, G_p$. G is evidently full rank provided each ρ_j satisfies the foregoing eigenvalue condition. Then straightforward calculation reveals that $G^{-1}(\Sigma^{-1} \otimes T_{\rho_1})(G^{-1})^T$ equals

$$\begin{pmatrix}
\Sigma_{11}^{-1} T_{\rho_1} & \Sigma_{12}^{-1} A_1 A_2^T & \cdots & \Sigma_{1p}^{-1} A_1 A_p^T \\
\Sigma_{21}^{-1} A_2 A_1^T & \Sigma_{22}^{-1} T_{\rho_2} & \cdots & \Sigma_{2p}^{-1} A_2 A_p^T \\
\vdots & \vdots & \cdots & \vdots \\
\Sigma_{p1}^{-1} A_p A_1^T & \Sigma_{p2}^{-1} A_p A_2^T & \cdots & \Sigma_{pp}^{-1} T_{\rho_p}
\end{pmatrix}. \tag{7.36}$$

The matrix in (7.36) is immediately positive definite and can be viewed as the inverse covariance matrix associated with $\boldsymbol{\Psi}^T = G\boldsymbol{\phi}^T$, where $\boldsymbol{\phi}$ has the inverse covariance matrix in (7.35) at $\rho = \rho_1$. Finally, the distribution of $\boldsymbol{\Psi} = (\boldsymbol{\Psi}_1, \ldots, \boldsymbol{\Psi}_n)$ where $\boldsymbol{\Psi}_i$ is $p \times 1$, with $\boldsymbol{\Psi} = P^T \boldsymbol{\Psi}' = P^T G \boldsymbol{\phi}' = P^T G P \boldsymbol{\phi}$ provides a new multivariate CAR specification, which we denote by $MCAR(\boldsymbol{\rho}, \Sigma)$. Note that the linear transformation relating $\boldsymbol{\Psi}$ and $\boldsymbol{\phi}$ is parametric, i.e., it involves the unknown $\boldsymbol{\rho}$. This class of models has $\binom{p+1}{2} + p$ parameters, and reduces to $MCAR(\rho_1, \Sigma)$ when $\rho_j = \rho_1$, $j = 2, \ldots, p$. Also note the interpretation of the diagonal blocks in (7.36): $\Sigma_{jj}^{-1} T_{\rho_j}$ is the inverse of the conditional covariance matrix of $\boldsymbol{\Psi}'_j$ given $\boldsymbol{\Psi}'_l$ for $l = 1, \ldots, p$ and $l \neq j$. Hence if $\rho_j = 0$, then $\Sigma_{jj}^{-1} T_{\rho_j} = \Sigma_{jj}^{-1} D_W$, i.e. $\{\Psi_{1j}, \Psi_{2j}, \ldots, \Psi_{nj}\}$ are conditionally independent given all of the other $\boldsymbol{\Psi}'$. This is analogous to the interpretation of $\rho = 0$ in the univariate "proper CAR" model from Exercise 5. Such conditional independence is the anticipated conclusion, since we are modeling through the inverse covariance matrix.

As a final comment, we clarify a distinction between these MCAR models and the twofold CAR model (for the case $p = 2$) of Kim, Sun, and Tsutakawa (2001). Rather than specifying the joint distribution through the conditional distributions $p(\boldsymbol{\phi}_i \mid \boldsymbol{\phi}_{j \neq i})$ as we (and Mardia, 1988) do, they instead specify the (univariate) *full* conditional distributions, $p(\phi_{il} \mid \boldsymbol{\phi}_{-(il)})$, where $\boldsymbol{\phi}_{-(il)}$ denotes all of the remaining ϕ's in $\boldsymbol{\phi}$.

7.4.3 Modeling issues

The MCAR specifications of the previous section are employed in models for spatial random effects arising in a hierarchical model. For instance, suppose we have a linear model with continuous data $\mathbf{Y}_{ik}$, $i = 1, \ldots, n$, $k = 1, \ldots, m_i$, where $\mathbf{Y}_{ik}$ is a $p \times 1$ vector denoting the kth response at the ith areal unit. The mean of the $\mathbf{Y}_{ik}$ is $\boldsymbol{\mu}_{ik}$ where $\mu_{ikj} = (\mathbf{X}_{ik})_j \boldsymbol{\beta}^{(j)} + \phi_{ij}$, $j = 1, \ldots, p$. Here $\mathbf{X}_{ik}$ is a $p \times s$ matrix with covariates associated with $\mathbf{Y}_{ik}$ having jth row $(\mathbf{X}_{ik})_j$, $\boldsymbol{\beta}^{(j)}$ is an $s \times 1$ coefficient vector associated with the jth component of the $\mathbf{Y}_{ik}$'s, and ϕ_{ij} is the jth component of the $p \times 1$ vector $\boldsymbol{\phi}_i$. Given $\left\{\boldsymbol{\beta}^{(j)}\right\}$, $\{\boldsymbol{\phi}_i\}$ and V, the $\mathbf{Y}_{ik}$ are conditionally independent $N(\mu_{ik}, V)$ variables. Adding a prior for $\left\{\beta^{(j)}\right\}$ and V and one of the MCAR models from Subsection 7.4.2 for the ϕ_i completes the second stage of the specification. Finally, a hyperprior on the MCAR parameters completes the model.

Alternatively, we might change the first stage to a multinomial. Here k disappears and $\mathbf{Y}_i$ is assumed to follow a multinomial distribution with sample size n_i and with $(p + 1) \times 1$ probability vector $\boldsymbol{\pi}_i$. Working on the logit scale, using cell $p + 1$ as the baseline, we could set $\log\left(\frac{\pi_{ij}}{\pi_{i,p+1}}\right) = \mathbf{X}_i^T \boldsymbol{\beta}^{(j)} + \phi_{ij}$, $j = 1, \ldots, p$, with $\mathbf{X}$'s, $\boldsymbol{\beta}$'s and $\boldsymbol{\phi}$'s interpreted as in the previous paragraph. Many other multivariate first stages could also be used, such as other multivariate exponential family models.

Regardless, model-fitting is most easily implemented using a Gibbs sampler with Metropolis updates where needed. The full conditionals for the $\boldsymbol{\beta}$'s will typically be normal (under a normal first-stage model) or else require Metropolis, slice, or adaptive rejection sampling (Gilks and Wild, 1992). For the $MCAR(1, \Sigma)$ and $MCAR(\rho, \Sigma)$ models, the full conditionals for the ϕ_i's will be likelihood-adjusted versions of the conditional distributions that define the MCAR, and are updated as a block. For the $MCAR(\rho, \Sigma)$ model, we can work with either the $\boldsymbol{\phi}$ or the $\boldsymbol{\psi}$ parametrization. With a non-Gaussian first stage, it will be awkward to pull the transformed effects out of the likelihood in order to do the updating. However, with a Gaussian first stage, it may well be more efficient to work on the transformed scale. Under the Gaussian first stage, the full conditional for V will be seen to follow an inverse Wishart, as will Σ. The ρ's do not follow standard distributions; in fact, discretization expedites computation, avoiding Metropolis steps.

We have chosen an illustrative prior for ρ in the ensuing example following three criteria. First, we insist that $\rho < 1$ to ensure propriety but allow $\rho = 0.99$. Second, we do not allow $\rho < 0$ since this would violate the similarity of spatial neighbors that we seek. Third, since even moderate spatial dependence requires values of ρ near 1 (recall the discussion in Subsection 3.3.1) we place prior mass that favors the upper range of ρ. In particular, we put equal mass on the following 31 values: $0, 0.05, 0.1, \ldots, 0.8, 0.82, 0.84, \ldots, 0.90, 0.91, 0.92, \ldots, 0.99$.

Finally, model choice arises here only in selecting among MCAR specifications. That is, we do not alter the mean vector in these investigations; our interest here lies solely in comparing the spatial explanations. Multivariate versions of the Gelfand and Ghosh (1998) criterion (4.13) for multivariate Gaussian data are employed.

Example 7.4 *(Analysis of the child growth data)*. Recalling the discussion of Subsection 7.4.2, it may be helpful to provide explicit expressions, with obvious notation, for the modeling and the resulting association structure.

Model	G	P	D_∞
$MCAR(1, \Sigma)$	34300.69	33013.10	67313.79
$MCAR(\rho, \Sigma)$	34251.25	33202.86	67454.11
$MCAR(\rho, \Sigma)$	34014.46	33271.97	67286.43

Table 7.5 *Model comparison for child growth data.*

We have, for the jth child in the ith RMU,

$$\mathbf{Y}_{ij} = \begin{pmatrix} (HAZ)_{ij} \\ (WAZ)_{ij} \end{pmatrix} = \mathbf{X}_{ij}^T \begin{pmatrix} \boldsymbol{\beta}^{(H)} \\ \boldsymbol{\beta}^{(W)} \end{pmatrix} + \begin{pmatrix} \phi_i^{(H)} \\ \phi_i^{(W)} \end{pmatrix} + \begin{pmatrix} \epsilon_{ij}^{(H)} \\ \epsilon_{ij}^{(W)} \end{pmatrix}.$$

In this setting, under say the $MCAR(\rho, \Sigma)$ model,

$$cov((HAZ)_{ij}, (HAZ)_{i'j'} \mid \boldsymbol{\beta}^{(H)}, \boldsymbol{\beta}^{(W)}, \rho, \Sigma, V)$$
$$= cov(\phi_i^{(H)}, \phi_{i'}^{(H)}) + V_{11}I_{i=i', j=j'} ,$$
$$cov((WAZ)_{ij}, (WAZ)_{i'j'} \mid \boldsymbol{\beta}^{(H)}, \boldsymbol{\beta}^{(W)}, \rho, \Sigma, V)$$
$$= cov(\phi_i^{(W)}, \phi_{i'}^{(W)}) + V_{22}I_{i=i', j=j'} ,$$

and $\quad cov((HAZ)_{ij}, (WAZ)_{i'j'} \mid \boldsymbol{\beta}^{(H)}, \boldsymbol{\beta}^{(W)}, \rho, \Sigma, V)$

$$= cov(\phi_i^{(H)}, \phi_{i'}^{(W)}) + V_{12}I_{i=i', j=j'} ,$$

where $cov(\phi_i^{(H)}, \phi_{i'}^{(H)}) = (D_W - \rho W)_{ii'}\Sigma_{11}$, $cov(\phi_i^{(W)}, \phi_{i'}^{(W)}) = (D_W - \rho W)_{ii'}\Sigma_{22}$, and $cov(\phi_i^{(H)}, \phi_{i'}^{(W)}) = (D_W - \rho W)_{ii'}\Sigma_{12}$. The interpretation of the components of Σ and V (particularly Σ_{12} and V_{12}) is now clarified.

We adopted noninformative uniform prior specifications on $\boldsymbol{\beta}^{(H)}$ and $\boldsymbol{\beta}^{(W)}$. For Σ and V we use inverse Wishart priors, i.e., $\Sigma^{-1} \sim W(\Omega_1, c_1)$, $V^{-1} \sim W(\Omega_2, c_2)$ where Ω_1, Ω_2 are $p \times p$ matrices and c_1, c_2 are shape parameters. Since we have no prior knowledge regarding the nature or extent of dependence, we choose Ω_1 and Ω_2 diagonal; the data will inform about the dependence *a posteriori*. Since the $\mathbf{Y}_{ij}$'s are centered and scaled on each dimension, setting $\Omega_1 = \Omega_2 = I$ seems appropriate. Finally, we set $c_1 = c_2 = 4$ to provide low precision for these priors. We adopted for ρ_1 and ρ_2 the prior discussed in the previous section. Simulation from the full conditional distributions of the $\boldsymbol{\beta}$'s and the $\boldsymbol{\psi}_i, i = 1, \ldots, n$ is straightforward as they are standard normal distributions. Similarly, the full conditionals for V^{-1} and Σ^{-1} are Wishart distributions. We implemented the Gibbs sampler with 10 parallel chains.

Table 7.5 offers a comparison of three MCAR models using (4.13), the

	Height (HAZ)			Weight (WAZ)		
Covariate	2.5%	50%	97.5%	2.5%	50%	97.5%
Global mean	−0.35	−0.16	−0.01	−0.48	−0.25	−0.15
Coconut	0.13	0.20	0.29	0.04	0.14	0.24
Sago	−0.16	−0.07	−0.00	−0.07	0.03	0.12
Sweet potato	−0.11	−0.03	0.05	−0.08	0.01	0.12
Taro	−0.09	0.01	0.10	−0.19	−0.09	0.00
Yams	−0.16	−0.04	0.07	−0.19	−0.05	0.08
Rice	0.30	0.40	0.51	0.26	0.38	0.49
Tinned fish	0.00	0.12	0.24	0.04	0.17	0.29
Fresh fish	0.13	0.23	0.32	0.08	0.18	0.28
Vegetables	−0.08	0.08	0.25	0.02	0.19	0.35
V_{11}, V_{22}	0.85	0.87	0.88	0.85	0.87	0.88
V_{12}	0.60	0.61	0.63			
Σ_{11}, Σ_{22}	0.30	0.37	0.47	0.30	0.39	0.52
Σ_{12}	0.19	0.25	0.35			
ρ_1, ρ_2	0.95	0.97	0.97	0.10	0.80	0.97

Table 7.6 *Posterior summaries of the dietary covariate coefficients, covariance components, and autoregression parameters for the child growth data using the most complex MCAR model.*

Gelfand and Ghosh (1998) criterion. The most complex model is preferred, offering sufficient improvement in goodness of fit to offset the increased complexity penalty. Summaries of the posterior quantities under this model are shown in Table 7.6. These were obtained from a posterior sample of size 1,000, obtained after running a 10-chain Gibbs sampler for 30,000 iterations with a burn-in of 5,000 iterations and a thinning interval of 30 iterations. Among the dietary factors, high consumption of sago and taro are correlated with lighter and shorter children, while high consumption of rice, fresh fish, and coconut are associated with both heavier and taller children. Children from villages with high consumption of vegetables or tinned fish are heavier.

The posterior for the correlation associated with Σ, $\Sigma_{12}/\sqrt{\Sigma_{11}\Sigma_{22}}$, has mean 0.67 with 95% credible interval $(0.57, 0.75)$, while the posterior for the correlation associated with V, $V_{12}/\sqrt{V_{11}V_{22}}$, has mean 0.71 with 95% credible interval $(0.70, 0.72)$. In addition, ρ_1 and ρ_2 differ. ∎

7.5 Exercises

1. Compute the coherence (generalized correlation) in (7.3):

(a) for the cross-covariance in (7.15), and

(b) for the cross-covariance in (7.19).

2. Let $Y(\mathbf{s}) = (Y_1(\mathbf{s}), Y_2(\mathbf{s}))^T$ be a bivariate process with a stationary cross-covariance matrix function

$$C(\mathbf{s} - \mathbf{s}') = \begin{pmatrix} c_{11}(\mathbf{s} - \mathbf{s}') & c_{12}(\mathbf{s} - \mathbf{s}') \\ c_{12}(\mathbf{s}' - \mathbf{s}) & c_{22}(\mathbf{s} - \mathbf{s}') \end{pmatrix},$$

and a set of covariates $\mathbf{x}(\mathbf{s})$. Let $\mathbf{y} = \left(\mathbf{y}_1^T, \mathbf{y}_2^T\right)^T$ be the $2n \times 1$ data vector, with $\mathbf{y}_1 = (y_1(\mathbf{s}_1), \dots, y_1(\mathbf{s}_n))^T$ and $\mathbf{y}_2 = (y_2(\mathbf{s}_1), \dots, y_2(\mathbf{s}_n))^T$.

(a) Show that the cokriging predictor has the form

$$E[Y_1(\mathbf{s}_0)|\mathbf{y}] = \mathbf{x}^T(\mathbf{s}_0)\boldsymbol{\beta} + \boldsymbol{\gamma}^T\Sigma^{-1}(\mathbf{y} - X\boldsymbol{\beta}),$$

i.e., as in (2.18), but with appropriate definitions of $\boldsymbol{\gamma}$ and Σ.

(b) Show further that if $\mathbf{s}_k$ is a site where $y_l(\mathbf{s}_k)$ is observed, then for $l = 1, 2$, $E[Y_l(\mathbf{s}_k)|\mathbf{y}] = y_l(\mathbf{s}_k)$ if and only if $\tau_l^2 = 0$.

3. Suppose $\mathbf{Y}(\mathbf{s})$ is a bivariate spatial process as in Exercise 2. In fact, suppose $\mathbf{Y}(\mathbf{s})$ is a Gaussian process. Let $Z_1(\mathbf{s}) = I(Y_1(\mathbf{s}) > 0)$, and $Z_2(\mathbf{s}) = I(Y_2(\mathbf{s}) > 0)$. Approximate the cross-covariance matrix of $\mathbf{Z}(\mathbf{s}) = (Z_1(\mathbf{s}), Z_2(\mathbf{s}))^T$.

4. The data in `www.biostat.umn.edu/~brad/data/ColoradoLMC.dat` record maximum temperature (in tenths of a degree Celsius) and precipitation (in cm) during the month of January 1997 at 50 locations in the U.S. state of Colorado.

(a) Let X denote temperature and Y denote precipitation. Following the model of Example 7.3, fit an LMC model to these data using the conditional approach, fitting X and then $Y|X$.

(b) Repeat this analysis, but this time fitting Y and then $X|Y$. Show that your new results agree with those from part (a) up to simulation variability.

5. If C_l and $C_{l'}$ are isotropic, obtain $C_{ll'}(\mathbf{s})$ in (7.31) by transformation to polar coordinates.

6. The usual and generalized (but still proper) MCAR models may be constructed using linear transformations of some nonspatially correlated variables. Consider a vector blocked by components, say $\boldsymbol{\phi} = \left(\boldsymbol{\phi}_1^T, \boldsymbol{\phi}_2^T\right)^T$, where each $\boldsymbol{\phi}_i$ is $n \times 1$, n being the number of areal units. Suppose we look upon these vectors as arising from linear transformations

$$\boldsymbol{\phi}_1 = A_1\mathbf{v}_1 \text{ and } \boldsymbol{\phi}_2 = A_2\mathbf{v}_2,$$

where A_1 and A_2 are any $n \times n$ matrices, $\mathbf{v}_1 = (v_{11}, \ldots, v_{1n})^T$ and $\mathbf{v}_2 = (v_{21}, \ldots, v_{2n})^T$ with covariance structure

$$Cov\,(v_{1i}, v_{1j}) = \lambda_{11} I_{[i=j]}, \; Cov\,(v_{1i}, v_{2j}) = \lambda_{12} I_{[i=j]},$$
$$\text{and } Cov\,(v_{2i}, v_{2j}) = \lambda_{22} I_{[i=j]} \, ,$$

where $I_{[i=j]} = 1$ if $i = j$ and 0 otherwise. Thus, although $\mathbf{v}_1$ and $\mathbf{v}_2$ are associated, their nature of association is nonspatial in that covariances remain same for every areal unit, and there is no association between variables in different units.

(a) Show that the dispersion matrix $\Sigma_{(\mathbf{v}_1, \mathbf{v}_2)}$ equals $\Lambda \otimes I$, where $\Lambda = (\lambda_{ij})_{i,j=1,2}$.

(b) Show that setting $A_1 = A_2 = A$ yields a separable covariance structure for ϕ. What choice of A would render a separable MCAR model, analogous to (7.34)?

(c) Show that appropriate (different) choices of A_1 and A_2 yield the generalized MCAR model, as in (7.36).

CHAPTER 8

Spatiotemporal modeling

In both theoretical and applied work, spatiotemporal modeling has received dramatically increased attention in the past few years. This reason behind this increase is easy to see: the proliferation of data sets that are both spatially and temporally indexed, and the attendant need to understand them. For example, in studies of air pollution, we are interested not only in the spatial nature of a pollutant surface, but also in how this surface changes over time. Customarily, temporal measurements (e.g., hourly, daily, three-day average, etc.) are collected at monitoring sites over several years. Similarly, with climate data we may be interested in spatial patterns of temperature or precipitation at a given time, but also in dynamic patterns in weather. With real estate markets, we might be interested in how the single-family home sales market changes on a quarterly or annual basis. Here an additional wrinkle arises in that we do not observe the *same* locations for each time period; the data are cross-sectional, rather than longitudinal.

Applications with areal unit data are also commonplace. For instance, we may look at annual lung cancer rates by county for a given state over a number of years to judge the effectiveness of a cancer control program. Or we might consider daily asthma hospitalization rates by zip code, over a period of several months.

From a methodological point of view, the introduction of time into spatial modeling brings a substantial increase in the scope of our work, as we must make separate decisions regarding spatial correlation, temporal correlation, and how space and time interact in our data. Such modeling will also carry an obvious associated increase in notational and computational complexity.

As in previous chapters, we make a distinction between the cases where the geographical aspect of the data is point level versus areal unit level. Again the former case is typically handled via Gaussian process models, while the latter often uses CAR specifications. A parallel distinction could be drawn for the temporal scale: is time viewed as continuous (say, over $\Re^+$ or some subinterval thereof) or discrete (daily, quarterly, etc.)? In the former case there is a conceptual measurement at each moment t. But in the latter case, we must determine whether each measurement should be interpreted as a block average over some time interval (analogous to block averaging in space), or whether it should be viewed merely as a measurement, e.g., a count attached to an associated time interval (and thus analogous to an areal unit measurement). Relatedly, when time is discretized, are we observing a time series of spatial data, e.g., the same points or areal units in each time period (as would be the case in our climate and pollution examples)? Or are we observing cross-sectional data, where the locations change with time period (as in our real estate setting)? In the case of time series, we could regard the data as a multivariate measurement vector at each location or areal unit. We could then employ multivariate spatial data models as in the previous chapter. With short series, this would be reasonable; with longer series, we would likely want to introduce aspects of usual time series modeling.

The nature and location of missing data is another issue that we have faced before, yet becomes doubly complicated in the spatiotemporal setting. The major goal of traditional kriging methods is to impute missing values at locations for which no data have been observed. Now we may encounter time points for which we lack spatial information, locations for which information is lacking for certain (possibly future) time points, or combinations thereof. Some of these combinations will be extrapolations (e.g., predicting future values at locations for which no data have been observed) that are statistically riskier than others (e.g., filling in missing values at locations for which we have data at some times but not others). Here the Bayesian hierarchical approach is particularly useful, since it not only helps organize our thinking about the model, but also fully accounts for all sources of uncertainty, and properly delivers wider confidence intervals for predictions that are "farther" from the observed data (in either space or time).

Our Atlanta data set (Figure 6.1) illustrates the sort of misalignment problem we face in many spatiotemporal settings. Here the number of ozone monitoring stations is small (just 8 or 10), but the amount of data collected from these stations over time (92 summer days for each of three years) is substantial. In this case, under suitable modeling assumptions, we may not only learn about the temporal nature of the data, but also enhance our understanding of the spatial process. Moreover, the additional computational burden to analyze the much larger data set within the Bayesian framework still turns out to be manageable.

In the next few sections we consider the case of point-level spatial data, so that point-point and point-block realignment can be contemplated as in Section 6.1. We initially focus on relatively simple *separable* forms for the space-time correlation, but also consider more complex forms that do not impose the separable models' rather severe restrictions on space-time interaction. We subsequently move on to spatiotemporal modeling for data where the spatial component can only be thought of as areal (block) level.

8.1 General modeling formulation

8.1.1 Preliminary analysis

Before embarking on a general spatiotemporal modeling formulation, consider the case of point-referenced data where time is discretized to customary integer-spaced intervals. We may look at a spatiotemporally indexed datum $Y(\mathbf{s}, t)$ in two ways. Writing $Y(\mathbf{s}, t) = Y_{\mathbf{s}}(t)$, it is evident that we have a spatially varying time series model. Writing $Y(\mathbf{s}, t) = Y_t(\mathbf{s})$, we instead have a temporally varying spatial model.

In fact, with locations $\mathbf{s}_i$, $i = 1, \ldots, n$ and time points $t = 1, \ldots, T$, we can collect the data into Y, an $n \times T$ matrix. Column averages of Y produce a time-averaged spatial process, while row averages yield a domain-averaged time series. In fact, suppose we center each column of Y by the vector of row averages and call the resulting matrix $\widetilde{Y}_{rows}$. Then clearly $\widetilde{Y}_{rows} \mathbf{1}_T = \mathbf{0}$, but also $\frac{1}{T} \widetilde{Y}_{rows} \widetilde{Y}_{rows}^T$ is an $n \times n$ matrix that is the sample spatial covariance matrix. Similarly, suppose we center each row of Y by the vector of column averages and call the resulting matrix $\widetilde{Y}_{cols}$. Now $\mathbf{1}_n^T \widetilde{Y}_{cols} = \mathbf{0}$ and $\frac{1}{n} \widetilde{Y}_{cols}^T \widetilde{Y}_{cols}$ is the $T \times T$ sample autocorrelation matrix.

One could also center Y by the grand mean of the $Y(\mathbf{s}, t)$. Indeed, to examine residual spatiotemporal structure, adjusted for the mean, one could fit a suitable OLS regression to the $Y(\mathbf{s}, t)$ and examine $\widehat{E}$, the matrix of residuals $\widehat{e}(\mathbf{s}, t)$. As above, $\frac{1}{T} \widehat{E}\widehat{E}^T$ is the residual spatial covariance matrix while $\frac{1}{n} \widehat{E}^T \widehat{E}$ is the residual autocorrelation matrix.

We can create the singular value decomposition (Harville, 1997) for any of the foregoing matrices. Say for Y, assuming $T < n$ for illustration, we can write

$$Y = UDV^T = \sum_{l=1}^{T} d_l \mathbf{u}_l \mathbf{v}_l^T , \qquad (8.1)$$

where U is an $n \times n$ orthogonal matrix with columns $\mathbf{u}_l$, V is a $T \times T$ orthogonal matrix with columns $\mathbf{v}_l$, and D is an $n \times T$ matrix of the form $\binom{\Delta}{0}$ where Δ is $T \times T$ diagonal with diagonal entries d_l, $l = 1, \ldots, T$. Without loss of generality, we can assume the d_l's are arranged in decreasing order of their absolute values. Then $\mathbf{u}_l \mathbf{v}_l^T$ is referred to as the lth *empirical orthogonal function* (EOF).

Thinking of $\mathbf{u}_l = (u_l(\mathbf{s}_1), \ldots, u_l(\mathbf{s}_n))^T$ and $\mathbf{v}_l = (v_l(1), \ldots, v_l(T))^T$, the expression in (8.1) represents the observed data as a sum of products of spatial and temporal variables, i.e., $Y(\mathbf{s}_i, t) = \sum d_l u_l(\mathbf{s}_i) v_l(t)$. Evidently, the expansion in (8.1) introduces redundant variables; there are already nT variables in $\mathbf{u}_1, \ldots, \mathbf{u}_T$. Suppose we approximate Y by its first EOF, that is, $Y \approx d_1 \mathbf{u}_1 \mathbf{v}_1^T$. Then we are saying that $Y(\mathbf{s}_i, t) \approx d_1 u_1(\mathbf{s}_i) v_1(t)$, i.e., the spatiotemporal process can be approximated by a product of a spatial process and a temporal process. Note this does *not* imply a *separable* covariance function for $Y(\mathbf{s}, t)$ (see (8.18)) since the u_1 process and the

v_1 process need not be independent. However, it does yield a reduction in dimension, introducing $n + T$ variables to represent Y, rather than nT. Adding the second EOF yields the approximation $Y(\mathbf{s}_i, t) \approx d_1 u_1(\mathbf{s}_i) v_1(t) + d_2 u_2(\mathbf{s}_i) v_2(t)$, a representation involving only $2(n + T)$ variables, and so on.

Note that

$$YY^T = UDD^TU^T = U \left(\begin{array}{cc} \Delta^2 & 0 \\ 0 & 0 \end{array} \right) U^T = \sum_{l=1}^{T} d_l^2 \mathbf{u}_l \mathbf{u}_l^T \ ,$$

clarifying the interpretation of the d_l's. (Of course, $YY^T = V^T D^T DV = V^T \Delta^2 V$ as well.) For instance, applied to the residual spatial covariance matrix, $\frac{1}{T} \widehat{E} \widehat{E}^T$, suppose the first two terms of the expansion explain most of the spatial covariance structure. This suggests that the first two EOFs provide a good explanation of $\widehat{E}$.

If $T > n$ we would just exchange T and n in the foregoing. Regardless, EOFs provide an exploratory tool for learning about spatial structure and suggesting models, in the spirit of the tools described in Section 2.3. For full inference, however, we require a full spatiotemporal model specification, the subject to which we now turn.

8.1.2 Model formulation

Modeling for spatiotemporal data can be given a fairly general formulation that naturally extends that of Chapter 5. Consider point-referenced locations and continuous time. Let $Y(\mathbf{s}, t)$ denote the measurement at location $\mathbf{s}$ at time t. Extending (5.1), for continuous data that we can assume to be roughly normally distributed, we can write the general form

$$Y(\mathbf{s}, t) = \mu(\mathbf{s}, t) + e(\mathbf{s}, t) \ , \tag{8.2}$$

where $\mu(\mathbf{s}, t)$ denotes the mean structure and $e(\mathbf{s}, t)$ denotes the residual. If $\mathbf{x}(\mathbf{s}, t)$ is a vector of covariates associated with $Y(\mathbf{s}, t)$ then we can set $\mu(\mathbf{s}, t) = \mathbf{x}(\mathbf{s}, t)^T \boldsymbol{\beta}(\mathbf{s}, t)$. Note that this form allows spatiotemporally varying coefficients, which is likely to be more generality than we would need or want; $\boldsymbol{\beta}(\mathbf{s}, t) = \boldsymbol{\beta}$ is frequently adopted. If t is discretized, $\boldsymbol{\beta}(\mathbf{s}, t) = \boldsymbol{\beta}(t)$ might be appropriate if there were enough time points to suggest a temporal change in the coefficient vector. Similarly, setting $\boldsymbol{\beta}(\mathbf{s}, t) = \boldsymbol{\beta}(\mathbf{s})$ yields spatially varying coefficients, the topic of Section 10.2. Finally, $e(\mathbf{s}, t)$ would typically be rewritten as $w(\mathbf{s}, t) + \epsilon(\mathbf{s}, t)$, where $\epsilon(\mathbf{s}, t)$ is a Gaussian white noise process and $w(\mathbf{s}, t)$ is a mean-zero spatiotemporal process.

We can therefore view (8.2) as a hierarchical model with a conditionally independent first stage given $\{\mu(\mathbf{s}, t)\}$ and $\{w(\mathbf{s}, t)\}$. But then, in the spirit of Section 5.2, we can replace the Gaussian first stage with another first-stage model (say, an exponential family model) and write

$Y(\mathbf{s}, t) \sim f(y(\mathbf{s}, t) \mid \mu(\mathbf{s}, t), w(\mathbf{s}, t))$, where

$$f(y(\mathbf{s}, t) \mid \mu(\mathbf{s}, t), w(\mathbf{s}, t)) = h(y(\mathbf{s}, t)) \exp\{\gamma[\eta(\mathbf{s}, t)y(\mathbf{s}, t) - \chi(\eta(\mathbf{s}, t))]\}, \tag{8.3}$$

where γ is a positive dispersion parameter. In (8.3), $g(\eta(\mathbf{s}, t)) = \mu(\mathbf{s}, t) + w(\mathbf{s}, t)$ for some link function g.

For areal unit data with discrete time, let Y_{it} denote the measurement for unit i at time period t. (In some cases we might obtain replications at i or t, e.g., the jth cancer case in county i, or the the jth property sold in school district i.) Analogous to (8.2) we can write

$$Y_{it} = \mu_{it} + e_{it}. \tag{8.4}$$

Now $\mu_{it} = \mathbf{x}_{it}^T \boldsymbol{\beta}_t$ (or perhaps just $\boldsymbol{\beta}$), and $e_{it} = w_{it} + \epsilon_{it}$ where the ϵ_{it} are unstructured heterogeneity terms and the w_{it} are spatiotemporal random effects, typically associated with a spatiotemporal CAR specification. Choices for this latter part of the model will be presented in Section 8.5.

Since areal unit data are often non-Gaussian (e.g., sparse counts), again we would view (8.4) as a hierarchical model and replace the first stage Gaussian specification with, say, a Poisson model. We could then write $Y_{it} \sim f(y_{it} | \mu_{it}, w_{it})$, where

$$f(y_{it} \mid \mu_{it}, w_{it}) = h(y_{it}) \exp\{\gamma[\eta_{it} y_{it} - \chi(\eta_{it})]\}, \tag{8.5}$$

with γ again a dispersion parameter, and $g(\eta_{it}) = \mu_{it} + w_{it}$ for some suitable link function g. With replications, we obtain Y_{ijt} hence $\mathbf{x}_{ijt}, \mu_{ijt}$, and η_{ijt}. Now we can write $g(\eta_{ijt}) = \mu_{ijt} + w_{ijt} + \epsilon_{ijt}$, enabling separation of spatial and heterogeneity effects.

Returning to the point-referenced data model (8.2), spatiotemporal richness is captured by extending $e(\mathbf{s}, t)$ beyond a white noise process. Below, α's denote temporal effects and w's denote spatial effects. Following Gelfand, Ecker, Knight, and Sirmans (2003), with t discretized, consider the following partitions for the error $e(\mathbf{s}, t)$:

$$e(\mathbf{s}, t) = \alpha(t) + w(\mathbf{s}) + \epsilon(\mathbf{s}, t), \tag{8.6}$$

$$e(\mathbf{s}, t) = \alpha_{\mathbf{s}}(t) + \epsilon(\mathbf{s}, t), \tag{8.7}$$

$$\text{and } e(\mathbf{s}, t) = w_t(\mathbf{s}) + \epsilon(\mathbf{s}, t). \tag{8.8}$$

The given forms avoid specification of space-time interactions. In each of (8.6), (8.7), and (8.8), the $\epsilon(\mathbf{s}, t)$ are i.i.d. $N(0, \sigma_\epsilon^2)$ and independent of the other processes. This pure error is viewed as a residual adjustment to the spatiotemporal explanation. (One could allow $Var(\epsilon(\mathbf{s}, t)) = \sigma_\epsilon^{2(t)}$, i.e., an error variance that changes with time. Modification to the details below is straightforward.)

Expression (8.6) provides an additive form in temporal and spatial effects (multiplicative on the original scale if the $Y(\mathbf{s}, t)$ are on the log scale). Expression (8.7) provides temporal evolution at each site; temporal effects are nested within sites. Expression (8.8) provides spatial evolution over time; spatial effects are nested within time. Spatiotemporal modeling beyond (8.6), (8.7), and (8.8) (particularly if t is continuous) necessitates the choice of a specification to connect the space and time scales; this is the topic of Section 8.2.

Next, we consider the components in (8.6), (8.7), and (8.8) in more detail. In (8.6), if t were continuous we could model $\alpha(t)$ as a one-dimensional stationary Gaussian process. In particular, for the set of actual sale times, $\{t_1, t_2, ..., t_m\}$, $\boldsymbol{\alpha} = (\alpha(t_1), ..., \alpha(t_m))' \sim N(\mathbf{0}, \sigma_\alpha^2 \Sigma(\phi))$ where $(\Sigma(\phi))_{rs} = Corr(\alpha(t_r), \alpha(t_s)) = \rho(|\ t_r - t_s\ |; \phi)$ for ρ a valid one-dimensional correlation function. Typical choices for ρ include the exponential, $\exp(-\phi\ |\ t_r - t_s\ |)$, and Gaussian, $\exp(-\phi(t_r - t_s)^2)$, forms, analogous to the spatial forms in Table 2.1.

With t confined to an indexing set, $t = 1, 2, ... T$, we can simply view $\alpha(1), ..., \alpha(T)$ as the coefficients associated with a set of time dummy variables. With this assumption for the $\alpha(t)$'s, if in (8.6), $w(\mathbf{s})$ is set to zero, $\boldsymbol{\beta}(t)$ is assumed constant over time and $\mathbf{X}(\mathbf{s}, t)$ is assumed constant over t, then upon differencing we obtain the seminal model for repeat property sales given in Bailey, Muth, and Nourse (1963). Also within these assumptions but restoring $\boldsymbol{\beta}$ to $\boldsymbol{\beta}(t)$, we obtain the extension of Knight, Dombrow, and Sirmans (1995). Alternatively, we might set $\alpha(t + 1) = \rho\alpha(t) + \eta(t)$ where $\eta(t)$ are i.i.d. $N(0, \sigma_\alpha^2)$. If $\rho < 1$ we have the familiar stationary $AR(1)$ time series, a special case of the continuous time model of the previous paragraph. If $\rho = 1$ the $\alpha(t)$ follow a random walk. With a finite set of times, time-dependent coefficients are handled analogously to the survival analysis setting (see, e.g., Cox and Oakes, 1984, Ch. 8).

The autoregressive and random walk specifications are naturally extended to provide a model for the $\alpha_{\mathbf{s}}(t)$ in (8.7). That is, we assume $\alpha_{\mathbf{s}}(t + 1) = \rho\alpha_{\mathbf{s}}(t) + \eta_{\mathbf{s}}(t)$ where again the $\eta_{\mathbf{s}}(t)$ are all i.i.d. Thus, there is no spatial modeling, only independent conceptual time series at each location. With spatial time series we can fit this model. With cross-sectional data, there is no information in the data about ρ so the likelihood can only identify the stationary variance $\sigma_\alpha^2/(1 - \rho^2)$ but not σ_α^2 or ρ. The case $\rho < 1$ with $\boldsymbol{\beta}(t)$ constant over time provides the models proposed in Hill, Knight, and Sirmans (1997) and in Hill, Sirmans, and Knight (1999). If $\rho = 1$ with $\boldsymbol{\beta}(t)$ and $\mathbf{X}(\mathbf{s}, t)$ constant over time, upon differencing we obtain the widely used model of Case and Shiller (1989). In application, it will be difficult to learn about the $\alpha_{\mathbf{s}}$ processes with typically one or at most two observations for each $\mathbf{s}$. The $w(\mathbf{s})$ are modeled as a Gaussian process following Section 2.2.

For $w_t(\mathbf{s})$ in (8.8), assuming t restricted to an index set, we can view the $w_t(\mathbf{s})$ as a collection of independent spatial processes. That is, rather than defining a dummy variable at each t, we conceptualize a separate spatial dummy *process* at each t. The components of $\mathbf{w}_t$ correspond to the sites at which measurements were observed in the time interval denoted by t. Thus, we capture the dynamics of location in a very general fashion. In particular, comparison of the respective process parameters reveals the nature of spatial evolution over time.

With a single time dummy variable at each t, assessment of temporal effects would be provided through inference associated with these variables. For example, a plot of the point estimates against time would clarify size and trend for the effects. With distinct spatial processes, how can we see such temporal patterns? A convenient reduction of each spatial process to a univariate random variable is the block average (see expression (6.1)).

To shed the independence assumption for the $w_t(\mathbf{s})$, we could instead assume that $w_t(\mathbf{s}) = \sum_{j=1}^{t} v_j(\mathbf{s})$ where the $v_j(\mathbf{s})$ are i.i.d. processes, again of one of the foregoing forms. Now, for $t < t^*$, $\mathbf{w}_t$ and $\mathbf{w}_{t^*}$ are not independent but $\mathbf{w}_t$ and $\mathbf{w}_{t^*} - \mathbf{w}_t$ are. This leads us to dynamic spatiotemporal models that are the focus of Section 8.4.

8.1.3 Associated distributional results

We begin by developing the likelihood under model (8.2) using (8.6), (8.7), or (8.8). Assuming $t \in \{1, 2, ..., T\}$, it is convenient to first obtain the joint distribution for $\mathbf{Y}' = (\mathbf{Y}_1', ..., \mathbf{Y}_T')$ where $\mathbf{Y}_t' = (Y(\mathbf{s}_1, t), ..., Y(\mathbf{s}_n, t))$. That is, each $\mathbf{Y}_t$ is $n \times 1$ and $\mathbf{Y}$ is $Tn \times 1$. This joint distribution will be multivariate normal. Thus, the joint distribution for the observed $Y(\mathbf{s}, t)$ requires only pulling off the appropriate entries from the mean vector and appropriate rows and columns from the covariance matrix. This simplifies the computational bookkeeping, though care is still required.

In the constant $\boldsymbol{\beta}$ case, associate with $\mathbf{Y}_t$ the matrix X_t whose ith row is $\mathbf{X}(\mathbf{s}_i, t)'$. Let $\boldsymbol{\mu}_t = X_t \boldsymbol{\beta}$ and $\boldsymbol{\mu}' = (\boldsymbol{\mu}_1', ..., \boldsymbol{\mu}_T')$. In the time-dependent parameter case we merely set $\boldsymbol{\mu}_t = X_t \boldsymbol{\beta}(t)$.

Under (8.6), let $\boldsymbol{\alpha}' = (\alpha(1), ..., \alpha(T))$, $\mathbf{w}' = (\mathbf{w}(\mathbf{s}_1), ..., \mathbf{w}(\mathbf{s}_n))$ and $\boldsymbol{\epsilon}' = (\epsilon(\mathbf{s}_1, 1), \epsilon(\mathbf{s}_1, 2), ..., \epsilon(\mathbf{s}_n, T))$. Then,

$$\mathbf{Y} = \boldsymbol{\mu} + \boldsymbol{\alpha} \otimes \mathbf{1}_{n \times 1} + \mathbf{1}_{T \times 1} \otimes \mathbf{w} + \boldsymbol{\epsilon} \qquad (8.9)$$

where $\otimes$ denotes the Kronecker product. Hence, given $\boldsymbol{\beta}$ along with the temporal and spatial effects,

$$\mathbf{Y} \mid \boldsymbol{\beta}, \boldsymbol{\alpha}, \mathbf{w}, \sigma_\epsilon^2 \sim N(\boldsymbol{\mu} + \boldsymbol{\alpha} \otimes \mathbf{1}_{n \times 1} + \mathbf{1}_{T \times 1} \otimes \mathbf{w} \,,\, \sigma_\epsilon^2 I_{Tn \times Tn}) . \qquad (8.10)$$

Let $\mathbf{w} \sim N(\mathbf{0}, \sigma_w^2 H(\delta))$. Suppose the $\alpha(t)$ follow an $AR(1)$ model, so that $\boldsymbol{\alpha} \sim N(\mathbf{0}, \sigma_\alpha^2 A(\rho))$ where $(A(\rho))_{ij} = \rho^{|i-j|}/(1 - \rho^2)$. Hence, if $\boldsymbol{\alpha}, \mathbf{w}$ and $\boldsymbol{\epsilon}$ are independent, marginalizing over $\boldsymbol{\alpha}$ and $\mathbf{w}$, i.e., integrating (8.10) with regard to the prior distribution of $\boldsymbol{\alpha}$ and $\mathbf{w}$, we obtain

$$\begin{aligned} \mathbf{Y} &\mid \boldsymbol{\beta}, \sigma_\epsilon^2, \sigma_\alpha^2, \rho, \sigma_w^2, \delta \\ &\sim N\left(\boldsymbol{\mu} \,,\, \sigma_\alpha^2 A(\rho) \otimes \mathbf{1}_{n \times 1} \mathbf{1}_{n \times 1}' + \sigma_w^2 \mathbf{1}_{T \times 1} \mathbf{1}_{T \times 1}' \otimes H(\delta) + \sigma_\epsilon^2 I_{Tn \times Tn}\right) . \end{aligned} \qquad (8.11)$$

262 SPATIOTEMPORAL MODELING

If the $\alpha(t)$ are coefficients associated with dummy variables (now $\boldsymbol{\beta}$ does not contain an intercept) we only marginalize over $\mathbf{w}$ to obtain

$$\mathbf{Y} \mid \boldsymbol{\beta}, \boldsymbol{\alpha}, \sigma_\epsilon^2, \sigma_w^2, \delta$$
$$\sim N\left(\boldsymbol{\mu} + \boldsymbol{\alpha} \otimes \mathbf{1}_{n\times 1} , \; \sigma_w^2 \mathbf{1}_{T\times 1} \mathbf{1}'_{T\times 1} \otimes H(\delta) + \sigma_\epsilon^2 I_{Tn\times Tn}\right) . \qquad (8.12)$$

The likelihood resulting from (8.10) arises as a product of independent normal densities by virtue of the conditional independence. This can facilitate model fitting but at the expense of a very high-dimensional posterior distribution. Marginalizing to (8.11) or (8.12) results in a much lower-dimensional posterior. Note, however, that while the distributions in (8.11) and (8.12) can be determined, evaluating the likelihood (joint density) requires a high-dimensional quadratic form and determinant calculation.

Turning to (8.7), if $\boldsymbol{\alpha}'(t) = (\alpha_{s_1}(t), ..., \alpha_{s_n}(t))$ and now we also define $\boldsymbol{\alpha}' = (\boldsymbol{\alpha}'(1), \ldots, \boldsymbol{\alpha}'(T))$ with $\boldsymbol{\epsilon}$ as above, then

$$\mathbf{Y} = \boldsymbol{\mu} + \boldsymbol{\alpha} + \boldsymbol{\epsilon} .$$

Now

$$\mathbf{Y} \mid \boldsymbol{\beta}, \boldsymbol{\alpha}, \sigma_\epsilon^2 \sim N\left(\boldsymbol{\mu} + \boldsymbol{\alpha} , \; \sigma_\epsilon^2 I_{Tn\times Tn}\right) .$$

If the $\alpha_{s_i}(t)$ follow an $AR(1)$ model independently across i, then marginalizing over $\boldsymbol{\alpha}$,

$$\mathbf{Y} \mid \boldsymbol{\beta}, \sigma_\epsilon^2, \sigma_\alpha^2, \rho \sim N(\boldsymbol{\mu} , \; A(\rho) \otimes I_{Tn\times Tn} + \sigma_\epsilon^2 I_{Tn\times Tn}). \qquad (8.13)$$

For (8.8), let $\mathbf{w}'_t = (w_t(\mathbf{s}_1), \ldots, w_t(\mathbf{s}_n))$ and $\mathbf{w}' = (\mathbf{w}'_1, \ldots, \mathbf{w}'_T)$. Then with $\boldsymbol{\epsilon}$ as above,

$$\mathbf{Y} = \boldsymbol{\mu} + \mathbf{w} + \boldsymbol{\epsilon} \qquad (8.14)$$

and

$$\mathbf{Y} \mid \boldsymbol{\beta}, \mathbf{w}, \sigma_\epsilon^2 \sim N\left(\boldsymbol{\mu} + \mathbf{w} , \; \sigma_\epsilon^2 I_{Tn\times Tn}\right) . \qquad (8.15)$$

If $\mathbf{w}_t \sim N(\mathbf{0}, \sigma_w^{2(t)} H(\delta^{(t)}))$ independently for $t = 1, \ldots, T$, then, marginalizing over $\mathbf{w}$,

$$\mathbf{Y} \mid \boldsymbol{\beta}, \sigma_\epsilon^2, \boldsymbol{\sigma}_w^2, \boldsymbol{\delta} \sim N(\boldsymbol{\mu} , \; D(\boldsymbol{\sigma}_w^2, \boldsymbol{\delta}) + \sigma_\epsilon^2 I_{Tn\times Tn}) , \qquad (8.16)$$

where $\boldsymbol{\sigma}_w^{2'} = (\sigma_w^{2(1)}, ..., \sigma_w^{2(T)})$, $\boldsymbol{\delta}' = (\delta^{(1)}, \ldots, \delta^{(T)})$, and $D(\boldsymbol{\sigma}_w^2, \boldsymbol{\delta})$ is block diagonal with the tth block being $\sigma_w^{2(t)}(H(\delta^{(t)}))$. Because D is block diagonal, likelihood evaluation associated with (8.16) is less of an issue than for (8.11) and (8.12).

We note that with either (8.7) or (8.8), $e(\mathbf{s}, t)$ is comprised of two sources of error that the data cannot directly separate. However, by incorporating a stochastic assumption on the $\alpha_\mathbf{s}(t)$ or on the $w_t(\mathbf{s})$, we can learn about the processes that guide the error components, as (8.13) and (8.16) reveal.

8.1.4 Prediction and forecasting

We now turn to forecasting under (8.2) with models (8.6), (8.7), or (8.8). Such forecasting involves prediction at location s_0 and time t_0, i.e., of $Y(s_0, t_0)$. Here s_0 may correspond to an already observed location, perhaps to a new location. However, typically $t_0 > T$ is of interest. Such prediction requires specification of an associated vector of characteristics $X(s_0, t_0)$. Also, prediction for $t_0 > T$ is available in the fixed coefficients case. For the time-varying coefficients case, we would need to specify a temporal model for $\beta(t)$.

In general, within the Bayesian framework, prediction at (s_0, t_0) follows from the posterior predictive distribution of $f(Y(s_0, t_0) \mid Y)$ where Y denotes the observed data vector. Assuming s_0 and t_0 are new, and for illustration, taking $U(s, t)$ as in (8.6),

$$f(Y(s_0, t_0) \mid Y) = \int f(Y(s_0, t_0) \mid \beta, \sigma_\epsilon^2, \alpha(t_0), w(s_0)) \\ \times dF(\beta, \alpha, w, \sigma_\epsilon^2, \sigma_\alpha^2, \rho, \sigma_w^2, \delta, \alpha(t_0), w(s_0) \mid Y) . \quad (8.17)$$

Using (8.17), given a random draw $(\beta^*, \sigma_\epsilon^{2*}, \alpha(t_0)^*, w(s_0)^*)$ from the posterior $f(\beta, \sigma_\epsilon^2, \alpha(t_0), w(s_0) \mid Y)$, if we draw $Y^*(s_0, t_0)$ from $N(X'(s_0, t_0)\beta^* + \alpha(t_0)^* + w(s_0)^*, \sigma_\epsilon^{2*})$, marginally, $Y^*(s_0, t_0) \sim f(Y(s_0, t_0) \mid Y)$.

Using sampling-based model fitting and working with (8.10), we obtain samples $(\beta^*, \sigma_\epsilon^{2*}, \sigma_\alpha^{2*}, \rho^*, \sigma_w^{2*}, \delta^*, \alpha^*, w^*)$ from the posterior distribution, $p(\beta, \sigma_\epsilon^2, \sigma_\alpha^2, \rho, \sigma_w^2, \delta, \alpha, w \mid Y)$. But $f(\beta, \sigma_\epsilon^2, \sigma_\alpha^2, \rho, \sigma_w^2, \delta, \alpha, w, \alpha(t_0), w(s_0) \mid Y) = f(\alpha(t_0) \mid \alpha, \sigma_\alpha^2, \rho) \cdot f(w(s_0) \mid w, \sigma_w^2, \delta) \cdot f(\beta, \sigma_\epsilon^2, \sigma_\alpha^2, \rho, \sigma_w^2, \delta, \alpha, w \mid Y)$. If, e.g., $t_0 = T+1$, and $\alpha(t)$ is modeled as a time series, $f(\alpha(T+1) \mid \alpha, \sigma_\alpha^2, \rho)$ is $N(\rho\alpha(T), \sigma_\alpha^2)$. If the $\alpha(t)$ are coefficients associated with dummy variables, setting $\alpha(T+1) = \alpha(T)$ is, arguably, the best one can do. The joint distribution of w and $w(s_0)$ is a multivariate normal from which $f(w(s_0) \mid w, \sigma_w^2, \delta)$ is a univariate normal. So if $\alpha(t_0)^* \sim f(\alpha(t_0) \mid \alpha^*, \sigma_\alpha^{2*}, \rho^*)$ and $w(s_0)^* \sim f(w(s_0) \mid w^*, \sigma_w^{2*}, \delta^*)$, along with β^* and σ_ϵ^{2*} we obtain a draw from $f(\beta, \sigma_\epsilon^2, \alpha(t_0), w(s_0) \mid Y)$. (If $t_0 \epsilon\{1, 2, ..., T\}$, $\alpha(t_0)$ is a component of α, then $\alpha(t_0)^*$ is a component of α^*. If s_0 is one of the $s_1, s_2, ..., s_n$, $w^*(s_0)$ is a component of w^*.) Alternatively, one can work with (8.13). Now, having marginalized over α and w, $Y(s, t)$ and Y are no longer independent. They have a multivariate normal distribution from which $f(Y(s, t) \mid Y, \beta, \sigma_\epsilon^2, \sigma_\alpha^2, \rho, \sigma_w^2, \delta)$ must be obtained. Note that for multiple predictions, $w(s_0)$ is replaced by a vector, say w_0. Now $f(w_0 \mid w, \sigma_w^2, \delta)$ is a multivariate normal distribution. No additional complications arise.

Example 8.1 *(Baton Rouge home sales).* We present a portion of the data analysis developed in Gelfand et al. (2003) for sales of single-family homes drawn from two regions in the city of Baton Rouge, LA. The two areas are known as Sherwood Forest and Highland Road. These regions are approximately the same size and have similar levels of transaction activity; they differ chiefly in the range of neighborhood characteristics and house

	Highland		Sherwood	
Year	Repeat	Single	Repeat	Single
1985	25	40	32	29
1986	20	35	32	39
1987	27	32	27	37
1988	16	26	20	34
1989	21	25	24	35
1990	42	29	27	37
1991	29	30	25	31
1992	33	38	39	27
1993	24	40	31	40
1994	26	35	20	34
1995	26	35	21	32
Total	289	365	298	375

Table 8.1 *Sample size by region, type of sale, and year.*

amenities found within. Sherwood Forest is a large, fairly homogeneous neighborhood located east, southeast of downtown Baton Rouge. Highland Road, on the other hand, is a major thoroughfare connecting downtown with the residential area to the southeast. Rather than being one homogeneous neighborhood, the Highland Road area consists, instead, of heterogeneous subdivisions. Employing two regions makes a local isotropy assumption more comfortable and allows investigation of possibly differing time effects and location dynamics.

For these regions, a subsample of all homes sold only once during the period 1985 through 1995 (single-sale transactions) and a second subsample of homes sold more than once (repeat-sale transactions) were drawn. These two samples can be studied separately to assess whether the population of single-sale houses differs from that of repeat-sale houses. The sample sizes are provided by year in Table 8.1. The location of each property is defined by its latitude and longitude coordinates, rescaled to UTM projection. In addition, a variety of house characteristics, to control for physical differences among the properties, are recorded at the time of sale. We use age, living area, other area (e.g., patios, garages, and carports) and number of bathrooms as covariates in our analysis. Summary statistics for these attributes appear in Table 8.2. We see that the homes in the Highland Road area are somewhat newer and slightly larger than those in the Sherwood area. The greater heterogeneity of the Highland Road homes is borne out

	Highland		Sherwood	
Variable	Repeat	Single	Repeat	Single
Age	11.10	12.49	14.21	14.75
	(8.15)	(11.37)	(8.32)	(10.16)
Bathrooms	2.18	2.16	2.05	2.02
	(0.46)	(0.56)	(0.36)	(0.40)
Living area	2265.4	2075.8	1996.0	1941.5
	(642.9)	(718.9)	(566.8)	(616.2)
Other area	815.1	706.0	726.0	670.6
	(337.7)	(363.6)	(258.1)	(289.2)

Table 8.2 *Mean (standard deviation) for house characteristics by region and type of sale.*

by the almost uniformly higher standard deviations for each covariate. In fact, we have more than 20 house characteristics in our data set, but elaborating the mean with additional features provides little improvement in R^2 and introduces multicollinearity problems. So, we confine ourselves to the four explanatory variables above and turn to spatial modeling to explain a portion of the remaining variability. Empirical semivariograms (2.9) offer evidence of spatial association, after adjusting for house characteristics.

We describe the results of fitting the model with mean $\mu(\mathbf{s}) = \mathbf{x}(\mathbf{s})^T \boldsymbol{\beta}$ and the error structure in (8.8). This is also the preferred model using the predictive model choice approach of Gelfand and Ghosh (4.13); we omit details. Fixed coefficients were justified by the shortness of the observation period. Again, an exponential isotropic correlation function was adopted.

To complete the Bayesian specification, we adopt rather noninformative priors in order to resemble a likelihood/least squares analysis. In particular, we assume a flat prior on the regression parameter $\boldsymbol{\beta}$ and inverse gamma (a, b) priors for $\sigma_\epsilon^2, \sigma_w^{2(t)}$ and $\delta^{(t)}, t = 1, \ldots T$. The shape parameter for these inverse gamma priors was fixed at two, implying an infinite prior variance. We choose the inverse gamma scale parameter for all $\delta^{(t)}$'s to be equal, i.e., $b_{\delta(1)} = b_{\delta(2)} = \ldots = b_{\delta(T)} = b_\delta$, say, and likewise for $\sigma_w^{2(t)}$. Furthermore, we set $b_{\sigma_\epsilon} = b_{\sigma_w^2}$ reflecting uncertain prior contribution from the nugget to the sill. Finally, the exact values of b_{σ_ϵ}, $b_{\sigma_w^2}$ and b_δ vary between region and type of sale reflecting different prior beliefs about these characteristics.

Variable	Repeat	Single
Highland region:		
intercept (β_0)	11.63 (11.59, 11.66)	11.45 (11.40, 11.50)
age (β_1)	−0.04 (−0.07, −0.02)	−0.08 (−0.11, −0.06)
bathrooms (β_2)	0.02 (−0.01, 0.04)	0.02 (−0.01, 0.05)
living area (β_3)	0.28 (0.25, 0.31)	0.33 (0.29, 0.37)
other area (β_4)	0.08 (0.06, 0.11)	0.07 (0.04, 0.09)
Sherwood region:		
intercept (β_0)	11.33 (11.30, 11.36)	11.30 (11.27, 11.34)
age (β_1)	−0.06 (−0.07, −0.04)	−0.05 (−0.07, −0.03)
bathrooms (β_2)	0.05 (0.03, 0.07)	0.00 (−0.02, 0.02)
living area (β_3)	0.19 (0.17, 0.21)	0.22 (0.19, 0.24)
other area (β_4)	0.02 (0.01, 0.04)	0.06 (0.04, 0.08)

Table 8.3 *Parameter estimates (median and 95% interval estimates) for house characteristics.*

Inference for the house characteristic coefficients is provided in Table 8.3 (point and 95% interval estimates). Age, living area, and other area are significant in all cases; number of bathrooms is significant only in Sherwood repeat sales. Significance of living area is much stronger in Highland than in Sherwood. The Highland sample is composed of homes from several heterogeneous neighborhoods. As such, living area not only measures differences in house size, but may also serve as a partial proxy for construction quality and for neighborhood location within the sample. The greater homogeneity of homes in Sherwood implies less variability in living area (as seen in Table 8.2) and reduces the importance of these variables in explaining house price.

Turning to the error structure, the parameters of interest for each region are the $\sigma_w^{2(t)}$, the $\delta^{(t)}$, and σ_e^2. The sill at time t is $Var(Y(s,t)) = \sigma_w^{2(t)} + \sigma_e^2$. Figure 8.1 plots the posterior medians of these sills. We see considerable difference in variability over the groups and over time, providing support for distinct spatial models at each t. Variability is highest for Highland single sales, lowest for Sherwood repeats. The additional insight is the effect of time. Variability is generally increasing over time.

We can obtain posterior median and interval estimates for $\sigma_w^{2(t)}/(\sigma_e^2 + \sigma_w^{2(t)})$, the proportion of spatial variance to total. The strength of the spatial story is considerable; 40 to 80% of the variability is spatial.

In Figure 8.2 we provide point and interval estimates for the range. The

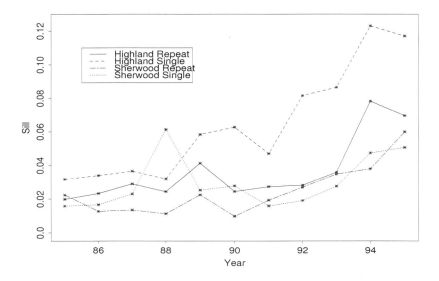

Figure 8.1 *Posterior median sill by year.*

ranges for the repeat sales are quite similar for the two regions, showing some tendency to increase in the later years of observation. By contrast, the range for the Highland single sales is much different from that for Sherwood. It is typically greater and much more variable. The latter again is a refection of the high variability in the single-sale home prices in Highland. The resulting posteriors are more dispersed.

Finally, in Figure 8.3, we present the posterior distribution of the block averages, mentioned at the end of Subsection 8.1.2, for each of the four analyses. Again, these block averages are viewed as analogues of more familiar time dummy variables. Time effects are evident. In all cases, we witness somewhat of a decline in magnitude in the 1980s and an increasing trend in the 1990s. ∎

8.2 Point-level modeling with continuous time

Suppose now that $\mathbf{s} \in \Re^2$ and $t \in \Re^+$ and we seek to define a spatiotemporal process $Y(\mathbf{s}, t)$. As in Subsection 2.2.1 we have to provide a joint distribution for an uncountable number of random variables. Again, we do this through arbitrary finite dimensional distributions. Confining ourselves to the Gaussian case, we only need to specify a valid spatiotemporal covariance function. Here, "valid" means that for any set of locations and any set of time points, the covariance matrix for the resulting set of ran-

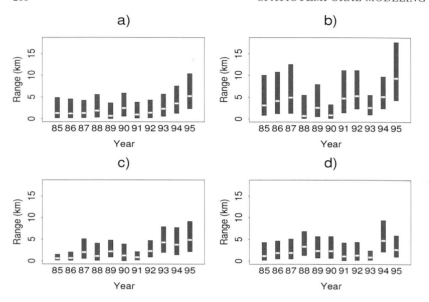

Figure 8.2 *Posterior median and 95% interval estimates for the range by year for (a) Highland repeat sales, (b) Highland single sales, (c) Sherwood repeat sales, and (d) Sherwood single sales.*

dom variables is positive definite. An important point here is that it is not sensible to combine **s** and t and propose a valid correlation function on $\Re^3$. This is because distance in space has nothing to do with "distance" on the time scale.

As a result, a stationary spatiotemporal covariance specification is assumed to take the form $cov(Y(\mathbf{s}, t), Y(\mathbf{s}', t')) = c(\mathbf{s} - \mathbf{s}', t - t')$. An isotropic form sets $cov(Y(\mathbf{s}, t), Y(\mathbf{s}', t')) = c(\|\mathbf{s} - \mathbf{s}'\|, |t - t'|)$. A frequently used choice is the *separable* form

$$cov(Y(\mathbf{s}, t), Y(\mathbf{s}', t')) = \sigma^2 \rho^{(1)}(\mathbf{s} - \mathbf{s}'; \boldsymbol{\phi}) \, \rho^{(2)}(t - t'; \boldsymbol{\psi}) , \qquad (8.18)$$

where $\rho^{(1)}$ is a valid two-dimensional correlation function and $\rho^{(2)}$ is a valid one-dimensional correlation function. Expression (8.18) shows that dependence attenuates in a multiplicative manner across space and time. Forms such as (8.18) have a history in spatiotemporal modeling; see, e.g., Mardia and Goodall (1993) and references therein.

Why is (8.18) valid? For locations $\mathbf{s}_1, \ldots, \mathbf{s}_I$ and times $t_1, \ldots, t_J$, collecting the variables a vector $\mathbf{Y}_s^T = (\mathbf{Y}^T(\mathbf{s}_1), \ldots, \mathbf{Y}^T(\mathbf{s}_I))$ where $\mathbf{Y}(\mathbf{s}_i) = (Y(\mathbf{s}_i, t_1), \ldots, Y(\mathbf{s}_i, t_J))^T$, the covariance matrix of $\mathbf{Y}_s$ is

$$\Sigma_{\mathbf{Y}_s}(\sigma^2, \phi, \psi) = \sigma^2 H_s(\boldsymbol{\phi}) \otimes H_t(\boldsymbol{\psi}) , \qquad (8.19)$$

where "$\otimes$" again denotes the Kronecker product. In (8.19), $H_s(\boldsymbol{\phi})$ is $I \times I$

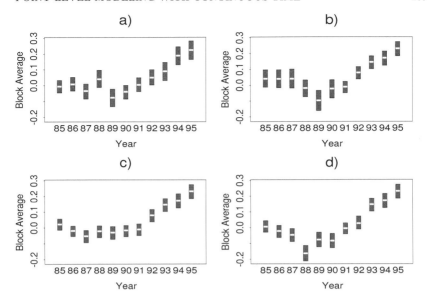

Figure 8.3 *Posterior median and 95% interval estimates for the block averages by year for (a) Highland repeat sales, (b) Highland single sales, (c) Sherwood repeat sales, and (d) Sherwood single sales.*

with $(H_s(\phi))_{ii'} = \rho^{(1)}(\mathbf{s}_i - \mathbf{s}_i'; \boldsymbol{\theta})$, and $H_t(\boldsymbol{\psi})$ is $J \times J$ with $(H_t(\boldsymbol{\psi}))_{jj'} = \rho^{(2)}(t_j - t_{j'}; \boldsymbol{\psi})$. Expression (8.19) clarifies that $\Sigma_{\mathbf{Y}_s}$ is positive definite, following the argument below (7.2). So, $\mathbf{Y}_s$ will be IJ-dimensional multivariate normal with, in obvious notation, mean vector $\boldsymbol{\mu}_s(\boldsymbol{\beta})$ and covariance matrix (8.19).

Given a prior for $\boldsymbol{\beta}, \sigma^2, \phi$, and $\boldsymbol{\psi}$, the Bayesian model is completely specified. Simulation-based model fitting can be carried out similarly to the static spatial case by noting the following. The log-likelihood arising from $\mathbf{Y}_s$ is

$$-\tfrac{1}{2} \log \left| \sigma^2 H_s(\phi) \otimes H_t(\boldsymbol{\psi}) \right|$$
$$-\tfrac{1}{2\sigma^2}(\mathbf{Y}_s - \boldsymbol{\mu}_s(\boldsymbol{\beta}))^T (H_s(\phi) \otimes H_t(\boldsymbol{\psi}))^{-1}(\mathbf{Y}_s - \boldsymbol{\mu}_s(\boldsymbol{\beta})) \ .$$

But in fact $\left| \sigma^2 H_s(\phi) \otimes H_t(\boldsymbol{\psi}) \right| = (\sigma^2)^{IJ} \left| H_s(\phi) \right|^J \left| H_t(\boldsymbol{\psi}) \right|^I$ and $(H_s(\phi) \otimes H_t(\boldsymbol{\psi}))^{-1} = H_s^{-1}(\phi) \otimes H_t^{-1}(\boldsymbol{\psi})$ by properties of Kronecker products. In other words, even though (8.19) is $IJ \times IJ$, we need only the determinant and inverse for an $I \times I$ and a $J \times J$ matrix, expediting likelihood evaluation and hence Gibbs sampling.

With regard to prediction, first consider new locations $\mathbf{s}_1', ..., \mathbf{s}_k'$ with interest in inference for $Y(\mathbf{s}_k', t_j)$. As with the observed data, we collect the $Y(\mathbf{s}_k', t_j)$ into vectors $\mathbf{Y}(\mathbf{s}_k')$, and the $\mathbf{Y}(\mathbf{s}_k')$ into a single $KJ \times 1$ vector $\mathbf{Y}_{s'}$. Even though we may not necessarily be interested in every component of

$\mathbf{Y}_{s'}$, the simplifying forms that follow suggest that, with regard to programming, it may be easiest to simulate draws from the entire predictive distribution $f(\mathbf{Y}_{s'} \mid \mathbf{Y}_s)$ and then retain only the desired components.

Since $f(\mathbf{Y}_{s'}|\mathbf{Y}_s)$ has a form analogous to (6.3), given posterior samples $(\boldsymbol{\beta}_g^*, \sigma_g^{2*}, \boldsymbol{\phi}_g^*, \boldsymbol{\psi}_g^*)$, we draw $\mathbf{Y}_{s',g}^*$ from $f(\mathbf{Y}_{s'} \mid \mathbf{Y}_s, \boldsymbol{\beta}_g^*, \sigma_g^{2*}, \boldsymbol{\phi}_g^*, \boldsymbol{\psi}_g^*)$, $g = 1, \ldots, G$. Analogous to (6.4),

$$f\left(\begin{pmatrix}\mathbf{Y}_s \\ \mathbf{Y}_{s'}\end{pmatrix}\Bigg| \boldsymbol{\beta}, \sigma^2, \boldsymbol{\phi}, \boldsymbol{\psi}\right) = N\left(\begin{pmatrix}\boldsymbol{\mu}_s(\boldsymbol{\beta}) \\ \boldsymbol{\mu}_{s'}(\boldsymbol{\beta})\end{pmatrix}, \Sigma_{\mathbf{Y}_s,\mathbf{Y}_{s'}}\right) \tag{8.20}$$

where

$$\Sigma_{\mathbf{Y}_s,\mathbf{Y}_{s'}} = \sigma^2 \begin{pmatrix} H_s(\boldsymbol{\phi}) \otimes H_t(\boldsymbol{\psi}) & H_{s,s'}(\boldsymbol{\phi}) \otimes H_t(\boldsymbol{\psi}) \\ H_{s,s'}^T(\boldsymbol{\phi}) \otimes H_t(\boldsymbol{\psi}) & H_{s'}(\boldsymbol{\phi}) \otimes H_t(\boldsymbol{\psi}) \end{pmatrix},$$

with obvious definitions for $H_{s'}(\boldsymbol{\phi})$ and $H_{s,s'}(\boldsymbol{\phi})$. But then the conditional distribution $\mathbf{Y}_{s'} \mid \mathbf{Y}_s, \boldsymbol{\beta}, \sigma^2, \boldsymbol{\phi}, \boldsymbol{\psi}$ is also normal, with mean

$$\boldsymbol{\mu}_{s'}(\boldsymbol{\beta}) + (H_{s,s'}^T(\boldsymbol{\phi}) \otimes H_t(\boldsymbol{\phi}))(H_s(\boldsymbol{\phi}) \otimes H_t(\boldsymbol{\psi}))^{-1}(Y_s - \boldsymbol{\mu}_s(\boldsymbol{\beta}))$$
$$= \boldsymbol{\mu}_{s'}(\boldsymbol{\beta}) + (H_{s,s'}^T(\boldsymbol{\phi})H_s^{-1}(\boldsymbol{\phi}) \otimes I_{J \times J})(\mathbf{Y}_s - \boldsymbol{\mu}_s(\boldsymbol{\beta})), \tag{8.21}$$

and covariance matrix

$$H_{s'}(\boldsymbol{\phi}) \otimes H_t(\boldsymbol{\psi})$$
$$-(H_{s,s'}^T \otimes H_t(\boldsymbol{\psi}))(H_s(\boldsymbol{\phi}) \otimes H_t(\boldsymbol{\psi}))^{-1}(H_{s,s'}(\boldsymbol{\phi}) \otimes H_t(\boldsymbol{\psi})) \tag{8.22}$$
$$= (H_{s'}(\boldsymbol{\phi}) - H_{s,s'}^T(\boldsymbol{\phi})H_s^{-1}(\boldsymbol{\phi})H_{s,s}(\boldsymbol{\phi})) \otimes H_t(\boldsymbol{\psi}),$$

using standard properties of Kronecker products. In (8.21), time disappears apart from $\boldsymbol{\mu}_{s'}(\boldsymbol{\beta})$, while in (8.22), time "factors out" of the conditioning. Sampling from this normal distribution usually employs the inverse square root of the conditional covariance matrix, but conveniently, this is

$$(H_{s'}(\boldsymbol{\phi}) - H_{s,s'}^T(\boldsymbol{\phi})H_s^{-1}(\boldsymbol{\phi})H_{s,s'}(\boldsymbol{\phi}))^{-\frac{1}{2}} \otimes H_t^{-\frac{1}{2}}(\boldsymbol{\psi}),$$

so the only work required beyond that in (6.5) is obtaining $H_t^{-\frac{1}{2}}(\boldsymbol{\psi})$, since $H_t^{-1}(\boldsymbol{\psi})$ will already have been obtained in evaluating the likelihood, following the discussion above.

For prediction not for points but for areal units (blocks) $B_1, \ldots, B_K$, we would set $\mathbf{Y}^T(B_k) = (Y(B_k, t_1), \ldots, Y(B_k, t_J))$ and then further set $\mathbf{Y}_B^T = (\mathbf{Y}^T(B_1), \ldots, \mathbf{Y}^T(B_K))$. Analogous to (6.6) we seek to sample $f(\mathbf{Y}_B \mid \mathbf{Y}_s)$, so we require $f(\mathbf{Y}_B \mid \mathbf{Y}_s, \boldsymbol{\beta}, \sigma^2, \boldsymbol{\phi}, \boldsymbol{\psi})$. Analogous to (8.20), this can be derived from the joint distribution $f((\mathbf{Y}_s, \mathbf{Y}_B)^T \mid \boldsymbol{\beta}, \sigma^2, \boldsymbol{\phi}, \boldsymbol{\psi})$, which is

$$N\left(\begin{pmatrix}\boldsymbol{\mu}_s(\boldsymbol{\beta}) \\ \boldsymbol{\mu}_B(\boldsymbol{\beta})\end{pmatrix}, \sigma^2 \begin{pmatrix} H_s(\boldsymbol{\phi}) \otimes H_t(\boldsymbol{\psi}) & H_{s,B}(\boldsymbol{\phi}) \otimes H_t(\boldsymbol{\psi}) \\ H_{s,B}^T(\boldsymbol{\phi}) \otimes H_t(\boldsymbol{\psi}) & H_B(\boldsymbol{\phi}) \otimes H_t(\boldsymbol{\psi}) \end{pmatrix}\right),$$

with $\boldsymbol{\mu}_B(\boldsymbol{\beta})$, $H_B(\boldsymbol{\phi})$, and $H_{s,B}(\boldsymbol{\phi})$ defined as in Section 6.1.2. Thus the distribution $f(\mathbf{Y}_B|\mathbf{Y}_s, \boldsymbol{\beta}, \sigma^2, \boldsymbol{\phi}, \boldsymbol{\psi})$ is again normal with mean and covariance matrix as given in (8.21) and (8.22), but with $\boldsymbol{\mu}_B(\boldsymbol{\beta})$ replacing $\boldsymbol{\mu}_{s'}(\boldsymbol{\beta})$, $H_B(\boldsymbol{\phi})$ replacing $H_{s'}(\boldsymbol{\phi})$, and $H_{s,B}(\boldsymbol{\phi})$ replacing $H_{s,s'}(\boldsymbol{\phi})$. Using the same

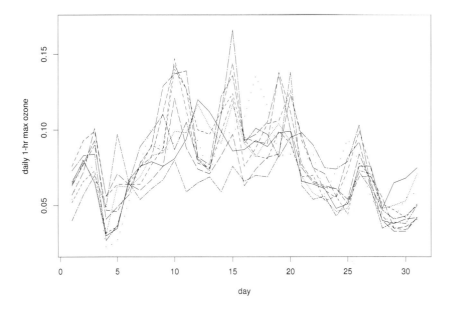

Figure 8.4 *Observed 1-hour maximum ozone measurement by day, July 1995, 10 Atlanta monitoring sites.*

Monte Carlo integrations as proposed in Section 6.1.2 leads to sampling the resultant $\widehat{f}(\mathbf{Y}_B \mid \mathbf{Y}_s, \boldsymbol{\beta}, \sigma^2, \boldsymbol{\phi}, \boldsymbol{\psi})$, and the same technical justification applies.

If we started with block data, $Y(B_i, t_j)$, then following (6.11) and (8.19),

$$f(\mathbf{Y}_B \mid \boldsymbol{\beta}, \sigma^2, \boldsymbol{\phi}, \boldsymbol{\psi}) = N(\mu_B(\boldsymbol{\beta}),\ \sigma^2(H_B(\boldsymbol{\phi}) \otimes H_t(\boldsymbol{\psi}))) . \qquad (8.23)$$

Given (8.23), the path for prediction at new points or at new blocks is clear, following the above and the end of Section 6.1.2; we omit the details.

Note that the association structure in (8.18) allows *forecasting* of the spatial process at time t_{J+1}. This can be done at observed or unobserved points or blocks following the foregoing development. To retain the above simplifying forms, we would first simulate the variables at t_{J+1} associated with observed points or blocks (with no change of support). We would then revise $H_t(\boldsymbol{\phi})$ to be $(J+1) \times (J+1)$ before proceeding as above.

Example 8.2 To illustrate the methods above, we use a spatiotemporal version of the Atlanta ozone data set. As mentioned in Section 6.1, we actually have ozone measurements at the 10 fixed monitoring stations shown in Figure 1.3 over the 92 summer days in 1995. Figure 8.4 shows the daily 1-hour maximum ozone reading for the sites during July of this same year.

| | Spatial only | | Spatiotemporal | |
	Point	95% Interval	Point	95% Interval
Point A	.125	(.040, .334)	.139	(.111, .169)
Point B	.116	(.031, .393)	.131	(.098, .169)
Zip 30317 (east-central)	.130	(.055, .270)	.138	(.121, .155)
Zip 30344 (south-central)	.123	(.055, .270)	.135	(.112, .161)
Zip 30350 (north)	.112	(.040, .283)	.109	(.084, .140)

Table 8.4 *Posterior medians and 95% equal-tail credible intervals for ozone levels at two points, and for average ozone levels over three blocks (zip codes), purely spatial model versus spatiotemporal model, Atlanta ozone data for July 15, 1995.*

There are several sharp peaks, but little evidence of a weekly (7-day) period in the data. The mean structure appears reasonably constant in space, with the ordering of the site measurements changing dramatically for different days. Moreover, with only 10 "design points" in the metro area, any spatial trend surface we fit would be quite speculative over much of the study region (e.g., the northwest and southwest metro; see Figure 1.3). The temporal evolution of the series is not inconsistent with a constant mean autoregressive error model; indeed, the lag 1 sample autocorrelation varies between .27 and .73 over the 10 sites, strongly suggesting the need for a model accounting for both spatial and temporal correlations.

We thus fit our spatiotemporal model with mean $\mu(\mathbf{s}, t; \boldsymbol{\beta}) = \mu$, but with spatial and temporal correlation functions $\rho^{(1)}(\mathbf{s}_i - \mathbf{s}_{i'}; \phi) = e^{-\phi\|\mathbf{s}_i - \mathbf{s}_{i'}\|}$ and $\rho^{(2)}(t_j - t_{j'}; \psi) = \psi^{|j-j'|}/(1 - \psi^2)$. Hence our model has four parameters: we use a flat prior for μ, an $IG(3, 0.5)$ prior for σ^2, a $G(0.003, 100)$ prior for ϕ, and a $U(0, 1)$ prior for ψ (thus eliminating the implausible possibility of *negative* autocorrelation in our data, but favoring no positive value over any other). To facilitate our Gibbs-Metropolis approach, we transform to $\theta = \log \phi$ and $\lambda = \log(\psi/(1-\psi))$, and subsequently use Gaussian proposals on these transformed parameters.

Running 3 parallel chains of 10,000 iterations each, sample traces (not shown) again indicate virtually immediate convergence of our algorithm. Posterior medians and 95% equal-tail credible intervals for the four parameters are as follows: for μ, 0.068 and (0.057, 0.080); for σ^2, 0.11 and (0.08, 0.17); for ϕ, 0.06 and (0.03, 0.08); and for ψ, 0.42 and (0.31, 0.52). The rather large value of ψ confirms the strong temporal autocorrelation suspected in the daily ozone readings.

Comparison of the posteriors for σ^2 and ϕ with those obtained for the static spatial model in Example 6.1 is not sensible, since these parameters have different meanings in the two models. Instead, we make this com-

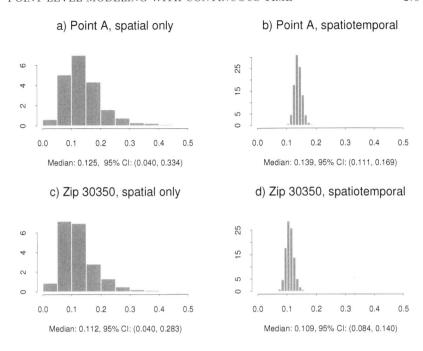

Figure 8.5 *Posterior predictive distributions for ozone concentration at point A and the block average over zip 30350, purely spatial model versus spatiotemporal model, Atlanta ozone data for July 15, 1995.*

parison in the context of point-point and point-block prediction. Table 8.4 provides posterior predictive summaries for the ozone concentrations for July 15, 1995, at points A and B (see Figure 1.3), as well as for the block averages over 3 selected Atlanta city zips: 30317, an east-central city zip very near to two monitoring sites; 30344, the south-central zip containing the points A and B; and 30350, the northernmost city zip. Results are shown for both the spatiotemporal model of this subsection and for the static spatial model previously fit in Example 6.1. Note that all the posterior medians are a bit higher under the spatiotemporal model, except for that for the northern zip, which remains low. Also note the significant increase in precision afforded by this model, which makes use of the data from all 31 days in July 1995, instead of only that from July 15. Figure 8.5 shows the estimated posteriors giving rise to the first and last rows in Table 8.4 (i.e., corresponding to the the July 15, 1995, ozone levels at point A and the block average over the northernmost city zip, 30350). The Bayesian approach's ability to reflect differing amounts of predictive uncertainty for the two models is clearly evident.

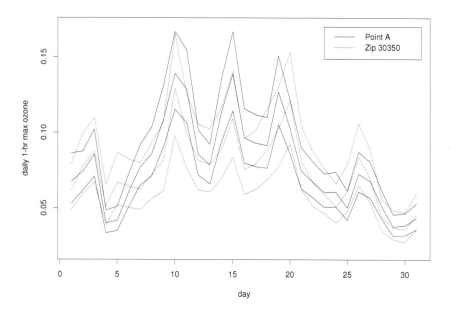

Figure 8.6 *Posterior medians and upper and lower .025 quantiles for the predicted 1-hour maximum ozone concentration by day, July 1995; solid lines, point A; dotted lines, block average over zip 30350 (northernmost Atlanta city zip).*

Finally, Figure 8.6 plots the posterior medians and upper and lower .025 quantiles produced by the spatiotemporal model by day for the ozone concentration at point A, as well as those for the block average in zip 30350. Note that the overall temporal pattern is quite similar to that for the data shown in Figure 8.4. Since point A is rather nearer to several data observation points, the confidence bands associated with it are often a bit narrower than those for the northern zip, but this pattern is not perfectly consistent over time. Also note that the relative positions of the bands for July 15 are consistent with the data pattern for this day seen in Figure 1.3, when downtown ozone exposures were higher than those in the northern metro. Finally, the day-to-day variability in the predicted series is substantially larger than the predictive variability associated with any given day. ∎

8.3 Nonseparable spatiotemporal models ⋆

The separable form for the spatiotemporal covariance function in (8.18) is convenient for computation and offers attractive interpretation. However, its form limits the nature of space-time interaction. Additive forms, arising from $w(\mathbf{s},t) = w(\mathbf{s}) + \alpha(t)$ with $w(\mathbf{s})$ and $\alpha(t)$ independent may be even more unsatisfying.

A simple way to extend (8.18) is through *mixing*. For instance, suppose $w(\mathbf{s},t) = w_1(\mathbf{s},t) + w_2(\mathbf{s},t)$ with w_1 and w_2 independent processes, each with a separable spatiotemporal covariance function, say $c_\ell(\mathbf{s} - \mathbf{s}', t - t) = \sigma_\ell^2 \rho_\ell^{(1)}(\mathbf{s} - \mathbf{s}') \rho_\ell^{(2)}(t - t')$, $\ell = 1, 2$. Then the covariance function for $w(\mathbf{s},t)$ is evidently the sum and is not separable. Building covariance functions in this way is easy to interpret but yields an explosion of parameters with finite mixing. Continuous parametric mixing, e.g.,

$$c(\mathbf{s} - \mathbf{s}', t - t) = \sigma^2 \int \rho^{(1)}(\mathbf{s} - \mathbf{s}', \boldsymbol{\phi}) \rho^{(2)}(t - t', \boldsymbol{\psi}) \, G_{\boldsymbol{\gamma}}(d\boldsymbol{\phi}, d\boldsymbol{\psi}) \,, \quad (8.24)$$

yields a function that depends only on σ^2 and $\boldsymbol{\gamma}$. Such forms have not received much attention in the literature to date.

Cressie and Huang (1999) introduce a flexible class of nonseparable stationary covariance functions that allow for space-time interaction. However, they work in the spectral domain and require that $c(\mathbf{s} - \mathbf{s}', t - t')$ can be computed explicitly, i.e., the Fourier inversion can be obtained in closed-form. Unfortunately this occurs only in very special cases. Recent work by Gneiting (2002) adopts a similar approach but obtains very general classes of valid space-time models that do not rely on closed form Fourier inversions. One simple example is the class $c(\mathbf{s} - \mathbf{s}', t - t') = \sigma^2(|t - t'| + 1)^{-1} \exp(-||\mathbf{s} - \mathbf{s}'||(|t - t'| + 1)^{-\beta/2})$. Here, β is a space-time interaction parameter; $\beta = 0$ provides a separable specification.

Stein (2003) also works in the spectral domain, providing a class of spectral densities whose resulting spatiotemporal covariance function is nonseparable with flexible analytic behavior. These spectral densities extend the Matérn form; see (2.13) or the discussion below equation (A.4) in Appendix A. In particular, the spectral density is

$$\widehat{c}(\mathbf{w}, v) \propto [c_1(\alpha_1^2 + ||\mathbf{w}||^2)^{\alpha_1} + c_2(\alpha_2 + v^2)^{\alpha_2}]^{-v} \,.$$

Unfortunately, the associated covariance function cannot be computed explicitly; fast Fourier transforms (see Appendix Section A.4) offer the best computational prospects. Also, unlike Gneiting's class, separability does not arise as a special or limiting case. We also mention related work using "blurring" discussed in Brown, Kåresen, Roberts, and Tonellato (2000).

8.4 Dynamic spatiotemporal models ⋆

In this section we follow the approach taken in Banerjee, Gamerman, and Gelfand, (2003), viewing the data as arising from a time series of spatial processes. In particular, we work in the setting of dynamic models (West and Harrison, 1997), describing the temporal evolution in a latent space. We achieve a class of dynamic models for spatiotemporal data.

Here, there is a growing literature. Non-Bayesian approaches include Huang and Cressie (1996), Wikle and Cressie (1999), and Mardia et al. (1998). Bayesian approaches include Tonellato (1997), Sanso and Guenni (1999), Stroud et al. (2001), and Huerta et al. (2003). The paper by Stroud et al. is attractive in being applicable to any data set that is continuous in space and discrete in time and allows straightforward computation using Kalman filtering.

8.4.1 Brief review of dynamic linear models

Dynamic linear models, often referred to as state-space models in the time-series literature, offer a versatile framework for fitting several time-varying models (West and Harrison, 1997). We briefly outline the general dynamic linear modeling framework. Thus, let $\mathbf{Y}_t$ be a $m \times 1$ vector of observables at time t. $\mathbf{Y}_t$ is related to a $p \times 1$ vector, $\boldsymbol{\theta}_t$, called the state vector, through a *measurement equation*. In general, the elements of $\boldsymbol{\theta}_t$ are not observable, but are generated by a first-order Markovian process, resulting in a *transition equation*. Therefore, we can describe the above framework as

$$\mathbf{Y}_t = F_t\boldsymbol{\theta}_t + \boldsymbol{\epsilon}_t, \ \boldsymbol{\epsilon}_t \sim N\left(\mathbf{0}, \Sigma_t^{\epsilon}\right).$$

$$\boldsymbol{\theta}_t = G_t\boldsymbol{\theta}_{t-1} + \boldsymbol{\eta}_t, \ \boldsymbol{\eta}_t \sim N\left(\mathbf{0}, \Sigma_t^{\eta}\right),$$

where F_t and G_t are $m \times p$ and $p \times p$ matrices, respectively. The first equation is the measurement equation, where $\boldsymbol{\epsilon}_t$ is a $m \times 1$ vector of serially uncorrelated Gaussian variables with mean $\mathbf{0}$ and an $m \times m$ covariance matrix, Σ_t^{ϵ}. The second equation is the transition equation with $\boldsymbol{\eta}_t$ being a $p \times 1$ vector of serially uncorrelated zero-centered Gaussian disturbances and Σ_t^{η} the corresponding $p \times p$ covariance matrix. Note that under (8.25), the association structure can be computed explicitly across time, e.g., $Cov\left(\boldsymbol{\theta}_t, \boldsymbol{\theta}_{t-1}\right) = G_t Var\left(\boldsymbol{\theta}_{t-1}\right)$ and $Cov\left(\mathbf{Y}_t, \mathbf{Y}_{t-1}\right) = F_t G_t Var\left(\boldsymbol{\theta}_{t-1}\right) F_t^T$.

F_t (in the measurement equation) and G_t (in the transition equation) are referred to as *system matrices* that may change over time. F_t and G_t may involve unknown parameters but, given the parameters, temporal evolution is in a predetermined manner. The matrix F_t is usually specified by the design of the problem at hand, while G_t is specified through modeling assumptions; for example, $G_t = I_p$, the $p \times p$ identity matrix would provide a random walk for $\boldsymbol{\theta}_t$. Regardless, the system is linear, and for any time point t, $\mathbf{Y}_t$ can be expressed as a linear combination of the present $\boldsymbol{\epsilon}_t$ and the present and past $\boldsymbol{\eta}_t$'s.

8.4.2 Formulation for spatiotemporal models

In this section we adapt the above dynamic modeling framework to univariate spatiotemporal models with spatially varying coefficients. For this we consider a collection of sites $S = \{s_1, ..., s_{N_s}\}$, and time-points $T = \{t_1, ..., t_{N_t}\}$, yielding observations $Y(s, t)$, and covariate vectors $x(s, t)$, for every $(s, t) \in S \times T$.

The response, $Y(s, t)$, is first modeled through a measurement equation, which incorporates the measurement error, $\epsilon(s, t)$, as serially and spatially uncorrelated zero-centered Gaussian disturbances. The transition equation now involves the regression parameters (slopes) of the covariates. The slope vector, say $\tilde{\beta}(s, t)$, is decomposed into a purely temporal component, β_t, and a spatiotemporal component, $\beta(s, t)$. Both these are generated through transition equations, capturing their Markovian dependence in time. While the transition equation of the purely temporal component is as in usual state-space modeling, the spatiotemporal component is generated by a multivariate Gaussian spatial process. Thus, we may write the spatiotemporal modeling framework as

$$
\begin{aligned}
Y(s, t) &= \mu(s, t) + \epsilon(s, t) \,;\, \epsilon(s, t) \overset{ind}{\sim} N\left(0, \sigma^2\right), &\text{(8.25)} \\
\mu(s, t) &= x^T(s, t)\tilde{\beta}(s, t), \\
\tilde{\beta}(s, t) &= \beta_t + \beta(s, t), &\text{(8.26)} \\
\beta_t &= \beta_{t-1} + \eta_t, \; \eta_t \overset{ind}{\sim} N_p\left(0, \Sigma_\eta\right), \\
\text{and } \beta(s, t) &= \beta(s, t-1) + w(s, t)\,.
\end{aligned}
$$

In (8.26), we introduce a linear model of coregionalization (Section 7.2) for $w(s, t)$, i.e., $w(s, t) = Av(s, t)$, with $v(s, t) = (v_1(s, t), ..., v_p(s, t))^T$. The $v_l(s, t)$ are serially independent replications of a Gaussian process with unit variance and correlation function $\rho_l(\cdot\,; \phi_l)$, henceforth denoted by $GP(0, \rho_l(\cdot\,; \phi_l))$, for $l = 1, \ldots, p$ and independent across l. In the current context, we assume that A does not depend upon (s, t). Nevertheless, this still allows flexible modeling for the spatial covariance structure, as we discuss below.

Moreover, allowing a spatially varying coefficient $\beta(s, t)$ to be associated with $x(s, t)$ provides an arbitrarily rich explanatory relationship for the x's with regard to the Y's (see Section 10.2 in this regard). By comparison, in Stroud et al. (2001), at a given t, a locally weighted mixture of linear regressions is proposed and only the purely temporal component of $\tilde{\beta}(s, t)$ is used. Such a specification requires both number of basis functions and number of mixture components.

Returning to our specification, note that if $v_l(\cdot, t) \overset{ind}{\sim} GP(0, \rho(\cdot; \phi))$, we have the intrinsic or separable model for $w(s, t)$. Allowing different correlation functions and decay parameters for the $v_l(\mathbf{s}, t)$, i.e., $v_l(\cdot, t) \overset{ind}{\sim}$ $GP(0, \rho_l(\cdot; \phi_l))$ yields the linear model of coregionalization (Section 7.2).

Following Subsection 8.4.1, we can compute the general association structure for the Y's under (8.25) and (8.26). For instance, we have the result that $Cov(Y(\mathbf{s}, t), Y(\mathbf{s}', t-1)) = \mathbf{x}^T(\mathbf{s}, t) \Sigma_{\tilde{\beta}(\mathbf{s}, t), \tilde{\beta}(\mathbf{s}', t-1)} \mathbf{x}(\mathbf{s}, t-1)$, where $\Sigma_{\tilde{\beta}(\mathbf{s}, t), \tilde{\beta}(\mathbf{s}', t-1)} = (t-1)\left(\Sigma_{\boldsymbol{\eta}} + \sum_{l=1}^{p} \rho_l(\mathbf{s} - \mathbf{s}'; \phi_l)\mathbf{a}_l\mathbf{a}_l^T\right)$. Furthermore, $var(Y(\mathbf{s}, t)) = \mathbf{x}^T(\mathbf{s}, t) t \left[\Sigma_{\boldsymbol{\eta}} + AA^T\right] \mathbf{x}(\mathbf{s}, t)$, with the result that $Corr(Y(\mathbf{s}, t), Y(\mathbf{s}', t-1))$ is $O(1)$ as $t \to \infty$.

A Bayesian hierarchical model for (8.25) and (8.26) may be completed by prior specifications such as

$$\boldsymbol{\beta}_0 \sim N(\mathbf{m}_0, C_0) \text{ and } \boldsymbol{\beta}(\cdot, 0) \equiv 0. \tag{8.27}$$
$$\Sigma_{\boldsymbol{\eta}} \sim IW\left(a_{\boldsymbol{\eta}}, B_{\boldsymbol{\eta}}\right), \ \Sigma_{\mathbf{w}} \sim IW\left(a_{\mathbf{w}}, B_{\mathbf{w}}\right) \text{ and } \sigma_\epsilon^2 \sim IG\left(a_\epsilon, b_\epsilon\right),$$
$$\mathbf{m}_0 \sim N(\mathbf{0}, \Sigma_0); \ \Sigma_0 = 10^5 \times I_p,$$

where $B_{\boldsymbol{\eta}}$ and $B_{\mathbf{w}}$ are $p \times p$ precision (hyperparameter) matrices for the inverted Wishart distribution.

Consider now data, in the form $(Y(\mathbf{s}_i, t_j))$ with $i = 1, 2, ..., N_s$ and $j = 1, 2, ..., N_t$. Let us collect, for each time point, the observations on all the sites. That is, we form, $\mathbf{Y}_t = (Y(\mathbf{s}_1, t), ..., Y(\mathbf{s}_{N_s}, t))^T$ and the $N_s \times N_s p$ block diagonal matrix $F_t = \left(\mathbf{x}^T(\mathbf{s}_1, t), \mathbf{x}^T(\mathbf{s}_2, t), ..., \mathbf{x}^T(\mathbf{s}_N, t)\right)$ for $t = t_1, ..., t_{N_t}$. ^{T}. Analogously we form the $N_s p \times 1$ vector $\boldsymbol{\theta}_t = \mathbf{1}_{N_s} \otimes \boldsymbol{\beta}_t + \boldsymbol{\beta}_t^*$, where $\boldsymbol{\beta}_t^* = (\boldsymbol{\beta}(\mathbf{s}_1, t), ..., \boldsymbol{\beta}(\mathbf{s}_{N_s}, t))^T$, $\boldsymbol{\beta}_t = \boldsymbol{\beta}_{t-1} + \boldsymbol{\eta}_t$, $\boldsymbol{\eta}_t \overset{ind}{\sim} N_p\left(\mathbf{0}, \Sigma_{\boldsymbol{\eta}}\right)$; and, with $\mathbf{w}_t = \left(\mathbf{w}^T(\mathbf{s}_1, t), ..., \mathbf{w}^T(\mathbf{s}_{N_s}, t)\right)^T$,

$$\boldsymbol{\beta}_t^* = \boldsymbol{\beta}_{t-1}^* + \mathbf{w}_t, \ \mathbf{w}_t \overset{ind}{\sim} N\left(\mathbf{0}, \sum_{l=1}^{p}(R_l(\phi_l) \otimes \Sigma_{\mathbf{w}, l})\right),$$

where $[R_l(\phi_l)]_{ij} = \rho_l(\mathbf{s}_i - \mathbf{s}_j; \phi_l)$ is the correlation matrix for $v_l(\cdot, t)$. We then write the data equation for a dynamic spatial model as

$$\mathbf{Y}_t = F_t \boldsymbol{\theta}_t + \boldsymbol{\epsilon}_t; \ t = 1, ..., N_t; \ \boldsymbol{\epsilon}_t \sim N\left(\mathbf{0}, \sigma_\epsilon^2 I_{N_s}\right).$$

With the prior specifications in (8.27), we can design a Gibbs sampler with Gaussian full conditionals for the temporal coefficients $\{\boldsymbol{\beta}_t\}$, the spatiotemporal coefficients $\{\boldsymbol{\beta}_t^*\}$, inverted Wishart for $\Sigma_{\boldsymbol{\eta}}$, and Metropolis steps for ϕ and the elements of $\Sigma_{\mathbf{w}, l}$. Updating of $\Sigma_{\mathbf{w}} = \sum_{l=1}^{p} \Sigma_{\mathbf{w}, l}$ is most efficiently done by reparametrizing the model in terms of the matrix square root of $\Sigma_{\mathbf{w}}$, say A, and updating the elements of the lower triangular matrix A. To be precise, consider the full conditional distribution,

$$f\left(\Sigma_{\mathbf{w}} | \boldsymbol{\gamma}, \phi_1, \phi_2\right) \propto f\left(\Sigma_{\mathbf{w}} | a_\gamma, B_\gamma\right) \frac{1}{|\sum_{l=1}^{p} R_l(\phi_l) \otimes \Sigma_{\mathbf{w}, l}|}$$
$$\times \exp\left(-\tfrac{1}{2}\boldsymbol{\beta}^{*T}\left(J^{-1} \otimes \left(\sum_{l=1}^{p} R_l(\phi_l) \otimes \Sigma_{\mathbf{w}, l}\right)^{-1}\right)\boldsymbol{\beta}^*\right).$$

The one-to-one relationship between elements of $\Sigma_{\mathbf{w}}$ and the Cholesky square root A is well known (see, e.g., Harville, 1997, p. 235). So, we reparametrize the above full conditional as

$$f\left(A|\boldsymbol{\gamma}, \phi_1, \phi_2\right) \propto f\left(h\left(A\right)|a_\gamma, B_\gamma\right) \left|\frac{\partial h}{\partial a_{ij}}\right| \frac{1}{\left|\sum_{l=1}^{p} R_l(\phi_l) \otimes \left(a_l a_l^T\right)\right|}$$
$$\times \exp\left(-\frac{1}{2}\boldsymbol{\beta}^{*T}\left(J^{-1} \otimes \left(\sum_{l=1}^{p} R_l\left(\phi_l\right) \otimes \left(a_l a_l^T\right)\right)^{-1}\right)\boldsymbol{\beta}^*\right).$$

Here, h is the function taking the elements of A, say a_{ij}, to those of the symmetric positive definite matrix $\Sigma_{\mathbf{w}}$. In the 2×2 case we have

$$h\left(a_{11}, a_{21}, a_{22}\right) = \left(a_{11}^2, a_{11}a_{21}, a_{21}^2 + a_{22}^2\right),$$

and the Jacobian is $4a_{11}^2 a_{22}$. Now, the elements of A are updated with univariate random-walk Metropolis proposals: lognormal or gamma for a_{11} and a_{22}, and normal for a_{21}. Additional computational burden is created, since now the likelihood needs to be computed for each of the three updates, but the chains are much better tuned (by controlling the scale of the univariate proposals) to move around the parameter space, thereby leading to better convergence behavior.

Example 8.3 (*Modeling temperature given precipitation*). Our spatial domain, shown in Figure 8.7 (see also color insert Figure C.6) along with elevation contours (in 100-m units), provides a sample of 50 locations (indicated by "+") in the state of Colorado. Each site provides information on monthly maximum temperature, and monthly mean precipitation. We denote the temperature summary in location $\mathbf{s}$ at time t, by $Y\left(\mathbf{s}, t\right)$, and the precipitation by $x\left(\mathbf{s}, t\right)$. Forming a covariate vector $\mathbf{x}^T\left(\mathbf{s}, t\right) = \left(1, x\left(\mathbf{s}, t\right)\right)$, we analyze the data using a coregionalized dynamic model, as outlined in Subsection 8.4.2. As a result, we have an intercept process $\tilde{\beta}_0\left(\mathbf{s}, t\right)$ and a slope process $\tilde{\beta}_1\left(\mathbf{s}, t\right)$, and the two processes are dependent.

Figure 8.8 displays the time-varying intercepts and slopes (coefficient of precipitation). As expected, the intercept is higher in the summer months and lower in the winter months, highest in July, lowest in December. In fact, the gradual increase from January to July, and the subsequent decrease toward December is evident from the plot. Precipitation seems to have a negative impact on temperature, although this seems to be significant only in the months of January, March, May, June, November, and December, i.e., seasonal pattern is retrieved although no such structure is imposed.

Table 8.5 displays the credible intervals for elements of the Σ_η matrix. Rows 1 and 2 show the medians and credible intervals for the respective *variances*; while Row 3 shows the *correlation*. The corresponding results for the elements of $\Sigma_{\mathbf{w}}$ are given in Table 8.6. A significant negative correlation is seen between the intercept and the slope processes, justifying our use of dependent processes. Next, in Table 8.7, we provide the measurement error variances for temperature along with the estimates of the spatial correlation parameters for the intercept and slope process. Also presented are the ranges implied by ϕ_1 and ϕ_2 for the marginal intercept process, $w_1\left(\mathbf{s}\right)$, and

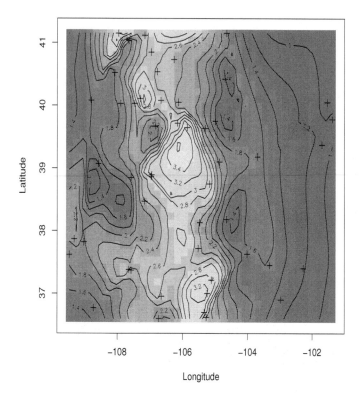

Longitude

Figure 8.7 *Map of the region in Colorado that forms the spatial domain (see also color insert). The data for the illustrations come from 50 locations, marked by "+" signs in this region.*

Σ_η	Median (2.5%, 97.5%)
$\Sigma_\eta\,[1,1]$	0.296 (0.130, 0.621)
$\Sigma_\eta\,[2,2]$	0.786 (0.198, 1.952)
$\Sigma_\eta\,[1,2]\,/\sqrt{\Sigma_\eta\,[1,1]\,\Sigma_\eta\,[2,2]}$	−0.562 (−0.807, −0.137)

Table 8.5 *Estimates of the variances and correlation from Σ_η, dynamic spatiotemporal modeling example.*

the marginal slope process, $w_2\,(\mathbf{s})$. The first range is computed by solving for the distance d, $\rho_1\,(\phi_1, d) = 0.05$, while the second range is obtained by solving $\left(a_{21}^2 \exp\left(-\phi_1 d\right) + a_{22}^2 \exp\left(-\phi_2 d\right)\right) / \left(a_{21}^2 + a_{22}^2\right) = 0.05$. The ranges are presented in units of 100 km with the maximum observed distance between our sites being approximately 742 km.

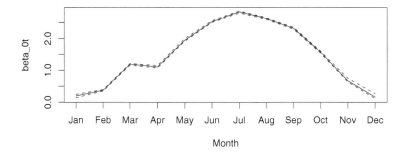

precipitation

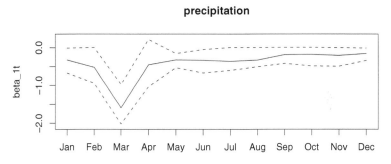

Figure 8.8 *Posterior distributions for the time-varying parameters in the temperature given precipitation example. The top graph corresponds to the intercept, while the lower one is the coefficient of precipitation. Solid lines represent the medians while the dashed lines correspond to the upper and lower credible intervals.*

$\Sigma_{\mathbf{w}}$	Median $(2.5\%, 97.5\%)$
$\Sigma_{\mathbf{w}}\,[1,1]$	$0.017\ (0.016,\ 0.019)$
$\Sigma_{\mathbf{w}}\,[2,2]$	$0.026\ (0.0065,\ 0.108)$
$\Sigma_{\mathbf{w}}\,[1,2]\,/\sqrt{\Sigma_{\mathbf{w}}\,[1,1]\,\Sigma_{\mathbf{w}}\,[2,2]}$	$-0.704\ (-0.843,\ -0.545)$

Table 8.6 *Estimates of the variances and correlation from $\Sigma_{\mathbf{w}}$, dynamic spatiotemporal modeling example.*

Finally, Figure 8.9 (see also color insert Figure C.7) displays the time-sliced image-contour plots for the slope process; similar figures can be drawn for the intercept process. For both processes, the spatial variation is better captured in the central and western edges of the domain. In Figure 8.9, all the months display broadly similar spatial patterns, with denser contour

Parameters	Median (2.5%, 97.5%)
σ_ϵ^2	0.134 (0.106, 0.185)
ϕ_1	1.09 (0.58, 2.04)
ϕ_2	0.58 (0.37, 1.97)
Range for intercept process	2.75 (1.47, 5.17)
Range for slope process	4.68 (1.60, 6.21)

Table 8.7 *Nugget effects and spatial correlation parameters, dynamic spatiotemporal modeling example.*

variations toward the west than the east. However, the spatial pattern does seem to be more pronounced in the months with more extreme weather, namely in the winter months of November through January and the summer months of June through August. ∎

8.5 Block-level modeling

We now return to spatiotemporal modeling for areal unit data, following the discussion of equations (8.4) and (8.5) in Section 8.1.

8.5.1 Aligned data

In the aligned data case, matters are relatively straightforward. Consider for example the spatiotemporal extension of the standard disease mapping setting described in Section 5.4.1. Here we would have $Y_{i\ell t}$ and $E_{i\ell t}$, the observed and expected disease counts in county i and demographic subgroup ℓ (race, gender, etc.) during time period t (without loss of generality we let t correspond to years in what follows). Again the issue of whether the $E_{i\ell t}$ are internally or externally standardized arises; in the more common former case we would use $n_{i\ell t}$, the number of persons at risk in county i during year t, to compute $E_{i\ell t} = n_{i\ell t}(\sum_{i\ell t} Y_{i\ell t} / \sum_{i\ell t} n_{i\ell t})$. That is, $E_{i\ell t}$ is the number of cases we would expect if the grand disease rate (all regions, subgroups, and years) were in operation throughout. The extension of the basic Section 5.4.1 Poisson regression model is then

$$Y_{i\ell t} \mid \mu_{i\ell t} \overset{ind}{\sim} Po\left(E_{i\ell t}\, e^{\mu_{i\ell t}}\right) ,$$

where $\mu_{i\ell t}$ is the log-relative risk of disease for region i, subgroup ℓ, and year t.

It now remains to specify the main effect and interaction components of $\mu_{i\ell t}$, and corresponding prior distributions. First the main effect for the demographic subgroups can be taken to have ordinary linear regression

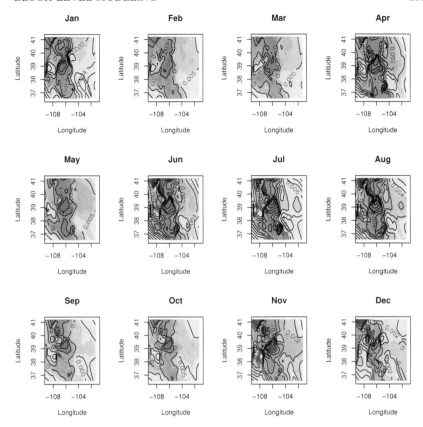

Figure 8.9 *Time-sliced image-contour plots displaying the posterior mean surface of the spatial residuals corresponding to the slope process in the temperature given precipitation model (see also color insert).*

structure, i.e., $\varepsilon_\ell = \mathbf{x}'_\ell \boldsymbol{\beta}$, with a flat prior for $\boldsymbol{\beta}$. Next, the main effects for time (say, δ_t) can be assigned flat priors (if we wish them to behave as fixed effects, i.e., temporal dummy variables), or an $AR(1)$ specification (if we wish them to reflect temporal autocorrelation). In some cases an even simpler structure (say, $\delta_t = \gamma t$) may be appropriate.

Finally, the main effects for space are similar to those assumed in the nontemporal case. Specifically, we might let

$$\psi_i = \mathbf{z}'_i \boldsymbol{\omega} + \theta_i + \phi_i \ ,$$

where $\boldsymbol{\omega}$ has a flat prior, the θ_i capture *heterogeneity* among the regions via the i.i.d. specification,

$$\theta_i \overset{iid}{\sim} N(0 \ , \ 1/\tau) \ ,$$

and the ϕ_i capture regional *clustering* via the CAR prior,

$$\phi_i \mid \phi_{j\neq i} \sim N(\bar{\phi}_i\,,\,1/(\lambda m_i))\,.$$

As usual, m_i is the number of neighbors of region i, and $\bar{\phi}_i = m_i^{-1}\Sigma_{j\in\partial_i}\,\phi_j$.

Turning to spatiotemporal interactions, suppose for the moment that demographic effects are not affected by region and year. Consider then the *nested* model,

$$\theta_{it} \overset{iid}{\sim} N(0\,,\,1/\tau_t) \text{ and } \phi_{it} \sim CAR(\lambda_t)\,, \tag{8.28}$$

where $\tau_t \overset{iid}{\sim} G(a,b)$ and $\lambda_t \overset{iid}{\sim} G(c,d)$. Provided these hyperpriors are not too informative, this allows "shrinkage" of the year-specific effects toward their grand mean, and in a way that allows the data to determine the amount of shrinkage.

Thus our most general model for $\mu_{i\ell t}$ is

$$\mu_{i\ell t} = \mathbf{x}'_\ell\boldsymbol{\beta} + \delta_t + \mathbf{z}'_i\boldsymbol{\omega} + \theta_{it} + \phi_{it}\,,$$

with corresponding joint posterior distribution proportional to

$$L(\boldsymbol{\beta},\delta,\boldsymbol{\omega},\boldsymbol{\theta},\boldsymbol{\phi};\mathbf{y})p(\delta)p(\boldsymbol{\theta}|\tau)p(\boldsymbol{\phi}|\lambda)p(\tau)p(\lambda)\,.$$

Computation via univariate Metropolis and Gibbs updating steps is relatively straightforward (and readily available in this aligned data setting in the WinBUGS language). However, convergence can be rather slow due to the weak identifiability of the joint parameter space. As a possible remedy, consider the the simple space-only case again for a moment. We may transform from $(\boldsymbol{\theta},\boldsymbol{\phi})$ to $(\boldsymbol{\theta},\boldsymbol{\eta})$ where $\eta_i = \theta_i + \phi_i$. Then $p(\boldsymbol{\theta},\boldsymbol{\eta}|\mathbf{y}) \propto L(\boldsymbol{\eta};\mathbf{y})p(\boldsymbol{\theta})p(\boldsymbol{\eta}-\boldsymbol{\theta})$, so that

$$p(\eta_i \mid \eta_{j\neq i},\boldsymbol{\theta},\mathbf{y}) \propto L(\eta_i;y_i)\,p(\eta_i-\theta_i \mid \{\eta_j-\theta_j\}_{j\neq i})$$

and

$$p(\theta_i \mid \theta_{j\neq i},\boldsymbol{\eta},\mathbf{y}) \propto p(\theta_i)\,p(\eta_i-\theta_i \mid \{\eta_j-\theta_j\}_{j\neq i})\,.$$

This simple transformation improves matters since each η_i full conditional is now well identified by the data point Y_i, while the weakly identified (indeed, "Bayesianly unidentified") θ_i now emerges in closed form as a normal distribution (since the nonconjugate Poisson likelihood no longer appears).

Example 8.4 The study of the trend of risk for a given disease in space and time may provide important clues in exploring underlying causes of the disease and helping to develop environmental health policy. Waller, Carlin, Xia, and Gelfand (1997) consider the following data set on lung cancer mortality in Ohio. Here Y_{ijkt} is the number of lung cancer deaths in county i during year t for gender j and race k in the state of Ohio. The data are recorded for $J = 2$ genders (male and female, indexed by s_j) and $K = 2$

Demographic subgroup	Contribution to ε_{jk}	Fitted relative risk
White males	0	1
White females	α	0.34
Nonwhite males	β	1.02
Nonwhite females	$\alpha + \beta + \xi$	0.28

Table 8.8 *Fitted relative risks, four sociodemographic subgroups in the Ohio lung cancer data.*

races (white and nonwhite, indexed by r_k) for each of the $I = 88$ Ohio counties over $T = 21$ years (1968–1988).

We adopt the model,

$$\mu_{ijkt} = s_j \alpha + r_k \beta + s_j r_k \xi + \theta_{it} + \phi_{it} \,, \qquad (8.29)$$

where $s_j = 1$ if $j = 2$ (female) and 0 otherwise, and $r_k = 1$ if $k = 2$ (nonwhite) and 0 otherwise. That is, there is one subgroup (white males) for which there is no contribution to the mean structure (8.38). For our prior specification, we select

$$\theta_{it} \overset{ind}{\sim} N\left(0 \,, \tfrac{1}{\tau_t}\right) \quad \text{and} \quad \phi_{it} \sim CAR(\lambda_t) \,;$$
$$\alpha, \beta, \xi \sim \text{flat} \,;$$
$$\tau_t \overset{iid}{\sim} G(1, 100) \quad \text{and} \quad \lambda_t \overset{iid}{\sim} G(1, 7) \,,$$

where the relative sizes of the hyperparameters in these two gamma distributions were selected following guidance given in Bernardinelli et al. (1995); see also Best et al. (1999) and Eberly and Carlin (2000).

Regarding implementation, five parallel, initially overdispersed MCMC chains were run for 500 iterations. Graphical monitoring of the chains for a representative subset of the parameters, along with sample autocorrelations and Gelman and Rubin (1992) diagnostics, indicated an acceptable degree of convergence by around the 100th iteration.

Histograms of the sampled values showed θ_{it} distributions centered near 0 in most cases, but ϕ_{it} distributions typically removed from 0, suggesting that the heterogeneity effects are not really needed in this model. Plots of $E(\tau_t|\mathbf{y})$ and $E(\lambda_t|\mathbf{y})$ versus t suggest increasing clustering and slightly increasing heterogeneity over time. The former might be the result of flight from the cities to suburban "collar counties" over time, while the latter is likely due to the elevated mean levels over time (for the Poisson, the variance increases with the mean).

Fitted relative risks obtained by Waller et al. (1997) for the four main demographic subgroups are shown in Table 8.8. The counterintuitively pos-

Demographic subgroup	Contribution to ε_{jk}	Fitted log-relative risk	Fitted relative risk
White males	0	0	1
White females	α	−1.06	0.35
Nonwhite males	β	0.18	1.20
Nonwhite females	$\alpha + \beta + \xi$	−1.07	0.34

Table 8.9 *Fitted relative risks, four sociodemographic subgroups in the Ohio lung cancer data*

itive fitted value for nonwhite females may be an artifact of the failure of this analysis to age-standardize the rates prior to modeling (or at least to incorporate age group as another demographic component in the model). To remedy this, consider the following revised and enhanced model, described by Xia and Carlin (1998), where we assume that

$$Y^*_{ijkt} \sim Poisson(E_{ijkt} \exp(\mu_{ijkt})) \,, \tag{8.30}$$

where again Y^*_{ijkt} denotes the observed age-adjusted deaths in county i for sex j, race k, and year t, and E_{ijkt} are the expected death counts. We also incorporate an ecological level smoking behavior covariate into our log-relative risk model, namely,

$$\mu_{ijkt} = \mu + s_j \alpha + r_k \beta + s_j r_k \xi + p_i \rho + \gamma t + \phi_{it} \,, \tag{8.31}$$

where p_i is the true smoking proportion in county i, γ represents the fixed time effect, and the ϕ_{it} capture the random spatial effects over time, wherein clustering effects are nested within time. That is, writing $\boldsymbol{\phi}_t = (\phi_{1t}, \ldots, \phi_{It})'$, we let $\boldsymbol{\phi}_t \sim CAR(\lambda_t)$ where $\lambda_t \overset{iid}{\sim} G(c, d)$. We assume that the sociodemographic covariates (sex and race) do not interact with time or space. Following the approach of Bernardinelli, Pascutto et al. (1997), we introduce both sampling error and spatial correlation into the smoking covariate. Let

$$q_i \mid p_i \sim N(p_i, \sigma_q^2), \ i = 1, \ldots, I, \ \text{and} \tag{8.32}$$

$$\mathbf{p} \sim CAR(\lambda_p) \iff p_i \mid p_{j \neq i} \sim N(\mu_{p_i}, \sigma_{p_i}^2), \ i = 1, \ldots, I \,, \tag{8.33}$$

where q_i is the current smoking proportion observed in a sample survey of county i (an imperfect measurement of p_i), $\mu_{p_i} = \sum_{j \neq i} w_{ij} p_j / \sum_{j \neq i} w_{ij}$, and $\sigma_{p_i}^2 = (\lambda_p \sum_{j \neq i} w_{ij})^{-1}$. Note that the amount of smoothing in the two CAR priors above may differ, since the smoothing is controlled by different parameters λ_ϕ and λ_p. Like λ_ϕ, λ_p is also assigned a gamma hyperprior, namely, a $G(e, f)$.

We ran 5 independent chains using our Gibbs-Metropolis algorithm for

2200 iterations each; plots suggested discarding the first 200 samples as an adequate burn-in period. We obtained the 95% posterior credible sets [–1.14, –0.98], [0.07, 0.28], and [–0.37, –0.01] for α, β, and ξ, respectively. Note that all 3 fixed effects are significantly different from 0, in contrast to our Table 8.8 results, which failed to uncover a main effect for race. The corresponding point estimates are translated into the fitted relative risks for the four sociodemographic subgroups in Table 8.9. Nonwhite males experience the highest risk, followed by white males, with females of both races having much lower risks. ∎

8.5.2 Misalignment across years

In this subsection we develop a spatiotemporal model to accommodate the situation of Figure 8.10, wherein the response variable and the covariate are spatially aligned within any given timepoint, but not across timepoints (due to periodic changes in the regional grid). Assuming that the observed disease count Y_{it} for zip i in year t is conditionally independent of the other zip-level disease counts given the covariate values, we have the model,

$$Y_{it} \mid \mu_{it} \overset{ind}{\sim} Po(E_{it} \exp(\mu_{it})), \; i = 1, \ldots, I_t, \; t = 1, \ldots, T,$$

where the expected count for zip i in year t, E_{it}, is proportional to the population count. In our case, we set $E_{it} = Rn_{it}$, where n_{it} is the population count in zip i at year t and $R = (\sum_{it} Y_{it})/(\sum_{it} n_{it})$, the grand asthma hospitalization rate (i.e., the expected counts assume homogeneity of disease rates across all zips and years). The log-relative risk is modeled as

$$\mu_{it} = x_{it}\beta_t + \delta_t + \theta_{it} + \phi_{it}, \tag{8.34}$$

where x_{it} is the zip-level exposure covariate (traffic density) depicted for 1983 in Figure 8.10, β_t is the corresponding main effect, δ_t is an overall intercept for year t, and θ_{it} and ϕ_{it} are zip- and year-specific heterogeneity and clustering random effects, analogous to those described in Section 8.5.1. The changes in the zip grid over time cloud the interpretation of these random effects (e.g., a particular region may be indexed by different i in different years), but this does not affect the interpretation of the main effects β_t and δ_t; it is simply the analogue of unbalanced data in a longitudinal setting. In the spatiotemporal case, the distributions on these effects become

$$\boldsymbol{\theta}_t \overset{ind}{\sim} N\left(0, \frac{1}{\tau_t}I\right) \quad \text{and} \quad \boldsymbol{\phi}_t \overset{ind}{\sim} CAR(\lambda_t), \tag{8.35}$$

where $\boldsymbol{\theta}_t = (\theta_1, \ldots, \theta_{I_t})'$, $\boldsymbol{\phi}_t = (\phi_1, \ldots, \phi_{I_t})'$, and we encourage similarity among these effects across years by assuming $\tau_t \overset{iid}{\sim} G(a, b)$ and $\lambda_t \overset{iid}{\sim} G(c, d)$, where G again denotes the gamma distribution. Placing flat (uniform) priors on the main effects β_t and δ_t completes the model specification. Note

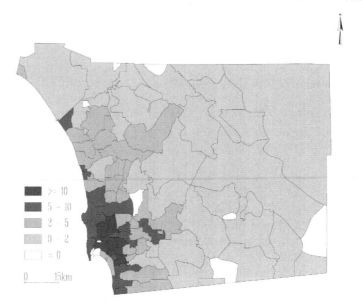

Figure 8.10 *Traffic density (average vehicles per km of major roadway) in thousands by zip code for 1983, San Diego County.*

that the constraints $\sum_i \phi_{it} = 0$, $t = 1, \ldots, T$ must be added to identify the year effects δ_t, due to the location invariance of the CAR prior.

Example 8.5 Asthma is the most common chronic disease diagnosis for children in the U.S. (National Center for Environmental Health, 1996). A large number of studies have shown a correlation between known products and byproducts of auto exhaust (such as ozone, nitrogen dioxide, and particulate matter) and pediatric asthma ER visits or hospitalizations. Several recent studies (e.g., Tolbert et al., 2000; Zidek et al., 1998; Best et al., 2000) have used hierarchical Bayesian methods in such investigations. An approach taken by some authors is to use proximity to major roadways (or some more refined measure of closeness to automobile traffic) as an omnibus measure of exposure to various asthma-inducing pollutants. We too adopt this approach and use the phrase "exposure" in what follows, even though in fact our traffic measures are really surrogates for the true exposure.

Our data set arises from San Diego County, CA, the region pictured in Figure 8.10. The city of San Diego is located near the southwestern corner of the map; the map's western boundary is the Pacific Ocean, while Mexico forms its southern boundary. The subregions pictured are the zip codes as defined in 1983; as mentioned earlier this grid changes over time.

Specifically, during the course of our 8-year (1983–1990) study period, the zip code boundaries changed four times: in 1984, 1987, 1988, and 1990.

The components of our data set are as follows. First, for a given year, we have the number of discharges from hospitalizations due to asthma for children aged 14 and younger by zip code (California Office of Statewide Health Planning and Development, 1997). The primary diagnosis was asthma based on the International Classification of Diseases, code 493 (U.S. Department of Health and Human Services, 1989). Assuming that patient records accurately report the correct zip code of residence, these data can be thought of as error-free.

Second, we have zip-level population estimates (numbers of residents aged 14 and younger) for each of these years, as computed by Scalf and English (1996). These estimates were obtained in ARC/INFO using the following process. First, a land-use covariate was used to assist in a linear interpolation between the 1980 and 1990 U.S. Census figures, to obtain estimates at the census block group level. Digitized hard-copy U.S. Postal Service maps or suitably modified street network files provided by the San Diego Association of Governments (SANDAG) were then used to reallocate these counts to the zip code grid for the year in question. To do this, the GIS first created a subregional grid by intersecting the block group and zip code grids. The block group population totals were allocated to the subregions per a combination of subregional area and population density (the latter again based on the land-use covariate). Finally, these imputed subregional counts were reaggregated to the zip grid. While there are several possible sources of uncertainty in these calculations, we ignore them in our initial round of modeling, assuming these population counts to be fixed and known.

Finally, for each of the major roads in San Diego County, we have mean yearly traffic counts on each road segment in our map. Here "major" roads are defined by SANDAG to include interstate highways or equivalent, major highways, access or minor highways, and arterial or collector routes. The sum of these numbers within a given zip divided by the total length of its major roads provides an aggregate measure of traffic exposure for the zip. These zip-level *traffic densities* are plotted for 1983 in Figure 8.10; this is the exposure measure we use in the following text.

We set $a = 1$, $b = 10$ (i.e., the τ_t have prior mean and standard deviation both equal to 10) and $c = 0.1$, $d = 10$ (i.e., the λ_t have prior mean 1, standard deviation $\sqrt{10}$). These are fairly vague priors designed to let the data dominate the allocation of excess spatial variability to heterogeneity and clustering. (As mentioned near equation (5.48), simply setting these two priors equal to each other would not achieve this, since the prior for the θ_{it} is specified *marginally*, while that for the ϕ_{it} is specified *conditionally* given the neighboring ϕ_{jt}.) Our MCMC implementation ran 3 parallel sampling chains for 5000 iterations each, and discarded the first 500 iterations as preconvergence "burn-in."

Plots of the posterior medians and 95% equal-tail Bayesian confidence intervals for β_t (not shown) makes clear that, with the exception of that for 1986, all of the β_t's are significantly greater than 0. Hence, the traffic exposure covariate in Figure 8.10 is positively associated with increased pediatric asthma hospitalization in seven of the eight years of our study. To interpret these posterior summaries, recall that their values are on the *log*-relative risk scale. Thus a zip having a 1983 traffic density of 10 thousand cars per km of roadway would have median relative risk $e^{10(.065)} = 1.92$ times higher than a zip with essentially no traffic exposure, with a corresponding 95% confidence interval of $(e^{10(.000)}, e^{10(.120)}) = (1.00, 3.32)$. There also appears to be a slight weakening of the traffic-asthma association over time.

Figure 8.11 provides ARC/INFO maps of the crude and fitted asthma rates (per thousand) in each of the zips for 1983. The crude rates are of course given by $r_{it} = Y_{it}/n_{it}$, while the fitted rates are given by $R \exp(\hat{\mu}_{it})$, where R is again the grand asthma rate across all zips and years and $\hat{\mu}_{it}$ is obtained by plugging in the estimated posterior means for the various components in equation (8.34). The figure clearly shows the characteristic Bayesian shrinkage of the crude rates toward the grand rate. In particular, no zip is now assigned a rate of exactly zero, and the rather high rates in the thinly populated eastern part of the map have been substantially reduced. However, the high observed rates in urban San Diego continue to be high, as the method properly recognizes the much higher sample sizes in these zips. There also appears to be some tendency for clusters of similar crude rates to be preserved, the probable outcome of the CAR portion of our model. ∎

8.5.3 Nested misalignment both within and across years

In this subsection we extend our spatiotemporal model to accommodate the situation of Figure 8.12, wherein the covariate is available on a grid that is a refinement of the grid for which the response variable is available (i.e., nested misalignment within years, as well as misalignment across years). Letting the subscript j index the subregions (which we also refer to as *atoms*) of zip i, our model now becomes

$$Y_{ijt} \mid \mu_{ijt} \sim Po(E_{ijt} \exp(\mu_{ijt})), \ i = 1, \ldots, I_t, \ j = 1, \ldots, J_{it}, \ t = 1, \ldots, T,$$

where the expected counts E_{ijt} are now Rn_{ijt}, with the grand rate R as before. The population of atom ijt is not known, and so we determine it by areal interpolation as $n_{ijt} = n_{it}(\text{area of atom } ijt)/(\text{area of zip } it)$. The

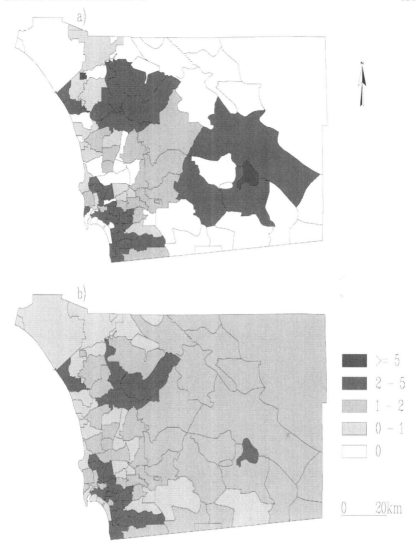

Figure 8.11 *Pediatric asthma hospitalization rate (per thousand children) by zip code for 1983, San Diego County: (a) crude rate, (b) temporally misaligned model fitted rate.*

log-relative risk in atom ijt is then modeled as

$$\mu_{ijt} = x_{ijt}\beta_t + \delta_t + \theta_{it} + \phi_{it}, \qquad (8.36)$$

where x_{ijt} is now the atom-level exposure covariate (depicted for 1983 in Figure 8.12), but β_t, δ_t, θ_{it} and ϕ_{it} are as before. Thus our prior specifica-

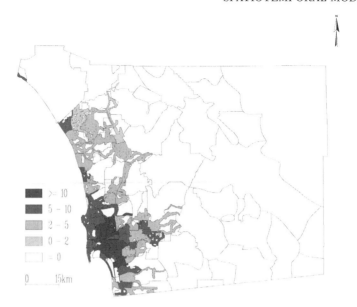

Figure 8.12 *Adjusted traffic density (average vehicles per km of major roadway) in thousands by zip code subregion for 1983, San Diego County.*

tion is exactly that of the previous subsection; priors for the θ_t and ϕ_t as given in equation (8.35), exchangeable gamma hyperpriors for the τ_t and λ_t with $a = 1$, $b = 10$, $c = 0.1$, and $d = 10$, and flat priors for the main effects β_t and δ_t.

Since only the zip-level hospitalization totals Y_{it} (and not the atom-level totals Y_{ijt}) are observed, we use the additivity of conditionally independent Poisson distributions to obtain

$$Y_{it} \mid \beta_t, \delta_t, \theta_{it}, \phi_{it} \sim Po\left(\sum_{j=1}^{J_{it}} E_{ijt} \exp(\mu_{ijt})\right), \ i = 1, \ldots, I_t, \ t = 1, \ldots, T.$$

(8.37)

Using expression (8.37), we can obtain the full Bayesian model specification for the observed data as

$$\left[\prod_{t=1}^{T} \prod_{i=1}^{I_t} p\left(y_{it} \mid \beta_t, \delta_t, \theta_{it}, \phi_{it}\right)\right] \left[\prod_{t=1}^{T} p(\boldsymbol{\theta}_t \mid \tau_t) p(\boldsymbol{\phi}_t \mid \lambda_t) p(\tau_t) p(\lambda_t)\right]$$

(8.38)

As in the previous section, only the τ_t and λ_t parameters may be updated via ordinary Gibbs steps, with Metropolis steps required for the rest.

Note that model specification (8.38) makes use of the atom-level covariate values x_{ijt}, but only the zip-level hospitalization counts Y_{it}. Of course, we might well be interested in *imputing* the values of the missing subregional counts Y_{ijt}, whose full conditional distribution is multinomial, namely,

$$(Y_{i1t}, \ldots, Y_{iJ_it}) \mid Y_{it}, \beta_t, \delta_t, \theta_{it}, \phi_{it} \sim Mult(Y_{it}, \{q_{ijt}\}) , \qquad (8.39)$$

$$\text{where} \quad q_{ijt} = \frac{E_{ijt}e^{\mu_{ijt}}}{\sum_{j=1}^{J_{it}} E_{ijt}e^{\mu_{ijt}}} .$$

Since this is a purely predictive calculation, Y_{ijt} values need not be drawn as part of the MCMC sampling order, but instead at the very end, conditional on the post-convergence samples.

Zhu, Carlin, English, and Scalf (2000) use Figure 8.12 to refine the definition of exposure used in Example 8.5 by subdividing each zip into subregions based on whether or not they are closer than 500 m to a major road. This process involves creating "buffers" around each road and subsequently overlaying them in a GIS, and has been previously used in several studies of vehicle emissions. This definition leads to some urban zips becoming "entirely exposed," as they contain no point further than 500 m from a major road; these are roughly the zips with the darkest shading in Figure 8.10 (i.e., those having traffic densities greater than 10,000 cars per year per km of major roadway). Analogously, many zips in the thinly populated eastern part of the county contained at most one major road, suggestive of little or no traffic exposure. As a result, we (somewhat arbitrarily) defined those zips in the two lightest shadings (i.e., those having traffic densities less than 2,000 cars per year per km of roadway) as being "entirely unexposed." This typically left slightly less than half the zips (47 for the year shown, 1983) in the middle range, having some exposed and some unexposed subregions, as determined by the intersection of the road proximity buffers. These subregions are apparent as the lightly shaded regions in Figure 8.12; the "entirely exposed" regions continue to be those with the darkest shading, while the "entirely unexposed" regions have no shading.

The fitted rates obtained by Zhu et al. (2000) provide a similar overall impression as those in Figure 8.11, except that the newer map is able to show subtle differences within several "partially exposed" regions. These authors also illustrate the interpolation of missing subregional counts Y_{ijt} using equation (8.39). Analogous to the block-block FMPC imputation in Subsection 6.2, the sampling-based hierarchical Bayesian method produces more realistic estimates of the subregional hospitalization counts, with associated confidence limits emerging as an automatic byproduct.

8.5.4 Nonnested misalignment and regression

In this subsection we consider spatiotemporal *regression* in the misaligned data setting motivated by our Atlanta ozone data set. Recall that the first component of this data set provides ozone measurements X_{itr} at between

8 and 10 fixed monitoring sites i for day t of year r, where $t = 1, \ldots, 92$ (the summer days from June 1 through August 31) and $r = 1, 2, 3$, corresponding to years 1993, 1994, and 1995 For example, Figure 1.3 shows the 8-hour daily maximum ozone measurements (in parts per million) at the 10 monitoring sites for a particular day (July 15, 1995), along with the boundaries of the 162 zip codes in the Atlanta metropolitan area.

A *second* component of this data set (about which we so far have said far less) provides relevant health outcomes, but only at the zip code level. Specifically, for each zip l, day t, and year r, we have the number of pediatric emergency room (ER) visits for asthma, Y_{ltr}, as well as the total number of pediatric ER visits, n_{ltr}. These data come from a historical records-based investigation of pediatric asthma emergency room visits to seven major emergency care centers in the Atlanta metropolitan statistical area during the same three summers. Our main substantive goal is an investigation of the relationship between ozone and pediatric ER visits for asthma in Atlanta, controlling for a range of sociodemographic covariates. Potential covariates (available only as zip-level summaries in our data set) include average age, percent male, percent black, and percent using Medicaid for payment (a crude surrogate for socioeconomic status). Clearly an investigation of the relationship between ozone exposure and pediatric ER visit count cannot be undertaken until the mismatch in the support of the (point-level) predictor and (zip-level) response variables is resolved.

A naive approach would be to average the ozone measurements belonging to a specific zip code, then relate this average ozone measurement to the pediatric asthma ER visit count in this zip. In fact, there are few monitoring sites relative to the number of zip codes; Figure 1.3 shows most of the zip codes contain no sites at all, so that most of the zip-level ER visit count data would be discarded. An alternative would be to aggregate the ER visits over the entire area and model them as a function of the average of the ozone measurements (that is, eliminate the spatial aspect of the data and fit a temporal-only model). Using this idea in a Poisson regression, we obtained a coefficient for ozone of 2.48 with asymptotic standard error 0.71 (i.e., significant positive effect of high ozone on ER visit rates). While this result is generally consistent with our findings, precise comparison is impossible for a number of reasons. First, this approach requires use of data from the entire Atlanta metro area (due to the widely dispersed locations of the monitoring stations), not data from the city only as our approach allows. Second, it does not permit use of available covariates (such as race and SES) that were spatially but not temporally resolved in our data set. Third, standardizing using expected counts E_i (as in equation (8.40) below) must be done only over days (not regions), so the effect of including them is now merely to adjust the model's intercept.

We now describe the disease component of our model, and subsequently assemble the full Bayesian hierarchical modeling specification for our spatially misaligned regression. Similar to the model of Subsection 8.5.3, we assume the zip-level asthma ER visit counts, Y_{ltr} for zip l during day t of summer r, follow a Poisson distribution,

$$Y_{ltr} \sim Poisson\left(E_{ltr}\exp(\lambda_{ltr})\right) , \tag{8.40}$$

where the E_{ltr} are expected asthma visit counts, determined via internal standardization as $E_{ltr} = n_{ltr}(\sum_{ltr} Y_{ltr} / \sum_{ltr} n_{ltr})$, where n_{ltr} is the total number of pediatric ER visits in zip code l on day t of year r. Thus E_{ltr} is the number of pediatric ER asthma visits we would expect from the given zip and day if the proportion of such visits relative to the total pediatric ER visit rate was homogeneous across all zips, days, and years. Hence λ_{ltr} in (8.40) can be interpreted as a log-relative risk of asthma among those children visiting the ER in group ltr. Our study design is thus a *proportional admissions model* (Breslow and Day, 1987, pp. 153–155).

We do not take n_{ltr} equal to the total number of children *residing* in zip l on day t of year r, since this standardization would implicitly presume a constant usage of the ER for pediatric asthma management across all zips, which seems unlikely (children from more affluent zips are more likely to have the help of family doctors or specialists in managing their asthma, and so would not need to rely on the ER; see Congdon and Best, 2000, for a solution to the related problem of adjusting for patient referral practices). Note however that this in turn means that our disease (pediatric asthma visits) is not particularly "rare" relative to the total (all pediatric visits). As such, our use of the Poisson distribution in (8.40) should not be thought of as an approximation to a binomial distribution for a rare event, but merely as a convenient and sensible model for a discrete variable.

For the log-relative risks in group ltr, we begin with the model,

$$\lambda_{ltr} = \beta_0 + \beta_1 X_{l,t-1,r} + \sum_{c=1}^{C}\alpha_c Z_{cl} + \sum_{d=1}^{D}\delta_d W_{dt} + \theta_l . \tag{8.41}$$

Here, β_0 is an intercept term, and β_1 denotes the effect of ozone exposure $X_{l,t-1,r}$ in zip l during day $t-1$ of year r. Note that we model pediatric asthma ER visit counts as a function of the ozone level on the *previous* day, in keeping with the most common practice in the epidemiological literature (see, e.g., Tolbert et al., 2000). This facilitates next-day predictions for pediatric ER visits given the current day's ozone level, with our Bayesian approach permitting full posterior inference (e.g., 95% prediction limits). However, it also means we have only $(J-1) \times 3 = 273$ days worth of usable data in our sample. Also, $\mathbf{Z}_l = (Z_{1l},\ldots,Z_{Cl})^T$ is a vector of C zip-level (but not time-varying) sociodemographic covariates with corresponding coefficient vector $\boldsymbol{\alpha} = (\alpha_1,\ldots,\alpha_C)^T$, and $\mathbf{W}_t = (W_{1t},\ldots,W_{Dt})^T$ is a vector of D day-level (but not spatially varying) temporal covariates with corresponding coefficient vector $\boldsymbol{\delta} = (\delta_1,\ldots,\delta_D)^T$. Finally, θ_l is a zip-specific

random effect designed to capture extra-Poisson variability in the observed
ER visitation rates. These random effects may simply be assumed to be
exchangeable draws from a $N(0, 1/\tau)$ distribution (thus modeling overall
heterogeneity), or may instead be assumed to vary spatially using a conditionally autoregressive (CAR) specification.

Of course, model (8.40)–(8.41) is not fittable as stated, since the zip-level
previous-day ozone values $X_{l,t-1,r}$ are not observed. Fortunately, we may
use the methods of Section 6.1 to perform the necessary point-block realignment. To connect our equation (8.41) notation with that used in Section 6.1,
let us write $\mathbf{X}_{B,r} \equiv \{X_{l,t-1,r}, l = 1, \ldots, L, t = 2, \ldots, J\}$ for the unobserved
block-level data from year r, and $\mathbf{X}_{s,r} \equiv \{X_{itr}, i = 1, \ldots, I, t = 1, \ldots, J\}$
for the observed site-level data from year r. Then, from equations (8.21)
and (8.22) and assuming no missing ozone station data for the moment, we
can find the conditional predictive distribution $f(\mathbf{X}_{B,r}|\mathbf{X}_{s,r}, \boldsymbol{\gamma}_r, \sigma_r^2, \boldsymbol{\phi}_r, \boldsymbol{\rho}_r)$
for year r. However, for these data some components of the $\mathbf{X}_{s,r}$ will be
missing, and thus replaced with imputed values $\mathbf{X}_{s,r}^{(m)}$, $m = 1, \ldots, M$, for
some modest number of imputations M (say, $M = 3$). (In a slight abuse of
notation here, we assume that any *observed* component of $\mathbf{X}_{s,r}^{(m)}$ is simply
set equal to that observed value for all m.)

Thus, the full Bayesian hierarchical model specification is given by

$$
\begin{aligned}
&[\textstyle\prod_r \prod_t \prod_l f(Y_{ltr}|\boldsymbol{\beta}, \boldsymbol{\alpha}, \boldsymbol{\delta}, \boldsymbol{\theta}, X_{l,t-1,r})]\, p(\boldsymbol{\beta}, \boldsymbol{\alpha}, \boldsymbol{\delta}, \boldsymbol{\theta}) \\
&\times \Big[\textstyle\prod_r f(\mathbf{X}_{B,r}|\mathbf{X}_{s,r}^{(m)}, \boldsymbol{\gamma}_r, \sigma_r^2, \boldsymbol{\phi}_r, \boldsymbol{\rho}_r) \\
&\qquad \times f(\mathbf{X}_{s,r}^{(m)}|\boldsymbol{\gamma}_r, \sigma_r^2, \boldsymbol{\phi}_r, \boldsymbol{\rho}_r) p(\boldsymbol{\gamma}_r, \sigma_r^2, \boldsymbol{\phi}_r, \boldsymbol{\rho}_r)\Big] \;,
\end{aligned} \tag{8.42}
$$

where $\boldsymbol{\beta} = (\beta_0, \beta_1)^T$, and $\boldsymbol{\gamma}_r, \sigma_r^2, \boldsymbol{\phi}_r$ and $\boldsymbol{\rho}_r$ are year-specific versions of the
parameters in (6.6). Note that there is a posterior distribution for each of
the M imputations. Model (8.42) assumes the asthma-ozone relationship
does not depend on year; the misalignment parameters are year-specific
only to permit year-by-year realignment.

Zhu, Carlin, and Gelfand (2003) offer a reanalysis of the Atlanta ozone
and asthma data by fitting a version of model (8.41), namely,

$$
\lambda_{ltr} = \beta_0 + \beta_1 X_{l,t-1,r}^{*(m,v)} + \alpha_1 Z_{1l} + \alpha_2 Z_{2l} + \delta_1 W_{1t} + \delta_2 W_{2t} + \delta_3 W_{3t} + \delta_4 W_{4t} \;,
\tag{8.43}
$$

where $X_{l,t-1,r}^{*(m,v)}$ denotes the (m,v)th imputed value for the zip-level estimate of the 8-hour daily maximum ozone measurement on the previous
day $(t-1)$. Our zip-specific covariates are Z_{1l} and Z_{2l}, the percent high
socioeconomic status and percent black race of those pediatric asthma ER
visitors from zip l, respectively. Of the day-specific covariates, W_{1t} indexes
day of summer ($W_{1t} = t \bmod 91$) and $W_{2t} = W_{1t}^2$, while W_{3t} and W_{4t} are
indicator variables for days in 1994 and 1995, respectively (so that 1993 is
taken as the reference year). We include both linear and quadratic terms

Parameter	Effect	Posterior median	95% Posterior credible set	Fitted relative risk
β_0	intercept	−0.4815	(−0.5761, −0.3813)	—
β_1	ozone	0.7860	(−0.7921, 2.3867)	1.016†
α_1	high SES	−0.5754	(−0.9839, −0.1644)	0.562
α_2	black	0.5682	(0.3093, 0.8243)	1.765
δ_1	day	−0.0131	(−0.0190, −0.0078)	—
δ_2	day^2	0.00017	(0.0001, 0.0002)	—
δ_3	year 1994	0.1352	(0.0081, 0.2478)	1.145
δ_4	year 1995	0.4969	(0.3932, 0.5962)	1.644

Table 8.10 *Fitted relative risks for the parameters of interest in the Atlanta pediatric asthma ER visit data, full model. (†This is the posterior median relative risk predicted to arise from a .02 ppm increase in ozone.)*

for day of summer in order to capture the rough U-shape in pediatric ER asthma visits, with June and August higher than July.

The analysis of Zhu et al. (2003) is only approximate, in that they run *separate* MCMC algorithms on the portions of the model corresponding to the two lines of model (8.42). In the spirit of the multiple imputation approach to the missing (point-level) ozone observations, they also retain $V = 3$ post-convergence draws from each of our $M = 3$ imputed data sets, resulting in $MV = 9$ zip-level approximately imputed ozone vectors $\mathbf{X}_{B,r}^{*(m,v)}$.

The results of this approach are shown in Table 8.10. The posterior median of β_1 (.7860) is positive, as expected. An increase of .02 ppm in 8-hour maximum ozone concentration (a relatively modest increase, as seen from Figure 1.3) thus corresponds to a fitted relative risk of exp(.7860 × .02) ≈ 1.016, or a 1.6% increase in relative risk of a pediatric asthma ER visit. However, the 95% credible set for β_1 does include 0, meaning that this positive association between ozone level and ER visits is not "Bayesianly significant" at the 0.05 level. Using a more naive approach but data from all 162 zips in the Atlanta metro area, Carlin et al. (1999) estimate the above relative risk as 1.026, marginally significant at the .05 level (that is, the lower limit of the 95% credible set for β_1 was precisely 0).

Regarding the demographic variables, the effects of both percent high SES and percent black emerge as significantly different from 0. The relative risk for a zip made entirely of high SES residents would be slightly more than half that of a comparable all-low SES zip, while a zip with a 100% black population would have a relative risk nearly 1.8 times that of a 100% nonblack zip. As for the temporal variables, day of summer is significantly negative and its square is significantly positive, confirming the U-shape of asthma relative risks over a given summer. Both year 1994 and year 1995 show higher relative risk compared with year 1993, with estimated increases in relative risk of about 15% and 64%, respectively.

8.6 Exercises

1. Suppose $Var(\epsilon(\mathbf{s}, t))$ in (8.6), (8.7), and (8.8) is revised to $\sigma_\epsilon^{2(t)}$.

   (a) Revise expressions (8.11), (8.13), and (8.16), respectively.

   (b) How would these changes affect simulation-based model fitting?

2. The data www.biostat.umn.edu/~brad/data/ColoradoS-T.dat contain the maximum monthly temperatures (in tenths of a degree Celcius) for 50 locations over 12 months in 1997. The elevation at each of the 50 sites is also given.

   (a) Treating month as the discrete time unit, temperature as the dependent variable, and elevation as a covariate, fit the additive space-time model (8.6) to this data. Provide posterior estimates of the important model parameters, and draw image-contour plots for each month.

   (*Hint:* Modify the WinBUGS code in Example 5.1 to fit a simple, nested spatiotemporal model. That is, use either the "direct" approach or the spatial.exp command to build an exponential kriging model for the data for a given month t with a range parameter ϕ_t, and then assume these parameters are in turn i.i.d. from (say) a $U(0, 10)$ distribution.)

   (b) Compare a few sensible models (changing the prior for the ϕ_t, including/excluding the covariate, etc.) using the DIC tool in WinBUGS. How does DIC seem to perform in this setting?

   (c) Repeat part (a) assuming the error structures (8.7) and (8.8). Can these models still be fit in WinBUGS, or must you now resort to your own C, Fortran, or R code?

3. Suppose $Y(\mathbf{s}_i, t_j)$, $i = 1, \ldots, n$, $j = 1, \ldots, m$ arise from a mean-zero stationary spatiotemporal process. Let $a_{ii'} = \sum_{j=1}^m Y(\mathbf{s}_i, t_j) Y(\mathbf{s}_{i'}, t_j)/m$, let $b_{jj'} = \sum_{i=1}^n Y(\mathbf{s}_i, t_j) Y(\mathbf{s}_i, t_{j'})/n$, and let $c_{ii',jj'} = Y(\mathbf{s}_i, t_j) Y(\mathbf{s}_{i'}, t_{j'})$.

   (a) Obtain $E(a_{ii'})$, $E(b_{jj'})$, and $E(c_{ii',jj'})$.

   (b) Argue that if we plot $c_{ii',jj'}$ versus $a_{ii'} \cdot b_{jj'}$, under a separable covariance structure, we can expect the plotted points to roughly lie along a straight line. (As a result, we might call this a *separability plot*.) What is the slope of this theoretical line?

   (c) Create a separability plot for the data in Exercise 2. Was the separability assumption there justified?

4. Consider again the data and model of Example 8.4, the former located at www.biostat.umn.edu/~brad/data2.html. Fit the Poisson spatiotemporal disease mapping model (8.30), but where we discard the smoking covariate, and also reverse the gender scores ($s_j = 1$ if male, 0 if female) so that the log-relative risk (8.31) is reparametrized as

$$\mu_{ijkt} = \mu + s_j \alpha + r_k \beta + s_j r_k (\xi - \alpha - \beta) + \gamma t + \phi_{it} .$$

Under this model, β now unequivocally captures the difference in log-relative risk between white and nonwhite females.

(a) Use either WinBUGS or your own R, C++, or Fortran code to find point and 95% interval estimates of β. Is there any real difference between the two female groups?

(b) Use either the mapping tool within WinBUGS or your own ArcView or other GIS code to map the fitted median nonwhite female lung cancer death rates per 1000 population for the years 1968, 1978, and 1988. Interpret your results. Is a temporal trend apparent?

5. In the following, let C_1 be a valid two-dimensional isotropic covariance function and let C_2 be a valid one-dimensional isotropic covariance function. Let $C_A(\mathbf{s}, t) = C_1(\mathbf{s}) + C_2(t)$ and $C_M(\mathbf{s}, t) = C_1(\mathbf{s})C_2(t)$. C_A is referred to as an *additive* (or *linear*) space-time covariance function, while C_M is referred to as a *multiplicative* space-time covariance function.

(a) Why are C_A and C_M valid?

(b) Comment on the behavior of C_A and C_M as $||\mathbf{s} - \mathbf{s}', t - t'|| \to 0$ (local limit), and as $||\mathbf{s} - \mathbf{s}', t - t'|| \to \infty$ (global limit).

CHAPTER 9

Spatial survival models

The use of survival models involving a random effect or "frailty" term is becoming more common. Usually the random effects are assumed to represent different clusters, and clusters are assumed to be independent. In this chapter, we consider random effects corresponding to clusters that are spatially arranged, such as clinical sites or geographical regions. That is, we might suspect that random effects corresponding to strata in closer proximity to each other might also be similar in magnitude.

Survival models have a long history in the biostatistical and medical literature (see, e.g., Cox and Oakes, 1984). Very often, time-to-event data will be grouped into *strata* (or *clusters*), such as clinical sites, geographic regions, and so on. In this setting, a hierarchical modeling approach using stratum-specific parameters called *frailties* is often appropriate. Introduced by Vaupel, Manton, and Stallard (1979), this is a mixed model with random effects (the frailties) that correspond to a stratum's overall health status.

To illustrate, let t_{ij} be the time to death or censoring for subject j in stratum i, $j = 1, \ldots, n_i$, $i = 1, \ldots, I$. Let $\mathbf{x}_{ij}$ be a vector of individual-specific covariates. The usual assumption of proportional hazards $h(t_{ij}; \mathbf{x}_{ij})$ enables models of the form

$$h(t_{ij}; \mathbf{x}_{ij}) = h_0(t_{ij}) \exp(\boldsymbol{\beta}^T \mathbf{x}_{ij}) \,, \qquad (9.1)$$

where h_0 is the *baseline hazard*, which is affected only multiplicatively by the exponential term involving the covariates. In the frailty setting, model (9.1) is extended to

$$
\begin{aligned}
h(t_{ij}; x_{ij}) &= h_0(t_{ij}) \, \omega_i \exp(\boldsymbol{\beta}^T \mathbf{x}_{ij}) \\
&= h_0(t_{ij}) \exp(\boldsymbol{\beta}^T \mathbf{x}_{ij} + W_i) \,, \qquad (9.2)
\end{aligned}
$$

where $W_i \equiv \log \omega_i$ is the stratum-specific frailty term, designed to capture differences among the strata. Typically a simple i.i.d. specification for the W_i is assumed, e.g.,

$$W_i \overset{iid}{\sim} N(0, \sigma^2) \,. \qquad (9.3)$$

With the advent of MCMC computational methods, the Bayesian approach to fitting hierarchical frailty models such as these has become increasingly popular (see, e.g., Carlin and Louis, 2000, Sec. 7.6). Perhaps the simplest approach is to assume a *parametric* form for the baseline hazard h_0. While a variety of choices (gamma, lognormal, etc.) have been explored in the literature, in Section 9.1 we adopt the Weibull, which seems to represent a good tradeoff between simplicity and flexibility. This then produces

$$h(t_{ij}; x_{ij}) = \rho t_{ij}^{\rho-1} \exp(\boldsymbol{\beta}^T \mathbf{x}_{ij} + W_i) . \tag{9.4}$$

Now, placing prior distributions on $\rho, \boldsymbol{\beta}$, and σ^2 completes the Bayesian model specification. Such models are by now a standard part of the literature, and easily fit (at least in the univariate case) using WinBUGS. Carlin and Hodges (1999) consider further extending model (9.4) to allow stratum-specific baseline hazards, i.e., by replacing ρ by ρ_i. MCMC fitting is again routine given a distribution for these new random effects, say, $\rho_i \overset{iid}{\sim} Gamma(\alpha, 1/\alpha)$, so that the ρ_i have mean 1 (corresponding to a constant hazard over time) but variance $1/\alpha$.

A richer but somewhat more complex alternative is to model the baseline hazard *nonparametrically*. In this case, letting γ_{ij} be a death indicator (0 if alive, 1 if dead) for patient ij, we may write the likelihood for our model $L(\beta, \mathbf{W}; \mathbf{t}, \mathbf{x}, \boldsymbol{\gamma})$ generically as

$$\prod_{i=1}^{I} \prod_{j=1}^{n_i} \{h(t_{ij}; \mathbf{x}_{ij})\}^{\gamma_{ij}} \exp\left\{-H_{0i}(t_{ij}) \exp\left(\boldsymbol{\beta}^T \mathbf{x}_{ij} + W_i\right)\right\} ,$$

where $H_{0i}(t) = \int_0^t h_{0i}(u) \, du$, the integrated baseline hazard. A frailty distribution parametrized by λ, $p(\mathbf{W}|\lambda)$, coupled with prior distributions for λ, $\boldsymbol{\beta}$, and the hazard function h complete the hierarchical Bayesian model specification.

In this chapter we consider both parametric and semiparametric hierarchical survival models for data sets that are spatially arranged. Such models might be appropriate anytime we suspect that frailties W_i corresponding to strata in closer proximity to each other might also be similar in magnitude. This could arise if, say, the strata corresponded to hospitals in a given region, to counties in a given state, and so on. The basic assumption here is that "expected" survival times (or hazard rates) will be more similar in proximate regions, due to underlying factors (access to care, willingness of the population to seek care, etc.) that vary spatially. We hasten to remind the reader that this does not imply that the observed survival times from subjects in proximate regions must be similar, since they include an extra level of randomness arising from their variability around their (spatially correlated) underlying model quantities.

9.1 Parametric models

9.1.1 Univariate spatial frailty modeling

While it is possible to identify centroids of geographic regions and employ spatial process modeling for these locations, the effects in our examples are more naturally associated with areal units. As such we work exclusively with CAR models for these effects, i.e., we assume that

$$\mathbf{W} \mid \lambda \sim CAR(\lambda) \ . \tag{9.5}$$

Also, we note that the resulting model for, say, (9.2) is an extended example of a generalized linear model for areal spatial data (Section 5.5). That is, (9.2) implies that

$$f(t_{ij}|\boldsymbol{\beta}, x_{ij}, W_i) = h_0(t_{ij})e^{\boldsymbol{\beta}^T \mathbf{x}_{ij} + W_i} e^{-H_0(t_{ij}) \exp(\boldsymbol{\beta}^T \mathbf{x}_{ij} + W_i)} \ . \tag{9.6}$$

In other words, $U_{ij} = H_0(t_{ij}) \sim \text{Exponential}\Big(\exp[-(\boldsymbol{\beta}^T \mathbf{x}_{ij} + W_i)]\Big)$ so $-\log EH_0(t_{ij}) = \boldsymbol{\beta}^T \mathbf{x}_{ij} + W_i$. The analogy with (5.51) and $g(\eta_i)$ is clear. The critical difference is that in Section 5.5 the link g is assumed known; here the link to the linear scale requires h_0, which is unknown (and will be modeled parametrically or nonparametrically).

Finally, we remark that it would certainly be possible to include both spatial and nonspatial frailties, which as already seen (Subsection 5.4.3) is now common practice in areal data modeling. Here, this would mean supplementing our spatial frailties W_i with a collection of nonspatial frailties, say, $V_i \overset{iid}{\sim} N(0, 1/\tau)$. The main problem with this approach is again that the frailties now become identified only by the prior, and so the proper choice of priors for τ and λ (or $\boldsymbol{\theta}$) becomes problematic. Another problem is the resultant decrease in algorithm performance wrought by the addition of so many additional, weakly identified parameters.

Bayesian implementation

As already mentioned, the models outlined above are straightforwardly implemented in a Bayesian framework using MCMC methods. In the parametric case, say (9.4), the joint posterior distribution of interest is

$$p(\boldsymbol{\beta}, \mathbf{W}, \rho, \lambda \mid \mathbf{t}, \mathbf{x}, \boldsymbol{\gamma}) \propto L(\boldsymbol{\beta}, \mathbf{W}, \rho \, ; \, \mathbf{t}, \mathbf{x}, \boldsymbol{\gamma}) \, p(\mathbf{W}|\lambda) \, p(\boldsymbol{\beta}) p(\rho) \, p(\lambda) \ , \tag{9.7}$$

where the first term on the right-hand side is the Weibull likelihood, the second is the CAR distribution of the random frailties, and the remaining terms are prior distributions. In (9.7), $\mathbf{t} = \{t_{ij}\}$ denotes the collection of times to death, $\mathbf{x} = \{\mathbf{x}_{ij}\}$ the collection of covariate vectors, and $\boldsymbol{\gamma} = \{\gamma_{ij}\}$ the collection of death indicators for all subjects in all strata.

For our investigations, we retain the parametric form of the baseline hazard given in (9.4). Thus $L(\beta, \mathbf{W}, \rho \,;\, \mathbf{t}, \mathbf{x}, \boldsymbol{\gamma})$ is proportional to

$$\prod_{i=1}^{I}\prod_{j=1}^{n_i}\left\{\rho t_{ij}^{\rho-1}\exp\left(\beta^T\mathbf{x}_{ij}+W_i\right)\right\}^{\gamma_{ij}}\exp\left\{-t_{ij}^{\rho}\exp\left(\beta^T\mathbf{x}_{ij}+W_i\right)\right\}.$$

$$(9.8)$$

The model specification in the Bayesian setup is completed by assigning prior distributions for β, ρ, and λ. Typically, a flat (improper uniform) prior is chosen for β, while vague but proper priors are chosen for ρ and λ, such as a $G(\alpha, 1/\alpha)$ prior for ρ and a $G(a, b)$ prior for λ. Hence the only extension beyond the disease mapping illustrations of Section 5.4 is the need to update ρ.

Example 9.1 *(Application to Minnesota infant mortality data).* We apply the methodology above to the analysis of infant mortality in Minnesota, originally considered by Banerjee, Wall, and Carlin (2003). The data were obtained from the linked birth-death records data registry kept by the Minnesota Department of Health. The data comprise 267,646 live births occurring during the years 1992–1996 followed through the first year of life, together with relevant covariate information such as birth weight, sex, race, mother's age, and the mother's total number of previous births. Because of the careful linkage connecting infant death certificates with birth certificates (even when the death occurs in a separate state), we assume that each baby in the data set that is not linked with a death must have been alive at the end of one year. Of the live births, only 1,547 babies died before the end of their first year. The number of days they lived is treated as the response t_{ij} in our models, while the remaining survivors were treated as "censored," or in other words, alive at the end of the study period. In addition to this information, the mother's Minnesota county of residence prior to the birth is provided. We implement the areal frailty model (9.5), the nonspatial frailty model (9.3), and a simple nonhierarchical ("no-frailty") model that sets $W_i = 0$ for all i.

For all of our models, we adopt a flat prior for β, and a $G(\alpha, 1/\alpha)$ prior for ρ, setting $\alpha = 0.01$. Metropolis random walk steps with Gaussian proposals were used for sampling from the full conditionals for β, while Hastings independence steps with gamma proposals were used for updating ρ. As for λ, in our case we are fortunate to have a data set that is large relative to the number of random effects to be estimated. As such, we simply select a vague (mean 1, variance 1000) gamma specification for λ, and rely on the data to overwhelm the priors.

Table 9.1 compares our three models in terms of two of the criteria discussed in Subsection 4.2.3, DIC and effective model size p_D. For the no-frailty model, we see a p_D of 8.72, very close to the actual number of parameters, 9 (8 components of β plus the Weibull parameter ρ). The random effects models have substantially larger p_D values, though much

Model	p_D	DIC
No-frailty	8.72	511
Nonspatial frailty	39.35	392
CAR frailty	34.52	371

Table 9.1 *DIC and effective number of parameters p_D for competing parametric survival models.*

Covariate	2.5%	50%	97.5%
Intercept	−2.135	−2.024	−1.976
Sex (boys = 0)			
girls	−0.271	−0.189	−0.105
Race (white = 0)			
black	−0.209	−0.104	−0.003
Native American	0.457	0.776	1.004
unknown	0.303	0.871	1.381
Mother's age	−0.005	−0.003	−0.001
Birth weight in kg	−1.820	−1.731	−1.640
Total births	0.064	0.121	0.184
ρ	0.411	0.431	0.480
σ	0.083	0.175	0.298

Table 9.2 *Posterior summaries for the nonspatial frailty model.*

smaller than their actual parameter counts (which would include the 87 random frailties W_i); apparently there is substantial shrinkage of the frailties toward their grand mean. The DIC values suggest that each of these models is substantially better than the no-frailty model, despite their increased size. Though the spatial frailty model has the best DIC value, plots of the full estimated posterior deviance distributions (not shown) suggest substantial overlap. On the whole we seem to have modest support for the spatial frailty model over the ordinary frailty model.

Tables 9.2 and 9.3 provide 2.5, 50, and 97.5 posterior percentiles for the main effects in our two frailty models. In both cases, all of the predictors are significant at the .05 level. Since the reference group for the sex variable is boys, we see that girls have a lower hazard of death during the first year of life. The reference group for the race variables is white; the Native American beta coefficient is rather striking. In the CAR model, this covariate increases the posterior median hazard rate by a factor of $e^{0.782} = 2.19$. The

Covariate	2.5%	50%	97.5%
Intercept	−2.585	−2.461	−2.405
Sex (boys = 0)			
girls	−0.224	−0.183	−0.096
Race (white = 0)			
black	−0.219	−0.105	−0.007
Native American	0.455	0.782	0.975
unknown	0.351	0.831	1.165
Mother's age	−0.005	−0.004	−0.003
Birth weight in kg	−1.953	−1.932	−1.898
Total births	0.088	0.119	0.151
ρ	0.470	0.484	0.497
λ	12.62	46.07	100.4

Table 9.3 *Posterior summaries for the CAR frailty model.*

effect of "unknown" race is also significant, but more difficult to interpret: in this group, the race of the infant was not recorded on the birth certificate. Separate terms for Hispanics, Asians, and Pacific Islanders were also originally included in the model, but were eliminated after emerging as not significantly different from zero. Note that the estimate of ρ is quite similar across models, and suggests a decreasing baseline hazard over time. This is consistent with the fact that a high proportion (495, or 32%) of the infant deaths in our data set occurred in the first *day* of life: the force of mortality (hazard rate) is very high initially, but drops quickly and continues to decrease throughout the first year.

A benefit of fitting the spatial CAR structure is seen in the reduction of the length of the 95% credible intervals for the covariates in the spatial models compared to the i.i.d. model. As we might expect, there are modest efficiency gains when the model that better specifies the covariance structure of its random effects is used. That is, since the spatial dependence priors for the frailties are in better agreement with the likelihood than is the independence prior, the prior-to-posterior learning afforded by Bayes' Rule leads to smaller posterior variances in the former cases. Most notably, the 95% credible set for the effect of "unknown" race is (0.303, 1.381) under the nonspatial frailty model (Table 9.2), but (0.351, 1.165) under the CAR frailty model (Table 9.3), a reduction in length of roughly 25%.

Figures 9.1 and 9.2 (see also color insert Figures C.8 and C.9) map the posterior medians of the W_i under the nonspatial (i.i.d. frailties) and CAR models, respectively, where the models include all of the covariates listed in Tables 9.2 and 9.3. As expected, no clear spatial pattern is evident in

IID model (with covariates)

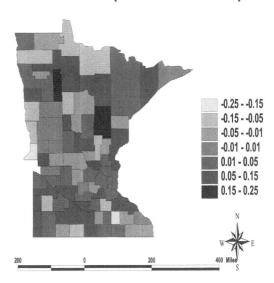

Figure 9.1 *Posterior median frailties, i.i.d. model with covariates, Minnesota county-level infant mortality data (see also color insert).*

the i.i.d. map, but from the CAR map we are able to identify two clusters of counties having somewhat higher hazards (in the southwest following the Minnesota River, and in the northeast "arrowhead" region), and two clusters with somewhat lower hazards (in the northwest, and the southeastern corner). Thus, despite the significance of the covariates now in these models, Figure 9.2 suggests the presence of some still-missing, spatially varying covariate(s) relevant for infant mortality. Such covariates might include location of birth (home or hospital), overall quality of available health or hospital care, mother's economic status, and mother's number of prior abortions or miscarriages.

In addition to the improved appearance and epidemiological interpretation of Figure 9.2, another reason to prefer the CAR model is provided in Figure 9.3, which shows boxplots of the posterior median frailties for the two cases corresponding to Figures 9.1 and 9.2, plus two preliminary models in which *no* covariates **x** are included. The tightness of the full CAR boxplot suggests this model is best at reducing the need for the frailty terms. This is as it should be, since these terms are essentially spatial residuals, and represent lingering lack of fit in our spatial model (although they may well also account for some excess *non*spatial variability, since our current models do not include nonspatial frailty terms). Note that all of the full

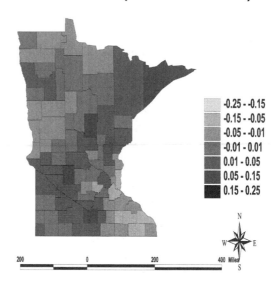

Figure 9.2 *Posterior median frailties, CAR model with covariates, Minnesota county-level infant mortality data (see also color insert).*

CAR residuals are in the range (–0.15, 0.10), or (0.86, 1.11) on the hazard scale, suggesting that missing spatially varying covariates have only a modest (10 to 15%) impact on the hazard; from a practical standpoint, this model fits quite well. ∎

9.1.2 Spatial frailty versus logistic regression models

In many contexts (say, a clinical trial enrolling and following patients at spatially proximate clinical centers), a spatial survival model like ours may be the only appropriate model. However, since the Minnesota infant mortality data does not have any babies censored because of loss to followup, competing risks, or any reason other than the end of the study, there is no ambiguity in defining a *binary* survival outcome for use in a random effects logistic regression model. That is, we replace the event time data t_{ij} with an indicator of whether the subject did ($Y_{ij} = 0$) or did not ($Y_{ij} = 1$) survive the first year. Letting $p_{ij} = Pr(Y_{ij} = 1)$, our model is then

$$logit(p_{ij}) = \widetilde{\beta}^T \mathbf{x}_{ij} + \widetilde{W}_i \,, \tag{9.9}$$

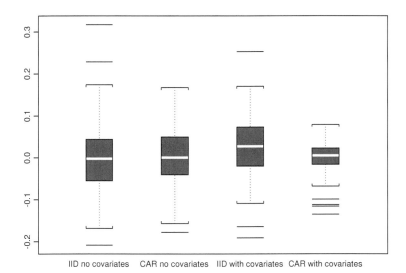

IID no covariates   CAR no covariates   IID with covariates   CAR with covariates

Figure 9.3 *Boxplots of posterior median frailties, i.i.d. and CAR models with and without covariates.*

with the usual flat prior for $\widetilde{\beta}$ and an i.i.d. or CAR prior for the $\widetilde{W}_i$. As a result, (9.9) is exactly an example of a generalized linear model for areal spatial data.

Other authors (Doksum and Gasko, 1990; Ingram and Kleinman, 1989) have shown that in this case of no censoring before followup (and even in cases of equal censoring across groups), it is possible to get results for the $\widetilde{\beta}$ parameters in the logistic regression model very similar to those obtained in the proportional hazards model (9.1), except of course for the differing interpretations (log odds versus log relative risk, respectively). Moreover when the probability of death is very small, as it is in the case of infant mortality, the log odds and log relative risk become even more similar. Since it uses more information (i.e., time to death rather than just a survival indicator), intuitively, the proportional hazards model should make gains over the logistic model in terms of power to detect significant covariate effects. Yet, consistent with the simulation studies performed by Ingram and Kleinman (1989), our experience with the infant mortality data indicate that only a marginal increase in efficiency (decrease in variance) is exhibited by the posterior distributions of the parameters.

On the other hand, we did find some difference in terms of the estimated random effects in the logistic model compared to the proportional hazards model. Figure 9.4 shows a scatterplot of the estimated posterior medians of W_i versus $\widetilde{W}_i$ for each county obtained from the models where there were

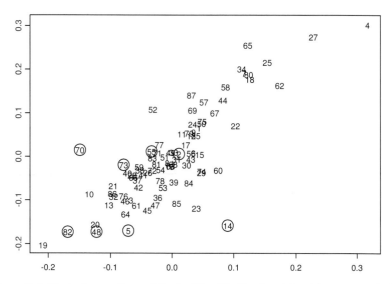

Figure 9.4 *Posterior medians of the frailties W_i (horizontal axis) versus posterior medians of the logistic random effects $\widetilde{W}_i$ (vertical axis). Plotting character is county number; significance of circled counties is described in the text.*

no covariates, and the random effects were assumed to i.i.d. The sample correlation of these estimated random effects is 0.81, clearly indicating that they are quite similar. Yet there are still some particular counties that result in rather different values under the two models. One way to explain this difference is that the hazard functions are not exactly proportional across the 87 counties of Minnesota. A close examination of the counties that had differing $\widetilde{W}_i$ versus W_i shows that they had different average times at death compared to other counties with similar overall death rates. Consider for example County 70, an outlier circled in Figure 9.4, and its comparison to circled Counties 73, 55, and 2, which have similar death rates (and hence roughly the same horizontal position in Figure 9.4). We find County 70 has the smallest mean age at death, implying that it has more early deaths, explaining its smaller frailty estimate. Conversely, County 14 has a higher average time at death but overall death rates similar to Counties 82, 48, and 5 (again note the horizontal alignment in Figure 9.4), and as a result has higher estimated frailty. A lack of proportionality in the baseline hazard rates across counties thus appears to manifest as a departure from linearity in Figure 9.4.

   We conclude this subsection by noting that previous work by Carlin and Hodges (1999) suggests a generalization of our basic model (9.4) to

$$h(t_{ij}; x_{ij}) = \rho_i t_{ij}^{\rho_i - 1} \exp(\beta^T \mathbf{x}_{ij} + W_i) \, .$$

That is, we allow two sets of random effects: the existing frailty parameters W_i, and a new set of shape parameters ρ_i. This then allows both the overall level and the shape of the hazard function over time to vary from county to county. Either i.i.d. or CAR priors could be assigned to these two sets of random effects, which could themselves be correlated within county. In the latter case, this might be fit using the MCAR model of Section 7.4; see Jin and Carlin (2003), as well as Section 9.4.

9.2 Semiparametric models

While parametric models are easily interpretable and often afford a surprisingly good fit to survival data, many practitioners continue to prefer the additional richness of the nonparametric baseline hazard offered by the celebrated Cox model. In this section we turn to nonparametric models for the baseline hazard. Such models are often referred to as *semiparametric*, since we continue to assume proportional hazards of the form (9.1) in which the covariate effects are still modeled parametrically, While Li and Ryan (2002) address this problem from a classical perspective, in this section we follow the hierarchical Bayesian approach of Banerjee and Carlin (2002).

Within the Bayesian framework, several authors have proposed treating the Cox partial likelihood as a full likelihood, to obtain a posterior distribution for the treatment effect. However, this approach does not allow fully hierarchical modeling of stratum-specific baseline hazards (with stratum-specific frailties) because the baseline hazard is implicit in the partial likelihood computation. In the remainder of this section, we describe two possible methodological approaches to modeling the baseline hazard in Cox regression, which thus lead to two semiparametric spatial frailty techniques. We subsequently revisit the Minnesota infant mortality data.

9.2.1 Beta mixture approach

Our first approach uses an idea of Gelfand and Mallick (1995) that flexibly models the integrated baseline hazard as a mixture of monotone functions. In particular, these authors use a simple transformation to map the integrated baseline hazard onto the interval $[0, 1]$, and subsequently approximate this function by a weighted mixture of incomplete beta functions. Implementation issues are discussed in detail by Gelfand and Mallick (1995) and also by Carlin and Hodges (1999) for stratum-specific baseline hazards. The likelihood and Bayesian hierarchical setup remain exactly as above.

Thus, we let $h_{0i}(t)$ be the baseline hazard in the ith region and $H_{0i}(t)$ be the corresponding integrated baseline hazard, and define

$$J_{0i}(t) = a_0 H_{0i}(t) / [a_0 H_{0i}(t) + b_0] \ ,$$

which conveniently takes values in $[0, 1]$. We discuss below the choice of a_0

and b_0 but note that this is not as much a modeling issue as a computational one, important only to ensure appropriate coverage of the interval $[0, 1]$. We next model $J_{0i}(t)$ as a mixture of $Beta(r_l, s_l)$ cdfs, for $l = 1, \ldots, m$. The r_l and s_l are chosen so that the beta cdfs have evenly spaced means and are centered around $\widetilde{J_0}(t)$, a suitable function transforming the time scale to $[0, 1]$. We thus have

$$
J_{0i}(t) = \sum_{l=1}^{m} v_{il} \, IB\left(\widetilde{J_0}(t); r_l, s_l\right) ,
$$

where $\sum_{l=1}^{m} v_{il} = 1$ for all i, and $IB(\cdot \, ; a, b)$ denotes the incomplete beta function (i.e., the cdf of a $Beta\,(a, b)$ distribution). Since any distribution function on $[0, 1]$ can be approximated arbitrarily well by a finite mixture of beta cdfs, the same is true for J_{0i}, an increasing function that maps $[0, 1]$ onto itself. Thus, working backward, we find the following expression for the cumulative hazard in terms of the above parameters:

$$
H_{0i}(t) = \frac{b_0 \sum_{l=1}^{m} v_{il} \, IB\left(\widetilde{J_0}(t); r_l, s_l\right)}{a_0 \left\{1 - \sum_{l=1}^{m} v_{il} \, IB\left(\widetilde{J_0}(t); r_l, s_l\right)\right\}} .
$$

Taking derivatives, we have for the hazard function,

$$
h_{0i}(t) = \frac{b_0 \frac{\partial}{\partial t} \widetilde{J_0}(t) \sum_{l=1}^{m} v_{il} Beta\left(\widetilde{J_0}(t); r_l, s_l\right)}{a_0 \left\{1 - \sum_{l=1}^{m} v_{il} \, IB\left(\widetilde{J_0}(t); r_l, s_l\right)\right\}^2} .
$$

Typically m, the number of mixands of the beta cdfs, is fixed, as are the $\{(r_l, s_l)\}_{l=1}^{m}$, so chosen that the resulting beta densities cover the interval $[0, 1]$. For example, we might fix $m = 5$, $\{r_l\} = (1, 2, 3, 4, 5)$ and $\{s_l\} = (5, 4, 3, 2, 1)$, producing five evenly-spaced beta cdfs.

Regarding the choice of a_0 and b_0, we note that it is intuitive to specify $\widetilde{J_0}(t)$ to represent a plausible central function around which the J_{0i}'s are distributed. Thus, if we consider the cumulative hazard function of an exponential distribution to specify $\widetilde{J_0}(t)$, then we get $\widetilde{J_0}(t) = a_0 t / (a_0 t + b_0)$. In our Minnesota infant mortality data set, since the survival times ranged between 1 day and 365 days, we found $a_0 = 5$ and $b_0 = 100$ lead to values for $\widetilde{J_0}(t)$ that largely cover the interval $[0, 1]$, and so fixed them as such. The likelihood is thus a function of the regression coefficients β, the stratum-specific weight vectors $\mathbf{v}_i = (v_{i1}, \ldots, v_{im})^T$, and the spatial effects W_i. It is natural to model the $\mathbf{v}_i$'s as draws from a Dirichlet$(\phi_1, \ldots, \phi_m)$ distribution, where for simplicity we often take $\phi_1 = \cdots = \phi_m = \phi$.

9.2.2 Counting process approach

The second nonparametric baseline hazard modeling approach we investigate is that of Clayton (1991, 1994). While the method is less transparent theoretically, it is gaining popularity among Bayesian practitioners due to its ready availability within WinBUGS. Here we give only the essential ideas, referring the reader to Andersen and Gill (1982) or Clayton (1991) for a more complete treatment. The underlying idea is that the number of failures up to time t is assumed to arise from a *counting process* $N(t)$. The corresponding *intensity process* is defined as

$$I(t)\, dt = E\left(dN(t)\,|F_{t-}\right)\ ,$$

where $dN(t)$ is the increment of N over the time interval $[t, t+dt)$, and F_{t-} represents the available data up to time t. For each individual, $dN(t)$ therefore takes the value 1 if the subject fails in that interval, and 0 otherwise. Thus $dN(t)$ may be thought of as the "death indicator process," analogous to γ in the model of the previous subsection. For the jth subject in the ith region, under the proportional hazards assumption, the intensity process (analogous to our hazard function $h\left(t_{ij}; \mathbf{x}_{ij}\right)$) is modeled as

$$I_{ij}(t) = Y_{ij}(t)\lambda_0(t) \exp\left(\boldsymbol{\beta}^T \mathbf{x}_{ij} + W_i\right)\ ,$$

where $\lambda_0(t)$ is the baseline hazard function and $Y_{ij}(t)$ is an indicator process taking the value 1 or 0 according to whether or not subject i is observed at time t. Under the above formulation and keeping the same notation as above for $\mathbf{W}$ and $\mathbf{x}$, a Bayesian hierarchical model may be formulated as:

$$
\begin{aligned}
dN_{ij}(t) &\sim\ Poisson\left(I_{ij}(t)\, dt\right)\ , \\
I_{ij}(t)\, dt &=\ Y_{ij}(t)\exp\left(\boldsymbol{\beta}^T \mathbf{x}_{ij} + W_i\right) d\Lambda_0(t)\ , \\
d\Lambda_0(t) &\sim\ Gamma\left(c\, d\Lambda_0^*(t), c\right)\ .
\end{aligned}
$$

As before, priors $p\left(\mathbf{W}|\lambda\right)$, $p\left(\lambda\right)$, and $p(\boldsymbol{\beta})$ are required to completely specify the Bayesian hierarchical model. Here, $d\Lambda_0(t) = \lambda_0(t)\, dt$ may be looked upon as the increment or jump in the integrated baseline hazard function occurring during the time interval $[t, t+dt)$. Since the conjugate prior for the Poisson mean is the gamma distribution, $\Lambda_0(t)$ is conveniently modeled as a process whose increments $d\Lambda_0(t)$ are distributed according to gamma distributions. The parameter c in the above setup represents the degree of confidence in our prior guess for $d\Lambda_0(t)$, given by $d\Lambda_0^*(t)$. Typically, the prior guess $d\Lambda_0^*(t)$ is modeled as $r\, dt$, where r is a guess at the failure rate per unit time. The LeukFr example in the WinBUGS examples manual offers an illustration of how to code the above formulation.

Example 9.2 *(Application to Minnesota infant mortality data, continued)*. We now apply the methodology above to the reanalysis of our Minnesota infant mortality data set. For both the CAR and nonspatial models

Model	p_D	DIC
No-frailty	6.82	507
Nonspatial frailty	27.46	391
CAR frailty	32.52	367

Table 9.4 *DIC and effective number of parameters p_D for competing nonparametric survival models.*

we implemented the Cox model with the two semiparametric approaches outlined above. We found very similar results, and so in our subsequent analysis we present only the results with the beta mixture approach (Subsection 9.2.1). For all of our models, we adopt vague Gaussian priors for $\boldsymbol{\beta}$. Since the full conditionals for each component of $\boldsymbol{\beta}$ are log-concave, adaptive rejection sampling was used for sampling from the $\boldsymbol{\beta}$ full conditionals. As in Section 9.1, we again simply select a vague $G(0.001, 1000)$ (mean 1, variance 1000) specification for CAR smoothness parameter λ, though we maintain more informative priors on the other variance components.

Table 9.4 compares our three models in terms of DIC and effective model size p_D. For the no-frailty model, we see a p_D of 6.82, reasonably close to the actual number of parameters, 8 (the components of $\boldsymbol{\beta}$). The other two models have substantially larger p_D values, though much smaller than their actual parameter counts (which would include the 87 random frailties W_i); apparently there is substantial shrinkage of the frailties toward their grand mean. The DIC values suggest that both of these models are substantially better than the no-frailty model, despite their increased size. As in Table 9.1, the spatial frailty model has the best DIC value.

Tables 9.5 and 9.6 provide 2.5, 50, and 97.5 posterior percentiles for the main effects in our two frailty models, respectively. In both tables, all of the predictors are significant at the .05 level. Overall, the results are broadly similar to those from our earlier parametric analysis in Tables 9.2 and 9.3. For instance, the effect of being in the Native American group is again noteworthy. Under the CAR model, this covariate increases the posterior median hazard rate by a factor of $e^{0.599} = 1.82$. The benefit of fitting the spatial CAR structure is also seen again in the reduction of the length of the 95% credible intervals for the spatial model compared to the i.i.d. model. Most notably, the 95% credible set for the effect of "mother's age" is $(-0.054, -0.014)$ under the nonspatial frailty model (Table 9.5), but $(-0.042, -0.013)$ under the CAR frailty model (Table 9.6), a reduction in length of roughly 28%. Thus overall, adding spatial structure to the frailty terms appears to be reasonable and beneficial. Maps analogous to Figures 9.1 and 9.2 (not shown) reveal a very similar story. ■

Covariate	2.5%	50%	97.5%
Intercept	−2.524	−1.673	−0.832
Sex (boys = 0)			
girls	−0.274	−0.189	−0.104
Race (white = 0)			
black	−0.365	−0.186	−0.012
Native American	0.427	0.737	1.034
unknown	0.295	0.841	1.381
Mother's age	−0.054	−0.035	−0.014
Birth weight in kg	−1.324	−1.301	−1.280
Total births	0.064	0.121	0.184

Table 9.5 *Posterior summaries for the nonspatial semiparametric frailty model.*

Covariate	2.5%	50%	97.5%
Intercept	−1.961	−1.532	−0.845
Sex (boys = 0)			
girls	−0.351	−0.290	−0.217
Race (white = 0)			
black	−0.359	−0.217	−0.014
Native American	0.324	0.599	0.919
unknown	0.365	0.863	1.316
Mother's age	−0.042	−0.026	−0.013
Birth weight in kg	−1.325	−1.301	−1.283
Total births	0.088	0.135	0.193

Table 9.6 *Posterior summaries for the CAR semiparametric frailty model.*

9.3 Spatiotemporal models

In this section we follow Banerjee and Carlin (2003) to develop a semiparametric (Cox) hierarchical Bayesian frailty model for capturing spatiotemporal heterogeneity in survival data. We then use these models to describe the pattern of breast cancer in the 99 counties of Iowa while accounting for important covariates, spatially correlated differences in the hazards among the counties, and possible space-time interactions.

We begin by extending the framework of the preceding section to incorporate temporal dependence. Here we have t_{ijk} as the response (time to death) for the jth subject residing in the ith county who was diagnosed in the kth year, while the individual-specific vector of covariates is now denoted by $\mathbf{x}_{ijk}$, for $i = 1, 2, ..., I$, $k = 1, ..., K$, and $j = 1, 2, ..., n_{ik}$. We note that "time" is now being used in two ways. The measurement or response is a survival time, but these responses are themselves observed at different areal units *and* different times (years). Furthermore, the spatial random effects W_i in the preceding section are now modified to W_{ik}, to represent spatiotemporal frailties corresponding to the ith county for the kth diagnosis year. Our spatial frailty specification in (9.1) now becomes

$$h\left(t_{ijk}; \mathbf{x}_{ijk}\right) = h_{0i}\left(t_{ijk}\right) \exp\left(\boldsymbol{\beta}^T \mathbf{x}_{ijk} + W_{ik}\right). \quad (9.10)$$

Our CAR prior would now have conditional representation $W_{ik} \mid W_{(i' \neq i)k} \sim N(\overline{W}_{ik}, 1/(\lambda_k m_i))$.

Note that we can account for temporal correlation in the frailties by assuming that the λ_k are themselves identically distributed from a common hyperprior (Subsection 8.5.1). A gamma prior (usually vague but proper) is often selected here, since this is particularly convenient for MCMC implementation. A flat prior for $\boldsymbol{\beta}$ is typically chosen, since this still admits a proper posterior distribution. Adaptive rejection (Gilks and Wild, 1992) or Metropolis-Hastings sampling are usually required to update the $\mathbf{W}_k$ and $\boldsymbol{\beta}$ parameters in a hybrid Gibbs sampler.

We remark that it would certainly be possible to include both spatial and nonspatial frailties, as mentioned in Subsection 9.1.1. This would mean supplementing our spatial frailties W_{ik} with a collection of nonspatial frailties, say $V_{ik} \overset{iid}{\sim} N(0, 1/\tau_k)$. We summarize our full hierarchical model as follows:

$$L\left(\boldsymbol{\beta}, \mathbf{W}; \mathbf{t}, \mathbf{x}, \gamma\right) \propto \prod_{k=1}^{K}\prod_{i=1}^{I}\prod_{j=1}^{n_{ik}} \{h_{0i}\left(t_{ijk}; \mathbf{x}_{ijk}\right)\}^{\gamma_{ijk}}$$

$$\times \exp\left\{-H_{0i}\left(t_{ijk}\right)\exp\left(\boldsymbol{\beta}^T \mathbf{x}_{ijk} + W_{ik} + V_{ik}\right)\right\},$$

where $p(\mathbf{W}_k|\lambda_k) \sim CAR(\lambda_k)$ $p(\mathbf{V}_k|\tau_k) \sim N_I(\mathbf{0}, \tau_k \mathbf{I})$
and $\lambda_k \sim G(a, b)$, $\tau_k \sim G(c, d)$ for $k = 1, 2, ..., K$.

In the sequel we adopt the beta mixture approach of Subsection 9.2.1 to model the baseline hazard functions $H_{0i}(t_{ijk})$ nonparametrically.

Example 9.3 *(Analysis of Iowa SEER breast cancer data).* The National Cancer Institute's SEER program (seer.cancer.gov) is the most authoritative source of cancer data in the U.S., offering county-level summaries on a yearly basis for several states in various parts of the country. In particular, the database provides a cohort of 15,375 women in Iowa who were diagnosed with breast cancer starting in 1973, and have been undergoing treatment and have been progressively monitored since. Only those who have been identified as having died from metastasis of cancerous nodes in the breast are considered to have failed, while the rest (including those

Covariate	2.5%	50%	97.5%
Age at diagnosis	0.0135	0.0148	0.0163
Number of primaries	–0.43	–0.40	–0.36
Race (white = 0)			
black	–0.14	0.21	0.53
other	–2.25	–0.30	0.97
Stage (local = 0)			
regional	0.30	0.34	0.38
distant	1.45	1.51	1.58

Table 9.7 *Posterior summaries for the spatiotemporal frailty model.*

who might have died from metastasis of other types of cancer, or from other causes of death) are considered censored. By the end of 1998, 11,912 of the patients had died of breast cancer while the remaining were censored, either because they survived until the end of the study period, dropped out of the study, or died of causes other than breast cancer. For each individual, the data set records the time in months (1 to 312) that the patient survived, and her county of residence at diagnosis. Several individual-level covariates are also available, including race (black, white, or other), age at diagnosis, number of primaries (i.e., the number of other types of cancer diagnosed for this patient), and the stage of the disease (local, regional, or distant).

Results for the full model

We begin by summarizing our results for the spatiotemporal frailty model described above, i.e., the full model having both spatial frailties W_{ik} and nonspatial frailties V_{ik}. We chose vague $G(0.01, 0.01)$ hyperpriors for the λ_k and τ_k (having mean 1 but variance 100) in order to allow maximum flexibility in the partitioning of the frailties into spatial and nonspatial components. Best et al. (1999) suggest that a higher variance prior for the τ_k (say, a $G(0.001, 0.001)$) may lead to better prior "balance" between the spatial and nonspatial random effects, but there is controversy on this point and so we do not pursue it here. While overly diffuse priors (as measured for example as in Weiss, 1996) may result in weak identifiability of these parameters, their posteriors remain proper, and the impact of these priors on the posterior for the well-identified subset of parameters (including β and the log-relative hazards themselves) should be minimal (Daniels and Kass, 1999; Eberly and Carlin, 2000).

Table 9.7 provides 2.5, 50, and 97.5 posterior percentiles for the main effects (components of β) in our model. All of the predictors *except* those having to do with race are significant at the .05 level. Since the reference

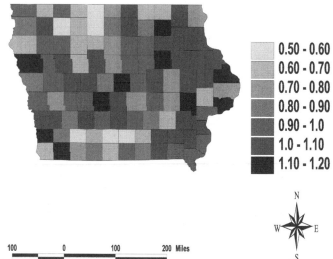

▦ 0.50 - 0.60
▦ 0.60 - 0.70
▦ 0.70 - 0.80
▦ 0.80 - 0.90
▦ 0.90 - 1.0
▦ 1.0 - 1.10
▦ 1.10 - 1.20

Figure 9.5 *Fitted spatiotemporal frailties, Iowa counties, 1986 (see also color insert).*

group for the stage variable is local, we see that women with regional and distant (metastasized) diagnoses have higher and much higher hazard of death, respectively; the posterior median hazard rate increases by a factor of $e^{1.51} = 4.53$ for the latter group. Higher age at diagnosis also increases the hazard, but a larger number of primaries (the number of other types of cancer a patient is be suffering from) actually leads to a *lower* hazard, presumably due to the competing risk of dying from one of these other cancers.

Figure 9.5 (see also color insert Figure C.10) maps the posterior medians of the frailties $W_{ik} + V_{ik}$ for the representative year 1986. We see clusters of counties with lower median frailties in the north-central and south-central parts of the state, and also clusters of counties with higher median frailties in the central, northeastern, and southeastern parts of the state.

Maps for other representative years showed very similar patterns, as well as an overall decreasing pattern in the frailties over time (see Banerjee and Carlin, 2003, for details). Figure 9.6 clarifies this pattern by showing box-plots of the posterior medians of the W_{ik} over time (recall our full model does not have year-specific intercepts; the average of the W_{ik} for year k plays this role). We see an essentially horizontal trend during roughly the first half of our observation period, followed by a decreasing trend that seems to be accelerating. Overall the total decrease in median log hazard is about 0.7 units, or about a 50% reduction in hazard over the observation period. A cancer epidemiologist would likely be unsurprised by this

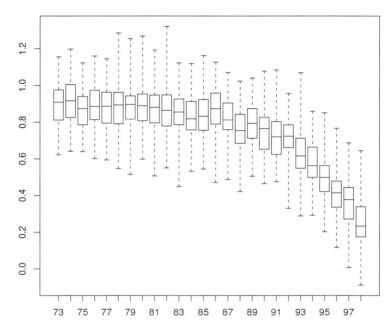

Figure 9.6 *Boxplots of posterior medians for the spatial frailties* W_{ik} *over the Iowa counties for each year,* $k=1973, \ldots, 1998$.

decline, since it coincides with the recent rise in the use of mammography by American women.

Bayesian model choice

For model choice, we again turn to the DIC criterion. The first six lines of Table 9.8 provide p_D and DIC values for our full model and several simplications thereof. Note the full model (sixth line) is estimated to have only just over 150 effective parameters, a substantial reduction (recall there are $2 \times 99 \times 26 = 5148$ random frailty parameters alone). Removing the spatial frailties W_{ik} from the log-relative hazard has little impact on p_D, but substantial negative impact on the DIC score. By contrast, removing the nonspatial frailties V_{ik} reduces (i.e., improves) both p_D and DIC, consistent with our findings in the previous subsection. Further simplifying the model to having a single set of spatial frailties W_i that do not vary with time (but now also reinserting year-specific intercepts α_k) has little effect on p_D but does improve DIC a bit more (though this improvement appears only slightly larger than the order of Monte Carlo error in our calculations). Even more drastic simplifications (eliminating the W_i, and perhaps even the α_k) lead to further drops in p_D, but at the cost of unacceptably large

Baseline hazard	Log-relative hazard	p_D	DIC
Semiparametric mixture	$\boldsymbol{\beta}^T \mathbf{x}_{ijk}$	6.17	780
Semiparametric mixture	$\boldsymbol{\beta}^T \mathbf{x}_{ijk} + \alpha_k$	33.16	743
Semiparametric mixture	$\boldsymbol{\beta}^T \mathbf{x}_{ijk} + \alpha_k + W_i$	80.02	187
Semiparametric mixture	$\boldsymbol{\beta}^T \mathbf{x}_{ijk} + W_{ik}$	81.13	208
Semiparametric mixture	$\boldsymbol{\beta}^T \mathbf{x}_{ijk} + V_{ik}$	149.45	732
Semiparametric mixture	$\boldsymbol{\beta}^T \mathbf{x}_{ijk} + W_{ik} + V_{ik}$	151.62	280
Weibull	$\boldsymbol{\beta}^T \mathbf{x}_{ijk} + \alpha_k + W_i$	79.22	221
Weibull	$\boldsymbol{\beta}^T \mathbf{x}_{ijk} + W_{ik}$	80.75	239
Weibull	$\boldsymbol{\beta}^T \mathbf{x}_{ijk} + W_{ik} + V_{ik}$	141.67	315

Table 9.8 *DIC and effective number of parameters p_D for the competing models.*

increases in DIC. Thus our county-level breast cancer survival data seem to have strong spatial structure that is still unaccounted for by the covariates in Table 9.7, but structure that is fairly similar for all diagnosis years.

The last three lines of Table 9.8 reconsider the best three log-relative hazard models above, but where we now replace the semiparametric mixture baseline hazard with a Weibull hazard having region-specific baseline hazards $h_{0i}(t_{ijk}; \rho_i) = \rho_i t_{ijk}^{\rho_i - 1}$ (note the spatial frailties play the role of the second parameter customarily associated with the Weibull model). These fully parametric models offer small advantages in terms of parsimony (smaller p_D), but these gains are apparently more than outweighed by a corresponding degradation in fit (much larger DIC score). ∎

9.4 Multivariate models ⋆

In this section we extend to multivariate spatial frailty modeling, using the MCAR model introduced in Subsection 7.4. In particular, we use a semiparametric model, and consider MCAR structure on both residual (spatial frailty) and regression (space-varying coefficient) terms. We also extend to the spatiotemporal case by including temporally correlated cohort effects (say, one for each year of initial disease diagnosis) that can be summarized and plotted over time. Example 9.4 illustrates the utility of our approach in an analysis of survival times of patients suffering from one or more types of cancer. We obtain posterior estimates of key fixed effects, smoothed maps of both frailties and spatially varying coefficients, and compare models using the DIC criterion.

Static spatial survival data with multiple causes of death

Consider the following multivariate survival setting. Let t_{ijk} denote the time to death or censoring for the kth patient having the jth type of primary cancer living in the ith county, $i = 1, \ldots, n$, $j = 1, \ldots, p$, $k = 1, \ldots, s_{ij}$, and let γ_{ijk} be the corresponding death indicator. Let us write $\mathbf{x}_{ijk}$ as the vector of covariates for the above individual, and let $\mathbf{z}_{ijk}$ denote the vector of cancer indicators for this individual. That is, $\mathbf{z}_{ijk} = (z_{ijk1}, z_{ijk2}, \ldots, z_{ijkp})^T$ where $z_{ijkl} = 1$ if patient ijk suffers from cancer type l, and 0 otherwise (note that $z_{ijkj} = 1$ by definition). Then we can write the likelihood of our proportional hazards model $L(\boldsymbol{\beta}, \boldsymbol{\theta}, \boldsymbol{\Phi}; \mathbf{t}, \mathbf{x}, \boldsymbol{\gamma})$ as

$$\prod_{i=1}^{n} \prod_{j=1}^{p} \prod_{k=1}^{s_{ij}} \{h(t_{ijk}; \mathbf{x}_{ijk}, \mathbf{z}_{ijk})\}^{\gamma_{ijk}}$$
$$\times \exp\left\{-H_{0i}(t_{ijk}) \exp\left(\mathbf{x}_{ijk}^T \boldsymbol{\beta} + \mathbf{z}_{ijk}^T \boldsymbol{\theta} + \phi_{ij}\right)\right\}, \tag{9.11}$$

where

$$h(t_{ijk}; \mathbf{x}_{ijk}, \mathbf{z}_{ijk}) = h_{0i}(t_{ijk}) \exp\left(\mathbf{x}_{ijk}^T \boldsymbol{\beta} + \mathbf{z}_{ijk}^T \boldsymbol{\theta} + \phi_{ij}\right). \tag{9.12}$$

Here, $H_{0i}(t_{ijk}) = \int_0^{t_{ijk}} h_{0i}(u)\, du$, $\boldsymbol{\phi}_i = (\phi_{i1}, \phi_{i2}, \ldots, \phi_{in})^T$, $\boldsymbol{\beta}$ and $\boldsymbol{\theta}$ are given flat priors, and

$$\boldsymbol{\Phi} \equiv \left(\boldsymbol{\phi}_1^T, \ldots, \boldsymbol{\phi}_n^T\right)^T \sim MCAR(\rho, \boldsymbol{\Sigma}),$$

using the notation of Subsection 7.4.2. The region-specific baseline hazard functions $h_{0i}(t_{ijk})$ are modeled using the beta mixture approach (Subsection 9.2.1) in such a way that the intercept in $\boldsymbol{\beta}$ remains estimable. We note that we could extend to a county *and* cancer-specific baseline hazard h_{0ij}; however, preliminary exploratory analyses of our data suggest such generality is not needed here.

Several alternatives to model formulation (9.12) immediately present themselves. For example, we could convert to a space-varying coefficients model (Assunção, 2003), replacing the log-relative hazard $\mathbf{x}_{ijk}^T \boldsymbol{\beta} + \mathbf{z}_{ijk}^T \boldsymbol{\theta} + \phi_{ij}$ in (9.12) with

$$\mathbf{x}_{ijk}^T \boldsymbol{\beta} + \mathbf{z}_{ijk}^T \boldsymbol{\theta}_i, \tag{9.13}$$

where $\boldsymbol{\beta}$ again has a flat prior, but $\boldsymbol{\Theta} \equiv \left(\boldsymbol{\theta}_1^T, \ldots, \boldsymbol{\theta}_n^T\right)^T \sim MCAR(\rho, \boldsymbol{\Sigma})$. In Example 9.4 we apply this method to our cancer data set; we defer mention of still other log-relative hazard modeling possibilities until after this illustration.

MCAR specification, simplification, and computing

To efficiently implement the $MCAR(\rho, \boldsymbol{\Sigma})$ as a prior distribution for our spatial process, suppose that we are using the usual 0-1 adjacency weights in W. Then recall from equation (7.34) that we may express the MCAR precision matrix $\mathbf{B} \equiv \Sigma_{\phi}^{-1}$ in terms of the $n \times n$ adjacency matrix W as

$$\mathbf{B} = (Diag(m_i) - \rho W) \otimes \Lambda ,$$

where we have added a propriety parameter ρ. Note that this is a Kronecker product of an $n \times n$ and a $p \times p$ matrix, thereby rendering $\mathbf{B}$ as $np \times np$ as required. In fact, $\mathbf{B}$ may be looked upon as the Kronecker product of two partial precision matrices: one for the spatial components, $(Diag(m_i) - \rho W)$ (depending upon their adjacency structure and number of neighbors), and another for the variation across diseases, given by Λ. We thus alter our notation slightly to $MCAR(\rho, \Lambda)$.

Also as a consequence of this form, a sufficient condition for positive definiteness of the dispersion matrix for this MCAR model becomes $|\rho| < 1$ (as in the univariate case). Negative smoothness parameters are not desirable, so we typically take $0 < \rho < 1$. We can now complete the Bayesian hierarchical formulation by placing appropriate priors on ρ (say, a $Unif(0,1)$ or $Beta(18,2)$) and Λ (say, a $Wishart(\rho, \Lambda_0)$).

The Gibbs sampler is the MCMC method of choice here, particularly because, as in the univariate case, it takes advantage of the MCAR's conditional specification. Adaptive rejection sampling may be used to sample the regression coefficients $\boldsymbol{\beta}$ and $\boldsymbol{\theta}$, while Metropolis steps with (possibly multivariate) Gaussian proposals may be employed for the spatial effects $\boldsymbol{\Phi}$. The full conditional for ρ is nicely suited for slice sampling (see Subsection 4.3.3), given its bounded support. Finally, the full conditional for Λ^{-1} emerges in closed form as an inverted Wishart distribution.

We conclude this subsection by recalling that our model can be generalized to admit different propriety parameters ρ_j for different diseases (c.f. the discussion surrounding equation (7.35)). We notate this model as $MCAR(\boldsymbol{\rho}, \Lambda)$, where $\boldsymbol{\rho} = (\rho_1, \ldots, \rho_p)^T$.

Spatiotemporal survival data

Here we extend our model to allow for cohort effects. Let r index the year in which patient ijk entered the study (i.e., the year in which the patient's primary cancer was diagnosed). Extending model (9.12) we obtain the log-relative hazard,

$$\mathbf{x}_{ijkr}^T \boldsymbol{\beta} + \mathbf{z}_{ijkr}^T \boldsymbol{\theta} + \phi_{ijr} , \tag{9.14}$$

with the obvious corresponding modifications to the likelihood (9.11). Here, $\phi_{ir} = (\phi_{i1r}, \phi_{i2r}, ..., \phi_{ipr})^T$ and $\boldsymbol{\Phi}_r = \left(\phi_{1r}^T, \ldots, \phi_{nr}^T\right)^T \stackrel{ind}{\sim} MCAR(\rho_r, \Lambda_r)$. This permits addition of an exchangeable prior structure,

$$\rho_r \stackrel{iid}{\sim} Beta(a,b) \text{ and } \Lambda_r \stackrel{iid}{\sim} Wishart(\rho, \Lambda_0) ,$$

where we may choose fixed values for a, b, ρ, and Λ_0, or place hyperpriors on them and estimate them from the data. Note also the obvious extension to disease-specific ρ_{jr}, as mentioned at the end of the previous subsection.

Example 9.4 *(Application to Iowa SEER multiple cancer survival data).* We illustrate the approach with an analysis of SEER data on 17,146 patients from the 99 counties of the state of Iowa who have been diagnosed with cancer between 1992 and 1998, and who have a well-identified primary cancer. Our covariate vector $\mathbf{x}_{ijk}$ consists of a constant (intercept), a gender indicator, the age of the patient, indicators for race with "white" as the baseline, indicators for the stage of the primary cancer with "local" as the baseline, and indicators for year of primary cancer diagnosis (cohort) with the first year (1992) as the baseline. The vector $\mathbf{z}_{ijk}$ comprises the indicators of which cancers the patient has; the corresponding parameters will thus capture the effect of these cancers on the hazards regardless of whether they emerge as primary or secondary.

With regard to modeling details, we used five separate (cancer-specific) propriety parameters ρ_j having an exchangeable $Beta(18, 2)$ prior, and a vague $Wishart\,(\rho = 5, \Lambda_0 = Diag(.01, .01, .01, .01, .01))$ for Λ. (Results for $\boldsymbol{\beta}, \boldsymbol{\theta}$, and $\boldsymbol{\Phi}$ under a $U(0, 1)$ prior for the ρ_j were broadly similar.) Table 9.9 gives posterior summaries for the main effects $\boldsymbol{\beta}$ and $\boldsymbol{\theta}$; note that $\boldsymbol{\theta}$ is estimable despite the presence of the intercept since many individuals have more than one cancer. No race or cohort effects emerged as significantly different from zero, so they have been deleted; all remaining effects are shown here. All of these effects are significant and in the directions one would expect. In particular, the five cancer effects are consistent with results of previous modeling of this and similar data sets, with pancreatic cancer emerging as the most deadly (posterior median log relative hazard 1.701) and colorectal and small intestinal cancer relatively less so (.252 and .287, respectively).

Table 9.10 gives posterior variance and correlation summaries for the frailties ϕ_{ij} among the five cancers for two representative counties, Dallas (urban; Des Moines area) and Clay (rural northwest). Note that the correlations are as high as 0.528 (pancreas and stomach in Dallas County), suggesting the need for the multivariate structure inherent in our MCAR frailty model. Note also that summarizing the posterior distribution of Λ^{-1} would be inappropriate here, since despite the Kronecker structure here (as in (7.35)), Λ^{-1} cannot be directly interpreted as a primary cancer covariance matrix across counties.

Turning to geographic summaries, Figure 9.7 (see also color insert Figure C.11) shows `ArcView` maps of the posterior means of the MCAR spatial frailties ϕ_{ij}. Recall that in this model, the ϕ_{ij} play the role of spatial residuals, capturing any spatial variation not already accounted for by the spatial main effects $\boldsymbol{\beta}$ and $\boldsymbol{\theta}$. The lack of spatial pattern in these maps

Variable	2.5%	50%	97.5%
Intercept	0.102	0.265	0.421
Sex (female = 0)	0.097	0.136	0.182
Age	0.028	0.029	0.030
Stage of primary cancer (local = 0)			
regional	0.322	0.373	0.421
distant	1.527	1.580	1.654
Type of primary cancer			
colorectal	0.112	0.252	0.453
gallbladder	1.074	1.201	1.330
pancreas	1.603	1.701	1.807
small intestine	0.128	0.287	0.445
stomach	1.005	1.072	1.141

Table 9.9 *Posterior quantiles for the fixed effects in the MCAR frailty model.*

suggest there is little additional spatial "story" in the data beyond what is already being told by the fixed effects. However, the map scales reveal that one cancer (gallbladder) is markedly different from the others, both in terms of total range of the mean frailties (rather broad) and their center (negative; the other four are centered near 0).

Next, we change from the MCAR spatial frailty model to the MCAR spatially varying coefficients model (9.13). This model required a longer burn-in period (20,000 instead of 10,000), but otherwise our prior and MCMC control parameters remain unchanged. Figure 9.8 (see also color insert Figure C.12) shows ArcView maps of the resulting posterior means of the spatially varying coefficients θ_{ij}. Unlike the ϕ_{ij} in the previous model, these parameters are not "residuals," but the effects of the presence of the primary cancer indicated on the death rate in each county. Clearly these maps show a strong spatial pattern, with (for example) southwestern Iowa counties having relatively high fitted values for pancreatic and stomach cancer, while southeastern counties fare relatively poorly with respect to colorectal and small intestinal cancer. The overall levels for each cancer are consistent with those given for the corresponding fixed effects θ in Table 9.9 for the spatial frailty model.

Table 9.11 gives the effective model sizes p_D and DIC scores for a variety of spatial survival models. The first two listed (fixed effects only and standard CAR frailty) have few effective parameters, but also poor (large) DIC scores. The MCAR spatial frailty models (which place the MCAR on Φ) fare better, especially when we add the disease-specific ρ_j (the model

Dallas County	Colo-rectal	Gall-bladder	Pancreas	Small intestine	Stomach
Colorectal	0.852	0.262	0.294	0.413	0.464
Gallbladder		1.151	0.314	0.187	0.175
Pancreas			0.846	0.454	0.528
Small intestine				1.47	0.413
Stomach					0.908

Clay County	Colo-rectal	Gall-bladder	Pancreas	Small intestine	Stomach
Colorectal	0.903	0.215	0.273	0.342	0.352
Gallbladder		1.196	0.274	0.128	0.150
Pancreas			0.852	0.322	0.402
Small intestine				1.515	0.371
Stomach					1.068

Table 9.10 *Posterior variances and correlation summaries, Dallas and Clay Counties, MCAR spatial frailty model. Diagonal elements are estimated variances, while off-diagonal elements are estimated correlations.*

Log-relative hazard model	p_D	DIC
$\mathbf{x}_{ijk}^T\boldsymbol{\beta} + \mathbf{z}_{ijk}^T\boldsymbol{\theta}$	10.97	642
$\mathbf{x}_{ijk}^T\boldsymbol{\beta} + \mathbf{z}_{ijk}^T\boldsymbol{\theta} + \phi_i,\ \phi \sim CAR(\rho, \lambda)$	103.95	358
$\mathbf{x}_{ijk}^T\boldsymbol{\beta} + \mathbf{z}_{ijk}^T\boldsymbol{\theta} + \phi_{ij},\ \boldsymbol{\Phi} \sim MCAR(\rho = 1, \Lambda)$	172.75	247
$\mathbf{x}_{ijk}^T\boldsymbol{\beta} + \mathbf{z}_{ijk}^T\boldsymbol{\theta} + \phi_{ij},\ \boldsymbol{\Phi} \sim MCAR(\rho, \Lambda)$	172.40	246
$\mathbf{x}_{ijk}^T\boldsymbol{\beta} + \mathbf{z}_{ijk}^T\boldsymbol{\theta} + \phi_{ij},\ \boldsymbol{\Phi} \sim MCAR(\rho_1, \ldots, \rho_5, \Lambda)$	175.71	237
$\mathbf{x}_{ijk}^T\boldsymbol{\beta} + \mathbf{z}_{ijk}^T\boldsymbol{\theta} + \phi_{ij} + \epsilon_{ij},\ \boldsymbol{\Phi} \sim MCAR(\rho_1, \ldots, \rho_5, \Lambda),$	177.25	255
$\quad \epsilon_{ij} \overset{iid}{\sim} N(0, \tau^2)$		
$\mathbf{x}_{ijk}^T\boldsymbol{\beta} + \mathbf{z}_{ijk}^T\boldsymbol{\theta}_i,\ \boldsymbol{\Theta} \sim MCAR(\rho, \Lambda)$	169.42	235
$\mathbf{x}_{ijk}^T\boldsymbol{\beta} + \mathbf{z}_{ijk}^T\boldsymbol{\theta}_i,\ \boldsymbol{\Theta} \sim MCAR(\rho_1, \ldots, \rho_5, \Lambda)$	171.46	229

Table 9.11 *DIC comparison, spatial survival models for the Iowa cancer data.*

summarized in Table 9.9, Table 9.10, and Figure 9.7). However, adding heterogeneity effects ϵ_{ij} to this model adds essentially no extra effective parameters, and is actually harmful to the overall DIC score (since we are adding complexity for little or no benefit in terms of fit). Finally, the two

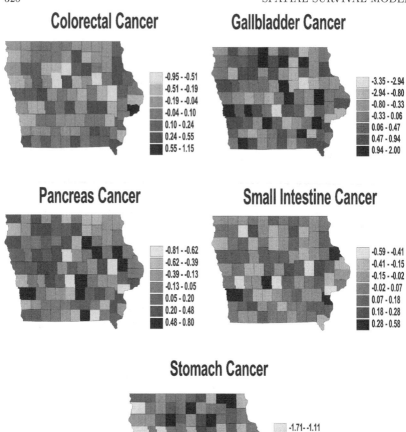

Figure 9.7 *Posterior mean spatial frailties, Iowa cancer data, static spatial MCAR model (see also color insert).*

spatially varying coefficients models enjoy the best (smallest) DIC scores, but only by a small margin over the best spatial frailty model.

Finally, we fit the spatiotemporal extension (9.14) of our MCAR frailty model to the data where the cohort effect (year of study entry r) is taken into account. Year-by-year boxplots of the posterior median frailties (Figure 9.9) reveal the expected steadily decreasing trend for all five cancers, though it is not clear how much of this decrease is simply an artifact of the censoring of survival times for patients in more recent cohorts. The

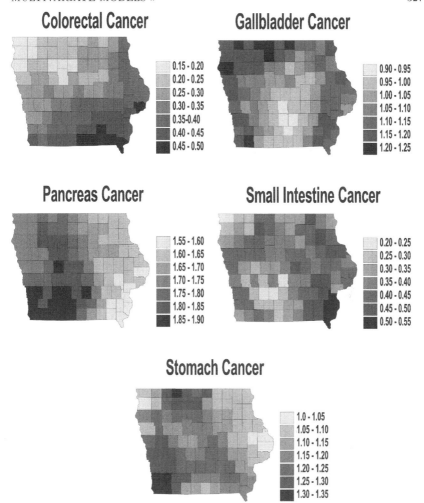

Figure 9.8 *Posterior mean spatially varying coefficients, Iowa cancer data, static spatial MCAR model (see also color insert).*

spatiotemporal extension of the spatially varying coefficients model (9.13) (i.e., $\mathbf{x}_{ijkr}^T \boldsymbol{\beta} + \mathbf{z}_{ijkr}^T \boldsymbol{\theta}_{ir}$) might well produce results that are temporally more interesting in this case. Incorporating change points, cancer start date measurement errors, and other model enhancements (say, interval censoring) might also be practically important model enhancements here. ∎

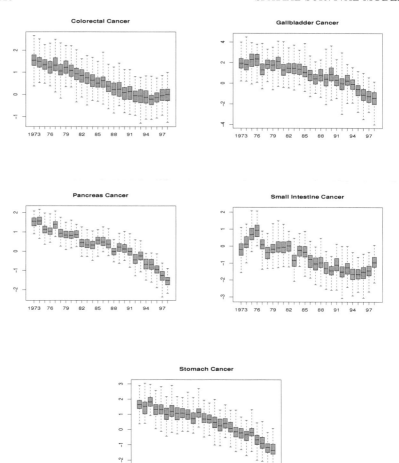

Figure 9.9 *Boxplots of posterior medians for the spatial frailties* ϕ_{ijr} *over the 99 Iowa counties for each year,* $r = 1973, \ldots, 1998$.

9.5 Spatial cure rate models ⋆

In Section 9.1 we investigated spatially correlated frailties in traditional parametric survival models, choosing a random effects distribution to reflect the spatial structure in the problem. Sections 9.2 and 9.3 extended this approach to spatial and spatiotemporal settings within a semiparametric model.

In this section our ultimate goal is the proper analysis of a geographically referenced smoking cessation study, in which we observe subjects periodically through time to check for relapse following an initial quit attempt. Each patient is observed once each year for five consecutive years, whereupon the current average number of cigarettes smoked at each visit is recorded, along with the zip code of residence and several other potential explanatory variables. This data set requires us to extend the work of Carlin and Hodges (1999) in a number of ways. The primary extension involves the incorporation of a *cure fraction* in our models. In investigating the effectiveness of quitting programs, data typically reveal many former smokers having successfully given up smoking, and as such may be thought of as "cured" of the deleterious habit. Incorporating such cure fractions in survival models leads to *cure rate models*, which are often applied in survival settings where the endpoint is a particular disease (say, breast cancer) which the subject may never reexperience. These models have a long history in the biostatistical literature, with the most popular perhaps being that of Berkson and Gage (1952). This model has been extensively studied in the statistical literature by a number of authors, including Farewell (1982, 1986), Goldman (1984), and Ewell and Ibrahim (1997). Recently, cure rates have been studied in somewhat more general settings by Chen, Ibrahim, and Sinha (1999) following earlier work by Yakovlev and Tsodikov (1996).

In addition, while this design can be analyzed as an ordinary right-censored survival model (with relapse to smoking as the endpoint), the data are perhaps more accurately viewed as *interval-censored*, since we actually observe only approximately annual intervals within which a failed quitter resumed smoking. We will consider both right- and interval-censored models, where in the former case we simply approximate the time of relapse by the midpoint of the corresponding time interval. Finally, we capture spatial variation through zip code-specific spatial random effects in the cure fraction or the hazard function, which in either case may act as spatial frailties. We find that incorporating the covariates and frailties into the hazard function is most natural (both intuitively and methodologically), especially after adopting a Weibull form for the baseline hazard.

9.5.1 Models for right- and interval-censored data

Right-censored data

Our cure rate models are based on those of Chen et al. (1999) and derived assuming that some latent biological process is generating the observed data. Suppose there are I regions and n_i patients in the ith region. We denote by T_{ij} the random variable for time to event (relapse, in our case) of the jth person in the ith region, where $j = 1, 2, ..., n_i$ and $i = 1, 2, ..., I$. (While acknowledging the presence of the regions in our notation, we postpone explicit spatial modeling to the next section.) Suppose that the (i, j)th individual has N_{ij} potential latent (unobserved) risk factors, the presence of any of which (i.e., $N_{ij} \geq 1$) will ultimately lead to the event. For example, in cancer settings these factors may correspond to metastasis-competent tumor cells within the individual. Typically, there will be a number of subjects who do not undergo the event during the observation period, and are therefore considered censored. Thus, letting U_{ijk}, $k = 1, 2, ..., N_{ij}$ be the time to an event arising from the kth latent factor for the (i, j)th individual, the observed time to event for an uncensored individual is generated by $T_{ij} = \min\{U_{ijk}, \ k = 1, 2, ..., N_{ij}\}$. If the (i, j)th individual is right-censored at time t_{ij}, none of the latent factors have led to an event by that time, and clearly $T_{ij} > t_{ij}$ (and in fact $T_{ij} = \infty$ if $N_{ij} = 0$).

Given N_{ij}, the U_{ijk}'s are independent with survival function $S\left(t | \Psi_{ij}\right)$ and corresponding density function $f\left(t | \Psi_{ij}\right)$. The parameter Ψ_{ij} is a collection of all the parameters (including possible regression parameters) that may be involved in a parametric specification for the survival function S. In this section we will work with a two-parameter Weibull distribution specification for the density function $f\left(t | \Psi_{ij}\right)$, where we allow the Weibull scale parameter ρ to vary across the regions, and η, which may serve as a link to covariates in a regression setup, to vary across individuals. Therefore $f\left(t | \rho_i, \eta_{ij}\right) = \rho_i t^{\rho_i - 1} \exp\left(\eta_{ij} - t^{\rho_i} \exp\left(\eta_{ij}\right)\right)$.

In terms of the hazard function h, $f\left(t | \rho_i, \eta_{ij}\right) = h\left(t | \rho_i, \eta_{ij}\right) S\left(t | \rho_i, \eta_{ij}\right)$, with $h\left(t; \rho_i, \eta_{ij}\right) = \rho_i t^{\rho_i - 1} \exp\left(\eta_{ij}\right)$ and $S\left(t | \rho_i, \eta_{ij}\right) = \exp\left(-t^{\rho_i} \exp\left(\eta_{ij}\right)\right)$. Note we implicitly assume proportional hazards, with baseline hazard function $h_0\left(t | \rho_i\right) = \rho_i t^{\rho_i - 1}$. Thus an individual ij who is censored at time t_{ij} before undergoing the event contributes $\left(S\left(t_{ij} | \rho_i, \eta_{ij}\right)\right)^{N_{ij}}$ to the likelihood, while an individual who experiences the event at time t_{ij} contributes $N_{ij}\left(S\left(t_{ij} | \rho_i, \eta_{ij}\right)\right)^{N_{ij}-1} f\left(t_{ij} | \rho_i, \eta_{ij}\right)$. The latter expression follows from the fact that the event is experienced when any one of the latent factors occurs. Letting ν_{ij} be the observed event indicator for individual ij, this person contributes

$$
\begin{aligned}
&L\left(t_{ij} | N_{ij}, \rho_i, \eta_{ij}, \nu_{ij}\right) \\
&= \left(S\left(t_{ij} | \rho_i, \eta_{ij}\right)\right)^{N_{ij}(1-\nu_{ij})} \left(N_{ij} S\left(t_{ij} | \rho_i, \eta_{ij}\right)^{N_{ij}-1} f\left(t_{ij} | \rho_i, \eta_{ij}\right)\right)^{\nu_{ij}},
\end{aligned}
$$

and the joint likelihood for all the patients can now be expressed as

$$
\begin{aligned}
&L\left(\{t_{ij}\} \,|\, \{N_{ij}\}, \{\rho_i\}, \{\eta_{ij}\}, \{\nu_{ij}\}\right) \\
&= \prod_{i=1}^{I} \prod_{j=1}^{n_i} L\left(t_{ij} | N_{ij}, \rho_i, \eta_{ij}, \nu_{ij}\right) \\
&= \prod_{i=1}^{I} \prod_{j=1}^{n_i} \left(S\left(t_{ij} | \rho_i, \eta_{ij}\right)\right)^{N_{ij}(1-\nu_{ij})} \\
&\qquad \times \left(N_{ij} S\left(t_{ij} \rho_i, \eta_{ij}\right)^{N_{ij}-1} f\left(t_{ij} | \rho_i, \eta_{ij}\right)\right)^{\nu_{ij}} \\
&= \prod_{i=1}^{I} \prod_{j=1}^{n_i} \left(S\left(t_{ij} | \rho_i, \eta_{ij}\right)\right)^{N_{ij}-\nu_{ij}} \left(N_{ij} f\left(t_{ij} | \rho_i, \eta_{ij}\right)\right)^{\nu_{ij}}.
\end{aligned}
$$

This expression can be rewritten in terms of the hazard function as

$$\prod_{i=1}^{I}\prod_{j=1}^{n_i} \left(S\left(t_{ij}|\rho_i, \eta_{ij}\right)\right)^{N_{ij}} \left(N_{ij} h\left(t_{ij}|\rho_i, \eta_{ij}\right)\right)^{\nu_{ij}}. \qquad (9.15)$$

A Bayesian hierarchical formulation is completed by introducing prior distributions on the parameters. We will specify independent prior distributions $p\left(N_{ij}|\theta_{ij}\right)$, $p\left(\rho_i|\psi_\rho\right)$ and $p\left(\eta_{ij}|\psi_\eta\right)$ for $\{N_{ij}\}$, $\{\rho_i\}$, and $\{\eta_{ij}\}$, respectively. Here, ψ_ρ, ψ_η, and $\{\theta_{ij}\}$ are appropriate hyperparameters. Assigning independent hyperpriors $p\left(\theta_{ij}|\psi_\theta\right)$ for $\{\theta_{ij}\}$ and assuming the hyperparameters $\psi = \left(\psi_\rho, \psi_\eta, \psi_\theta\right)$ to be fixed, the posterior distribution for the parameters, $p\left(\{\theta_{ij}\}, \{\eta_{ij}\}, \{N_{ij}\}, \{\rho_i\} \mid \{t_{ij}\}, \{\nu_{ij}\}\right)$, is easily found (up to a proportionality constant) using (9.15) as

$$\prod_{i=1}^{I} \left\{ p\left(\rho_i|\psi_\rho\right) \prod_{j=1}^{n_i} \left[S\left(t_{ij}|\rho_i, \eta_{ij}\right)\right]^{N_{ij}} \left[N_{ij} h\left(t_{ij}|\rho_i, \eta_{ij}\right)\right]^{\nu_{ij}} \right.$$
$$\left. \times p\left(N_{ij}|\theta_{ij}\right) p\left(\eta_{ij}|\psi_\eta\right) p\left(\theta_{ij}|\psi_\theta\right) \right\}.$$

Chen et al. (1999) assume that the N_{ij} are distributed as independent Poisson random variables with mean θ_{ij}, i.e., $p\left(N_{ij}|\theta_{ij}\right)$ is $Poisson\left(\theta_{ij}\right)$. In this setting it is easily seen that the survival distribution for the (i, j)th patient, $P\left(T_{ij} \geq t_{ij}|\rho_i, \eta_{ij}\right)$, is given by $\exp\left\{-\theta_{ij}\left(1 - S\left(t_{ij}|\rho_i, \eta_{ij}\right)\right)\right\}$. Since $S\left(t_{ij}|\rho_i, \eta_{ij}\right)$ is a proper survival function (corresponding to the latent factor times U_{ijk}), as $t_{ij} \to \infty$, $P\left(T_{ij} \geq t_{ij}|\rho_i, \eta_{ij}\right) \to \exp\left(-\theta_{ij}\right) > 0$. Thus we have a subdistribution for T_{ij} with a *cure fraction* given by $\exp\left(-\theta_{ij}\right)$. Here a hyperprior on the θ_{ij}'s would have support on the positive real line.

While there could certainly be multiple latent factors that increase the risk of smoking relapse (age started smoking, occupation, amount of time spent driving, tendency toward addictive behavior, etc.), this is rather speculative and certainly not as justifiable as in the cancer setting for which the multiple factor approach was developed (where $N_{ij} > 1$ is biologically motivated). As such, we instead form our model using a single, omnibus, "propensity for relapse" latent factor. In this case, we think of N_{ij} as a *binary* variable, and specify $p\left(N_{ij}|\theta_{ij}\right)$ as Bernoulli $\left(1 - \theta_{ij}\right)$. In this setting it is easier to look at the survival distribution after marginalizing out the N_{ij}. In particular, note that

$$P\left(T_{ij} \geq t_{ij}|\rho_i, \eta_{ij}, N_{ij}\right) = \begin{cases} S\left(t_{ij}|\rho_i, \eta_{ij}\right), & N_{ij} = 1 \\ 1, & N_{ij} = 0 \end{cases}.$$

That is, if the latent factor is absent, the subject is cured (does not experience the event). Marginalizing over the Bernoulli distribution for N_{ij}, we obtain for the (i, j)th patient the survival function $S^*\left(t_{ij}|\theta_{ij}, \rho_i, \eta_{ij}\right) \equiv P\left(T_{ij} \geq t_{ij}|\rho_i, \eta_{ij}\right) = \theta_{ij} + \left(1 - \theta_{ij}\right) S\left(t_{ij}|\rho_i, \eta_{ij}\right)$, which is the classic cure-rate model attributed to Berkson and Gage (1952) with cure fraction θ_{ij}. Now we can write the likelihood function for the data marginalized over

$\{N_{ij}\}$, $L\left(\{t_{ij}\} \mid \{\rho_i\}, \{\theta_{ij}\}, \{\eta_{ij}\}, \{\nu_{ij}\}\right)$, as

$$\prod_{i=1}^{I} \prod_{j=1}^{n_i} \left[S^*\left(t_{ij} \mid \theta_{ij}, \rho_i, \eta_{ij}\right)\right]^{1-\nu_{ij}} \left(-\frac{d}{dt_{ij}} S^*\left(t_{ij} \mid \theta_{ij}, \rho_i, \eta_{ij}\right)\right)^{\nu_{ij}}$$
$$= \prod_{i=1}^{I} \prod_{j=1}^{n_i} \left[S^*\left(t_{ij} \mid \theta_{ij}, \rho_i, \eta_{ij}\right)\right]^{1-\nu_{ij}} \left[(1-\theta_{ij}) f\left(t_{ij} \mid \rho_i, \eta_{ij}\right)\right]^{\nu_{ij}},$$

which in terms of the hazard function becomes

$$\prod_{i=1}^{I} \prod_{j=1}^{n_i} \left[S^*\left(t_{ij} \mid \theta_{ij}, \rho_i, \eta_{ij}\right)\right]^{1-\nu_{ij}} \left[(1-\theta_{ij}) S\left(t_{ij} \mid \rho_i, \eta_{ij}\right) h\left(t_{ij} \mid \rho_i, \eta_{ij}\right)\right]^{\nu_{ij}},$$

(9.16)

where the hyperprior for θ_{ij} has support on $(0, 1)$. Now the posterior distribution of the parameters is proportional to

$$L\left(\{t_{ij}\} \mid \{\rho_i\}, \{\theta_{ij}\}, \{\eta_{ij}\}, \{\nu_{ij}\}\right) \prod_{i=1}^{I} \left\{ p\left(\rho_i \mid \psi_\rho\right) \prod_{j=1}^{n_i} p\left(\eta_{ij} \mid \psi_\eta\right) p\left(\theta_{ij} \mid \psi_\theta\right) \right\}.$$

(9.17)

Turning to the issue of incorporating covariates, in the general setting with N_{ij} assumed to be distributed Poisson, Chen et al. (1999) propose their introduction in the cure fraction through a suitable link function g, so that $\theta_{ij} = g\left(\mathbf{x}_{ij}^T \widetilde{\boldsymbol{\beta}}\right)$, where g maps the entire real line to the positive axis. This is sensible when we believe that the risk factors affect the probability of an individual being cured. Proper posteriors arise for the regression coefficients $\widetilde{\boldsymbol{\beta}}$ even under improper priors. Unfortunately, this is no longer true when N_{ij} is Bernoulli (i.e., in the Berkson and Gage model). Vague but proper priors may still be used, but this makes the parameters difficult to interpret, and can often lead to poor MCMC convergence.

Since a binary N_{ij} seems most natural in our setting, we instead introduce covariates into $S\left(t_{ij} \mid \rho_i, \eta_{ij}\right)$ through the Weibull link η_{ij}, i.e., we let $\eta_{ij} = \mathbf{x}_{ij}^T \boldsymbol{\beta}$. This seems intuitively more reasonable anyway, since now the covariates influence the underlying factor that brings about the smoking relapse (and thus the rapidity of this event). Also, proper posteriors arise here for $\boldsymbol{\beta}$ under improper posteriors even though N_{ij} is binary. As such, henceforth we will only consider the situation where the covariates enter the model in this way (through the Weibull link function). This means we are unable to separately estimate the effect of the covariates on both the *rate* of relapse and the *ultimate level* of relapse, but "fair" estimation here (i.e., allocating the proper proportions of the covariates' effects to each component) is not clear anyway since flat priors could be selected for $\boldsymbol{\beta}$, but not for $\widetilde{\boldsymbol{\beta}}$. Finally, all of our subsequent models also assume a constant cure fraction for the entire population (i.e., we set $\theta_{ij} = \theta$ for all i, j).

Note that the posterior distribution in (9.17) is easily modified to incorporate covariates. For example, with $\eta_{ij} = \mathbf{x}_{ij}^T \boldsymbol{\beta}$, we replace $\prod_{ij} p\left(\eta_{ij} \mid \psi_\eta\right)$

in (9.17) with $p(\boldsymbol{\beta}|\psi_\beta)$, with ψ_β as a fixed hyperparameter. Typically a flat or vague Gaussian prior may be taken for $p(\boldsymbol{\beta}|\psi_\beta)$.

Interval-censored data

The formulation above assumes that our observed data are right-censored. This means that we are able to observe the actual relapse time t_{ij} when it occurs prior to the final office visit. In reality, our study (like many others of its kind) is only able to determine patient status at the office visits themselves, meaning we observe only a time *interval* (t_{ijL}, t_{ijU}) within which the event (in our case, smoking relapse) is known to have occurred. For patients who did not resume smoking prior to the end of the study we have $t_{ijU} = \infty$, returning us to the case of right-censoring at time point t_{ijL}. Thus we now set $\nu_{ij} = 1$ if subject ij is interval-censored (i.e., experienced the event), and $\nu_{ij} = 0$ if the subject is right-censored.

Following Finkelstein (1986), the general interval-censored cure rate likelihood, $L(\{(t_{ijL}, t_{ijU})\} \mid \{N_{ij}\}, \{\rho_i\}, \{\eta_{ij}\}, \{\nu_{ij}\})$, is given by

$$\prod_{i=1}^{I} \prod_{j=1}^{n_i} [S(t_{ijL}|\rho_i, \eta_{ij})]^{N_{ij} - \nu_{ij}} \{N_{ij} [S(t_{ijL}|\rho_i, \eta_{ij}) - S(t_{ijU}|\rho_i, \eta_{ij})]\}^{\nu_{ij}}$$

$$= \prod_{i=1}^{I} \prod_{j=1}^{n_i} [S(t_{ijL}|\rho_i, \eta_{ij})]^{N_{ij}} \left\{ N_{ij} \left(1 - \frac{S(t_{ijU}|\rho_i, \eta_{ij})}{S(t_{ijL}|\rho_i, \eta_{ij})} \right) \right\}^{\nu_{ij}}.$$

As in the previous section, in the Bernoulli setup after marginalizing out the $\{N_{ij}\}$ the foregoing becomes $L(\{(t_{ijL}, t_{ijU})\} \mid \{\rho_i\}, \{\theta_{ij}\}, \{\eta_{ij}\}, \{\nu_{ij}\})$, and can be written as

$$\prod_{i=1}^{I} \prod_{j=1}^{n_i} S^*(t_{ijL}|\theta_{ij}, \rho_i, \eta_{ij}) \left\{ 1 - \frac{S^*(t_{ijU}|\theta_{ij}, \rho_i, \eta_{ij})}{S^*(t_{ijL}|\theta_{ij}, \rho_i, \eta_{ij})} \right\}^{\nu_{ij}}. \tag{9.18}$$

We omit details (similar to those in the previous section) arising from the Weibull parametrization and subsequent incorporation of covariates through the link function η_{ij}.

9.5.2 Spatial frailties in cure rate models

The development of the hierarchical framework in the preceding section acknowledged the data as coming from I different geographical regions (clusters). Such clustered data are common in survival analysis and often modeled using cluster-specific frailties ϕ_i. As with the covariates, we will introduce the frailties ϕ_i through the Weibull link as intercept terms in the log-relative risk; that is, we set $\eta_{ij} = \mathbf{x}_{ij}^T \boldsymbol{\beta} + \phi_i$.

Here we allow the ϕ_i to be spatially correlated across the regions; similarly we would like to permit the Weibull baseline hazard parameters, ρ_i,

SPATIAL SURVIVAL MODELS

to be spatially correlated. A natural approach in both cases is to use a univariate CAR prior. While one may certainly employ separate, independent CAR priors on ϕ and $\zeta \equiv \{\log \rho_i\}$, another option is to allow these two spatial priors to themselves be correlated. In other words, we may want a bivariate spatial model for the $\delta_i = (\phi_i, \zeta_i)^T = (\phi_i, \log \rho_i)^T$. As mentioned in Sections 7.4 and 9.4, we may use the MCAR distribution for this purpose. In our setting, the MCAR distribution on the concatenated vector $\boldsymbol{\delta} = (\boldsymbol{\phi}^T, \boldsymbol{\zeta}^T)^T$ is Gaussian with mean $\mathbf{0}$ and precision matrix $\Lambda^{-1} \otimes (Diag\,(m_i) - \rho W)$, where Λ is a 2×2 symmetric and positive definite matrix, $\rho \in (0, 1)$, and m_i and W remain as above. In the current context, we may also wish to allow different smoothness parameters (say, ρ_1 and ρ_2) for ϕ and ζ, respectively, as in Section 9.4. Henceforth, in this section we will denote the proper MCAR with a common smoothness parameter by $MCAR\,(\rho, \Lambda)$, and the multiple smoothness parameter generalized MCAR by $MCAR\,(\rho_1, \rho_2, \Lambda)$. Combined with independent (univariate) CAR models for ϕ and ζ, these offer a broad range of potential spatial models.

9.5.3 Model comparison

Suppose we let Ω denote the set of all model parameters, so that the deviance statistic (4.9) becomes

$$D(\Omega) = -2 \log f(\mathbf{y}|\Omega) + 2 \log h(\mathbf{y}) \ . \tag{9.19}$$

When DIC is used to compare nested models in standard exponential family settings, the unnormalized likelihood $L(\Omega; \mathbf{y})$ is often used in place of the normalized form $f(\mathbf{y}|\Omega)$ in (9.19), since in this case the normalizing function $m(\Omega) = \int L(\Omega; \mathbf{y})d\mathbf{y}$ will be free of Ω and constant across models, hence contribute equally to the DIC scores of each (and thus have no impact on model selection). However, in settings where we require comparisons across different likelihood distributional forms, it appears one must be careful to use the properly scaled joint density $f(\mathbf{y}|\Omega)$ for each model.

We argue that use of the usual proportional hazards likelihood (which of course is not a joint density function) *is* in fact appropriate for DIC computation here, provided we make a fairly standard assumption regarding the relationship between the survival and censoring mechanisms generating the data. Specifically, suppose the distribution of the censoring times is independent of that of the survival times *and* does not depend upon the survival model parameters (i.e., independent, noninformative censoring). Let $g\,(t_{ij})$ denote the density of the censoring time for the ijth individual, with corresponding survival (1-cdf) function $R\,(t_{ij})$. Then the right-censored likelihood (9.16) can be extended to the joint likelihood specification,

$$\prod_{i=1}^{I} \prod_{j=1}^{n_i} [S^*\,(t_{ij}|\theta_{ij}, \rho_i, \eta_{ij})]^{1-\nu_{ij}}$$
$$\times [(1 - \theta_{ij})\,S\,(t_{ij}|\rho_i, \eta_{ij})\,h\,(t_{ij}|\rho_i, \eta_{ij})]^{\nu_{ij}}\,[R\,(t_{ij})]^{\nu_{ij}}\,[g\,(t_{ij})]^{1-\nu_{ij}} \ ,$$

as for example in Le (1997, pp. 69–70). While not a joint probability density, this likelihood is still an everywhere nonnegative and integrable function of the survival model parameters Ω, and thus suitable for use with the Kullback-Leibler divergences that underlie DIC (Spiegelhalter et al., 2002, p. 586). But by assumption, $R(t)$ and $g(t)$ do not depend upon Ω. Thus, like an $m(\Omega)$ that is free of Ω, they may be safely ignored in both the p_D and DIC calculations. Note this same argument implies that we can use the unnormalized likelihood (9.16) when comparing not only nonnested parametric survival models (say, Weibull versus gamma), but even parametric and semiparametric models (say, Weibull versus Cox) provided our definition of "likelihood" is comparable across models.

Note also that here our "focus" (in the nomenclature of Spiegelhalter et al., 2002) is solely on Ω. An alternative would be instead to use a missing data formulation, where we include the likelihood contribution of $\{s_{ij}\}$, the collection of latent survival times for the right-censored individuals. Values for both Ω and the $\{s_{ij}\}$ could then be imputed along the lines given by Cox and Oakes (1984, pp. 165–166) for the EM algorithm or Spiegelhalter et al. (1995b, the "mice" example) for the Gibbs sampler. This would alter our focus from Ω to $(\Omega, \{s_{ij}\})$, and p_D would reflect the correspondingly larger effective parameter count.

Turning to the interval censored case, here matters are only a bit more complicated. Converting the interval-censored likelihood (9.18) to a joint likelihood specification yields

$$\prod_{i=1}^{I} \prod_{j=1}^{n_i} S^* \left(t_{ijL} | \theta_{ij}, \rho_i, \eta_{ij} \right) \left(1 - \frac{S^* \left(t_{ijU} | \theta_{ij}, \rho_i, \eta_{ij} \right)}{S^* \left(t_{ijL} | \theta_{ij}, \rho_i, \eta_{ij} \right)} \right)^{\nu_{ij}}$$

$$\times \left[R \left(t_{ijL} \right) \right]^{\nu_{ij}} \left(1 - \frac{R \left(t_{ijU} \right)}{R \left(t_{ijL} \right)} \right)^{\nu_{ij}} \left[g \left(t_{ijL} \right) \right]^{1-\nu_{ij}} .$$

Now $\left[R \left(t_{ijL} \right) \right]^{\nu_{ij}} \left(1 - R \left(t_{ijU} \right) / R \left(t_{ijL} \right) \right)^{\nu_{ij}} \left[g \left(t_{ijL} \right) \right]^{1-\nu_{ij}}$ is the function absorbed into $m(\Omega)$, and is again free of Ω. Thus again, use of the usual form of the interval-censored likelihood presents no problems when comparing models within the interval-censored framework (including nonnested parametric models, or even parametric and semiparametric models).

Note that it does *not* make sense to compare a particular right-censored model with a particular interval-censored model. The form of the available data is different; model comparison is only appropriate to a given data set.

Example 9.5 *(Smoking cessation data)*. We illustrate our methods using the aforementioned study of smoking cessation, a subject of particular interest in studies of lung health and primary cancer control. Described more fully by Murray et al. (1998), the data consist of 223 subjects who reside in 53 zip codes in the southeastern corner of Minnesota. The subjects, all of whom were smokers at study entry, were randomized into either a

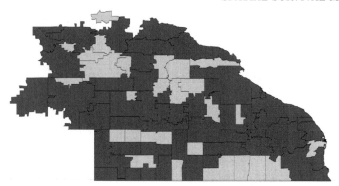

Figure 9.10 *Map showing missingness pattern for the smoking cessation data: lightly shaded regions are those having no responses.*

smoking intervention (SI) group, or a usual care (UC) group that received no special antismoking intervention. Each subject's smoking habits were monitored at roughly annual visits for five consecutive years. The subjects we analyze are actually the subset who are known to have quit smoking at least once during these five years, and our event of interest is whether they relapse (resume smoking) or not. Covariate information available for each subject includes sex, years as a smoker, and the average number of cigarettes smoked per day just prior to the quit attempt.

To simplify matters somewhat, we actually fit our spatial cure rate models over the 81 contiguous zip codes shown in Figure 9.10, of which only the 54 dark-shaded regions are those contributing patients to our data set. This enables our models to produce spatial predictions even for the 27 unshaded regions in which no study patients actually resided. All of our MCMC algorithms ran 5 initially overdispersed sampling chains, each for 20,000 iterations. Convergence was assessed using correlation plots, sample trace plots, and Gelman-Rubin (1992) statistics. In every case a burn-in period of 15,000 iterations appeared satisfactory. Retaining the remaining 5,000 samples from each chain yielded a final sample of 25,000 for posterior summarization.

Table 9.12 provides the DIC scores for a variety of random effects cure rate models in the interval-censored case. Models 1 and 2 have only random frailty terms ϕ_i with i.i.d. and CAR priors, respectively. Models 3 and 4 add random Weibull shape parameters $\zeta_i = \log \rho_i$, again with i.i.d. and CAR priors, respectively, independent of the priors for the ϕ_i. Finally, Models 5 and 6 consider the full MCAR structure for the (ϕ_i, ζ_i) pairs, assuming common and distinct spatial smoothing parameters, respectively. The DIC scores do not suggest that the more complex models are significantly better; apparently the data encourage a high degree of shrinkage in the random

Model	Log-relative risk	pD	DIC
1	$\mathbf{x}_{ij}^{T}\boldsymbol{\beta} + \phi_i;\ \phi_i \overset{iid}{\sim} N\left(0, \tau_\phi\right),\ \rho_i = \rho\ \forall\ i$	10.3	438
2	$\mathbf{x}_{ij}^{T}\boldsymbol{\beta} + \phi_i;\ \{\phi_i\} \sim CAR\left(\lambda_\phi\right),\ \rho_i = \rho\ \forall\ i$	9.4	435
3	$\mathbf{x}_{ij}^{T}\boldsymbol{\beta} + \phi_i;\ \phi_i \overset{iid}{\sim} N\left(0, \tau_\phi\right),\ \zeta_i \overset{iid}{\sim} N\left(0, \tau_\zeta\right)$	13.1	440
4	$\mathbf{x}_{ij}^{T}\boldsymbol{\beta} + \phi_i;\ \{\phi_i\} \sim CAR\left(\lambda_\phi\right),\ \{\zeta_i\} \sim CAR\left(\lambda_\zeta\right)$	10.4	439
5	$\mathbf{x}_{ij}^{T}\boldsymbol{\beta} + \phi_i;\ (\{\phi_i\}, \{\zeta_i\}) \sim MCAR\left(\rho, \Lambda\right)$	7.9	434
6	$\mathbf{x}_{ij}^{T}\boldsymbol{\beta} + \phi_i;\ (\{\phi_i\}, \{\zeta_i\}) \sim MCAR\left(\rho_\phi, \rho_\zeta, \Lambda\right)$	8.2	434

Table 9.12 *DIC and p_D values for various competing interval-censored models.*

Parameter	Median	(2.5%, 97.5%)
Intercept	–2.720	(–4.803, –0.648)
Sex (male = 0)	0.291	(–0.173, 0.754)
Duration as smoker	–0.025	(–0.059, 0.009)
SI/UC (usual care = 0)	–0.355	(–0.856, 0.146)
Cigarettes smoked per day	0.010	(–0.010, 0.030)
θ (cure fraction)	0.694	(0.602, 0.782)
ρ_ϕ	0.912	(0.869, 0.988)
ρ_ζ	0.927	(0.906, 0.982)
Λ_{11} (spatial variance component, ϕ_i)	0.005	(0.001, 0.029)
Λ_{22} (spatial variance component, ζ_i)	0.007	(0.002, 0.043)
$\Lambda_{12}/\sqrt{\Lambda_{11}\Lambda_{22}}$	0.323	(–0.746, 0.905)

Table 9.13 *Posterior quantiles, full model, interval-censored case.*

effects (note the low p_D scores). In what follows we present results for the "full" model (Model 6) in order to preserve complete generality, but emphasize that any of the models in Table 9.12 could be used with equal confidence.

Table 9.13 presents estimated posterior quantiles (medians, and upper and lower .025 points) for the fixed effects $\boldsymbol{\beta}$, cure fraction θ, and hyperparameters in the interval-censored case. The smoking intervention does appear to produce a decrease in the log relative risk of relapse, as expected. Patient sex is also marginally significant, with women more likely to relapse than men, a result often attributed to the (real or perceived) risk of weight gain following smoking cessation. The number of cigarettes smoked per day does not seem important, but duration as a smoker is significant, and in

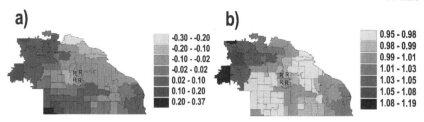

Figure 9.11 *Maps of posterior means for the* ϕ_i *(a) and the* ρ_i *(b) in the full spatial MCAR model, assuming the data to be interval-censored (see also color insert).*

a possibly counterintuitive direction: shorter-term smokers relapse sooner. This may be due to the fact that people are better able to quit smoking as they age (and are thus confronted more clearly with their own mortality).

The estimated cure fraction in Table 9.13 is roughly .70, indicating that roughly 70% of smokers in this study who attempted to quit have in fact been "cured." The spatial smoothness parameters ρ_ϕ and ρ_ζ are both close to 1, again suggesting we would lose little by simply setting them both equal to 1 (as in the standard CAR model). Finally, the last lines of both tables indicate only a moderate correlation between the two random effects, again consistent with the rather weak case for including them in the model at all.

We compared our results to those obtained from the R function `survreg` using a Weibull link, and also to Weibull regression models fit in a Bayesian fashion using the `WinBUGS` package. While neither of these alternatives featured a cure rate (and only the `WinBUGS` analysis included spatial random effects), both produced fixed effect estimates quite consistent with those in Table 9.13.

Turning to graphical summaries, Figure 9.11 (see also color insert Figure C.13) maps the posterior medians of the frailty (ϕ_i) and shape (ρ_i) parameters in the full spatial MCAR (Model 6) case. The maps reveal some interesting spatial patterns, though the magnitudes of the differences appear relatively small across zip codes. The south-central region seems to be of some concern, with its high values for both ϕ_i (high overall relapse rate) and ρ_i (increasing baseline hazard over time). By contrast, the four zip codes comprising the city of Rochester, MN (home of the Mayo Clinic, and marked with an "R" in each map) suggest slightly better than average cessation behavior. Note that a nonspatial model cannot impute anything other than the "null values" ($\phi_i = 0$ and $\rho_i = 1$) for any zip code contributing no data (all of the unshaded regions in Figure 9.10). Our spatial model however is able to impute nonnull values here, in accordance with the observed values in neighboring regions. ∎

EXERCISES 339

Unit	Drug	Time	Unit	Drug	Time	Unit	Drug	Time
A	1	74+	E	1	214	H	1	74+
A	2	248	E	2	228+	H	1	88+
A	1	272+	E	2	262	H	1	148+
A	2	344				H	2	162
			F	1	6			
B	2	4+	F	2	16+	I	2	8
B	1	156+	F	1	76	I	2	16+
			F	2	80	I	2	40
C	2	100+	F	2	202	I	1	120+
			F	1	258+	I	1	168+
D	2	20+	F	1	268+	I	2	174+
D	2	64	F	2	368+	I	1	268+
D	2	88	F	1	380+	I	2	276
D	2	148+	F	1	424+	I	1	286+
D	1	162+	F	2	428+	I	1	366
D	1	184+	F	2	436+	I	2	396+
D	1	188+				I	2	466+
D	1	198+	G	2	32+	I	1	468+
D	1	382+	G	1	64+			
D	1	436+	G	1	102	J	1	18+
			G	2	162+	J	1	36+
E	1	50+	G	2	182+	J	2	160+
E	2	64+	G	1	364+	J	2	254
E	2	82						
E	1	186+	H	2	22+	K	1	28+
E	1	214+	H	1	22+	K	1	70+
						K	2	106+

Table 9.14 *Survival times (in half-days) from the MAC treatment trial, from Carlin and Hodges (1999). Here, "+" indicates a censored observation.*

9.6 Exercises

1. The data located at www.biostat.umn.edu/~brad/data/MAC.dat, and also shown in Table 9.14, summarize a clinical trial comparing two treatments for *Mycobacterium avium* complex (MAC), a disease common in late-stage HIV-infected persons. Eleven clinical centers ("units") have enrolled a total of 69 patients in the trial, 18 of which have died; see Cohn et al. (1999) and Carlin and Hodges (1999) for full details regarding this trial.

As in Section 9.1, let t_{ij} be the time to death or censoring and x_{ij} be the treatment indicator for subject j in stratum i ($j = 1, \ldots, n_i$, $i = 1, \ldots, k$). With proportional hazards and a Weibull baseline hazard, stratum i's hazard is then

$$
\begin{aligned}
h(t_{ij}; x_{ij}) &= h_0(t_{ij}) \omega_i \exp(\beta_0 + \beta_1 x_{ij}) \\
&= \rho_i t_{ij}^{\rho_i - 1} \exp(\beta_0 + \beta_1 x_{ij} + W_i) \,,
\end{aligned}
$$

where $\rho_i > 0$, $\boldsymbol{\beta} = (\beta_0, \beta_1)' \in \Re^2$, and $W_i = \log \omega_i$ is a clinic-specific frailty term.

(a) Assume i.i.d. specifications for these random effects, i.e.,

$$
W_i \overset{iid}{\sim} N(0, 1/\tau) \quad \text{and} \quad \rho_i \overset{iid}{\sim} G(\alpha, \alpha) \,.
$$

Then as in the `mice` example (`WinBUGS` Examples Vol 1),

$$
\mu_{ij} = \exp(\beta_0 + \beta_1 x_{ij} + W_i) \,,
$$

so that $t_{ij} \sim Weibull(\rho_i, \mu_{ij})$. Use `WinBUGS` to obtain posterior summaries for the main and random effects in this model. Use vague priors on β_0 and β_1, a moderately informative $G(1,1)$ prior on τ, and set $\alpha = 10$. (You might also recode the drug covariate from (1,2) to (–1,1), in order to ease collinearity between the slope β_1 and the intercept β_0.)

(b) From Table 9.15, we can obtain the lattitude and longitude of each of the 11 sites, hence the distance d_{ij} between each pair. These distances are included in `www.biostat.umn.edu/~brad/data/MAC.dat`; note they have been scaled so that the largest (New York-San Francisco) equals 1. (Note that since sites F and H are virtually coincident (both in Detroit, MI), we have recoded them as a single clinic (#6) and now think of this as a 10-site model.) Refit the model in `WinBUGS` assuming the frailties to have spatial correlation following the isotropic exponential kriging model,

$$
\mathbf{W} \sim N_k(\mathbf{0}, H), \text{ where } H_{ij} = \sigma^2 \exp(-\phi d_{ij}) \,,
$$

where as usual $\sigma^2 = 1/\tau$, and where we place a $G(3, 0.1)$ (mean 30) prior on ϕ.

2. The file `www.biostat.umn.edu/~brad/data/smoking.dat` contains the southeastern Minnesota smoking cessation data discussed in Section 9.5. At each of up to five office visits, the smoking status of persons who had recently quit smoking was assessed. We define relapse to smoking as the endpoint, and denote the failure or censoring time of person j in county i by t_{ij}. The data set (already in `WinBUGS` format) also contains the adjacency matrix for the counties in question.

Unit	Number	City
A	1	Harlem (New York City), NY
B	2	New Orleans, LA
C	3	Washington, DC
D	4	San Francisco, CA
E	5	Portland, OR
F	6a	Detroit, MI (Henry Ford Hospital)
G	7	Atlanta, GA
H	6b	Detroit, MI (Wayne State University)
I	8	Richmond, VA
J	9	Camden, NJ
K	10	Albuquerque, NM

Table 9.15 *Locations of the clinical sites in the MAC treatment data set.*

(a) Assuming that smoking relapses occurred on the day of the office visit when they were detected, build a hierarchical spatial frailty model to analyze these data. Code your model in WinBUGS, run it, and summarize your results. Use the DIC tool to compare a few competing prior or likelihood specifications.

(b) When we observe a subject who has resumed smoking, all we really know is that his failure (relapse) point occurred somewhere between his last office visit and this one. As such, improve your model from part (a) by building an interval-censored version.

3. Consider the extension of the Section 9.4 model in the single endpoint, multiple cause case to the *multiple* endpoint, multiple cause case — say, for analyzing times until diagnosis of each cancer (if any), rather than merely a single time until death. Write down a model, likelihood, and prior specification (including an appropriately specified MCAR distribution) to handle this case.

CHAPTER 10

Special topics in spatial process modeling

Earlier chapters have developed the basic theory and the general hierarchical Bayesian modeling approach for handling spatial and spatiotemporal point-referenced data. In this chapter, we consider some special topics that are of interest in the context of such models.

The first returns to the notion of smoothness of process realizations, as initially discussed in Subsection 2.2.3. In Section 10.1 we rigorize the ideas of mean square continuity and differentiability. In particular, mean square differentiability suggests the definitions of finite difference processes and directional derivative processes. Such processes are useful in learning about gradients in spatial surfaces.

The second idea returns to our discussion of multivariate spatial processes (Chapter 7). However, we now think of their use not for modeling multivariate spatial data, but for generalizing the linear regression relationship between a response $Y(\mathbf{s})$ and an explanatory variable $X(\mathbf{s})$. In fact, with a model for $Y(\mathbf{s})$ that includes an intercept, the spatial random effect ($w(\mathbf{s})$ in equation (5.1)) could be viewed as a spatially varying adjustment to the intercept. This suggests that we might also incorporate a spatially varying adjustment to the *slope* associated with $X(\mathbf{s})$. Since the slope and intercept would be anticipated to be dependent, we would need a bivariate spatial process to jointly model these adjustments. This idea came up previously in the context of spatial frailty modeling in Section 9.4 and will be explored much more fully in Section 10.2.

The third idea considers a spatial process version of the familiar cumulative distribution function (CDF). Imagine a random realization of a spatial process restricted to locations in a region D having finite area. If the surface is then sliced by a horizontal plane at a specified height, there will be a portion of D where the surface will lie below this plane; for the remainder of D, the surface will lie above it. As the height is increased, the proportion of the surface lying below the plane will approach 1. This motivates the idea of a *spatial CDF* (SCDF). We consider properties of and inference for the SCDF in Section 10.3.

SPECIAL TOPICS IN SPATIAL PROCESS MODELING

10.1 Process smoothness revisited ⋆

10.1.1 Smoothness of a univariate spatial process

We confine ourselves to smoothness properties of a univariate spatial process, say, $\{Y(\mathbf{s}), \mathbf{s} \in \Re^d\}$; for a discussion of multivariate processes, see Banerjee and Gelfand (2003). In our investigation of smoothness properties we look at two types of continuity, continuity in the L_2 sense and continuity in the sense of process realizations. Unless otherwise noted, we assume the processes to have 0 mean and finite second-order moments.

Definition 10.1 A process $\{Y(\mathbf{s}), \mathbf{s} \in \Re^d\}$ is L_2 continuous at $\mathbf{s}_0$ if and only if $\lim_{\mathbf{s} \to \mathbf{s}_0} E[Y(\mathbf{s}) - Y(\mathbf{s}_0)]^2 = 0$. Continuity in the L_2 sense is also referred to as *mean square continuity*, and will be denoted by $Y(\mathbf{s}) \xrightarrow{L_2} Y(\mathbf{s}_0)$.

Definition 10.2 A process $\{Y(\mathbf{s}), \mathbf{s} \in \Re^d\}$ is *almost surely continuous* at $\mathbf{s}_0$ if $Y(\mathbf{s}) \longrightarrow Y(\mathbf{s}_0)$ *a.s.* as $\mathbf{s} \longrightarrow \mathbf{s}_0$. If the process is almost surely continuous for every $\mathbf{s}_0 \in \Re^d$ then the process is said to have continuous realizations.

In general, one form of continuity does not imply the other since one form of convergence does not imply the other. However, if $Y(\mathbf{s})$ is a bounded process then a.s. continuity implies L_2 continuity. Of course, each implies that $Y(\mathbf{s}) \xrightarrow{P} Y(\mathbf{s}_0)$.

Example 10.1 Almost sure continuity does not imply mean square continuity. To see this, let $t \in [0, 1]$ with $\omega \sim U(0, 1)$ and define

$$Y(t; \omega) = \begin{cases} \left(t - \frac{1}{2}\right)^{-1} I_{\left(\frac{1}{2}, t\right)}(\omega) & \text{if } t \in \left(\frac{1}{2}, 1\right] \\ 0 & \text{if } t \in \left[0, \frac{1}{2}\right] \end{cases}.$$

Then $Y(t; \omega) \longrightarrow 0$ *a.s.* as $t \longrightarrow \frac{1}{2}$. But $E[Y^2(t; \omega)] \longrightarrow \infty$ as $t \longrightarrow \frac{1}{2}$ if $t \in \left(\frac{1}{2}, 1\right]$ and $E\left[Y^2(t; \omega)\right] = 0$ if $t \in \left[0, \frac{1}{2}\right]$. Thus the process does not converge in L_2 although it does so almost surely. ∎

Example 10.2 Mean square continuity does not imply almost sure continuity. To see this, construct a process over $t \in \Re^+$ defined through $\omega \sim U(0, 1)$ as follows. Let $Y\left(\frac{1}{t}; \omega\right) = 0$, if t is not a positive integer, $Y(1; \omega) = I_{\left(0, \frac{1}{2}\right)}(\omega)$, $Y\left(\frac{1}{2}; \omega\right) = I_{\left(\frac{1}{2}, 1\right)}(\omega)$, $Y\left(\frac{1}{3}; \omega\right) = I_{\left(0, \frac{1}{3}\right)}(\omega)$, $Y\left(\frac{1}{4}; \omega\right) = I_{\left(\frac{1}{3}, \frac{2}{3}\right)}(\omega)$, $Y\left(\frac{1}{5}; \omega\right) = I_{\left(\frac{2}{3}, 1\right)}(\omega)$, and so on. That is, we construct the process as a sequence of moving indicators on successively finer arithmetic divisions of the unit interval. We see here that $E\left[Y^2\left(\frac{1}{t}; \omega\right)\right] \longrightarrow 0$ as $t \longrightarrow 0$, so that $Y\left(\frac{1}{t}; \omega\right) \xrightarrow{L_2} 0$. However the process is not continuous almost surely since $Y\left(\frac{1}{t}; \omega\right)$ is equal to one infinitely often. ∎

The above definitions apply to any stochastic process (possibly nonstationary). Cramér and Leadbetter (1967) and Hoel, Port, and Stone (1972) outline conditions on the covariance function for mean square continuity for processes on the real line. For a process on $\Re^d$, we denote the covariance function $C(\mathbf{s},\mathbf{s}') = cov(Y(\mathbf{s}),Y(\mathbf{s}'))$, so that the definition of mean square continuity is equivalent to $\lim_{\mathbf{s}' \to \mathbf{s}}[C(\mathbf{s}',\mathbf{s}') - 2C(\mathbf{s}',\mathbf{s}) + C(\mathbf{s},\mathbf{s})] = 0$. It follows that continuity in $\mathbf{s}$ and $\mathbf{s}'$ serve as sufficient conditions for mean square continuity. For a (weakly) stationary process, mean square continuity is equivalent to the covariance function $C(\mathbf{s})$ being continuous at $\mathbf{0}$. This follows easily since $E[Y(\mathbf{s}') - Y(\mathbf{s})]^2 = 2(C(\mathbf{0}) - C(\mathbf{s}' - \mathbf{s}))$ for a weakly stationary process and enables a simple practical check for mean square continuity.

Kent (1989) investigates continuous process realizations through a Taylor expansion of the covariance function. Let $\{Y(\mathbf{s}), \mathbf{s} \in \Re^d\}$ be a real-valued stationary spatial process on $\Re^d$. Kent proves that if $C(\mathbf{s})$ is d-times continuously differentiable and $C_d(\mathbf{s}) = C(\mathbf{s}) - P_d(\mathbf{s})$, where $P_d(\mathbf{s})$ is the Taylor polynomial of degree d for $C(\mathbf{s})$ about $\mathbf{0}$, satisfies the condition,

$$|C_d(\mathbf{s})| = O\left(||\mathbf{s}||^{d+\beta}\right)$$

for some $\beta > 0$, then there exists a version of the spatial process $\{Y(\mathbf{s}), \mathbf{s} \in \Re^d\}$ with continuous realizations. If $C(\mathbf{s})$ is d-times continuously differentiable then it is of course continuous at $\mathbf{0}$ and so, from the previous paragraph, the process is mean square continuous.

Let us suppose that $f : L_2 \longrightarrow \Re^1$ (L_2 is the usual Hilbert space of random variables induced by the L_2 metric) is a continuous function. Let $\{Y(\mathbf{s}), \mathbf{s} \in \Re^d\}$ be a process that is continuous almost surely. Then the process $Z(\mathbf{s}) = f(Y(\mathbf{s}))$ is almost surely continuous, being the composition of two continuous functions. The validity of this statement is direct and does not require checking Kent's conditions. Indeed, the process $Z(\mathbf{s})$ need not be stationary even if $Y(\mathbf{s})$ is. However the existence of the covariance function $C(\mathbf{s},\mathbf{s}') = E[f(Y(\mathbf{s}))f(Y(\mathbf{s}'))]$, via the Cauchy-Schwartz inequality, requires $Ef^2(Y(\mathbf{s})) < \infty$.

While almost sure continuity of the new process $Z(\mathbf{s})$ follows routinely, the mean square continuity of $Z(\mathbf{s})$ is not immediate. However, from the remark below Definition 10.2, if $f : \Re^1 \longrightarrow \Re^1$ is a continuous function that is bounded and $Y(\mathbf{s})$ is a process that is continuous almost surely, then the process $Y(\mathbf{s}) = f(Z(\mathbf{s}))$ (a process on $\Re^d$) is mean square continuous.

More generally suppose f is a continuous function that is Lipschitz of order 1, and $\{Y(\mathbf{s}), \mathbf{s} \in \Re^d\}$ is a process which is mean square continuous. Then the process $Z(\mathbf{s}) = f(Y(\mathbf{s}))$ is mean square continuous. To see this, note that since f is Lipschitz of order 1 we have $|f(Y(\mathbf{s}+\mathbf{h})) - f(Y(\mathbf{s}))| \le K|Y(\mathbf{s}+\mathbf{h}) - Y(\mathbf{s})|$ for some constant K. It follows that $E[f(Y(\mathbf{s}+\mathbf{h})) - f(Y(\mathbf{s}))]^2 \le K^2 E[Y(\mathbf{s}+\mathbf{h}) - Y(\mathbf{s})]^2$, and the mean square continuity of $Z(\mathbf{s})$ follows directly from the mean square continuity of $Y(\mathbf{s})$.

We next formalize the notion of a mean square differentiable process. Our definition is motivated by the analogous definition of total differentiability of a function of $\Re^d$ in a nonstochastic setting. In particular, $Y(\mathbf{s})$ is mean square differentiable at $\mathbf{s}_0$ if there exists a vector $\nabla_Y(\mathbf{s}_0)$, such that, for any scalar h and any unit vector $\mathbf{u}$,

$$Y(\mathbf{s}_0 + h\mathbf{u}) = Y(\mathbf{s}_0) + h\mathbf{u}^T \nabla_Y(\mathbf{s}_0) + r(\mathbf{s}_0, h\mathbf{u}) , \qquad (10.1)$$

where $r(\mathbf{s}_0, h\mathbf{u}) \to 0$ in the L_2 sense as $h \to 0$. That is, we require for any unit vector $\mathbf{u}$,

$$\lim_{h \to 0} E \left(\frac{Y(\mathbf{s}_0 + h\mathbf{u}) - Y(\mathbf{s}_0) - h\mathbf{u}^T \nabla_Y(\mathbf{s}_0)}{h} \right)^2 = 0 . \qquad (10.2)$$

The first-order linearity condition for the process is required to ensure that mean square differentiable processes are mean square continuous. A counterexample when this condition does not hold is given in Banerjee and Gelfand (2003).

10.1.2 Directional finite difference and derivative processes

The focus of this subsection is to address the problem of the rate of change of a spatial surface at a given point in a given direction. Such slopes or gradients are of interest in so-called digital terrain models for exploring surface roughness. They would also arise in meteorology to recognize temperature or rainfall gradients or in environmental monitoring to understand pollution gradients. With spatial computer models, where the process generating the $Y(\mathbf{s})$ is essentially a black box and realizations are costly to obtain, inference regarding local rates of change becomes important. The application we study here considers rates of change for unobservable or latent spatial processes. For instance, in understanding real estate markets for single-family homes, spatial modeling of residuals provides adjustment to reflect desirability of location, controlling for the characteristics of the home and property. Suppose we consider the rate of change of the residual surface in a given direction at, say, the central business district. Transportation costs to the central business district vary with direction. Increased costs are expected to reduce the price of housing. Since transportation cost information is not included in the mean, directional gradients to the residual surface can clarify this issue.

Spatial gradients are customarily defined as finite differences (see, e.g., Greenwood, 1984, and Meyer, Ericksson, and Maggio, 2001). Evidently the scale of resolution will affect the nature of the resulting gradient (as we illustrate in Example 10.2). To characterize local rates of change without having to specify a scale, infinitesimal gradients may be considered. Ultimately, the nature of the data collection and the scientific questions of interest would determine preference for an infinitesimal or a finite gradient.

For the former, gradients (derivatives) are quantities of basic importance in geometry and physics. Researchers in the physical sciences (e.g., geophysics, meteorology, oceanography) often formulate relationships in terms of gradients. For the latter, differences, viewed as discrete approximations to gradients, may initially seem less attractive. However, in applications involving spatial data, scale is usually a critical question (e.g., in environmental, ecological, or demographic settings). Infinitesimal local rates of change may be of less interest than finite differences at the scale of a map of interpoint distances.

Following the discussion of Subsection 10.1.1, with $\mathbf{u}$ a unit vector, let

$$Y_{\mathbf{u},h}\left(\mathbf{s}\right) = \frac{Y\left(\mathbf{s}+h\mathbf{u}\right) - Y\left(\mathbf{s}\right)}{h} \tag{10.3}$$

be the finite difference at $\mathbf{s}$ in direction $\mathbf{u}$ at scale h. Clearly, for a fixed $\mathbf{u}$ and h, $Y_{\mathbf{u},h}\left(\mathbf{s}\right)$ is a well-defined process on $\Re^d$, which we refer to as the finite difference process at scale h in direction $\mathbf{u}$.

Next, let $D_{\mathbf{u}}Y\left(\mathbf{s}\right) = \lim_{h \to 0} Y_{\mathbf{u},h}\left(\mathbf{s}\right)$ if the limit exists. We see that if $Y\left(\mathbf{s}\right)$ is a mean square differentiable process in $\Re^d$, i.e., (10.2) holds for every $\mathbf{s}_0$ in $\Re^d$, then for each $\mathbf{u}$,

$$\begin{aligned} D_{\mathbf{u}}Y\left(\mathbf{s}\right) &= \lim_{h \to 0} \frac{Y\left(\mathbf{s}+h\mathbf{u}\right) - Y\left(\mathbf{s}\right)}{h} \\ &= \lim_{h \to 0} \frac{h\mathbf{u}^T \nabla_Y\left(\mathbf{s}\right) + r\left(\mathbf{s},h\mathbf{u}\right)}{h} = \mathbf{u}^T \nabla_Y\left(\mathbf{s}\right) . \end{aligned}$$

So $D_{\mathbf{u}}Y\left(\mathbf{s}\right)$ is a well-defined process on $\Re^d$, which we refer to as the directional derivative process in the direction $\mathbf{u}$.

Note that if the unit vectors $\mathbf{e}_1, \mathbf{e}_2, ..., \mathbf{e}_d$ form an orthonormal basis set for $\Re^d$, any unit vector $\mathbf{u}$ in $\Re^d$ can be written as $\mathbf{u} = \sum_{i=1}^{d} w_i \mathbf{e}_i$ with $w_i = \mathbf{u}^T \mathbf{e}_i$ and $\sum_{i=1}^{d} w_i^2 = 1$. But then,

$$D_{\mathbf{u}}Y\left(\mathbf{s}\right) = \mathbf{u}^T \nabla_Y\left(\mathbf{s}\right) = \sum_{i=1}^{d} w_i \mathbf{e}_i^T \nabla_Y\left(\mathbf{s}\right) = \sum_{i=1}^{d} w_i D_{\mathbf{e}_i} Y\left(\mathbf{s}\right) . \tag{10.4}$$

Hence, to study directional derivative processes in arbitrary directions we need only work with a basis set of directional derivative processes. Also from (10.4) it is clear that $D_{-\mathbf{u}}Y\left(\mathbf{s}\right) = -D_{\mathbf{u}}Y\left(\mathbf{s}\right)$. Applying the Cauchy-Schwarz inequality to (10.4), for every unit vector $\mathbf{u}$, $D_{\mathbf{u}}^2 Y\left(\mathbf{s}\right) \leq \sum_{i=1}^{d} D_{\mathbf{e}_i}^2 Y\left(\mathbf{s}\right)$. Hence, $\sum_{i=1}^{d} D_{\mathbf{e}_i}^2 Y\left(\mathbf{s}\right)$ is the maximum over all directions of $D_{\mathbf{u}}^2 Y\left(\mathbf{s}\right)$. At location $\mathbf{s}$, this maximum is achieved in the direction $\mathbf{u} = \nabla_Y\left(\mathbf{s}\right) / \|\nabla_Y\left(\mathbf{s}\right)\|$, and the maximizing value is $\|\nabla_Y\left(\mathbf{s}\right)\|$. In the following text we work with the customary orthonormal basis defined by the coordinate axes so that $\mathbf{e}_i$ is a $d \times 1$ vector with all 0's except for a 1 in the ith row. In fact, with this basis, $\nabla_Y\left(\mathbf{s}\right) = \left(D_{\mathbf{e}_1} Y\left(\mathbf{s}\right), ..., D_{\mathbf{e}_d} Y\left(\mathbf{s}\right)\right)^T$. The result in (10.4) is a limiting result as $h \to 0$. From (10.4), the presence of h shows that to study finite difference processes at scale h in arbitrary directions we have no reduction to a basis set.

Formally, finite difference processes require less assumption for their existence. To compute differences we need not worry about a numerical degree of smoothness for the realized spatial surface. However, issues of numerical stability can arise if h is too small. Also, with directional derivatives in, say, two-dimensional space, following the discussion below (10.4), we only need work with north and east directional derivatives processes in order to study directional derivatives in arbitrary directions.

10.1.3 Distribution theory

If $E\left(Y\left(\mathbf{s}\right)\right) = 0$ for all $\mathbf{s} \in \Re^d$ then $E\left(Y_{\mathbf{u},h}\left(\mathbf{s}\right)\right) = 0$ and $E\left(D_{\mathbf{u}}Y\left(\mathbf{s}\right)\right) = 0$. Let $C_{\mathbf{u}}^{(h)}\left(\mathbf{s},\mathbf{s}'\right)$ and $C_{\mathbf{u}}\left(\mathbf{s},\mathbf{s}'\right)$ denote the covariance functions associated with the process $Y_{\mathbf{u},h}\left(\mathbf{s}\right)$ and $D_{\mathbf{u}}Y\left(\mathbf{s}\right)$, respectively. If $\boldsymbol{\Delta} = \mathbf{s}-\mathbf{s}'$ and $Y\left(\mathbf{s}\right)$ is (weakly) stationary we immediately have

$$C_{\mathbf{u}}^{(h)}(\mathbf{s},\mathbf{s}') = \frac{\left(2C\left(\boldsymbol{\Delta}\right) - C\left(\boldsymbol{\Delta}+h\mathbf{u}\right) - C\left(\boldsymbol{\Delta}-h\mathbf{u}\right)\right)}{h^2}\,, \qquad (10.5)$$

whence $Var\left(Y_{\mathbf{u},h}\left(\mathbf{s}\right)\right) = 2\left(C\left(\mathbf{0}\right) - C\left(h\mathbf{u}\right)\right)/h^2$. If $Y\left(\mathbf{s}\right)$ is isotropic and we replace $C(\mathbf{s},\mathbf{s}')$ by $\widetilde{C}(||\mathbf{s}-\mathbf{s}'||)$, we obtain

$$C_{\mathbf{u}}^{(h)}\left(\mathbf{s},\mathbf{s}'\right) = \frac{\left(2\widetilde{C}\left(||\boldsymbol{\Delta}||\right) - \widetilde{C}\left(||\boldsymbol{\Delta}+h\mathbf{u}||\right) - \widetilde{C}\left(||\boldsymbol{\Delta}-h\mathbf{u}||\right)\right)}{h^2}\,. \qquad (10.6)$$

Expression (10.6) shows that even if $Y\left(\mathbf{s}\right)$ is isotropic, $Y_{\mathbf{u},h}\left(\mathbf{s}\right)$ is only stationary. Also $Var\left(Y_{\mathbf{u},h}\left(\mathbf{s}\right)\right) = 2\left(\widetilde{C}\left(0\right) - \widetilde{C}\left(h\right)\right)/h^2 = \gamma\left(h\right)/h^2$ where $\gamma\left(h\right)$ is the familiar variogram of the $Y\left(\mathbf{s}\right)$ process (Subsection 2.1.2).

Similarly, if $Y\left(\mathbf{s}\right)$ is stationary we may show that if all second-order partial and mixed derivatives of C exist and are continuous, the limit of (10.5) as $h \to 0$ is

$$C_{\mathbf{u}}\left(\mathbf{s},\mathbf{s}'\right) = -\mathbf{u}^T\Omega\left(\boldsymbol{\Delta}\right)\mathbf{u}\,, \qquad (10.7)$$

where $\left(\Omega\left(\boldsymbol{\Delta}\right)\right)_{ij} = \partial^2 C\left(\boldsymbol{\Delta}\right)/\partial\boldsymbol{\Delta}_i\partial\boldsymbol{\Delta}_j$. By construction, (10.7) is a valid covariance function on $\Re^d$ for any $\mathbf{u}$. Also, $Var\left(D_{\mathbf{u}}Y\left(\mathbf{s}\right)\right) = -\mathbf{u}^T\Omega\left(0\right)\mathbf{u}$. If $Y\left(\mathbf{s}\right)$ is isotropic, using standard chain rule calculations we obtain

$$C_{\mathbf{u}}\left(\mathbf{s},\mathbf{s}'\right) = -\left\{\left(1 - \frac{\left(\mathbf{u}^T\boldsymbol{\Delta}\right)^2}{||\boldsymbol{\Delta}||^2}\right)\frac{\widetilde{C}'\left(||\boldsymbol{\Delta}||\right)}{||\boldsymbol{\Delta}||} + \frac{\left(\mathbf{u}^T\boldsymbol{\Delta}\right)^2}{||\boldsymbol{\Delta}||^2}\widetilde{C}''\left(||\boldsymbol{\Delta}||\right)\right\}\,. \qquad (10.8)$$

Again, if $Y\left(\mathbf{s}\right)$ is isotropic, $D_{\mathbf{u}}Y\left(\mathbf{s}\right)$ is only stationary. In addition, we have $Var\left(D_{\mathbf{u}}Y\left(\mathbf{s}\right)\right) = -\widetilde{C}''\left(0\right)$ which also shows that, provided $\widetilde{C}$ is twice differentiable at 0, $\lim_{h\to 0}\gamma\left(h\right)/h^2 = -\widetilde{C}''\left(0\right)$, i.e., $\gamma\left(h\right) = O\left(h^2\right)$ for h small.

For $Y(\mathbf{s})$ stationary we can also calculate

$$cov\left(Y(\mathbf{s}), Y_{\mathbf{u},h}(\mathbf{s}')\right) = \left(C(\boldsymbol{\Delta} - h\mathbf{u}) - C(\boldsymbol{\Delta})\right)/h\,,$$

from which $cov\left(Y(\mathbf{s}), Y_{\mathbf{u},h}(\mathbf{s})\right) = \left(C(h\mathbf{u}) - C(\mathbf{0})\right)/h$. But then,

$$\begin{aligned}
cov\left(Y(\mathbf{s}), D_{\mathbf{u}}Y(\mathbf{s}')\right) &= \lim_{h\to 0}\left(C(\boldsymbol{\Delta} - h\mathbf{u}) - C(\boldsymbol{\Delta})\right)/h \\
&= -D_{\mathbf{u}}C(\boldsymbol{\Delta}) = D_{\mathbf{u}}C(-\boldsymbol{\Delta})\,,
\end{aligned}$$

since $C(\boldsymbol{\Delta}) = C(-\boldsymbol{\Delta})$. In particular, we have that $cov\left(Y(\mathbf{s}), D_{\mathbf{u}}Y(\mathbf{s})\right) = \lim_{h\to 0}\left(C(h\mathbf{u}) - C(\mathbf{0})\right)/h = D_{\mathbf{u}}C(\mathbf{0})$. The existence of the directional derivative process ensures the existence of $D_{\mathbf{u}}C(\mathbf{0})$. Moreover, since $C(h\mathbf{u}) = C(-h\mathbf{u})$, $C(h\mathbf{u})$ (viewed as a function of h) is even, so $D_{\mathbf{u}}C(\mathbf{0}) = 0$. Thus, $Y(\mathbf{s})$ and $D_{\mathbf{u}}Y(\mathbf{s})$ are uncorrelated. Intuitively, this is sensible. The level of the process at a particular location is uncorrelated with the directional derivative in any direction at that location. This is not true for directional differences. Also, in general, $cov\left(Y(\mathbf{s}), D_{\mathbf{u}}Y(\mathbf{s}')\right)$ will not be 0.

Under isotropy,

$$cov\left(Y(\mathbf{s}), Y_{\mathbf{u},h}(\mathbf{s}')\right) = \frac{\widetilde{C}\left(||\boldsymbol{\Delta} - h\mathbf{u}||\right) - \widetilde{C}\left(||\boldsymbol{\Delta}||\right)}{h}\,.$$

Now $cov\left(Y(\mathbf{s}), Y_{\mathbf{u},h}(\mathbf{s})\right) = \left(\widetilde{C}(h) - \widetilde{C}(0)\right)/h = \gamma(h)/2h$, so this means $cov\left(Y(\mathbf{s}), D_{\mathbf{u}}Y(\mathbf{s})\right) = \widetilde{C}'(0) = \lim_{h\to 0}\gamma(h)/2h = 0$ since, as above, if $\widetilde{C}''(0)$ exists, $\gamma(h) = O(h^2)$.

Suppose we consider the bivariate process $\mathbf{Z}_{\mathbf{u}}^{(h)}(\mathbf{s}) = (Y(\mathbf{s}), Y_{\mathbf{u},h}(\mathbf{s}))^T$. It is clear that this process has mean 0 and, if $Y(\mathbf{s})$ is stationary, cross-covariance matrix $V_{\mathbf{u},h}(\boldsymbol{\Delta})$ given by

$$\begin{pmatrix} C(\boldsymbol{\Delta}) & \frac{C(\boldsymbol{\Delta}-h\mathbf{u})-C(\boldsymbol{\Delta})}{h} \\ \frac{C(\boldsymbol{\Delta}+h\mathbf{u})-C(\boldsymbol{\Delta})}{h} & \frac{2C(\boldsymbol{\Delta})-C(\boldsymbol{\Delta}+h\mathbf{u})-C(\boldsymbol{\Delta}-h\mathbf{u})}{h^2} \end{pmatrix}\,. \tag{10.9}$$

Since $\mathbf{Z}_{\mathbf{u}}^{(h)}(\mathbf{s})$ arises by linear transformation of $Y(\mathbf{s})$, (10.9) is a valid cross-covariance matrix in $\Re^d$. But since this is true for every h, letting $h \to 0$,

$$V_{\mathbf{u}}(\boldsymbol{\Delta}) = \begin{pmatrix} C(\boldsymbol{\Delta}) & -D_{\mathbf{u}}C(\boldsymbol{\Delta}) \\ D_{\mathbf{u}}C(\boldsymbol{\Delta}) & -\mathbf{u}^T\Omega(\boldsymbol{\Delta})\mathbf{u} \end{pmatrix} \tag{10.10}$$

is a valid cross-covariance matrix in $\Re^d$. In fact, $V_{\mathbf{u}}$ is the cross-covariance matrix for the bivariate process $\mathbf{Z}_{\mathbf{u}}(s) = \begin{pmatrix} Y(\mathbf{s}) \\ D_{\mathbf{u}}Y(\mathbf{s}) \end{pmatrix}$.

If, in addition, we assume that $Y(\mathbf{s})$ is a stationary Gaussian process it is clear, again by linearity, that $\mathbf{Z}_{\mathbf{u}}^h(\mathbf{s})$ is a stationary bivariate Gaussian process. But then, by a standard limiting moment generating function argument, $\mathbf{Z}_{\mathbf{u}}(\mathbf{s})$ is a stationary bivariate Gaussian process and thus $D_{\mathbf{u}}Y(\mathbf{s})$ is a stationary univariate Gaussian process. As an aside, we note that for a given $\mathbf{s}$, $D_{\frac{\nabla_Y(\mathbf{s})}{||\nabla_Y(\mathbf{s})||}}Y(\mathbf{s})$ is not normally distributed, and in fact the set $\{D_{\frac{\nabla_Y(\mathbf{s})}{||\nabla_Y(\mathbf{s})||}}Y(\mathbf{s}) : \mathbf{s} \in \Re^d\}$ is not a spatial process.

Extension to a pair of directions with associated unit vectors $\mathbf{u}_1$ and $\mathbf{u}_2$ results in a trivariate Gaussian process $\mathbf{Z}(\mathbf{s}) = (Y(\mathbf{s}), D_{\mathbf{u}_1}Y(\mathbf{s}), D_{\mathbf{u}_2}Y(\mathbf{s}))^T$ with associated cross-covariance matrix $V_{\mathbf{Z}}(\mathbf{\Delta})$ given by

$$\begin{pmatrix} C(\mathbf{\Delta}) & -(\nabla C(\mathbf{\Delta}))^T \\ \nabla C(\mathbf{\Delta}) & -\Omega(\mathbf{\Delta}) \end{pmatrix}. \qquad (10.11)$$

At $\mathbf{\Delta} = 0$, (10.11) becomes a diagonal matrix.

We conclude this subsection with a useful example. Recall the power exponential family of isotropic covariance functions of the previous subsection, $\tilde{C}(||\mathbf{\Delta}||) = \alpha \exp(-\phi ||\mathbf{\Delta}||^\nu)$, $0 < \nu \le 2$. It is apparent that $\tilde{C}''(0)$ exists only for $\nu = 2$. The Gaussian covariance function is the only member of the class for which directional derivative processes can be defined. However, as we have noted in Subsection 2.2.3, the Gaussian covariance function produces process realizations that are too smooth to be attractive for practical modeling.

Turning to the Matérn class, $\tilde{C}(||\mathbf{\Delta}||) = \alpha (\phi ||\mathbf{\Delta}||)^\nu K_\nu(\phi ||\mathbf{\Delta}||)$, ν is a smoothness parameter controlling the extent of mean square differentiability of process realizations (Stein, 1999a). At $\nu = 3/2$, $\tilde{C}(||\mathbf{\Delta}||)$ takes the closed form $\tilde{C}(||\mathbf{\Delta}||) = \sigma^2(1 + \phi ||\mathbf{\Delta}||) \exp(-\phi ||\mathbf{\Delta}||)$ where σ^2 is the process variance. This function is exactly twice differentiable at 0. We have a (once but not twice) mean square differentiable process, which therefore does not suffer the excessive smoothness implicit with the Gaussian covariance function.

For this choice one can show that $\nabla \tilde{C}(||\mathbf{\Delta}||) = -\sigma^2\phi^2 \exp(-\phi ||\mathbf{\Delta}||)\mathbf{\Delta}$, that $\left(H_{\tilde{C}}(||\mathbf{\Delta}||)\right)_{ii} = -\sigma^2\phi^2 \exp(-\phi ||\mathbf{\Delta}||)(1 - \phi\Delta_i^2/||\mathbf{\Delta}||)$, and also that $\left(H_{\tilde{C}}(||\mathbf{\Delta}||)\right)_{ij} = \sigma^2\phi^2 \exp(-\phi ||\mathbf{\Delta}||)\Delta_i\Delta_j/||\mathbf{\Delta}||$. In particular, $V_{\mathbf{u}}(0) = \sigma^2 BlockDiag(1, \phi^2 I)$.

10.1.4 Directional derivative processes in modeling

We work in $d = 2$-dimensional space and can envision the following types of modeling settings in which directional derivative processes would be of interest. For $Y(\mathbf{s})$ purely spatial with constant mean, we would seek $D_{\mathbf{u}}Y(\mathbf{s})$. In the customary formulation $Y(\mathbf{s}) = \mu(\mathbf{s}) + W(\mathbf{s}) + \epsilon(\mathbf{s})$ we would instead want $D_{\mathbf{u}}W(\mathbf{s})$. In the case of a spatially varying coefficient model $Y(\mathbf{s}) = \beta_0(\mathbf{s}) + \beta_1(\mathbf{s})X(\mathbf{s}) + \epsilon(\mathbf{s})$ such as in Section 10.2, we would examine $D_{\mathbf{u}}\beta_0(\mathbf{s})$, $D_{\mathbf{u}}\beta_1(\mathbf{s})$, and $D_{\mathbf{u}}EY(\mathbf{s})$ with $EY(\mathbf{s}) = \beta_0(\mathbf{s}) + \beta_1(\mathbf{s})X(\mathbf{s})$.

Consider the constant mean purely spatial process for illustration, where we have $Y(\mathbf{s})$ a stationary process with mean μ and covariance function $C(\mathbf{\Delta}) = \sigma^2\rho(\mathbf{\Delta})$ where ρ is a valid two-dimensional correlation function.

For illustration we work with the general Matérn class parametrized by ϕ and ν, constraining $\nu > 1$ to ensure the (mean square) existence of the directional derivative processes. Letting $\boldsymbol{\theta} = \left(\mu, \sigma^2, \phi, \nu\right)$, for locations $\mathbf{s}_1, \mathbf{s}_2, ..., \mathbf{s}_n$, the likelihood $L\left(\boldsymbol{\theta}; \mathbf{Y}\right)$ is proportional to

$$\left(\sigma^2\right)^{-n/2} \left|R\left(\phi, \nu\right)\right|^{-1/2} \exp\left\{-\frac{1}{2\sigma^2}\left(\mathbf{Y} - \mu\mathbf{1}\right)^T R^{-1}\left(\phi, \nu\right)\left(\mathbf{Y} - \mu\mathbf{1}\right)\right\}.$$
(10.12)

In (10.12), $\mathbf{Y}^T = \left(Y(\mathbf{s}_1), \ldots, Y(\mathbf{s}_n)\right)^T$ and $\left(R\left(\phi, \nu\right)\right)_{ij} = \rho\left(\mathbf{s}_i - \mathbf{s}_j; \phi, \nu\right)$.

In practice we would usually propose fairly noninformative priors for μ, σ^2, ϕ and ν, e.g., vague normal (perhaps flat), vague inverse gamma, vague gamma, and $U\left(1, 2\right)$, respectively. With regard to the prior on ν, we follow the suggestion of Stein (1999a) and others, who observe that, in practice, it will be very difficult to distinguish $\nu = 2$ from $\nu > 2$. Fitting of this low-dimensional model is easiest using slice sampling (Appendix Section A.6).

A contour or a grey-scale plot of the posterior mean surface is of primary interest in providing a smoothed display of spatial pattern and of areas where the process is elevated or depressed. To handle finite differences at scale h, in the sequel we work with the vector of eight compass directions, $N, NE, E, \ldots$. At $\mathbf{s}_i$, we denote this vector by $\mathbf{Y}_h\left(\mathbf{s}_i\right)$ and let $\mathbf{Y}_h = \left\{\mathbf{Y}_h\left(\mathbf{s}_i\right), i = 1, 2, ..., n\right\}$. With directional derivatives we only need $\mathbf{D}\left(\mathbf{s}_i\right)^T = \left(D_{10}Y(\mathbf{s}_i), D_{01}Y(\mathbf{s}_i)\right)$ and let $\mathbf{D} = \left\{\mathbf{D}\left(\mathbf{s}_i\right), i = 1, 2, ..., n\right\}$. We seek samples from the predictive distribution $f\left(\mathbf{Y}_h|\mathbf{Y}\right)$ and $f\left(\mathbf{D}|\mathbf{Y}\right)$. In $\mathbf{Y}_h\left(\mathbf{s}_i\right)$, $Y\left(\mathbf{s}_i\right)$ is observed, hence fixed in the predictive distribution. So we can replace $Y_{u,h}\left(\mathbf{s}_i\right)$ with $Y\left(\mathbf{s}_i + h\mathbf{u}\right)$; posterior predictive samples of $Y\left(\mathbf{s}_i + h\mathbf{u}\right)$ are immediately converted to posterior predictive samples of $Y_{u,h}\left(\mathbf{s}_i\right)$ by linear transformation. Hence, the directional finite differences problem is merely a large Bayesian kriging problem requiring spatial prediction at the set of $8n$ locations $\left\{Y\left(\mathbf{s}_i + h\mathbf{u}_r\right), i = 1, 2, ..., n; r = 1, 2, ..., 8\right\}$. Denoting this set by $\widetilde{\mathbf{Y}}_h$, we require samples from $f\left(\widetilde{\mathbf{Y}}_h|\mathbf{Y}\right)$. From the relationship, $f\left(\widetilde{\mathbf{Y}}_h|\mathbf{Y}\right) = \int f\left(\widetilde{\mathbf{Y}}_h|\mathbf{Y}, \boldsymbol{\theta}\right) f\left(\boldsymbol{\theta}|\mathbf{Y}\right) d\boldsymbol{\theta}$ this can be done one for one with the $\boldsymbol{\theta}_l^*$'s by drawing $\widetilde{\mathbf{Y}}_{h,l}^*$ from the multivariate normal distribution $f\left(\widetilde{\mathbf{Y}}_h|\mathbf{Y}, \boldsymbol{\theta}_l^*\right)$, as detailed in Section 5.1. Similarly $f\left(\mathbf{D}|\mathbf{Y}\right) = \int f\left(\mathbf{D}|\mathbf{Y}, \boldsymbol{\theta}\right) f\left(\boldsymbol{\theta}|\mathbf{Y}\right) d\boldsymbol{\theta}$. The cross-covariance function in (10.11) allows us to immediately write down the joint multivariate normal distribution of $\mathbf{Y}$ and $\mathbf{D}$ given $\boldsymbol{\theta}$ and thus, at $\boldsymbol{\theta}_l^*$, the conditional multivariate normal distribution $f\left(\mathbf{D}|\mathbf{Y}, \boldsymbol{\theta}_l^*\right)$. At a specified new location $\mathbf{s}_0$, with finite directional differences we need to add spatial prediction at the nine new locations, $Y\left(\mathbf{s}_0\right)$ and $Y\left(\mathbf{s}_0 + h\mathbf{u}_r\right), r = 1, 2, ..., 8$. With directional derivatives, we again can use (10.9) to obtain the joint distribution of $\mathbf{Y}, \mathbf{D}, Y\left(\mathbf{s}_0\right)$ and $\mathbf{D}\left(\mathbf{s}_0\right)$ given θ and thus the conditional distribution $f\left(\mathbf{D}, Y\left(\mathbf{s}_0\right), \mathbf{D}\left(\mathbf{s}_0\right)|\mathbf{Y}, \boldsymbol{\theta}\right)$.

Turning to the random spatial effects model we now assume that

$$Y(\mathbf{s}) = \mathbf{x}^T(\mathbf{s})\boldsymbol{\beta} + w(\mathbf{s}) + \epsilon(\mathbf{s}) . \quad (10.13)$$

In (10.13), $\mathbf{x}(\mathbf{s})$ is a vector of location characteristics, $w(\mathbf{s})$ is a mean 0 stationary Gaussian spatial process with parameters σ^2, ϕ, and ν as above, and $\epsilon(\mathbf{s})$ is a Gaussian white noise process with variance τ^2, intended to capture measurement error or microscale variability. Such a model is appropriate for the real estate example mentioned in Subsection 10.1.2, where $Y(\mathbf{s})$ is the log selling price and $\mathbf{x}(\mathbf{s})$ denotes associated house and property characteristics. Here $w(\mathbf{s})$ measures the spatial adjustment to log selling price at location $\mathbf{s}$ reflecting relative desirability of the location. $\epsilon(\mathbf{s})$ is needed to capture microscale variability. Here such variability arises because two identical houses arbitrarily close to each other need not sell for essentially the same price due to unobserved differences in buyers, sellers, and brokers across transactions.

For locations $\mathbf{s}_1, \mathbf{s}_2, ..., \mathbf{s}_n$, with $\boldsymbol{\theta} = \left(\boldsymbol{\beta}, \tau^2, \sigma^2, \phi, \nu\right)$ the model in (10.13) produces a marginal likelihood $L(\boldsymbol{\theta}; Y)$ (integrating over $\{w(\mathbf{s}_i)\}$) proportional to $\left|\sigma^2 R(\phi, \nu) + \tau^2 I\right|^{-1/2} e^{-\frac{1}{2}(Y-X\boldsymbol{\beta})^T \left(\sigma^2 R(\phi,\nu)+\tau^2 I\right)^{-1}(Y-X\boldsymbol{\beta})}$. Priors for $\boldsymbol{\beta}$, τ^2, σ^2, ϕ and ν are similar to the first illustration, vague choices that are normal, inverse gamma, inverse gamma, gamma, and uniform, respectively. Again slice sampling provides an efficient fitting algorithm.

Further inference with regard to (10.13) focuses on the spatial process itself. That is, we would be interested in the posterior spatial effect surface and in rates of change associated with this surface. The former is usually handled with samples of the set of $w(\mathbf{s}_i)$ given $\mathbf{Y}$ along with, perhaps, a grey-scaling or contouring routine. The latter would likely be examined at new locations. For instance in the real estate example, spatial gradients would be of interest at the central business district or at other externalities such as major roadway intersections, shopping malls, airports, or waste disposal sites but not likely at the locations of the individual houses.

As below (10.12) with $\mathbf{W}^T = (w(\mathbf{s}_1), ..., w(\mathbf{s}_n))$, we sample $f(\mathbf{W}|Y)$ one for one with the θ_l^*'s using $f(\mathbf{W}|\mathbf{Y}) = \int f(\mathbf{W}|\mathbf{Y}, \boldsymbol{\theta}) f(\boldsymbol{\theta}|\mathbf{Y}) d\boldsymbol{\theta}$, as described in Section 5.1. But also, given $\boldsymbol{\theta}$, the joint distribution of $\mathbf{W}$ and $\mathbf{V}(\mathbf{s}_0)$ where $\mathbf{V}(\mathbf{s}_0)$ is either $\mathbf{W}_h(\mathbf{s}_0)$ or $\mathbf{D}(\mathbf{s}_0)$ is multivariate normal. For instance, with $D(\mathbf{s}_0)$, the joint normal distribution can be obtained using (10.11) and as a result so can the conditional normal distribution $f(\mathbf{V}(\mathbf{s}_0)|\mathbf{W}, \boldsymbol{\theta})$. Lastly, since

$$f(\mathbf{V}(\mathbf{s}_0)|\mathbf{Y}) = \int f(\mathbf{V}(\mathbf{s}_0)|\mathbf{W}, \boldsymbol{\theta}) f(\mathbf{W}|\boldsymbol{\theta}, \mathbf{Y}) f(\boldsymbol{\theta}|\mathbf{Y}) d\boldsymbol{\theta} d\mathbf{W} ,$$

we can also obtain samples from $f(\mathbf{V}(\mathbf{s}_0)|\mathbf{Y})$ one for one with the θ_l^*'s.

Example 10.3 We illustrate with a simulated example. We generate from a Gaussian random field with constant mean μ and a covariance struc-

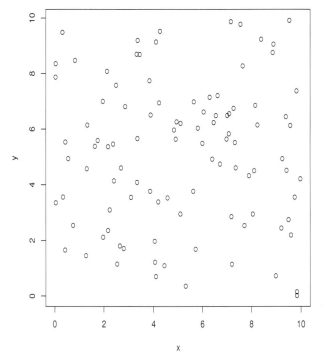

Figure 10.1 *Locations of the 100 sites where simulated draws from the random field have been observed.*

ture specified through the Matérn ($\nu = 3/2$) covariance function, $\sigma^2(1 + \phi d)\exp(-\phi d)$. The field is observed on a randomly sampled set of points within a 10×10 square. That is, the x and y coordinates lie between 0 and 10. For illustration, we set $\mu = 0$, $\sigma^2 = 1.0$ and $\phi = 1.05$. Our data consists of $n = 100$ observations at the randomly selected sites shown in Figure 10.1. The maximum observed distance in our generated field is approximately 13.25 units. The value of $\phi = 1.05$ provides an effective isotropic range of about 4.5 units. We also perform a Bayesian kriging on the data to develop a predicted field. Figure 10.2 shows a grey-scale plot with contour lines displaying the topography of the "kriged" field. We will see below that our predictions of the spatial gradients at selected points are consistent with the topography around those points, as depicted in Figure 10.2.

Adopting a flat prior for μ, an IG(2, 0.1) (mean = 10, infinite variance) prior for σ^2, a $G(2, 0.1)$ prior (mean = 20, variance = 200) for ϕ, and a uniform on $(1, 2)$ for ν, we obtain the posterior estimates for our parameters shown in Table 10.1, in good agreement with the true values.

We next examine the directional derivatives and directional finite differ-

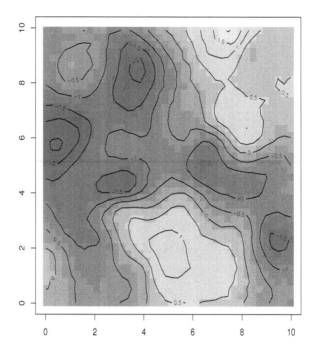

Figure 10.2 *Grey-scale plot with contour lines showing the topography of the random field in the simulated example.*

Parameter	50%	(2.5%, 97.5%)
μ	−0.39	(−0.91, 0.10)
σ^2	0.74	(0.50, 1.46)
ϕ	1.12	(0.85, 1.41)
ν	1.50	(1.24, 1.77)

Table 10.1 *Posterior estimates for model parameters, simulated example.*

ences for the unit vectors corresponding to angles of 0, 45, 90, 135, 180, 225, 270, and 315 degrees with the horizontal axis in a counterclockwise direction at the point. For the finite differences we consider $h = 1.0$ and 0.1. Recall that $D_{-\mathbf{u}}Y(\mathbf{s}) = -D_{\mathbf{u}}Y(\mathbf{s})$. Table 10.2 presents the resulting posterior predictive inference for the point $(3.5, 3.5)$ in Figure 10.2.

We see that $(3.5, 3.5)$ seems to be in an interesting portion of the surface, with many contour lines nearby. It is clear from the contour lines that there

Angle	$D_{\mathbf{u}}Y$ (s) ($h = 0$)	$h = 1.0$	$h = 0.1$
0	−0.06 (−1.12,1.09)	0.51 (−0.82,1.81)	−0.08 (−1.23,1.20)
45	−1.49 (−2.81,−0.34)	−0.01 (−1.29,1.32)	−1.55 (−2.93,−0.56)
90	−2.07 (−3.44,−0.66)	−0.46 (−1.71,0.84)	−2.13 (−3.40,−0.70)
135	−1.42 (−2.68,−0.23)	−0.43 (−1.69,0.82)	−1.44 (−2.64,−0.23)
180	0.06 (−1.09,1.12)	−0.48 (−1.74,0.80)	0.08 (−1.19,1.23)
225	1.49 (0.34,2.81)	0.16 (−1.05,1.41)	1.61 (0.52,3.03)
270	2.07 (0.66,3.44)	0.48 (−0.91,1.73)	2.12 (0.68,3.43)
315	1.42 (0.23,2.68)	1.12 (−0.09,2.41)	1.44 (0.24,2.68)

Table 10.2 *Posterior medians and (2.5%, 97.5%) predictive intervals for directional derivatives and finite differences at point (3.5, 3.5).*

Angle	50%	(2.5%, 97.5%)	angle	50%	(2.5%, 97.5%)
0	0.12	(−1.45, 1.80)	180	−0.12	(−1.80, 1.45)
45	−0.05	(−1.70, 1.51)	225	0.05	(−1.51, 1.70)
90	−0.18	(−1.66, 1.34)	270	0.18	(−1.34, 1.66)
135	−0.24	(−1.72, 1.18)	315	0.24	(−1.18, 1.72)

Table 10.3 *Posterior medians and (2.5%, 97.5%) predictive intervals for directional derivatives at point (5, 5).*

is a negative northern gradient (downhill) and a positive southern gradient (uphill) around the point (3.5, 3.5). On the other hand, there does not seem to be any significant E-W gradient around that point as seen from the contour lines through that point running E-W. This is brought out very clearly in column 1 of Table 10.2. The angles of 0 and 180 degrees that correspond to the E-W gradients are not at all significant. The N-S gradients are indeed pronounced as seen by the 90 and 270 degree gradients. The directional derivatives along the diagonals also indicate the presence of a gradient. There are significant downhill gradients toward the northeast, north, and northwest, and (therefore) significant uphill gradients toward the southwest, south, and southeast. Hence the directional derivative process provides inferential quantification consistent with features that are captured descriptively and visually in Figure 10.2. Note further that while the results for $h = 0.1$ are very close to those for $h = 0$, at the scale $h = 1.0$ the gradients disappear.

As an aside, we look at the point (5, 5) in Figure 10.2, which is located in an essentially "flat" or featureless portion of the region. Here we would

expect the directional derivative to be insignificantly different from 0 in all
directions because of the constant level, revealed by the constant shade and
the lack of contour lines in the region. Indeed, this is what emerges from
Table 10.3. ■

10.2 Spatially varying coefficient models

In Section 8.1 we introduced a spatially varying coefficient process in the
evolution equation of the spatiotemporal dynamic model. Similarly, in Section 9.4 we considered multiple spatial frailty models with regression coefficients that were allowed to vary spatially. We return to this topic here
to amplify the scope of possibilities for such modeling. In particular, in
the spatial-only case, we denote the value of the coefficient at location $\mathbf{s}$
by $\beta(\mathbf{s})$. This coefficient can be resolved at either areal unit or point level.
With the former, the $\beta(\mathbf{s})$ surface consists of "tiles" at various heights, one
tile per areal unit. For the latter, we achieve a more flexible spatial surface.

Using tiles, concern arises regarding the arbitrariness of the scale of resolution, the lack of smoothness of the surface, and the inability to interpolate the value of the surface to individual locations. When working with
point-referenced data it will be more attractive to allow the coefficients to
vary by location, to envision for a particular coefficient, a spatial surface.
For instance, in our example below we also model the (log) selling price
of single-family houses. Customary explanatory variables include the age
of the house, the square feet of living area, the square feet of other area,
and the number of bathrooms. If the region of interest is a city or greater
metropolitan area, it is evident that the capitalization rate (e.g., for age)
will vary across the region. In some parts of the region older houses will
be more valued than in other parts. By allowing the coefficient of age to
vary with location, we can remedy the foregoing concerns. With practical
interest in mind (say, real estate appraisal), we can predict the coefficient
for arbitrary properties, not just for those that sold during the period of
investigation. Similar issues arise in modeling environmental exposure to a
particular pollutant where covariates might include temperature and precipitation.

One possible approach would be to model the spatial surface for the
coefficient parametrically. In the simplest case this would require the rather
arbitrary specification of a polynomial surface function; a range of surfaces
too limited or inflexible might result. More flexibility could be introduced
using a spline surface over two- dimensional space; see, e.g., Luo and Wahba
(1998) and references therein. However, this requires selection of a spline
function and determination of the number of and locations of the knots in
the space. Also, with multiple coefficients, a multivariate specification of a
spline surface is required. The approach we adopt here is arguably more
natural and at least as flexible. We model the spatially varying coefficient

surface as a realization from a spatial process. For multiple coefficients we employ a multivariate spatial process model.

To clarify interpretation and implementations, we first develop our general approach in the case of a single covariate, hence two spatially varying coefficient processes, one for "intercept" and one for "slope." We then turn to the case of multiple covariates. Since even in the basic multiple regression setting, coefficient estimates typically reveal some strong correlations, it is expected that the collection of spatially varying coefficient processes will be dependent. Hence, we employ a multivariate process model. Indeed we present a further generalization to build a spatial analogue of a multilevel regression model (see, e.g., Goldstein, 1995). We also consider flexible spatiotemporal possibilities. The previously mentioned real estate setting provides site level covariates whose coefficients are of considerable practical interest and a data set of single-family home sales from Baton Rouge, LA, enables illustration. Except for regions exhibiting special topography, we anticipate that a spatially varying coefficient model will prove more useful than, for instance, a trend surface model. That is, incorporating a polynomial in latitude and longitude into the mean structure would not be expected to serve as a surrogate for allowing the variability across the region of a coefficient for, say, age or living area of a house.

10.2.1 Approach for a single covariate

Recall the usual Gaussian stationary spatial process model as in (5.1),

$$Y(\mathbf{s}) = \mu(\mathbf{s}) + w(\mathbf{s}) + \epsilon(\mathbf{s}) , \qquad (10.14)$$

where $\mu(\mathbf{s}) = \mathbf{x}(\mathbf{s})^T \beta$ and $\epsilon(\mathbf{s})$ is a white noise process, i.e., $E(\epsilon(\mathbf{s})) = 0$, $\mathrm{Var}(\epsilon(\mathbf{s})) = \tau^2$, $\mathrm{cov}(\epsilon(\mathbf{s}), \epsilon(\mathbf{s}')) = 0$, and $w(\mathbf{s})$ is a second-order stationary mean-zero process independent of the white noise process, i.e., $E(w(\mathbf{s})) = 0$, $\mathrm{Var}(w(\mathbf{s})) = \sigma^2$, $\mathrm{cov}(w(\mathbf{s}), w(\mathbf{s}')) = \sigma^2 \rho(\mathbf{s}, \mathbf{s}'; \phi)$, where ρ is a valid two-dimensional correlation function.

Letting $\mu(\mathbf{s}) = \beta_0 + \beta_1 x(\mathbf{s})$, write $w(\mathbf{s}) = \beta_0(\mathbf{s})$ and define $\tilde{\beta}_0(\mathbf{s}) = \beta_0 + \beta_0(\mathbf{s})$. Then $\beta_0(\mathbf{s})$ can be interpreted as a random spatial adjustment at location $\mathbf{s}$ to the overall intercept β_0. Equivalently, $\tilde{\beta}_0(\mathbf{s})$ can be viewed as a random intercept process. For an observed set of locations $\mathbf{s}_1, \mathbf{s}_2, \ldots, \mathbf{s}_n$ given $\beta_0, \beta_1, \{\beta_0(\mathbf{s}_i)\}$ and τ^2, the $Y(\mathbf{s}_i) = \beta_0 + \beta_1 x(\mathbf{s}_i) + \beta_0(\mathbf{s}_i) + \epsilon(\mathbf{s}_i), i = 1, \ldots, n$, are conditionally independent. Then $L(\beta_0, \beta_1, \{\beta_0(\mathbf{s}_i)\}, \tau^2; \mathbf{y})$, the first-stage likelihood, is

$$(\tau^2)^{-\frac{n}{2}} \exp\left\{ -\frac{1}{2\tau^2} \sum (Y(\mathbf{s}_i) - (\beta_0 + \beta_1 x(\mathbf{s}_i) + \beta_0(\mathbf{s}_i)))^2 \right\} . \qquad (10.15)$$

In obvious notation, the distribution of $\mathbf{B}_0 = (\beta_0(\mathbf{s}_1), \ldots, \beta_0(\mathbf{s}_n))^T$ is

$$f(\mathbf{B}_0 \mid \sigma_0{}^2, \phi_0) = N(\mathbf{0}, \sigma_0{}^2 H_0(\phi_0)) , \qquad (10.16)$$

where $(H_0(\phi_0))_{ij} = \rho_0(\mathbf{s}_i - \mathbf{s}_j; \phi_0)$. For all of the discussion and examples below, we adopt the Matérn correlation function, (2.8). With a prior on $\beta_0, \beta_1, \tau^2, \sigma_0^2$, and ϕ_0, specification of the Bayesian hierarchical model is completed. Under (10.15) and (10.16), we can integrate over $\mathbf{B}_0$, obtaining $L(\beta_0, \beta_1, \tau^2, {\sigma_0}^2, \phi_0; \mathbf{y})$, the marginal likelihood, as

$$|{\sigma_0}^2 H_0(\phi_0) + \tau^2 I|^{-\frac{1}{2}} e^{\{-\frac{1}{2}(\mathbf{y}-\beta_0\mathbf{1}-\beta_1\mathbf{x})^T (\sigma_0^2 H_0(\phi_0)+\tau^2 I)^{-1}(\mathbf{y}-\beta_0\mathbf{1}-\beta_1\mathbf{x})\}} ,$$

$$(10.17)$$

where $\mathbf{x} = (x(\mathbf{s}_1), \ldots, x(\mathbf{s}_n))^T$.

We note analogies with standard Gaussian random effects models, where $Y_{ij} = \beta_0 + \beta_1 x_{ij} + \alpha_i + e_{ij}$ with $\alpha_i \overset{iid}{\sim} N(0, \sigma_\alpha{}^2)$ and $\epsilon_{ij} \overset{iid}{\sim} N(0, \sigma_\epsilon{}^2)$. In this case, replications are needed to identify (separate) the variance components. Because of the dependence between the $\beta_0(s_i)$, replications are not needed in the spatial case, as (10.17) reveals. Also, (10.14) can be interpreted as partitioning the total error in the regression model into "intercept process" error and "pure" error.

The foregoing development immediately suggests how to formulate a spatially varying coefficient model. Suppose we write

$$Y(\mathbf{s}) = \beta_0 + \beta_1 x(\mathbf{s}) + \beta_1(\mathbf{s})x(\mathbf{s}) + \epsilon(\mathbf{s}) . \qquad (10.18)$$

In (10.18), $\beta_1(\mathbf{s})$ is a second-order stationary mean-zero Gaussian process with variance σ_1^2 and correlation function $\rho_1(\cdot; \phi_1)$. Also, let $\tilde{\beta}_1(\mathbf{s}) = \beta_1 + \beta_1(\mathbf{s})$. Now $\beta_1(\mathbf{s})$ can be interpreted as a random spatial adjustment at location $\mathbf{s}$ to the overall slope β_1. Equivalently, $\tilde{\beta}_1(\mathbf{s})$ can be viewed as a random slope process. In effect, we are employing an uncountable dimensional function to explain the relationship between $x(\mathbf{s})$ and $Y(\mathbf{s})$.

Expression (10.18) yields obvious modification of (10.15) and (10.16). In particular, the resulting marginalized likelihood becomes

$$L(\beta_0, \beta_1, \tau^2, {\sigma_1}^2, \phi_1; \mathbf{y}) = |{\sigma_1}^2 D_x H_1(\phi_1)D_x + \tau^2 I|^{-\frac{1}{2}}$$
$$\times e^{\{-\frac{1}{2}(\mathbf{y}-\beta_0\mathbf{1}-\beta_1\mathbf{x})^T ({\sigma_1}^2 D_x H_1(\phi_1)D_x+\tau^2 I)^{-1}(\mathbf{y}-\beta_0\mathbf{1}-\beta_1\mathbf{x})\}} , \qquad (10.19)$$

where D_x is diagonal with $(D_x)_{ii} = x(\mathbf{s}_i)$. With $\mathbf{B}_1 = (\beta_1(\mathbf{s}_1), \ldots, \beta_1(\mathbf{s}_n))^T$ we can sample $f(\mathbf{B}_1|\mathbf{y})$ and $f(\beta_1(\mathbf{s}_{new})|\mathbf{y})$ via composition.

Note that (10.18) provides a heterogeneous, nonstationary process for the data regardless of the choice of covariance function for the $\beta_1(\mathbf{s})$ process, since $\text{Var}(Y(\mathbf{s}) \mid \beta_0, \beta_1, \tau^2, {\sigma_1}^2, \phi_1) = x^2(\mathbf{s})\sigma_1{}^2 + \tau^2$ and $\text{cov}(Y(\mathbf{s}), Y(\mathbf{s}') \mid \beta_0, \beta_1, \tau^2, {\sigma_1}^2, \phi_1) = {\sigma_1}^2 x(\mathbf{s})x(\mathbf{s}')\rho_1(\mathbf{s}-\mathbf{s}'; \phi_1)$. As a result, we observe that in practice, (10.18) is sensible only if we have $x(\mathbf{s}) > 0$. In fact, centering and scaling, which is usually advocated for better behaved model fitting, is inappropriate here. With centered $x(\mathbf{s})$'s we would find the likely untenable behavior that $\text{Var}(Y(\mathbf{s}))$ decreases and then increases in $x(\mathbf{s})$. Worse, for an essentially central $x(\mathbf{s})$ we would find $Y(\mathbf{s})$ essentially independent of $Y(\mathbf{s}')$ for any $\mathbf{s}'$. Also, scaling the $x(\mathbf{s})$'s accomplishes nothing. $\beta_1(\mathbf{s})$ would be inversely rescaled since the model only identifies $\beta_1(\mathbf{s})x(\mathbf{s})$.

This leads to concerns regarding possible approximate collinearity of $\mathbf{x}$, the vector of $x(\mathbf{s}_i)$'s, with the vector $\mathbf{1}$. Expression (10.19) shows that a badly behaved likelihood will arise if $\mathbf{x} \approx c\mathbf{1}$. But, we can reparametrize (10.18) to $Y(\mathbf{s}) = \beta_0' + \beta_1'\tilde{x}(\mathbf{s}) + \beta_1(\mathbf{s})x(\mathbf{s}) + \epsilon(\mathbf{s})$ where $\tilde{x}(\mathbf{s})$ is centered and scaled with obvious definitions for β_0' and β_1'. Now $\tilde{\beta}_1(\mathbf{s}) = \beta_1'/s_x + \beta_1(\mathbf{s})$ where s_x is the sample standard deviation of the $x(\mathbf{s})$'s.

As below (10.17), we can draw an analogy with standard longitudinal linear growth curve modeling, where $Y_{ij} = \beta_0 + \beta_1 x_{ij} + \beta_{1i} x_{ij} + \epsilon_{ij}$, i.e., a random slope for each individual. Also, $U(\mathbf{s})$, the total error in the regression model (10.18) is now partitioned into "slope process" error and "pure" error.

The general specification incorporating both $\beta_0(\mathbf{s})$ and $\beta_1(\mathbf{s})$ would be

$$Y(\mathbf{s}) = \beta_0 + \beta_1 x(\mathbf{s}) + \beta_0(\mathbf{s}) + \beta_1(\mathbf{s})x(\mathbf{s}) + \epsilon(\mathbf{s}). \tag{10.20}$$

Expression (10.20) parallels the usual linear growth curve modeling by introducing both an intercept process and a slope process. The model in (10.20) requires a bivariate process specification in order to determine the joint distribution of $\mathbf{B}_0$ and $\mathbf{B}_1$. It also partitions the total error into intercept process error, slope error, and pure error.

10.2.2 Multivariate spatially varying coefficient models

For the case of a $p \times 1$ multivariate covariate vector $\mathbf{X}(\mathbf{s})$ at location $\mathbf{s}$ where, for convenience, $\mathbf{X}(\mathbf{s})$ includes a 1 as its first entry to accommodate an intercept, we generalize (10.20) to

$$Y(\mathbf{s}) = \mathbf{X}^T(\mathbf{s})\tilde{\boldsymbol{\beta}}(\mathbf{s}) + \epsilon(\mathbf{s}) , \tag{10.21}$$

where $\tilde{\boldsymbol{\beta}}(\mathbf{s})$ is assumed to follow a p-variate spatial process model. With observed locations $\mathbf{s}_1, \mathbf{s}_2, \ldots, \mathbf{s}_n$, let X be $n \times np$ block diagonal having as block for the ith row $\mathbf{X}^T(\mathbf{s}_i)$. Then we can write $\mathbf{Y} = X^T\tilde{\mathbf{B}} + \epsilon$ where $\tilde{\mathbf{B}}$ is $np \times 1$, the concatenated vector of the $\tilde{\boldsymbol{\beta}}(s)$, and $\epsilon \sim N(0, \tau^2 I)$.

In practice, to assume that the component processes of $\tilde{\boldsymbol{\beta}}(\mathbf{s})$ are independent is likely inappropriate. That is, in the simpler case of simple linear regression, negative association between slope and intercept is usually seen. (This is intuitive if one envisions overlaying random lines that are likely relative to a fixed scattergram of data points.) The dramatic improvement in model performance when dependence is incorporated is shown in Example 10.4. To formulate a multivariate Gaussian process for $\tilde{\boldsymbol{\beta}}(\mathbf{s})$ we require the mean and the cross-covariance function. For the former, following Subsection 10.2.1, we take this to be $\boldsymbol{\mu}_\beta = (\beta_1, \ldots, \beta_p)^T$. For the latter we require a valid p-variate choice. In the following paragraphs we work with a separable form (Section 7.1), yielding

$$\tilde{\mathbf{B}} \sim N(\mathbf{1}_{n \times 1} \otimes \boldsymbol{\mu}_\beta , \; H(\phi) \otimes T) . \tag{10.22}$$

If if $\tilde{\mathbf{B}} = \mathbf{B} + \mathbf{1}_{n\times 1} \otimes \boldsymbol{\mu}_\beta$, then we can write (10.21) as

$$Y(\mathbf{s}) = \mathbf{X}^T(\mathbf{s})\boldsymbol{\mu}_\beta + \mathbf{X}^T(\mathbf{s})\boldsymbol{\beta}(\mathbf{s}) + \epsilon(\mathbf{s}) . \qquad (10.23)$$

In (10.23) the total error in the regression model is partitioned into $p+1$ pieces, each with an obvious interpretation. Following Subsection 10.2.1, using (10.21) and (10.22) we can integrate over $\boldsymbol{\beta}$ to obtain

$$L(\boldsymbol{\mu}_\beta, \tau^2, T, \phi; \mathbf{y}) = |X(H(\phi) \otimes T)X^T + \tau^2 I|^{-\frac{1}{2}}$$
$$\times e^{\{-\frac{1}{2}(\mathbf{y}-X(1\otimes\boldsymbol{\mu}_\beta))^T (X(H(\phi)\otimes T)X^T+\tau^2 I)^{-1}(\mathbf{y}-X(1\otimes\boldsymbol{\mu}_\beta))\}} . \qquad (10.24)$$

This apparently daunting form still involves only $n \times n$ matrices.

The Bayesian model is completed with a prior $p\left(\boldsymbol{\mu}_\beta, \tau^2, T, \phi\right)$, which we assume to take the product form $p(\boldsymbol{\mu}_\beta)p(\tau^2)p(T)p(\phi)$. Below, these components will be normal, inverse gamma, inverse Wishart, and gamma, respectively.

With regard to prediction, $p(\tilde{\mathbf{B}}|\mathbf{y})$ can be sampled one for one with the posterior samples from $f\left(\boldsymbol{\mu}_\beta, \tau^2, T, \phi|\mathbf{y}\right)$ using $f\left(\tilde{\mathbf{B}}|\boldsymbol{\mu}_\beta, \tau^2, T, \phi, y\right)$, which is $N\left(A\mathbf{a}, A\right)$ where $A = (X^T X/\tau^2 + H^{-1}(\phi) \otimes T^{-1})^{-1}$ and $\mathbf{a} = X^T\mathbf{y}/\tau^2 + \left(H^{-1}(\phi) \otimes T^{-1}\right)\left(1 \otimes \boldsymbol{\mu}_\beta\right)$. Here A is $np \times np$ but, for sampling $\tilde{\boldsymbol{\beta}}$, only a Cholesky decomposition of A is needed, and only for the retained posterior samples. Prediction at a new location, say, $\mathbf{s}_{new}$, requires samples from $f\left(\tilde{\boldsymbol{\beta}}\left(\mathbf{s}_{new}\right)|\tilde{\mathbf{B}}, \boldsymbol{\mu}_\beta, \tau^2, T, \phi\right)$. Defining $\mathbf{h}_{new}(\phi)$ to be the $n \times 1$ vector with ith row entry $\rho\left(\mathbf{s}_i - \mathbf{s}_{new}; \phi\right)$, this distribution is normal with mean

$$\boldsymbol{\mu}_\beta + \left(\mathbf{h}_{new}^T(\phi) \otimes T\right)\left(H^{-1}(\phi) \otimes T^{-1}\right)\left(\tilde{\mathbf{B}} - \mathbf{1}_{nx1} \otimes \boldsymbol{\mu}_\beta\right)$$
$$= \boldsymbol{\mu}_\beta + \left(\mathbf{h}_{new}^T(\phi) H^{-1}(\phi) \otimes I\right)\left(\tilde{\mathbf{B}} - \mathbf{1}_{nx1} \otimes \boldsymbol{\mu}_\beta\right) ,$$

and covariance matrix $T - \left(\mathbf{h}_{new}^T(\phi) \otimes T\right)\left(H^{-1}(\phi) \otimes T^{-1}\right)\left(\mathbf{h}_{new}(\phi) \otimes T\right)$ $= \left(I - \mathbf{h}_{new}^T(\phi) H^{-1}(\phi) \mathbf{h}_{new}(\phi)\right) T$. Finally, the predictive distribution for $Y\left(\mathbf{s}_{new}\right)$, namely $f\left(Y\left(\mathbf{s}_{new}\right) | \mathbf{y}\right)$, is sampled by composition, as usual.

We conclude this subsection by noting an extension of (10.21) when we have repeated measurements at location s. That is, suppose we have

$$Y(\mathbf{s}, l) = \mathbf{X}^T(\mathbf{s}, l)\boldsymbol{\beta}(\mathbf{s}) + \epsilon(\mathbf{s}, l) , \qquad (10.25)$$

where $l = 1, \ldots, L_\mathbf{s}$ with $L_\mathbf{s}$ the number of measurements at $\mathbf{s}$ and the $\epsilon(\mathbf{s}, l)$ still white noise. As an illustration, in the real estate context, $\mathbf{s}$ might denote the location for an apartment block and l might index apartments in this block that have sold, with the lth apartment having characteristics $\mathbf{X}(\mathbf{s}, l)$. Suppose further that $\mathbf{Z}(\mathbf{s})$ denotes an $r \times 1$ vector of site-level characteristics. For an apartment block, these characteristics might include amenities provided or distance to the central business district. Then (10.25) can be extended to a multilevel model in the sense of Goldstein (1995) or

Raudenbush and Bryk (2002). In particular we can write

$$\boldsymbol{\beta}\left(s\right)=\begin{pmatrix}\mathbf{Z}^{T}\left(\mathbf{s}\right)\boldsymbol{\gamma}_{1}\\\vdots\\\mathbf{Z}^{T}\left(\mathbf{s}\right)\boldsymbol{\gamma}_{p}\end{pmatrix}+\mathbf{w}\left(\mathbf{s}\right)\ .\tag{10.26}$$

In (10.26), $\boldsymbol{\gamma}_{j}, j = 1, \ldots, p$, is an $r \times 1$ vector associated with $\tilde{\beta}_{j}\left(\mathbf{s}\right)$, and $\mathbf{w}\left(\mathbf{s}\right)$ is a mean-zero multivariate Gaussian spatial process, for example, as above. In (10.26), if the $\mathbf{w}\left(\mathbf{s}\right)$ were independent we would have a usual multilevel model specification. In the case where $\mathbf{Z}\left(\mathbf{s}\right)$ is a scalar capturing just an intercept, we return to the initial model of this subsection.

10.2.3 Spatiotemporal data

A natural extension of the modeling of the previous sections is to the case where we have data correlated at spatial locations across time. If, as in Section 8.2, we assume that time is discretized to a finite set of equally spaced points on a scale, we can conceptualize a time series of spatial processes that are observed only at the spatial locations $\mathbf{s}_{1}, \ldots, \mathbf{s}_{n}$.

Adopting a general notation that parallels (10.20), let

$$Y(\mathbf{s},t) = \mathbf{X}^{T}\left(\mathbf{s},t\right)\tilde{\boldsymbol{\beta}}\left(\mathbf{s},t\right)+\epsilon\left(\mathbf{s},t\right)\ ,\ \ t = 1, 2, \ldots, M\ .\tag{10.27}$$

That is, we introduce spatiotemporally varying intercepts and spatiotemporally varying slopes. Alternatively, if we write $\tilde{\boldsymbol{\beta}}\left(\mathbf{s},t\right) = \boldsymbol{\beta}\left(\mathbf{s},t\right)+\boldsymbol{\mu}_{\beta}$, we are partitioning the total error into $p + 1$ spatiotemporal intercept pieces including $\epsilon\left(\mathbf{s},t\right)$, each with an obvious interpretation. So we continue to assume that $\epsilon\left(\mathbf{s},t\right) \overset{iid}{\sim} N\left(0, \tau^{2}\right)$, but need to specify a model for $\tilde{\boldsymbol{\beta}}\left(\mathbf{s},t\right)$. Regardless, (10.27) defines a nonstationary process having moments $E(Y(\mathbf{s},t)) = \mathbf{X}^{T}(\mathbf{s},t)\tilde{\beta}(s,t)$, $Var(Y(\mathbf{s},t)) = \mathbf{X}^{T}(\mathbf{s},t)\Sigma_{\tilde{\beta}_{(,t)}}\mathbf{X}(\mathbf{s},t)+\tau^{2}$, and $Cov(Y(\mathbf{s},t),Y(\mathbf{s}',t')) = \mathbf{X}^{T}(\mathbf{s},t)\Sigma_{\tilde{\beta}(\mathbf{s},t),\tilde{\beta}(\mathbf{s}',t')}\mathbf{X}(\mathbf{s}',t')$.

Section 8.4 handled (10.27) using a dynamic model. Here we consider four alternative specifications for $\boldsymbol{\beta}\left(\mathbf{s},t\right)$. Paralleling the customary assumption from longitudinal data modeling (where the time series are usually short), we could set

- **Model 1:** $\boldsymbol{\beta}(\mathbf{s},t) = \boldsymbol{\beta}(\mathbf{s})$, where $\boldsymbol{\beta}(\mathbf{s})$ is modeled as in the previous sections. This model can be viewed as a local linear growth curve model.

- **Model 2:** $\boldsymbol{\beta}(\mathbf{s},t) = \boldsymbol{\beta}(\mathbf{s}) + \boldsymbol{\alpha}(t)$, where $\boldsymbol{\beta}(\mathbf{s})$ is again as in Model 1. In modeling $\boldsymbol{\alpha}\left(t\right)$, two possibilities are (i) treat the $\alpha_{k}\left(t\right)$ as time dummy variables, taking this set of pM variables to be a priori independent and identically distributed; and (ii) model the $\boldsymbol{\alpha}\left(t\right)$ as a random walk or autoregressive process. The components could be assumed independent across k, but for greater generality, we take them to be dependent, using a separable form that replaces $\mathbf{s}$ with t and takes ρ to be a valid correlation function in just one dimension.

- **Model 3:** $\boldsymbol{\beta}(\mathbf{s}, t) = \boldsymbol{\beta}^{(t)}(\mathbf{s})$, i.e., we have spatially varying coefficient processes nested within time. This model is an analogue of the nested effects areal unit specification in Waller et al. (1997); see also Gelfand, Ecker et al. (2003). The processes are assumed independent across t (essentially dummy time processes) and permit temporal evolution of the coefficient process. Following Subsection 10.2.2, the process $\boldsymbol{\beta}^{(t)}(\mathbf{s})$ would be mean-zero, second-order stationary Gaussian with cross-covariance specification at time t, $C^{(t)}(\mathbf{s}, \mathbf{s}')$ where $\left(C^{(t)}(\mathbf{s}, \mathbf{s}')\right)_{lm} = \rho\left(\mathbf{s} - \mathbf{s}'; \phi^{(t)}\right) \tau_{lm}^{(t)}$. We have specified Model 3 with a common $\boldsymbol{\mu}_\beta$ across time. This enables some comparability with the other models we have proposed. However, we can increase flexibility by replacing $\boldsymbol{\mu}_\beta$ with $\boldsymbol{\mu}_\beta^{(t)}$.

- **Model 4:** For $\rho^{(1)}$ a valid two-dimensional correlation function, $\rho^{(2)}$ a valid one-dimensional choice, and T positive definite symmetric, $\boldsymbol{\beta}(\mathbf{s}, t)$ such that $\Sigma_{\left[\boldsymbol{\beta}(\mathbf{s}, t), \boldsymbol{\beta}(\mathbf{s}', t')\right]} = \rho^{(1)}\left(\mathbf{s} - \mathbf{s}'; \phi\right) \rho^{(2)}\left(t - t'; \gamma\right) T$. This model proposes a separable covariance specification in space and time, as in Section 8.2. Here $\rho^{(1)}$ obtains spatial association as in earlier subsections that is attenuated across time by $\rho^{(2)}$. The resulting covariance matrix for the full vector $\boldsymbol{\beta}$, blocked by site and time within site has the convenient form $H_2(\gamma) \otimes H_1(\phi) \otimes T$.

In each of the above models we can marginalize over $\boldsymbol{\beta}(\mathbf{s}, t)$ as we did earlier in this section. Depending upon the model it may be more computationally convenient to block the data by site or by time. We omit the details and notice only that, with n sites and T time points, the resulting likelihood will involve the determinant and inverse of an $nT \times nT$ matrix (typically a large matrix; see Appendix Section A.6).

Note that all of the foregoing modeling can be applied to the case of cross-sectional data where the set of observed locations varies with t. This is the case, for instance, with our real estate data. We only observe a selling price at the time of a transaction. With n_t locations in year t, the likelihood for all but Model 3 will involve a $\sum n_t \times \sum n_t$ matrix.

Example 10.4 *(Baton Rouge housing prices).* We analyze a sample from a database of real estate transactions in Baton Rouge, LA, during the eight-year period 1985–1992. In particular, we focus on modeling the log selling price of single-family homes. In real estate modeling it is customary to work with log selling price in order to achieve better approximate normality. A range of house characteristics are available. We use four of the most common choices: age of house, square feet of living area, square feet of other area (e.g., garages, carports, storage), and number of bathrooms. For the static spatial case, a sample of 237 transactions was drawn from 1992.

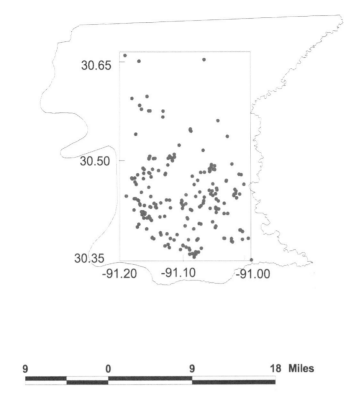

9   0   9   18 Miles

Figure 10.3 *Locations sampled within the parish of Baton Rouge for the static spatial models.*

Model	Fit	Variance penalty	D_K
Five-dimensional	42.21	36.01	78.22
Three-dimensional (best)	61.38	47.83	109.21
Two-dimensional (best)	69.87	46.24	116.11
Independent process	94.36	59.34	153.70

Table 10.4 *Values of posterior predictive model choice criterion (over all models).*

Figure 10.3 shows the parish of Baton Rouge and the locations contained in an encompassing rectangle within the parish.

We fit a variety of models, where in all cases the correlation function is from the Matérn class. We used priors that are fairly noninformative and comparable across models as sensible. First, we started with a spatially varying intercept and one spatially varying slope coefficient (the remaining coefficients do not vary), requiring a bivariate process model. There are four such models, and using D_K, the Gelfand and Ghosh (1998) criterion (4.13), the model with a spatially varying living area coefficient emerges as best. Next, we introduced two spatially varying slope coefficient processes along with a spatially varying intercept, requiring a trivariate process model. There are six models here; the one with spatially varying age and living area is best. Finally, we allowed five spatially varying processes: an intercept and all four coefficients, using a five-dimensional process model. We also fit a model with five independent processes. From Table 10.4 the five-dimensional dependent process model is far superior and the independent process model is a dismal last, supporting our earlier intuition.

The prior specification used for the five-dimensional dependent process model is as follows. We take vague $N\left(\mathbf{0}, 10^5 I\right)$ for $\boldsymbol{\mu}_\beta$, a five-dimensional inverse Wishart, $IW\left(5, Diag\left(0.001\right)\right)$, for T, and an inverse gamma $IG\left(2,1\right)$ for τ^2 (mean 1, infinite variance). For the Matérn correlation function parameters ϕ and ν we assume gamma priors $G\left(2, 0.1\right)$ (mean 20 and variance 200). For all the models three parallel chains were run to assess convergence. Satisfactory mixing was obtained within 3000 iterations for all the models; 2000 further samples were generated and retained for posterior inference.

The resulting posterior inference summary is provided in Table 10.5. We note a significant negative overall age coefficient with significant positive overall coefficients for the other three covariates, as expected. The contribution to spatial variability from the components of $\boldsymbol{\beta}$ is captured through the diagonal elements of the T matrix scaled by the corresponding covariates following the discussion at the end of Subsection 10.2.1. We see that the spatial intercept process contributes most to the error variability with, perhaps surprisingly, the "bathrooms" process second. Clearly spatial variability overwhelms the pure error variability τ^2, showing the importance of the spatial model. The dependence between the processes is evident in the posterior correlation between the components. We find the anticipated negative association between the intercept process and the slope processes (apart from that with the "other area" process). Under the Matérn correlation function, by inverting $\rho\left(\cdot; \phi\right) = 0.05$ for a given value of the decay parameter γ and the smoothing parameter ν, we obtain the range, i.e., the distance beyond which spatial association becomes negligible. Posterior samples of $\left(\gamma, \nu\right)$ produce posterior samples for the range. The resulting posterior median is roughly 4 km over a somewhat sprawling parish that is roughly 22 km $\times$ 33 km. The smoothness parameter suggests processes with mean square differentiable realizations $\left(\nu > 1\right)$. Contour plots of the posterior mean spatial surfaces for each of the processes (not shown) are quite different.

Turning to the dynamic models proposed in Subsection 10.2.3, from the

Parameter	2.5%	50%	97.5%
β_0 (intercept)	9.908	9.917	9.928
β_1 (age)	−0.008	−0.005	−0.002
β_2 (living area)	0.283	0.341	0.401
β_3 (other area)	0.133	0.313	0.497
β_4 (bathrooms)	0.183	0.292	0.401
T_{11}	0.167	0.322	0.514
$\bar{x}_1^2 T_{22}$	0.029	0.046	0.063
$\bar{x}_2^2 T_{33}$	0.013	0.028	0.047
$\bar{x}_3^2 T_{44}$	0.034	0.045	0.066
$\bar{x}_4^2 T_{55}$	0.151	0.183	0.232
$T_{12}/\sqrt{T_{11}T_{22}}$	−0.219	−0.203	−0.184
$T_{13}/\sqrt{T_{11}T_{33}}$	−0.205	−0.186	−0.167
$T_{14}/\sqrt{T_{11}T_{44}}$	0.213	0.234	0.257
$T_{15}/\sqrt{T_{11}T_{55}}$	−0.647	−0.583	−0.534
$T_{23}/\sqrt{T_{22}T_{33}}$	−0.008	0.011	0.030
$T_{24}/\sqrt{T_{22}T_{44}}$	0.061	0.077	0.098
$T_{25}/\sqrt{T_{22}T_{55}}$	−0.013	0.018	0.054
$T_{34}/\sqrt{T_{33}T_{44}}$	−0.885	−0.839	−0.789
$T_{35}/\sqrt{T_{33}T_{55}}$	−0.614	−0.560	−0.507
$T_{45}/\sqrt{T_{44}T_{55}}$	0.173	0.232	0.301
ϕ (decay)	0.51	1.14	2.32
ν (smoothness)	0.91	1.47	2.87
range (in km)	2.05	4.17	9.32
τ^2	0.033	0.049	0.077

Table 10.5 *Inference summary for the five-dimensional multivariate spatially varying coefficients model.*

Baton Rouge database we drew a sample of 120 transactions at distinct spatial locations for the years 1989, 1990, 1991, and 1992. We compare Models 1–4. In particular, we have two versions of Model 2; 2a has the $\boldsymbol{\alpha}(t)$ as four i.i.d. time dummies, while 2b uses the multivariate temporal process model for $\boldsymbol{\alpha}(t)$. We also have two versions of Model 3; 3a has a common $\boldsymbol{\mu}_\beta$ across t, while 3b uses $\boldsymbol{\mu}_\beta^{(t)}$. In all cases the five-dimensional spatially varying coefficient model for β's was employed. Table 10.6 shows the results. Model 3, where space is nested within time, turns out to be the best with Model 4 following closely behind. We omit the posterior inference summary for Model 3b, noting only that the overall coefficients $\left(\boldsymbol{\mu}_\beta^{(t)}\right)$ do

Model	Independent process			Dependent process		
	G	P	D_∞	G	P	D_∞
1	88.58	56.15	144.73	54.54	29.11	83.65
2a	77.79	50.65	128.44	47.92	26.95	74.87
2b	74.68	50.38	125.06	43.38	29.10	72.48
3a	59.46	48.55	108.01	43.74	20.63	64.37
3b	57.09	48.41	105.50	42.35	21.04	63.39
4	53.55	52.98	106.53	37.84	26.47	64.31

Table 10.6 *Model choice criteria for various spatiotemporal process models.*

not change much over time. However, there is some indication that spatial range is changing over time. ∎

10.2.4 Generalized linear model setting

We briefly consider a generalized linear model version of (10.21), replacing the Gaussian first stage with

$$f\left(y\left(\mathbf{s}_i\right) \mid \theta\left(\mathbf{s}_i\right)\right) = h\left(y\left(\mathbf{s}_i\right)\right) \exp\left(\theta\left(\mathbf{s}_i\right) y\left(\mathbf{s}_i\right) - b\left(\theta\left(\mathbf{s}_i\right)\right)\right) , \qquad (10.28)$$

where, using a canonical link, $\theta\left(\mathbf{s}_i\right) = \mathbf{X}^T\left(\mathbf{s}_i\right) \tilde{\boldsymbol{\beta}}\left(\mathbf{s}_i\right)$. In (10.28) we could include a dispersion parameter with little additional complication.

The resulting first-stage likelihood becomes

$$L\left(\tilde{\boldsymbol{\beta}}; \mathbf{y}\right) = \exp\left\{\sum y\left(\mathbf{s}_i\right) \mathbf{X}^T\left(\mathbf{s}_i\right) \tilde{\boldsymbol{\beta}}\left(\mathbf{s}_i\right) - b\left(\mathbf{X}^T\left(\mathbf{s}_i\right) \tilde{\boldsymbol{\beta}}\left(\mathbf{s}_i\right)\right)\right\} . \qquad (10.29)$$

Taking the prior on $\tilde{\boldsymbol{\beta}}$ in (10.22), the Bayesian model is completely specified with a prior on on ϕ, T and $\boldsymbol{\mu}_\beta$.

This model can be fit using a conceptually straightforward Gibbs sampling algorithm, which updates the components of $\boldsymbol{\mu}_\beta$ and $\tilde{\boldsymbol{\beta}}$ using adaptive rejection sampling. With an inverse Wishart prior on T, the resulting full conditional of T is again inverse Wishart. Updating ϕ is usually very awkward because it enters in the Kronecker form in (10.22). Slice sampling is not available here since we cannot marginalize over the spatial effects; Metropolis updates are difficult to design but offer perhaps the best possibility. Also problematic is the repeated componentwise updating of $\tilde{\boldsymbol{\beta}}$. This hierarchically centered parametrization (Gelfand, Sahu, and Carlin, 1995, 1996) is preferable to working with $\boldsymbol{\mu}_\beta$ and $\boldsymbol{\beta}$, but in our experience the algorithm still exhibits serious autocorrelation problems.

10.3 Spatial CDFs

In this section, we review the essentials of spatial cumulative distribution functions (SCDFs), including a hierarchical modeling approach for inference. We then extend the basic definition to allow covariate weighting of the SCDF estimate, as well as versions arising under a bivariate random process.

10.3.1 Basic definitions and motivating data sets

Suppose that $X(\mathbf{s})$ is the log-ozone concentration at location $\mathbf{s}$ over a particular time period. Thinking of $X(\mathbf{s})$, $\mathbf{s} \in D$ as a spatial process, we might wish to find the proportion of area in D that has ozone concentration below some level w (say, a level above which exposure is considered to be unhealthful). This proportion is the random variable,

$$F(w) = Pr\left[\mathbf{s} \in D : X(\mathbf{s}) \le w\right] = \frac{1}{|D|} \int_D Z_w(\mathbf{s})d\mathbf{s}\,, \qquad (10.30)$$

where $|D|$ is the area of D, and $Z_w(\mathbf{s}) = 1$ if $X(\mathbf{s}) \le w$, and 0 otherwise. Since $X(\mathbf{s})$, $\mathbf{s} \in D$ is random, (10.30) is a random function of $w \in \Re$ that increases from 0 to 1 and is right-continuous. Thus while $F(w)$ is not the usual cumulative distribution function (CDF) of X at $\mathbf{s}$ (which would be given by $Pr[X(\mathbf{s}) \le x]$, and is not random), it does have all the properties of a CDF, and so is referred to as the *spatial cumulative distribution function*, or SCDF. For a constant mean stationary process, all $X(\mathbf{s})$ have the same marginal distribution, whence $E[F(w)] = Pr\left(X(\mathbf{s}) \le w\right)$. It is also easy to show that $Var[F(w)] = \frac{1}{|D|^2} \int_D \int_D Pr\left(X(\mathbf{s}) \le w, X(\mathbf{s}') \le w\right) d\mathbf{s}d\mathbf{s}'$ $-[Pr\left(X(\mathbf{s}) \le w\right)]^2$. Overton (1989) introduced the idea of an SCDF, and used it to analyze data from the National Surface Water Surveys. Lahiri et al. (1999) developed a subsampling method that provides (among other things) large-sample prediction bands for the SCDF, which they show to be useful in assessing the foliage condition of red maple trees in the state of Maine.

The *empirical* SCDF based upon data $\mathbf{X}_s = (X(\mathbf{s}_1), \ldots, X(\mathbf{s}_n))'$ at w is the proportion of the $X(\mathbf{s}_i)$ that take values less than or equal to w. Large-sample investigation of the behavior of the empirical SCDF requires care to define the appropriate asymptotics; see Lahiri et al. (1999) and Zhu, Lahiri, and Cressie (2002) for details. When n is not large, as is the case in our applications, the empirical SCDF may become less attractive. Stronger inference can be achieved if one is willing to make stronger distributional assumptions regarding the process $X(\mathbf{s})$. For instance, if $X(\mathbf{s})$ is assumed to be a Gaussian process, the joint distribution of $\mathbf{X}_s$ is multivariate normal. Given a suitable prior specification, a Bayesian framework provides the predictive distribution of $F(w)$ given $X(\mathbf{s})$.

Though (10.30) can be studied analytically, it is difficult to work with in practice. However, approximation of (10.30) via Monte Carlo integration is natural (and may be more convenient than creating a grid of points over D), i.e., replacing $F(w)$ by

$$\widehat{F}(w) = \frac{1}{L} \sum_{\ell=1}^{L} Z_w(\tilde{\mathbf{s}}_\ell) , \qquad (10.31)$$

where the $\tilde{\mathbf{s}}_\ell$ are chosen randomly in D, and $Z_w(\tilde{\mathbf{s}}_\ell) = 1$ if $X(\tilde{\mathbf{s}}_\ell) \leq w$, and 0 otherwise. Suppose we seek a realization of $\widehat{F}(w)$ from the predictive distribution of $\widehat{F}(w)$ given $\mathbf{X}_s$. In Section 6.1 we showed how to sample from the predictive distribution $p(\mathbf{X}_{\tilde{s}} \mid \mathbf{X}_s)$ for $\mathbf{X}_{\tilde{s}}$ arising from new locations $\tilde{\mathbf{s}} = (\tilde{\mathbf{s}}_1, \dots, \tilde{\mathbf{s}}_L)'$. In fact, samples $\{\mathbf{X}_{\tilde{s}}^{(g)}, g = 1, \dots, G\}$ from the posterior predictive distribution,

$$p(\mathbf{X}_{\tilde{s}} \mid \mathbf{X}_s) = \int p(\mathbf{X}_{\tilde{s}} \mid \mathbf{X}_s, \boldsymbol{\beta}, \boldsymbol{\theta}) p(\boldsymbol{\beta}, \boldsymbol{\theta} \mid \mathbf{X}_s) d\boldsymbol{\beta} d\boldsymbol{\theta} ,$$

may be obtained one for one from posterior samples by composition.

The predictive distribution of (10.30) can be sampled at a given w by obtaining $X(\tilde{\mathbf{s}}_\ell)$ using the above algorithm, hence $Z_w(\tilde{\mathbf{s}}_\ell)$, and then calculating $\widehat{F}(w)$ using (10.31). Since interest is in the entire function $F(w)$, we would seek realizations of the approximate function $\widehat{F}(w)$. These are most easily obtained, up to an interpolation, using a grid of w values $\{w_1 < \cdots < w_k < \cdots < w_K\}$, whence each $(\boldsymbol{\beta}^{(g)}, \boldsymbol{\theta}^{(g)})$ gives a realization at grid point w_k,

$$\widehat{F}^{(g)}(w_k) = \frac{1}{L} \sum_{\ell=1}^{L} Z_{w_k}^{(g)}(\tilde{\mathbf{s}}_\ell) , \qquad (10.32)$$

where now $Z_{w_k}^{(g)}(\tilde{\mathbf{s}}_\ell) = 1$ if $X^{(g)}(\tilde{\mathbf{s}}_\ell) \leq w_k$, and 0 otherwise. Handcock (1999) describes a similar Monte Carlo Bayesian approach to estimating SCDFs in his discussion of Lahiri et al. (1999).

Expression (10.32) suggests placing all of our $\widehat{F}^{(g)}(w_k)$ values in a $K \times G$ matrix for easy summarization. For example, a histogram of all the $\widehat{F}^{(g)}(w_k)$ in a particular row (i.e., for a given grid point w_k) provides an estimate of the predictive distribution of $\widehat{F}(w_k)$. On the other hand, each column (i.e., for a given Gibbs draw g) provides (again up to, say, linear interpolation) an approximate draw from the predictive distribution of the SCDF. Hence, averaging these columns provides, with interpolation, essentially the posterior predictive mean for F and can be taken as an *estimated* SCDF. But also, each draw from the predictive distribution of the SCDF can be inverted to obtain any quantile of interest (e.g., the median exposure $d_{.50}^{(g)}$). A histogram of these inverted values in turn provides an estimate of the posterior distribution of this quantile (in this case, $\widehat{p}(d_{.50}|\mathbf{X}_s)$). While this algorithm provides general inference for SCDFs, for most data sets it will be computationally very demanding, since a large L will be required to make (10.32) sufficiently accurate.

Our interest in this methodology is motivated by two environmental data sets; we describe both here but only present the inference for the second. The first is the Atlanta eight-hour maximum ozone data, which exemplifies the case of an air pollution variable measured at points, with a demographic covariate measured at a block level. Recall that its first component is a collection of ambient ozone levels in the Atlanta, GA, metropolitan area, as reported by Tolbert et al. (2000). Ozone measurements X_{itr} are available at between 8 and 10 fixed monitoring sites i for day t of year r, where $t = 1, \ldots, 92$ (the summer days from June 1 through August 31) and $r = 1, 2, 3$, corresponding to years 1993, 1994, and 1995. The reader may wish to flip back to Figure 1.3, which shows the 8-hour daily maximum ozone measurements in parts per million at the 10 monitoring sites for one of the days (July 15, 1995). This figure also shows the boundaries of the 162 zip codes in the Atlanta metropolitan area, with the 36 zips falling within the city of Atlanta encircled by the darker boundary on the map. An environmental justice assessment of exposure to potentially harmful levels of ozone would be clarified by examination of the predictive distribution of a *weighted* SCDF that uses the racial makeups of these city zips as the weights. This requires generalizing our SCDF simulation in (10.32) to accommodate covariate weighting in the presence of misalignment between the response variable (at point-referenced level) and the covariate (at areal-unit level).

SCDFs adjusted with point-level covariates present similar challenges. Consider the spatial data setting of Figure 10.4, recently presented and analyzed by Gelfand, Schmidt, and Sirmans (2002). These are the locations of several air pollutant monitoring sites in central and southern California, all of which measure ozone, carbon monoxide, nitric oxide (NO), and nitrogen dioxide (NO_2). For a given day, suppose we wish to compute an SCDF for the log of the daily median NO exposure adjusted for the log of the daily median NO_2 level (since the health effects of exposure to high levels of one pollutant may be exacerbated by further exposure to high levels of the other). Here the data are all point level, so that Bayesian kriging methods of the sort described above may be used. However, we must still tackle the problem of *bivariate* kriging (for both NO and NO_2) in a computationally demanding setting (say, to the $L = 500$ randomly selected points shown as dots in Figure 10.4). In some settings, we must also resolve the misalignment in the data itself, which arises when NO or NO_2 values are missing at some of the source sites.

Figure 10.4 *Locations of 67 NO and NO$_2$ monitoring sites, California air quality data; 500 randomly selected target locations are also shown as dots.*

10.3.2 Derived-process spatial CDFs

Point- versus block-level spatial CDFs

The spatial CDF in (10.30) is customarily referred to as *the* SCDF associated with the spatial process $X(\mathbf{s})$. In fact, we can formulate many other useful SCDFs under this process. We proceed to elaborate choices of possible interest.

Suppose for instance that our data arrive at areal unit level, i.e., we observe $X(B_j)$, $j = 1, \ldots, J$ such that the B_j are disjoint with union D, the entire study region. Let $Z_w(B_j) = 1$ if $X(B_j) \leq w$, and 0 otherwise. Then

$$\widetilde{F}(w) = \frac{1}{|D|} \sum_{j=1}^{J} |B_j| \, Z_w(B_j) \qquad (10.33)$$

again has the properties of a CDF and thus can also be interpreted as a spatial CDF. In fact, this CDF is a step function recording the proportion of the area of D that (at block-level resolution) lies below w. Suppose in fact that the $X(B_j)$ can be viewed as block averages of the process

$X(\mathbf{s})$, i.e., $X(B_j) = \frac{1}{|B_j|} \int_{B_j} X(\mathbf{s}) ds$. Then (10.30) and (10.33) can be compared: write (10.30) as $\frac{1}{|D|} \sum_j |B_j| \left[\frac{1}{|B_j|} \int_{B_j} I(X(\mathbf{s}) \le w) ds \right]$ and (10.33) as $\frac{1}{|D|} \sum_j |B_j| I \left[\left(\frac{1}{|B_j|} \int_{B_j} X(\mathbf{s}) ds \right) \le w \right]$. Interpreting $\mathbf{s}$ to have a uniform distribution on B_j, the former is $\frac{1}{|D|} \sum_j |B_j| E_{B_j} [I(X(\mathbf{s}) \le w)]$ while the latter is $\frac{1}{|D|} \sum_j |B_j| I [E_{B_j}(X(\mathbf{s})) \le w]$. In fact, if $X(\mathbf{s})$ is stationary, while $E[F(w)] = P(X(\mathbf{s}) \le w)$, $E[\widetilde{F}(w)] = \frac{1}{|D|} \sum_j |B_j| P(X(B_j) \le w)$. For a Gaussian process, under weak conditions $X(B_j)$ is normally distributed with mean $E[X(\mathbf{s})]$ and variance $\frac{1}{|B_j|^2} \int_{B_j} \int_{B_j} c(\mathbf{s} - \mathbf{s}'; \boldsymbol{\theta}) ds ds'$, so $E[\widetilde{F}(w)]$ can be obtained explicitly. Note also that since $\frac{1}{|B_j|} \int_{B_j} I(X(\mathbf{s}) \le w) ds$ is the customary spatial CDF for region B_j, then by the alternate expression for (10.30) above, $F(w)$ is an areally weighted average of *local* SCDFs.

Thus (10.30) and (10.33) differ, but (10.33) should neither be viewed as "incorrect" nor as an approximation to (10.30). Rather, it is an alternative SCDF derived under the $X(\mathbf{s})$ process. Moreover, if only the $X(B_j)$ have been observed, it is arguably the most sensible empirical choice. Indeed, the Multiscale Advanced Raster Map (MARMAP) analysis system project (www.stat.psu.edu/~gpp/marmap_system_partnership.htm) is designed to work with "empirical cell intensity surfaces" (i.e., the tiled surface of the $X(B_j)$'s over D) and calculates the "upper level surfaces" (variants of (10.33)) for description and inference regarding multicategorical maps and cellular surfaces.

Next we seek to introduce covariate weights to the spatial CDF, as motivated in Subsection 10.3.1. For a nonnegative function $r(\mathbf{s})$ that is integrable over D, define the SCDF associated with $X(\mathbf{s})$ weighted by r as

$$F_r(w) = \frac{\int_D r(\mathbf{s}) Z_w(\mathbf{s}) ds}{\int_D r(\mathbf{s}) ds} . \qquad (10.34)$$

Evidently (10.34) satisfies the properties of a CDF and generalizes (10.30) (i.e., (10.30) is restored by taking $r(\mathbf{s}) \equiv 1$). But as (10.30) suggests expectation with respect to a uniform density for $\mathbf{s}$ over D, (10.34) suggests expectation with respect to the density $r(\mathbf{s}) / \int_D r(\mathbf{s}) ds$. Under a stationary process, $E[F(w)] = P(X(\mathbf{s}) \le w)$ and $Var[F(w)]$ is

$$\frac{1}{(\int_D r(\mathbf{s}) ds)^2} \int_D \int_D r(\mathbf{s}) r(\mathbf{s}') P(X(\mathbf{s}) \le w, X(\mathbf{s}') \le w) ds ds'$$
$$-[P(X(\mathbf{s}) \le w)]^2 .$$

There is an empirical SCDF associated with (10.34) that extends the empirical SCDF in Subsection 10.3.1 using weights $r(\mathbf{s}_i) / \sum_i r(\mathbf{s}_i)$ rather than $1/n$. This random variable is mentioned in Lahiri et al. (1999, p. 87). Following Subsection 10.3.1, we adopt a Bayesian approach and seek a predictive distribution for $F_r(w)$ given $\mathbf{X}_s$. This is facilitated by Monte

Carlo integration of (10.34), i.e.,

$$\widehat{F}_r(w) = \frac{\sum_{\ell=1}^{L} r(\mathbf{s}_\ell) Z_w(\mathbf{s}_\ell)}{\sum_{\ell=1}^{L} r(\mathbf{s}_\ell)} . \tag{10.35}$$

Covariate weighted SCDFs for misaligned data

In the environmental justice application described in Subsection 10.3.1, the covariate is only available (indeed, only meaningful) at an areal level, i.e., we observe only the population density associated with B_j. How can we construct a covariate weighted SCDF in this case? Suppose we make the assignment $r(\mathbf{s}) = r_j$ for all $\mathbf{s} \in B_j$, i.e., that the density surface is constant over the areal unit (so that $r_j |B_j|$ is the observed population density for B_j). Inserting this into (10.34) we obtain

$$F_r^*(w) = \frac{\sum_{j=1}^{J} r_j |B_j| \left[\frac{1}{|B_j|} \int_{B_j} Z_w(\mathbf{s}) ds \right]}{\sum_{j=1}^{J} r_j |B_j|} . \tag{10.36}$$

As a special case of (10.34), (10.36) again satisfies the properties of a CDF and again has mean $P(X(\mathbf{s}) \le w)$. Moreover, as below (10.33), the bracketed expression in (10.36) is the spatial CDF associated with $X(\mathbf{s})$ restricted to B_j. Monte Carlo integration applied to (10.36) can use the same set of $\mathbf{s}_\ell$'s chosen randomly over D as in Subsection 10.3.1 or as in (10.35). In fact (10.35) becomes

$$\widehat{F}_r(w) = \frac{\sum_{j=1}^{J} r_j L_j \left[\frac{1}{L_j} \sum_{\mathbf{s}_\ell \in B_j} Z_w(\mathbf{s}_\ell) \right]}{\sum_{j=1}^{J} r_j L_j} , \tag{10.37}$$

where L_j is the number of $\mathbf{s}_\ell$ falling in B_j. Equation (10.36) suggests the alternative expression,

$$\widehat{F}_r^*(w) = \frac{\sum_{j=1}^{J} r_j |B_j| \left[\frac{1}{L_j} \sum_{\mathbf{s}_\ell \in B_j} Z_w(\mathbf{s}_\ell) \right]}{\sum_{j=1}^{J} r_j |B_j|} . \tag{10.38}$$

Expression (10.38) may be preferable to (10.37), since it uses the exact $|B_j|$ rather than the random L_j.

10.3.3 Randomly weighted SCDFs

If we work solely with the r_j's, we can view (10.35)–(10.38) as *conditional* on the r_j's. However if we work with $r(\mathbf{s})$'s, then we will need a probability model for $r(\mathbf{s})$ in order to interpolate to $r(\mathbf{s}_\ell)$ in (10.35). Since $r(\mathbf{s})$ and $X(\mathbf{s})$ are expected to be associated, we may conceptualize them as arising from a spatial process, and develop, say, a bivariate Gaussian spatial process model for both $X(\mathbf{s})$ and $h(r(\mathbf{s}))$, where h maps the weights onto $\Re^1$.

Let $\mathbf{Y}(\mathbf{s}) = (X(\mathbf{s}), h(r(\mathbf{s}))^T$ and $\mathbf{Y} = (\mathbf{Y}(\mathbf{s}_1), \ldots, \mathbf{Y}(\mathbf{s}_n))^T$. Analogous to the univariate situation in Subsection 10.3.1, we need to draw samples $\mathbf{Y}_{\tilde{s}}^{(g)}$ from $p(\mathbf{Y}_{\tilde{s}} \mid \mathbf{Y}, \boldsymbol{\beta}^{(g)}, \boldsymbol{\theta}^{(g)}, T^{(g)})$. Again this is routinely done via composition from posterior samples. Since the $\mathbf{Y}_{\tilde{s}}^{(g)}$ samples have marginal distribution $p(\mathbf{Y}_{\tilde{s}} \mid \mathbf{Y})$, we may use them to obtain predictive realizations of the SCDF, using either the unweighted form (10.32) or the weighted form (10.35).

The bivariate structure also allows for the definition of a *bivariate SCDF*,

$$F_{U,V}(w_u, w_v) = \frac{1}{|D|} \int_D I(U(\mathbf{s}) \leq w_u, V(\mathbf{s}) \leq w_v) d\mathbf{s}, \qquad (10.39)$$

which gives $Pr\left[\mathbf{s} \in D : U(\mathbf{s}) \leq w_u, V(\mathbf{s}) \leq w_v\right]$, the proportion of the region having values below the given thresholds for, say, two pollutants. Finally, a sensible *conditional* SCDF might be

$$
\begin{aligned}
F_{U|V}(w_u|w_v) &= \frac{\int_D I(U(\mathbf{s}) \leq w_u, V(\mathbf{s}) \leq w_v) d\mathbf{s}}{\int_D I(V(\mathbf{s}) \leq w_v) d\mathbf{s}} \\
&= \frac{\int_D I(U(\mathbf{s}) \leq w_u) I(V(\mathbf{s}) \leq w_v) d\mathbf{s}}{\int_D I(V(\mathbf{s}) \leq w_v) d\mathbf{s}}. \qquad (10.40)
\end{aligned}
$$

This expression gives $Pr\left[\mathbf{s} \in D : U(\mathbf{s}) \leq w_u \mid V(\mathbf{s}) \leq w_v\right]$, the proportion of the region having second pollutant values below the threshold w_v that *also has* first pollutant values below the threshold w_u. Note that (10.40) is again a weighted SCDF, with $r(\mathbf{s}) = I(V(\mathbf{s}) \leq w_v)$. Note further that we could easily alter (10.40) by changing the directions of either or both of its inequalities, if conditional statements involving high (instead of low) levels of either pollutant were of interest.

Example 10.5 *(California air quality data)*. We illustrate in the case of a bivariate Gaussian process using data collected by the California Air Resources Board, available at www.arb.ca.gov/aqd/aqdcd/aqdcddld.htm. The particular subset we consider are the mean NO and NO_2 values for July 6, 1999, as observed at the 67 monitoring sites shown as solid dots in Figure 10.4. Recall that in our notation, U corresponds to log(mean NO) while V corresponds to log(mean NO_2). A WSCDF based on these two variables is of interest since persons already at high NO risk may be especially vulnerable to elevated NO_2 levels. Figure 10.5 shows interpolated perspective, image, and contour plots of the raw data. Association of the pollutant levels is apparent; in fact, the sample correlation coefficient over the 67 pairs is 0.74.

We fit a separable, Gaussian bivariate model using the simple exponential spatial covariance structure $\rho(d_{ii'}, \boldsymbol{\theta}) = \exp(-\lambda d_{ii'})$, so that $\boldsymbol{\theta} \equiv \lambda$; no σ^2 parameter is required (nor identifiable) here due to the multiplicative presence of the T matrix. For prior distributions, we first assumed $T^{-1} \sim W((\nu R)^{-1}, \nu)$ where $\nu = 2$ and $R = 4I$. This is a reasonably vague

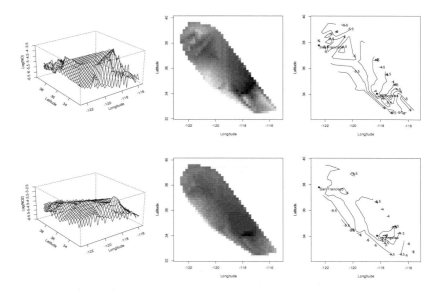

Figure 10.5 *Interpolated perspective, image, and contour plots of the raw log-NO (first row) and log-NO$_2$ (second row), California air quality data, July 6, 1999.*

specification, both in its small degrees of freedom ν and in the relative size of R (roughly the prior mean of T), since the entire range of the data (for both log-NO and log-NO$_2$) is only about 3 units. Next, we assume $\lambda \sim G(a, b)$, with the parameters chosen so that the effective spatial range is half the maximum diagonal distance M in Figure 10.4 (i.e., $3/E(\lambda) = .5M$), and the standard deviation is one half of this mean. Finally, we assume constant means $\mu_U(\mathbf{s}; \boldsymbol{\beta}) = \beta_U$ and $\mu_V(\mathbf{s}; \boldsymbol{\beta}) = \beta_V$, and let β_U and β_V have vague normal priors (mean 0, variance 1000).

Our initial Gibbs algorithm sampled over $\boldsymbol{\beta}$, T^{-1}, and λ. The first two of these may be sampled from closed-form full conditionals (normal and inverse Wishart, respectively) while λ is sampled using Hastings independence chains with $G(\frac{2}{3}, \frac{3}{2})$ proposals. We used 3 parallel chains to check convergence, followed by a "production run" of 2000 samples from a single chain for posterior summarization. Histograms (not shown) of the posterior samples for the bivariate kriging model are generally well behaved and consistent with the results in Figure 10.5.

Figure 10.6 shows perspective plots of raw and kriged log-NO and log-NO$_2$ surfaces, where the plots in the first column are the (interpolated) raw data (as in the first column of Figure 10.5), those in the second column are based on a single Gibbs sample, and those in the third column represent the average over 2000 post-convergence Gibbs samples. The plots in this

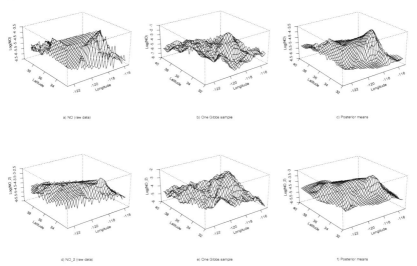

Figure 10.6 *Perspective plots of kriged log-NO and log-NO$_2$ surfaces, California air quality data. First column, raw data; second column, based on a single Gibbs sample; third column, average over 2000 post-convergence Gibbs samples.*

final column are generally consistent with those in the first, except that they exhibit the spatial smoothness we expect of our posterior means.

Figure 10.7 shows several SCDFs arising from samples from our bivariate kriging algorithm. First, the solid line shows the ordinary SCDF (10.32) for log-NO. Next, we computed the weighted SCDF for two choices of weight function in (10.35). In particular, we weight log-NO exposure U by $h^{-1}(V)$ using $h^{-1}(V) = \exp(V)$ and $h^{-1}(V) = \exp(V)/(\exp(V) + 1)$ (the exponential and inverse logit functions). Since V is log-NO$_2$ exposure, this amounts to weighting by NO$_2$ itself, and by NO$_2$/(NO$_2$+1). The results from these two h^{-1} functions turn out to be visually indistinguishable, and are shown as the dotted line in Figure 10.7. This line is shifted to the right from the unweighted version, indicating higher harmful exposure when the second (positively correlated) pollutant is accounted for.

Also shown as dashed lines in Figure 10.7 are several WSCDFs that result from using a particular indicator of whether log-NO$_2$ remains below a certain threshold. These WSCDFs are thus also conditional SCDFs, as in equation (10.40). Existing EPA guidelines and expertise could be used to inform the choice of clinically meaningful thresholds; here we simply demonstrate the procedure's behavior for a few illustrative thresholds. For example, when the threshold is set to –3.0 (a rather high value for this pollutant on the log scale), nearly all of the weights equal 1, and the WSCDF differs little from the unweighted SCDF. However, as this thresh-

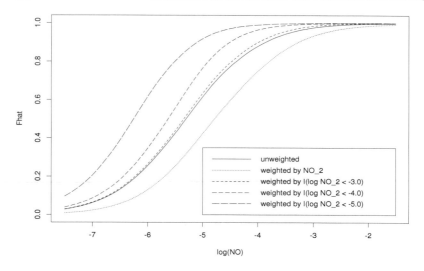

Figure 10.7 *Weighted SCDFs for the California air quality data: solid line, ordinary SCDF for log-NO; dotted line, weighted SCDF for log-NO using NO_2 as the weight; dashed lines, weighted SCDF for log-NO using various indicator functions of log-NO_2 as the weights.*

old moves lower (to −4.0 and −5.0), the WSCDF moves further to the left of its unweighted counterpart. This movement is understandable, since our indicator functions are *decreasing* functions of log-NO_2; movement to the right could be obtained simply by reversing the indicator inequality. ∎

Appendices

Matrix theory and spatial computing methods

In this appendix we discuss matrix theory and computing issues that are essential in the Bayesian implementation of the spatial and spatiotemporal process models described in this text. Specifically, repeated likelihood evaluation in MCMC implementation of hierarchical spatial models necessitates repeated matrix inversion and determinant evaluation.

Linear algebra lies at the core of the applied sciences, and matrix computations arise in a wide array of scenarios in engineering, applied mathematics, statistical computing, and many other scientific fields. As a result, the subject of numerical linear algebra has grown at a tremendous pace over the past three decades, bringing about new algorithms and techniques to efficiently perform matrix computations.

This brief appendix only skims the surface of this vast and very important subject. There are several excellent books on matrix computation; we would particularly recommend the book by Golub and van Loan (1996) for a more thorough and rigorous study. Our initial focus will be on matrix inversions, determinant computations, and Cholesky decompositions. Since the most efficient matrix inversion algorithms actually solve linear systems of equations, we start with this topic. After our matrix theory discussion, we present sections on special algorithms for point-referenced and areal data models. The former employs slice sampling (Neal, 2003), while the latter relies on structured MCMC (Sargent et al., 2000).

A.1 Gaussian elimination and LU decomposition

In this section we consider the solution to the linear system $A\mathbf{x} = \mathbf{b}$, where A is an $n \times n$ nonsingular matrix and $\mathbf{b}$ is a known vector. For illustration, let us start with $n = 3$, so that

$$A = \begin{pmatrix} a_{11} & a_{12} & a_{13} \\ a_{21} & a_{22} & a_{23} \\ a_{31} & a_{32} & a_{33} \end{pmatrix} \text{ and } \mathbf{b} = \begin{pmatrix} b_1 \\ b_2 \\ b_3 \end{pmatrix} . \tag{A.1}$$

The most popular method of solving a system as above is using a series of elementary (also called *echelon*) transformations that involve only multiplying a row (or column) by a constant, or adding one row (or column) to another. This approach allows us to reduce A to an upper triangular matrix U, thereby reducing a general system to an upper triangular system that is far easier to solve (using backsubstitution).

Specifically, we first *sweep out* the first column of A by premultiplying with an elementary matrix E_1, to obtain

$$E_1 A = \begin{pmatrix} a_{11} & a_{12} & a_{13} \\ 0 & a_{22}^{(1)} & a_{23}^{(1)} \\ 0 & a_{32}^{(1)} & a_{33}^{(1)} \end{pmatrix} . \tag{A.2}$$

The matrix E_1 is easily seen to be

$$E_1 = \begin{pmatrix} 1 & 0 & 0 \\ -a_{21}/a_{11} & 1 & 0 \\ -a_{31}/a_{11} & 0 & 1 \end{pmatrix} = I - \boldsymbol{\tau}_1 \mathbf{e}_1^T ,$$

where $\boldsymbol{\tau}_1 = (0, a_{21}/a_{11}, a_{31}/a_{11})^T$ and $\mathbf{e}_1 = (1, 0, 0)^T$. With a second elementary transformation, E_2, we proceed to sweep out the elements below the diagonal in the second column. This yields

$$E_2 E_1 A = \begin{pmatrix} a_{11} & a_{12} & a_{13} \\ 0 & a_{22}^{(1)} & a_{23}^{(1)} \\ 0 & 0 & a_{33}^{(2)} \end{pmatrix} ,$$

which is in upper triangular form U, as desired. The matrix E_2 is seen to be

$$E_2 = \begin{pmatrix} 1 & 0 & 0 \\ 0 & 1 & 0 \\ 0 & -a_{23}^{(1)}/a_{22}^{(1)} & 1 \end{pmatrix} = I - \boldsymbol{\tau}_2 \mathbf{e}_2^T ,$$

where $\boldsymbol{\tau}_1 = (0, 0, a_{23}^{(1)}/a_{22}^{(1)})^T$ and $\mathbf{e}_2 = (0, 1, 0)^T$. We therefore have that $E_2 E_1 A = U$, so that $A = E_1^{-1} E_2^{-1} U$. Note however that $(I - \boldsymbol{\tau}_i \mathbf{e}_i^T)(I + \boldsymbol{\tau}_i \mathbf{e}_i^T) = I - \boldsymbol{\tau}_i \left(\mathbf{e}_i^T \boldsymbol{\tau}_i \right) \mathbf{e}_i^T = I$ for $i = 1, 2$, since $\mathbf{e}_i^T \boldsymbol{\tau}_i = 0$ for $i = 1, 2$. Thus, $E_1^{-1} = (I + \boldsymbol{\tau}_1 \mathbf{e}_1^T)$ and $E_2^{-1} = (I + \boldsymbol{\tau}_2 \mathbf{e}_2^T)$. Clearly both E_1^{-1} and E_2^{-1} are unit lower triangular (lower triangular matrices with diagonals as unity), and so $L = E_1^{-1} E_2^{-1}$ is a unit lower triangular matrix. Thus, this sweeping strategy, known as *Gaussian elimination*, leads us to the LU decomposition of a nonsingular square matrix A as $A = LU$.

In the case of a general n, the same argument eventually results in

$$A = E_1^{-1} E_2^{-1} \cdots E_{n-1}^{-1} U ,$$

where $E_k = I - \boldsymbol{\tau}_k \mathbf{e}_k^T$. Again $\mathbf{e}_k$ is the coordinate vector with a 1 in the kth place and 0's elsewhere, $\tau_1^T = (0, a_{21}/a_{11}, ..., a_{n1}/a_{11})$, and $\tau_k^T =$

$\left(0, ..., 0, a_{k+1,k}^{(k-1)}/a_{kk}^{(k-1)}, ..., a_{nk}^{(k-1)}/a_{kk}^{(k-1)}\right)$ for $k = 2, \ldots, n - 1$. The important feature to note here is that $\boldsymbol{\tau}_k$ has a 0 in its kth position, rendering it orthogonal to $\mathbf{e}_k$, so that E_k^{-1} is still $I + \boldsymbol{\tau}_k \mathbf{e}_k^T$. Thus,

$$L = \prod_{k=1}^{n-1} \left(I + \boldsymbol{\tau}_k \mathbf{e}_k^T\right) = I + \sum_{k=1}^{n-1} \boldsymbol{\tau}_k \mathbf{e}_k^T,$$

and we have a nice closed-form expression for the unit lower triangular matrix. The above simplification follows easily since $\mathbf{e}_i^T \boldsymbol{\tau}_j = 0$ whenever $j \geq i$.

Returning to the system $A\mathbf{x} = \mathbf{b}$, we now rewrite our system as $LU\mathbf{x} = \mathbf{b}$. We now solve the system in two stages. Stage 1 solves a lower triangular system, recursively solved rowwise from the first row. Stage 2 solves an upper triangular system, again solved recursively starting with the last row. That is,

$$\text{Stage 1} \quad : \quad \text{Solve for } \mathbf{y} \text{ in } L\mathbf{y} = \mathbf{b};$$
$$\text{Stage 2} \quad : \quad \text{Solve for } \mathbf{x} \text{ in } U\mathbf{x} = \mathbf{y}.$$

An alternative way of formulating the LU decomposition is to in fact assume the existence of an LU decomposition (with a unit lower triangular L), and actually solve for the other elements of L and U. Here we present *Crout's algorithm* (see, e.g., Wilkinson, 1965), wherein we being by writing $A = (a_{ij})$ as

$$\begin{pmatrix} 1 & 0 & \cdots & 0 \\ l_{21} & 1 & \cdots & 0 \\ \vdots & \vdots & \ddots & \vdots \\ l_{n1} & l_{n2} & \cdots & 1 \end{pmatrix} \begin{pmatrix} u_{11} & u_{12} & \cdots & u_{1n} \\ 0 & u_{22} & \cdots & u_{2n} \\ \vdots & \vdots & \ddots & \vdots \\ 0 & 0 & \cdots & u_{nn} \end{pmatrix}.$$

Crout's algorithm proceeds by solving for the elements of L and U in the right order. We first consider the first row of the U matrix, $u_{1j} = a_{1j}$ for $j = 1, 2, \ldots, n$. Next we look at the first column of the L matrix, obtaining $l_{i1} = a_{i1}/u_{11}$ for $i = 2, 3, \ldots, n$. We then return to the rows; considering the second row of the U matrix, we have $l_{21}u_{1j} + u_{2j} = a_{2j}$, $j = 2, 3, \ldots, n$, so that $u_{2j} = a_{2j} - l_{21}u_{1j}$, $j = 2, 3, \ldots, n$. The next step would be to consider the second column of L, and so on. In general, Crout's algorithm takes the following form: for $i = 1, 2, \ldots, n$,

$$u_{ij} = a_{ij} - \sum_{k=1}^{i-1} l_{ik} u_{kj}, \quad j = i, i+1, \ldots, n;$$

$$l_{ji} = \frac{a_{ji} - \sum_{k=1}^{i-1} l_{jk} u_{ki}}{u_{ii}}, \quad j = i+1, \ldots, n.$$

In all of the above steps, $l_{ii} = 1$ for $i = 1, 2, \ldots, n$.

Crout's algorithm demonstrates an efficient way to compute the LU decomposition and also shows (through the equations) that the decomposition is unique. That is, if $A = L_1 U_1 = L_2 U_2$, then it must be that $L_1 = L_2$ and $U_1 = U_2$. Also note that Crout's algorithm will be able to detect when A is near singular and in such a situation can be designed to return an appropriate error message (or exception class in an object-oriented framework).

Gaussian elimination and Crout's algorithm differ only in the ordering of operations. Both algorithms are theoretically and numerically equivalent with complexity $O\left(n^3\right)$ (actually, the number of operations is approximately $n^3/3$, where an operation is defined as one multiplication and one addition). Also note that the LU decomposition may be converted to an LDU decomposition with both L and U being unit triangular matrices by taking D to be $Diag\left(u_{11}, \ldots, u_{nn}\right)$.

We hasten to add a few remarks concerning the important concept of *pivoting* for the above algorithms. Pivots are the scaling constants involved in the Gaussian elimination process. For example, in our preceding discussion, a_{11} and $a_{22}^{(1)}$. Pivoting is discussed at length in Golub and van Loan (1996) and deals with permuting the rows of matrix A by premultiplication with a permutation matrix P. Pivoting is vital for numerical stability and proper termination of the program. Pivoting generally results in the decomposition $PA = LU$.

We illustrate the importance of pivoting with a simple ill-conditioned example. Consider the matrix $A = \begin{pmatrix} 0 & 1 \\ 1 & 0 \end{pmatrix}$. A direct application of Crout's algorithm will clearly fail, since the leading diagonal element is 0. However, we can use pivoting by premultiplying by the permutation matrix $P = \begin{pmatrix} 0 & 1 \\ 1 & 0 \end{pmatrix}$, obtaining $PA = \begin{pmatrix} 1 & 0 \\ 0 & 1 \end{pmatrix}$. Now, Crout's algorithm is applied to the permuted matrix to produce $PA = LU$. Once the linear system is solved, if one needs to recover the original order (as, e.g., in getting A^{-1}), it is just a matter of permuting back the final result; that is, premultiplying by $P^{-1} \equiv P^T$ (since permutation matrices are orthogonal). This is easily verified here, with our final result being $A^{-1} = \begin{pmatrix} 0 & 1 \\ 1 & 0 \end{pmatrix}$. This admittedly pathological example shows that robust matrix LU algorithms must allow for pivoting. In practice, matrices such as $\begin{pmatrix} \epsilon & 1 \\ 1 & 0 \end{pmatrix}$, where the leading diagonal ϵ is extremely small, lead to numerical instability. Golub and van Loan (1996, pp. 110–121) provide a comprehensive review of different pivoting strategies, showing that even with seemingly well-conditioned problems, pivoting often leads to even more accurate algorithms.

A.2 Inverses and determinants

Among the advantages of Gaussian elimination and LU decomposition is the fact that the most numerically stable matrix inversion formulations are based upon them. Furthermore the determinant of a general square matrix is readily available from the LU decomposition. We have, since $\det(L) = 1$,

$$\det(A) = \det(LU) = \det(L)\det(U) = \det(U) = \prod_{i=1}^{n} u_{ii}.$$

Turning to the matrix inversion problem, we note that matrix inversion is equivalent to solving a set of linear systems of equations. That is, we need to solve for a matrix X such that $AX = I$. But writing $X = [\mathbf{x}_1, \mathbf{x}_2, ..., \mathbf{x}_n]$ and $I = [\mathbf{e}_1, \mathbf{e}_2, ..., \mathbf{e}_n]$, this is equivalent to solving for $\mathbf{x}_j$ such that

$$A\mathbf{x}_j = \mathbf{e}_j, \ j = 1, 2, \ldots, n \ .$$

Notice that we need to decompose the matrix just once and obtain the inverse columnwise. Therefore this matrix system can be solved as an $O\left(n^3\right)$ operation using Gaussian elimination or LU decomposition followed by backsubstitution. When applied to the inversion problem, Gaussian elimination is known as the *Gauss-Jordan* algorithm.

The LU (or LDU) decomposition technique is invaluable to us because a single decomposition allows us to obtain the determinant *and* the inverse of the matrix. We will see in the next section that further efficiency and savings can be accrued in situations where the matrix is symmetric and positive definite. This is what we will encounter since we will work with correlation matrices.

We conclude this section with a brief discussion about improving precision of matrix inverses. Since inverses are computed based upon simulated parameter values and have a crucial role to play in the proper movement of the Markov Chain, stability and numerical accuracy of the inversion process is vital. In certain cases, an iterative improvement to the initial result obtained is obtained.

Thus, suppose $A\mathbf{x}_j = \mathbf{e}_j$ has been solved via the above method using t-digit precision arithmetic. We first compute $\mathbf{r}_j = \mathbf{e}_j - A\hat{\mathbf{x}}_j$ in $2t$-digit (double precision) arithmetic, where $\hat{\mathbf{x}}_j$ is the initial solution. We then perform the following steps:

$$\begin{aligned}
\text{Solve for } \mathbf{y} \quad &: \quad L\mathbf{y} = \mathbf{r}_j; \\
\text{Solve for } \mathbf{z} \quad &: \quad U\mathbf{z} = \mathbf{y}; \\
\text{Set} \quad &: \quad \mathbf{x}_{j(new)} = \hat{\mathbf{x}}_j + \mathbf{z} \ .
\end{aligned}$$

This process is known as *mixed-precision iterative improvement* (MPIS). The motivation for the above scheme is immediately seen through the fixed point system (in exact arithmetic) $A\mathbf{x}_{j(new)} = A\hat{\mathbf{x}}_j + A\mathbf{z} = \mathbf{e}_j - \mathbf{r}_j + \mathbf{r}_j = \mathbf{e}_j$.

Notice that the process is relatively cheap, requiring $O(n^2)$ operations, compared to the $O(n^3)$ operations in the original decomposition.

It is important to note that the above method is sensitive to the accuracy of r_j's. The above double-precision computing strategy often works well. In fact, Skeel (1980) has performed an error analysis deriving sufficient conditions when MPIS is effective. Another drawback of this approach is that the implementation is somewhat machine-dependent, and therefore not widely used in distributed software.

A.3 Cholesky decomposition

In most statistical applications we need to work with covariance matrices. These matrices are symmetric and positive definite. Such matrices enjoy excellent stability properties and have a special decomposition, called the *Cholesky decomposition*, given by $A = G^T G$. Here G is upper triangular and is called the "square root" of the matrix A. The existence of such a matrix is immediately seen by applying the LDU decomposition to A. Since A is symmetric, we in fact obtain $A = LDL^T$, where L is unit lower triangular and D is a diagonal matrix with all entries positive. Thus we may write $G^T = LD^{1/2}$ to obtain the Cholesky square root. Also, if A is real, so is its square root.

Note that, for general matrices, the algorithms discussed in Section A.2 can fail if no pivot selection is carried out, i.e. if we naively take the diagonal elements in order as pivots. This is not the case with symmetric positive definite matrices. For these matrices we are assured of obtaining a positive diagonal element in order, thereby yielding a nonzero pivot. It is also numerically stable to use these pivots. This means that the subtleties associated with pivoting in Crout's algorithm can be avoided, thus yielding a simpler and faster algorithm.

Crout's method, when applied to symmetric positive definite matrices yield the following set of equations: for $i = 1, 2, \ldots, n$,

$$g_{ii} = \left(a_{ii} - \sum_{k=1}^{i-1} g_{ik}^2 \right)^{1/2} ;$$

$$g_{ji} = \frac{a_{ij} - \sum_{k=1}^{i-1} g_{jk} g_{ik}}{g_{ii}}, \quad j = i+1, \ldots, n .$$

These steps form what is commonly referred to as the Cholesky algorithm. Applying the above equations, one can easily see that the g's are easily determined by the time they are really needed. Also, only components of A in the upper triangle (i.e. a_{ij}'s with $j \geq i$) are referenced. Since A is symmetric, these hold all the information that will be needed. The operation count of this algorithm is $n^3/6$, which is a reduction from the general LU decomposition by a factor of two. The Cholesky decomposition is still a $O(n^3)$ algorithm.

A.4 Fast Fourier transforms

The Fast Fourier Transform (FFT) constitutes one of the major break-throughs in computational mathematics. The FFT can be looked upon as a fast algorithm to compute the Discrete Fourier Transform (DFT), which enjoys wide applications in physics, engineering, and the mathematical sciences. The DFT implements discrete Fourier analysis and is used in time series and periodogram analysis. Here we briefly discuss the computational framework for DFTs; further details may be found in Monahan (2001, pp. 386–400) or Press et al. (1992, pp. 496–532).

For computational purposes, we develop the DFT in terms of a matrix transformation of a vector. For this discussion we let all indices range from 0 to $N-1$ (as in the C programming language). Thus if $\mathbf{x} = (x_0, \ldots, x_{N-1})^T$ is a vector representing a sequence of order N, then the DFT of $\mathbf{x}$ is given by

$$y_j = \sum_{k=0}^{N-1} \exp\left(-2\pi i j k/N\right) x_k , \qquad (A.3)$$

where $i = \sqrt{-1}$, and j and k are indices. Let $w = \exp\left(-2\pi i/N\right)$ (the Nth root of unity) and let W be the $N \times N$ matrix with (j,k)th element given by w^{jk}. Then the relationship in (A.3) can be represented as the linear tranformation $\mathbf{y} = W\mathbf{x}$, with $\mathbf{y} = (y_0, \ldots, y_{N-1})^T$. The matrix of the inverse transformation is given by W^{-1}, whose (j,k)th element is easily verified to be w^{-jk}.

Direct computation of this linear transformation involves $O(N^2)$ arithmetic operations (additions, multiplications and complex exponentiations). The FFT (Cooley and Tukey, 1965) is a modified algorithm that computes the above in only $O(N \log N)$ operations. Note that the difference in these complexities can be immense in terms of CPU time. Press et al. (1992) report that with $N = 10^6$, this difference can be between 2 weeks and 30 seconds of CPU time on a microsecond-cycle computer.

To illustrate the above modification, let us consider a composite $N = N_1 N_2$, where N_1 and N_2 are integers. Using the remainder theorem, we write the indices $j = q_1 N_1 + r_1$ and $k = q_2 N_2 + r_2$ with $q_1, r_2 \in [0, N_2 - 1]$ and $q_2, r_1 \in [0, N_1 - 1]$. It then follows that

$$
\begin{aligned}
y_j &= \sum_{k=0}^{N-1} w^{jk} x_j = \sum_{k=0}^{N-1} w^{(q_1 N_1 + r_1)k} x_k \\
&= \sum_{q_2=0}^{N_1-1} \sum_{r_2=0}^{N_2-1} w^{(q_1 N_1 + r_1)(q_2 N_2 + r_2)} x_{q_2 N_2 + r_2}
\end{aligned}
$$

$$= \sum_{r_2}^{N_2-1} w^{(q_1 N_1 + r_1) r_2} \sum_{q_2=0}^{N_1-1} (w^{N_2})^{q_2 r_1} x_{q_2 N_2 + r_2}$$

$$= \sum_{r_2}^{N_2-1} (w^{N_1})^{q_1 r_2} \left(w^{r_1 r_2} \sum_{q_2=0}^{N_1-1} (w^{N_2})^{q_2 r_1} x_{q_2 N_2 + r_2} \right) ,$$

where the equality in the third line arises from the fact that $w^{N_1 N_2 q_1 q_2} = 1$. This shows that each inner sum is a DFT of length N_1, while each of the N_2 outer sums is a DFT of length N_2. Therefore, to compute the above, we perform a DFT of length $N = N_1 N_2$ by first performing N_2 DFTs of length N_1 to obtain the inner sum, and then N_1 DFTs of length N_2 to obtain the outer sum. Effectively, the new algorithm involves $N_2 O(N_1) + N_1 O(N_2)$ arithmetic operations, which, when N is a power of 2, boils down to an $O(N \log_2 N)$ algorithm. The details of this setting may be found in Monahan (2001).

In spatial statistics, the FFT is often used for computing covariance functions and their spectral densities (Stein, 1999a). Recall that valid correlation functions are related to probability densities via Bochner's theorem. Restricting our attention to isotropic functions on $\Re^1$, we have

$$f(u) = \frac{1}{2\pi} \int_{-\infty}^{\infty} \exp(-itu) C(t) dt , \qquad (A.4)$$

where $f(u)$ is the *spectral density* obtained by a Fourier transform of the correlation function, $C(t)$. For example, the Matérn correlation function arises as a transform of the spectral density $f(u) = (\phi_2 + |u|^2)^{-(\nu+r/2)}$, up to a proportionality constant; see also equation (2.13).

To take advantage of the FFT in computing the Matérn correlation function, we first replace the continuous version of Bochner's integral in (A.4) by a finite sum over an evenly spaced grid of points between $-T$ and T, with T large enough so that

$$f(u) \approx \frac{\Delta_t}{2\pi} \sum_{j=0}^{N-1} \exp(-it_j u) C(t_j) , \qquad (A.5)$$

where $t_j = j \Delta_t$. Note that we have used the fact that C is an even function, so $\Delta_t = 2T/N$. Next, we select evenly spaced evaluation points for the spectral density $f(u)$, and rewrite equation (A.5) as

$$f(u_k) \approx \frac{\Delta_t}{2\pi} \sum_{j=0}^{N-1} \exp(-it_j u_k) K(t_j) .$$

This, in fact, is a DFT and is cast in the matrix equation, $\mathbf{y} = W\mathbf{x}$, with $\mathbf{y} = (f(u_0), ..., f(u_{M-1}))^T$, $\mathbf{x} = (C(t_0), ..., C(t_{N-1}))^T$, and W is the matrix of the transformation with (j, k)th element given by $\exp(-it_j u_k)$. Now this DFT is made "fast" (into an FFT) by appropriate choice of the evaluation points for f and C. Let Δ_u denote the spacings of u. The FFT implementation is obtained by ensuring the product of the two spacings to equal $2\pi/N$. That is, we ensure that $\Delta_t \Delta_u = 2\pi/N$. This results in $W_{jk} = \exp(-ijk2\pi/N)$, which is the exactly the FFT matrix.

The FFT enables fast conversion between the spectral domain and the frequency domain. Since the Matérn function has an easily computed spectral density, the inverse FFT is used to approximate $C(t)$ from $f(t)$, using W^{-1}. Note that $W_{jk}^{-1} = \exp(ijk2\pi/N)$, which enables a direct efficient computation of the inverse, instead of the usual $O(n^3)$ inversion algorithms.

A.5 Strategies for large spatial and spatiotemporal data sets

Implementing Gibbs sampling or other MCMC algorithms requires repeated evaluation of various full conditional density functions. In the case of hierarchical models built from random effects using Gaussian processes, this requires repeated evaluation of the likelihood and/or joint or conditional densities arising under the Gaussian process; see Section A.6. In particular, such computation requires evaluation of quadratic forms involving the inverse of covariance matrix and also the determinant of that matrix. Strictly speaking, we do not have to obtain the inverse in order to compute the quadratic form. Letting $\mathbf{z}^T A^{-1} \mathbf{z}$ denote a general object of this sort, if we obtain $A^{\frac{1}{2}}$ and solve $z = A^{\frac{1}{2}}v$ for v, then $v^T v = z^T A^{-1} z$. Still, with large n, computation associated with resulting $n \times n$ matrices can be unstable, and repeated computation (as for simulation-based model fitting) can be very slow, perhaps infeasible. We refer to this situation informally as "the big n problem."

Extension to multivariate models with, say, p measurements at a location leads to $np \times np$ matrices (see Section 7.2). Extension to spatiotemporal models (say, spatial time series at T time points) leads to $nT \times nT$ matrices (see Section 8.2). Of course, there may be modeling strategies that will simplify to, say, T $n \times n$ matrices or an $n \times n$ and a $T \times T$ matrix, but the problem will still persist if n is large. The objective of this section is thus to offer some suggestions and approaches for handling spatial process models in this case.

A.5.1 Subsampling

As a first approach, one could employ a subsample of the sampled locations, resulting in a computationally more tractable n. Though it may be unattractive to ignore some of the available data, it may be argued that the incremental inferential gain with regard to the process unknowns given, say, $4n$ points may be small. The issue is that we have dependent measurements. In the case of independent data, quadrupling sample size doubles precision. In the spatial setting, if we add a location very close to an existing location, the data from the new location may help with regard to learning about the noise or measurement error component, but will not add much to the inference about the spatial model. Indeed, with a purely spatial model, including such a new location may make the associated covariance matrix nearly singular, obviously leading to possible stability problems. With increasing sample size this will surely be the case eventually.

These remarks apply to prediction as well. Prediction at a new spatial location will improve very slowly with increasing sample size. To see this, consider the simplest kriging situation, i.e., ordinary kriging (constant mean, no pure error). Then, given process parameters, the conditional variance of $Y(s_0)$ for a new s_0 given data $\mathbf{Y}$ will be, in obvious notation, $\sigma^2(1 - R_{Y_0,Y}^T R_Y^{-1} R_{Y_0,Y})$. The quadratic form will initially grow quickly in n, but will then increase very slowly in n toward its asymptote of 1. Further, as in the previous paragraph, instability with regard to R_Y^{-1} will eventually arise.

This subsampling strategy can be formalized into a model-fitting approach following the ideas of Pardo-Igúzquiza and Dowd (1997). Specifically, for observations $Y(\mathbf{s}_i)$ arising from a Gaussian process with parameters $\boldsymbol{\theta}$, they propose replacing the joint density of $\mathbf{Y} = (Y(\mathbf{s}_1), \ldots, Y(\mathbf{s}_n))^T$, $f(\mathbf{y}|\boldsymbol{\theta})$, by

$$\prod_{i=1}^{n} f(y(\mathbf{s}_i) \mid y(\mathbf{s}_j),\ \mathbf{s}_j \in \partial\mathbf{s}_i)\,, \tag{A.6}$$

where $\partial\mathbf{s}_i$ defines some neighborhood of $\mathbf{s}_i$. For instance, it might be all $\mathbf{s}_j$ within some specified distance of $\mathbf{s}_i$, or perhaps the m $\mathbf{s}_j$'s closest to $\mathbf{s}_i$ for some integer m. Pardo-Igúzquiza and Dowd (1997) suggest the latter, propose $m = 10$ to 15, and check for stability of the inference about $\boldsymbol{\theta}$.

The justification for approximating $f(\mathbf{y}|\boldsymbol{\theta})$ by (A.6) is essentially that above, but a more formal argument is given by Vecchia (1988). Regardless, evaluation of (A.6) will involve n $m \times m$ matrices, rather than one $n \times n$ matrix.

A.5.2 Spectral methods

Another option is to work in the spectral domain (as advocated by Stein, 1999a, and Fuentes, 2002a). The idea is to transform to the space of frequencies, develop a periodogram (an estimate of the spectral density), and utilize the Whittle likelihood (Whittle, 1954; Guyon, 1995) in the spectral domain as an approximation to the data likelihood in the original space. The Whittle likelihood requires no matrix inversion so, as a result, computation is very rapid. In principle, inversion back to the original space is straightforward.

The practical concerns here are the following. First, there is discretization to implement a fast Fourier transform (see Section A.4). Then, there is a certain arbitrariness to the development of a periodogram. Empirical experience is employed to suggest how many low frequencies should be discarded. Also, there is concern regarding the performance of the Whittle likelihood as an approximation to the exact likelihood. Some empirical investigation suggests that this approximation is reasonably well centered, but does a less than satisfactory job in the tails (thus leading to poor estimation of model variances). Lastly, with non-Gaussian first stages, we will be doing all of this with random spatial effects that are never observed, making the implementation impossible. In summary, use of the spectral domain with regard to handling large n is limited in its application, and requires considerable familiarity with spectral analysis (discussed briefly in Subsection 2.2.2).

A.5.3 Lattice methods

Though Gaussian Markov random fields have received a great deal of recent attention for modeling areal unit data, they were originally introduced for points on a regular lattice. In fact, using inverse distance to create a proximity matrix, we can immediately supply a joint spatial distribution for variables at an arbitrary set of locations. As in Section 3.2, this joint distribution will be defined through its full conditional distribution. The joint density is recaptured using Brook's Lemma (3.7). The inverse of the covariance matrix is directly available, and the joint distribution can be made proper through the inclusion of an autocorrelation parameter. Other than the need to sample a large number of full conditional distributions, there is no big n problem. Indeed, many practitioners immediately adopt Gaussian Markov random field models as the spatial specification due to the computational convenience.

The disadvantages arising with the use of Gaussian Markov random fields should by now be familiar. First, and perhaps most importantly, we do not model association directly, which precludes the specification of models exhibiting certain correlation behavior. The joint distribution of the variables at two locations depends not only on their joint distribution given the rest of the variables, but also on the joint distribution of the rest of the variables. In fact, the relationship between entries in the inverse covariance matrix and the actual covariance matrix is very complex and highly nonlinear. Besag and Kooperberg (1995) showed, using a fairly small n that entries in the covariance matrix resulting from a Gaussian Markov random field specification need not behave as desired. They need not be positive nor decay with distance. With large n, the implicit transformation from inverse covariance matrix to covariance matrix is even more ill behaved (Conlon and Waller, 1999; Wall, 2003).

In addition, with a Gaussian Markov random field there is no notion of a stochastic process, i.e., a collection of variables at all locations in the region of interest with joint distributions determined through finite dimensional distributions. In particular, we cannot write down the distribution of the variable at a selected location in the region. Rather, the best we can do is determine a conditional distribution for this variable given the variables at some prespecified number of and set of locations. Also, introduction of nonspatial error is confusing. The conditional variance in the Gaussian Markov random field cannot be aligned in magnitude with the marginal variance associated with a white noise process.

Some authors have proposed approximating a Gaussian process with a Gaussian Markov random field. More precisely, a given set of spatial locations $s_1, ..., s_n$ along with a choice of correlation function yields an $n \times n$ covariance matrix Σ_1. How might we specify a Gaussian Markov random field with full rank inverse matrix Σ_2^{-1} such that $\Sigma_2 \approx \Sigma_1$? That is, unlike the previous paragraph where we start with a Gaussian Markov random field, here we start with the Gaussian spatial process.

A natural metric in this setting is Kullback-Liebler distance (see Besag and Kooperberg, 1995). If $f_1 \sim N(0, \Sigma_1)$ and $f_2 \sim N(0, \Sigma_2)$, the Kullback-Leibler distance of f_2 from f_1 is

$$KL(f_1, f_2) = \int f_1 \log(f_1/f_2) = -\frac{1}{2} \log \left| \Sigma_2^{-1} \Sigma_1 \right| + \frac{1}{2} tr(\Sigma_2^{-1} \Sigma_1 - I). \quad (A.7)$$

Hence, we only need Σ_1 and Σ_2^{-1} to compute (A.7). Using an algorithm originally proposed by Dempster (1972), Besag and Kooperberg provide approximation based upon making (A.7) small. Rue and Tjelmeland (2002) note that this approach does not well approximate the correlation function of the Gaussian process. In particular, it will not do well when spatial association decays slowly. Rue and Tjelmeland propose a "matched correlation" criterion that accommodates both local and global behavior.

A.5.4 Dimension reduction

Another strategy is to employ a dimension reduction approach. For instance, recall the idea of kernel convolution (see Subsection 5.3.2) where we represent the process $Y(s)$ by

$$Y(\mathbf{s}) = \int k(\mathbf{s} - \mathbf{s}')z(\mathbf{s}')d\mathbf{s}' , \quad (A.8)$$

where k is a kernel function (which might be parametric, and might be spatially varying) and $z(\mathbf{s})$ is a stationary spatial process (which might be white noise, that is, $\int_A z(\mathbf{s})d\mathbf{s} \sim N(0, \sigma^2 A)$ and $cov(\int_A z(\mathbf{s})d\mathbf{s}, \int_B z(\mathbf{s})d\mathbf{s}) = \sigma^2 |A \cap B|$). Finite approximation to (A.8) yields

$$Y(\mathbf{s}) = \sum_{j=1}^{J} k(\mathbf{s} - \mathbf{s}_j^*)z(\mathbf{s}_j^*) . \quad (A.9)$$

Expression (A.9) shows that given k, every variable in the region is expressible as a linear combination of the set $\{z(\mathbf{s}),\ j = 1, ..., J\}$. Hence, no matter how large n is, working with the z's, we never have to handle more than a $J \times J$ matrix. The richness associated with the class in (A.8) suggests reasonably good richness associated with (A.9). Versions of (A.9) to accommodate multivariate processes and spatiotemporal processes can be readily envisioned.

Concerns regarding the use of (A.9) involve two issues. First, how does one determine the number of and choice of the $\mathbf{s}_j^*$'s? How sensitive will inference be to these choices? Also, the joint distribution of $\{Y(\mathbf{s}_i),\ i = 1, ..., n\}$ will be singular for $n > J$. While this does not mean that $Y(\mathbf{s}_i)$ and $Y(\mathbf{s}_i')$ are perfectly associated, it does mean that specifying $Y(\cdot)$ at J distinct locations determines the value of the process at all other locations. As a result, such modeling may be more attractive for spatial random effects than for the data itself.

A variant of this strategy is a conditioning idea. Suppose we partition the region of interest into M subregions so that we have the total of n points partitioned into n_m in subregion m with $\sum_{m=1}^{M} n_m = n$. Suppose we assume that $Y(\mathbf{s})$ and $Y(\mathbf{s}')$ are conditionally independent given $\mathbf{s}$ lies in subregion m and $\mathbf{s}'$ lies in subregion m'. However, suppose we assign random effects $\gamma(\mathbf{s}_1^*), ..., \gamma(\mathbf{s}_M^*)$ with $\gamma(\mathbf{s}_m^*)$ assigned to subregion m. Suppose the $\mathbf{s}_M^*$'s are "centers" of the subregions (using an appropriate definition) and that the $\gamma(\mathbf{s}_M^*)$ follows a spatial process that we can envision as a *hyper*spatial process. There are obviously many ways to build such multilevel spatial structures, achieving a variety of spatial association behaviors. We do not elaborate here but note that matrices will now be $n_m \times n_m$ and $M \times M$ rather than $n \times n$.

A.5.5 Coarse-fine coupling

Lastly, particularly for hierarchical models with a non-Gaussian first stage, a version of the coarse-fine idea as in Higdon, Lee, and Holloman (2003) may be successful. The idea here is, with a non-Gaussian first stage, if spatial random effects (say, $\theta(\mathbf{s}_1), ..., \theta(\mathbf{s}_n)$) are introduced at the second stage, then, as in Subsection 5.2, the set of $\theta(\mathbf{s}_i)$ will have to be updated at each iteration of a Gibbs sampling algorithm.

Suppose n is large and that the "fine" chain does such updating. This chain will proceed very slowly. But now suppose that concurrently we run a "coarse" chain using a much smaller subset n' of the $\mathbf{s}_i$'s. The coarse chain will update very rapidly. Since the process for $\theta(\cdot)$ is the same in both chains it will be the case that the coarse one will explore the posterior more rapidly. However, we need realizations from the fine chain to fit the model using all of the data.

The coupling idea is to let both the fine and coarse chains run, and after a specified number of updates of the fine chain (and many more updates of the coarse chain, of course) we attempt a "swap;" i.e., we propose to swap the current value of the fine chain with that of the coarse chain. The swap attempt ensures that the equilibrium distributions for both chains are not compromised (see Higdon, Lee, and Holloman, 2003). For instance, given the values of the θ's for the fine iteration, we might just use the subset of θ's at the locations for the coarse chain. Given the values of the θ's for the coarse chain, we might do an appropriate kriging to obtain the θ's for the fine chain. Such coupling strategies have yet to be thoroughly investigated.

A.6 Slice Gibbs sampling for spatial process model fitting

Auxiliary variable methods are receiving increased attention among those who use MCMC algorithms to simulate from complex nonnormalized multivariate densities. Recent work in the statistical literature includes Tanner and Wong (1987), Besag and Green (1993), Besag et al. (1995), and Higdon (1998). The particular version we focus on here introduces a single auxiliary variable to "knock out" or "slice" the likelihood. Employed in the context of spatial modeling for georeferenced data using a Bayesian formulation with commonly used proper priors, in this section we show that convenient Gibbs sampling algorithms result. Our approach thus finds itself as a special case of recent work by Damien, Wakefield, and Walker (1999), who view methods based on multiple auxiliary variables as a general approach to constructing Markov chain samplers for Bayesian inference problems. We are also close in spirit to recent work of Neal (2003), who also employs a single auxiliary variable, but prefers to slice the entire nonnormalized joint density and then do a single multivariate updating of all the variables. Such updating requires sampling from a possibly high-dimensional uniform distribution with support over a very irregular region. Usually, a bounding rectangle is created and then rejection sampling is used. As a result, a single updating step will often be inefficient in practice.

Currently, with the wide availability of cheap computing power, Bayesian spatial model fitting typically turns to MCMC methods. However, most of these algorithms are hard to automate since they involve tuning tailored to each application. In this section we demonstrate that a *slice Gibbs sampler*, done by knocking out the likelihood and implemented with a Gibbs updating, enables essentially an automatic MCMC algorithm for fitting Gaussian spatial process models. Additional advantages over other simulation-based model fitting schemes accrue, as we explain below. In this regard, we could instead slice the product of the likelihood and the prior, yielding uniform draws to implement the Gibbs updates. However, the support for these conditional uniform updates changes with iteration. The conditional interval arises through matrix inverse and determinant functions of model parameters with matrices of dimension equal to the sample size. Slicing only the likelihood and doing Gibbs updates using draws from the prior along with rejection sampling is truly "off the shelf," requiring no tuning at all. Approaches that require first and second derivatives of the log likelihood or likelihood times prior, e.g., the MLE approach of Mardia and Marshall (1984) or Metropolis-Hastings proposal approaches within Gibbs samplers will be very difficult to compute, particularly with correlation functions such as those in the Matérn class.

Formally, if $L(\boldsymbol{\theta}; \boldsymbol{Y})$ denotes the likelihood and $\pi(\boldsymbol{\theta})$ is a proper prior, we introduce the single auxiliary variable U, which, given $\boldsymbol{\theta}$ and $\boldsymbol{Y}$, is distributed uniformly on $(0, L(\boldsymbol{\theta}; \boldsymbol{Y}))$. Hence the joint posterior distribution of $\boldsymbol{\theta}$ and U is given by

$$p(\boldsymbol{\theta}, U | \boldsymbol{Y}) \propto \pi(\boldsymbol{\theta})\, I(U < L(\boldsymbol{\theta}; \boldsymbol{Y}))\,, \qquad (A.10)$$

where I denotes the indicator function. The Gibbs sampler updates U according to its full conditional distribution, which is the above uniform. A component θ_i of $\boldsymbol{\theta}$ is updated by drawing from its prior subject to the indicator restriction given the other θ's and U. A standard distribution is sampled and only L needs to be evaluated. Notice that, if hyperparameters are introduced into the model, i.e., $\pi(\boldsymbol{\theta})$ is replaced with $\pi(\boldsymbol{\theta}|\boldsymbol{\eta})\pi(\boldsymbol{\eta})$, the foregoing still applies and $\boldsymbol{\eta}$ is updated without restriction. Though our emphasis here is spatial model fitting, it is evident that slice Gibbs sampling algorithms are more broadly applicable. With regard to computation, for large data sets often evaluation of $L(\boldsymbol{\theta}; \boldsymbol{Y})$ will produce an underflow, preventing sampling from the uniform distribution for U given $\boldsymbol{\theta}$ and $\boldsymbol{Y}$. However, $\log L(\boldsymbol{\theta}; \boldsymbol{Y})$ will typically not be a problem to compute. So, if $V = -\log U$, given $\boldsymbol{\theta}$ and $\boldsymbol{Y}$, $V + \log L(\boldsymbol{\theta}; \boldsymbol{Y}) \sim Exp(mean = 1.0)$, and we can transform (A.10) to $p(\boldsymbol{\theta}, V | \boldsymbol{Y}) \propto exp(-V)\, I(-\log L(\boldsymbol{\theta}; \boldsymbol{Y}) < V < \infty)$.

In fact, in some cases we can implement a more efficient slice sampling algorithm than the slice Gibbs sampler. We need only impose constrained sampling on a subset of the components of $\boldsymbol{\theta}$. In particular, suppose we write $\boldsymbol{\theta} = (\boldsymbol{\theta_1}, \boldsymbol{\theta_2})$ and suppose that the full conditional distribution for $\boldsymbol{\theta_1}$, $p(\boldsymbol{\theta_1}|\boldsymbol{\theta_2}, \boldsymbol{Y}) \propto L(\boldsymbol{\theta_1}, \boldsymbol{\theta_2}; \boldsymbol{Y})\pi(\boldsymbol{\theta_1}|\boldsymbol{\theta_2})$, is a standard distribution. Then consider the following iterative updating scheme: sample U given $\boldsymbol{\theta_1}$ and $\boldsymbol{\theta_2}$ as above; then, update $\boldsymbol{\theta_2}$ given $\boldsymbol{\theta_1}$ and U with a draw from $\pi(\boldsymbol{\theta_2}|\boldsymbol{\theta_1})$ subject to the constraint $U < L(\boldsymbol{\theta_1}, \boldsymbol{\theta_2}; \boldsymbol{Y})$; finally, update $\boldsymbol{\theta_1}$ with an unconditional draw from $p(\boldsymbol{\theta_1}|\boldsymbol{\theta_2}, \boldsymbol{Y})$. Formally, this scheme is not a Gibbs sampler. Suppressing $\boldsymbol{Y}$, we are updating $p(U|\boldsymbol{\theta_1}, \boldsymbol{\theta_2})$, then $p(\boldsymbol{\theta_2}|\boldsymbol{\theta_1}, U)$, and finally $p(\boldsymbol{\theta_1}|\boldsymbol{\theta_2})$. However, the first and third distribution uniquely determine $p(U, \boldsymbol{\theta_1}|\boldsymbol{\theta_2})$ and, this, combined with the second, uniquely determine the joint distribution. The Markov chain iterated in this fashion still has $p(\boldsymbol{\theta}, U | \boldsymbol{Y})$ as its stationary distribution. In fact, if $p(\boldsymbol{\theta_1}|\boldsymbol{\theta_2}, \boldsymbol{Y})$ is a standard distribution, this implies that we can marginalize over $\boldsymbol{\theta_1}$ and run the slice Gibbs sampler on $\boldsymbol{\theta_2}$ with U. Given posterior draws of $\boldsymbol{\theta_2}$, we can sample $\boldsymbol{\theta_1}$ one for one from its posterior using $p(\boldsymbol{\theta_1}|\boldsymbol{\theta_2}, \boldsymbol{Y})$ and the fact that $p(\boldsymbol{\theta}_1|\boldsymbol{Y}) = \int p(\boldsymbol{\theta}_1|\boldsymbol{\theta}_2, \boldsymbol{Y})p(\boldsymbol{\theta}_2|\boldsymbol{Y})$. Moreover, if $p(\boldsymbol{\theta}_1|\boldsymbol{\theta}_2, \boldsymbol{Y})$ is not a standard distribution, we can add Metropolis updating of $\boldsymbol{\theta}_1$ either in its entirety or through its components (we can also use Gibbs updating here). We employ these modified schemes for different choices of $(\boldsymbol{\theta}_1, \boldsymbol{\theta}_2)$ in the remainder of this section.

We note that the performance of the algorithm depends critically on the distribution of the number of draws needed from $\pi(\boldsymbol{\theta}_2|\boldsymbol{\theta}_1)$ to update $\boldsymbol{\theta}_2$ given $\boldsymbol{\theta}_1$ and U subject to the constraint $U < L(\boldsymbol{\theta}_1, \boldsymbol{\theta}_2; \boldsymbol{Y})$. Henceforth, this will be referred to as "getting a point in the slice." A naive rejection sampling scheme (repeatedly sample from $\pi(\boldsymbol{\theta}_2|\boldsymbol{\theta}_1)$ until we get to a point in the slice) may not always give good results. An algorithm that shrinks the support of $\pi(\boldsymbol{\theta}_2|\boldsymbol{\theta}_1)$ so that it gives a better approximation to the slice whenever there is a rejection is more appropriate.

We propose one such scheme called "shrinkage sampling" described in Neal (2003). In this context, it results in the following algorithm. For simplicity, let us assume $\boldsymbol{\theta}_2$ is one-dimensional. If a point $\hat{\theta}_2$ drawn from $\pi(\boldsymbol{\theta}_2|\boldsymbol{\theta}_1)$ is not in the slice and is larger (smaller) than the current value θ_2 (which is of course in the slice), the next draw is made from $\pi(\boldsymbol{\theta}_2|\boldsymbol{\theta}_1)$ truncated with the upper (lower) bound being $\hat{\theta}_2$. The truncated interval keeps shrinking with each rejection until a point in the slice is found. The multidimensional case works by shrinking hyperrectangles. As mentioned in Neal (2003), this ensures that the expected number of points drawn will not be too large, making it a more appropriate method for general use. However, intuitively it might result in higher autocorrelations compared to the simple rejection sampling scheme. In our experience, the shrinkage sampling scheme has performed better than the naive version in most cases.

Supressing $\boldsymbol{Y}$ in our notation, we summarize the main steps in our slice Gibbs sampling algorithm as follows:

(a) Partition $\boldsymbol{\theta} = (\boldsymbol{\theta}_1, \boldsymbol{\theta}_2)$ so that samples from $p(\boldsymbol{\theta}_1|\boldsymbol{\theta}_2)$ are easy to obtain;

(b) Draw $V = -\log L(\boldsymbol{\theta}) + Z$, where $Z \sim Exp(\text{mean} = 1)$;

(c) Draw $\boldsymbol{\theta}_2$ from $p(\boldsymbol{\theta}_2|\boldsymbol{\theta}_1, V) I(-\log L(\boldsymbol{\theta}) < V < \infty)$ using shrinkage sampling;

(d) Draw $\boldsymbol{\theta}_1$ from $p(\boldsymbol{\theta}_1|\boldsymbol{\theta}_2)$;

(e) Iterate (b) through (d) until we get the appropriate number of MCMC samples.

The spatial models on which we focus arise through the specification of a Gaussian process for the data. With, for example, an isotropic covariance function, proposals for simulating the range parameter for, say, an exponential choice, or the range and smoothness parameters for a Matérn choice can be difficult to develop. That is, these parameters appear in the covariance matrix for $\boldsymbol{Y}$ in a nonstructured way (unless the spatial locations are on a regular grid). They enter the likelihood through the determinant and inverse of this matrix. And, for large n, the fewer matrix inversion and determinant computations, the better. As a result, for a noniterative sampling algorithm, it is very difficult to develop an effective importance sampling distribution for all of the model parameters. Moreover, as over-all model dimension increases, resampling typically yields a very "spiked" discrete distribution.

Alternative Metropolis algorithms require effective proposal densities with careful tuning. Again, these densities are difficult to obtain for pa-rameters in the correlation function. Morover, in general, such algorithms will suffer slower convergence than the Gibbs samplers we suggest, since full conditional distributions are not sampled. Furthermore, in our expe-rience, with customary proposal distributions we often encounter serious autocorrelation problems. When thinning to obtain a sample of roughly uncorrelated values, high autocorrelation necessitates an increased num-ber of iterations. Additional iterations require additional matrix inversion and determinant calculation and can substantially increase run times. Dis-cretizing the parameter spaces has been proposed to expedite computation in this regard, but it too has problems. The support set is arbitrary, which may be unsatisfying, and the support will almost certainly be adaptive across iterations, diminishing any computational advantage.

A.6.1 Constant mean process with nugget

Suppose $Y(\mathbf{s}_1), \ldots, Y(\mathbf{s}_n)$ are observations from a constant mean spatial process over $s \in D$ with a nugget. That is,

$$Y(\mathbf{s}_i) = \mu + w(\mathbf{s}_i) + \epsilon(\mathbf{s}_i)\,, \qquad (\text{A.11})$$

where the $\epsilon(\mathbf{s}_i)$ are realizations of a white noise process with mean 0 and variance τ^2. In (A.11), the $w(\mathbf{s}_i)$ are realizations from a second-order sta-tionary Gaussian process with covariance function $\sigma^2 C(\mathbf{h}; \boldsymbol{\rho})$ where C is a valid two-dimensional correlation function with parameters $\boldsymbol{\rho}$ and sepa-ration vector $\mathbf{h}$. Below we work with the Matérn class (2.8), so that $\boldsymbol{\rho} = (\phi, \nu)$. Thus (A.11) becomes a five-parameter model: $\boldsymbol{\theta} = (\mu, \sigma^2, \tau^2, \phi, \nu)^T$.

Note that though the $Y(\mathbf{s}_i)$ are conditionally independent given the $w(\mathbf{s}_i)$, a Gibbs sampler that also updates the latent $w(\mathbf{s}_i)$'s will be sampling an $(n+5)$-dimensional posterior density. However, it is possible to marginal-ize explicitly over the $w(\mathbf{s}_i)$'s (see Section 5.1), and it is almost always preferable to implement iterative simulation with a lower-dimensional dis-tribution. The marginal likelihood associated with $\mathbf{Y} = (Y(\mathbf{s}_1), \ldots, Y(\mathbf{s}_n))$ is

$$\begin{aligned} L(\mu, \sigma^2, \tau^2, \phi, \nu; \mathbf{Y}) &= \mid \sigma^2 H(\boldsymbol{\rho}) + \tau^2 I \mid^{-\frac{1}{2}} \\ &\times \exp\{-(\mathbf{Y} - \mu \mathbf{1})^T (\sigma^2 H(\boldsymbol{\rho}) + \tau^2 I)^{-1} (\mathbf{Y} - \mu \mathbf{1})/2\}\,, \end{aligned} \qquad (\text{A.12})$$

where $(H(\boldsymbol{\rho}))_{ij} = \sigma^2 C(d_{ij}; \boldsymbol{\rho})$ (d_{ij} being the distance between $\mathbf{s}_i$ and $\mathbf{s}_j$).

Suppose we adopt a prior of the form $\pi_1(\mu)\pi_2(\tau^2)\pi_3(\sigma^2)\pi_4(\phi)\pi_5(\nu)$. Then (A.10) becomes $\pi_1(\mu)\pi_2(\tau^2)\pi_3(\sigma^2)\pi_4(\phi)\pi_5(\nu)\, I(U < L(\mu, \sigma^2, \tau^2, \phi, \nu; \boldsymbol{Y}))$. The Gibbs sampler is most easily implemented if, given ϕ and ν, we diagonalize $H(\boldsymbol{\rho})$, i.e., $H(\boldsymbol{\rho}) = P(\boldsymbol{\rho})D(\boldsymbol{\rho})(P(\boldsymbol{\rho}))^T$ where $P(\boldsymbol{\rho})$ is orthogonal with the columns of $P(\boldsymbol{\rho})$ giving the eigenvectors of $H(\boldsymbol{\rho})$ and $D(\boldsymbol{\rho})$ is a diagonal matrix with diagonal elements λ_i, the eigenvalues of $H(\boldsymbol{\rho})$. Then (A.12) simplifies to

$$\prod_{i=1}^{n}(\sigma^2\lambda_i+\tau^2))^{-\frac{1}{2}}\exp\{-\frac{1}{2}(\boldsymbol{Y}-\mu\mathbf{1})^T P(\boldsymbol{\rho})(\sigma^2 D(\boldsymbol{\rho})+\tau^2 I)^{-1}P^T(\boldsymbol{\rho})(\boldsymbol{Y}-\mu\mathbf{1})\}.$$

As a result, the constrained updating of σ^2 and τ^2 at a given iteration does not require repeated calculation of a matrix inverse and determinant. To minimize the number of diagonalizations of $H(\boldsymbol{\rho})$ we update ϕ and ν together. If there is interest in the $w(\mathbf{s}_i)$, their posteriors can be sampled straightforwardly after the marginalized model is fitted. For instance, $p(w(\mathbf{s}_i)|\boldsymbol{Y}) = \int p(w(\mathbf{s}_i)|\boldsymbol{\theta}, \boldsymbol{Y})p(\boldsymbol{\theta}|\boldsymbol{Y})d\boldsymbol{\theta}$ so each posterior sample $\boldsymbol{\theta}^\star$, using a draw from $p(w(\mathbf{s}_i)|\boldsymbol{\theta}^\star, \boldsymbol{Y})$ (which is a normal distribution), yields a sample from the posterior for $w(\mathbf{s}_i)$.

We remark that (A.11) can also include a parametric transformation of $Y(s)$. For instance, we could employ a power transformation to find a scale on which the Gaussian process assumption is comfortable. This only requires replacing $Y(\mathbf{s})$ with $Y^p(\mathbf{s})$ and adds one more parameter to the likelihood in (A.12). Lastly, we note that other dependence structures for $\boldsymbol{Y}$ can be handled in this fashion, e.g., equicorrelated forms, Toeplitz forms, and circulants.

A.6.2 Mean structure process with no pure error component

Now suppose $Y(\mathbf{s}_1), \ldots, Y(\mathbf{s}_n)$ are observations from a spatial process over $\mathbf{s} \in D$ such that

$$Y(\mathbf{s}_i) = \boldsymbol{X}^T(\mathbf{s}_i)\boldsymbol{\beta} + w(\mathbf{s}_i)\,. \tag{A.13}$$

Again, the $w(\mathbf{s}_i)$ are realizations from a second-order stationary Gaussian process with covariance parameters σ^2 and $\boldsymbol{\rho}$. In (A.13), $\boldsymbol{X}(\mathbf{s}_i)$ could arise as a vector of site level covariates or $\boldsymbol{X}^T(\mathbf{s}_i)\boldsymbol{\beta}$ could be a trend surface specification as in the illustration below. To complete the Bayesian specification we adopt a prior of the form $\pi_1(\boldsymbol{\beta})\pi_2(\sigma^2)\pi_3(\boldsymbol{\rho})$ where $\pi_1(\boldsymbol{\beta})$ is $N(\boldsymbol{\mu}_\beta, \boldsymbol{\Sigma}_\beta)$ with $\boldsymbol{\mu}_\beta$ and $\boldsymbol{\Sigma}_\beta$ known.

This model is not hierarchical in the sense of our earlier forms, but we can marginalize explicitly over $\boldsymbol{\beta}$, obtaining

$$L(\sigma^2, \boldsymbol{\rho}; \boldsymbol{Y}) = |\sigma^2 H(\boldsymbol{\rho}) + X\boldsymbol{\Sigma}_\beta X^T|^{-\frac{1}{2}}$$
$$\times \exp\{-(\boldsymbol{Y} - X\boldsymbol{\mu}_\beta)^T(\sigma^2 H(\boldsymbol{\rho}) + X\boldsymbol{\Sigma}_\beta X^T)^{-1}(\boldsymbol{Y} - X\boldsymbol{\mu}_\beta)/2\}\,, \tag{A.14}$$

where the rows of X are the $\boldsymbol{X}^T(\mathbf{s}_i)$. Here, $H(\boldsymbol{\rho})$ is positive definite while $X\Sigma_{\boldsymbol{\beta}}X^T$ is symmetric positive semidefinite. Hence, there exists a nonsingular matrix $Q(\rho)$ such that $(Q^{-1}(\rho))^T Q^{-1}(\rho) = H(\rho)$ and also satisfying $(Q^{-1}(\rho))^T \Omega Q^{-1}(\rho) = X\Sigma_{\boldsymbol{\beta}}X^T$, where Ω is diagonal with diagonal elements that are eigenvalues of $X\Sigma_{\boldsymbol{\beta}}X^T H^{-1}(\rho)$. Therefore, (A.14) simplifies to

$$|Q(\rho)| \; \prod_{i=1}^n (\sigma^2 + \lambda_i))^{-\frac{1}{2}}$$
$$\times \exp\left\{-\tfrac{1}{2}(\boldsymbol{Y} - X\mu_{\boldsymbol{\beta}})^T Q(\rho)^T (\sigma^2 I + \Omega)^{-1} Q(\rho)(\boldsymbol{Y} - X\mu_{\boldsymbol{\beta}})\right\} \; .$$

As in the previous section, we run a Gibbs sampler to update U given σ^2, ρ, and $\boldsymbol{Y}$, then σ^2 given ρ, U, and $\boldsymbol{Y}$, and finally ρ given σ^2, U, and $\boldsymbol{Y}$. Then, given posterior samples $\{\sigma_l^{2*}, \rho_l^*, \; l = 1, \ldots, L\}$ we can obtain posterior samples for $\boldsymbol{\beta}$ one for one given σ_l^{2*} and ρ_l^* by drawing β_l^* from a $N(A\mathbf{a}, A)$ distribution, where

$$A^{-1} = \tfrac{1}{\sigma_l^{2*}} X^T H^{-1}(\rho_l^*) X + \Sigma_{\boldsymbol{\beta}}$$
$$\text{and } \mathbf{a} = \tfrac{1}{\sigma_l^{2*}} X^T H^{-1}(\rho_l^*) \boldsymbol{Y} + \Sigma_{\boldsymbol{\beta}}^{-1} \mu_{\boldsymbol{\beta}} \; . \tag{A.15}$$

In fact, using standard identities (see, e.g., Rao, 1973, p. 29),

$$\left(\frac{1}{\sigma^2} X^T H^{-1}(\rho) X + \Sigma_{\boldsymbol{\beta}}^{-1}\right)^{-1} = \Sigma_{\boldsymbol{\beta}} - \Sigma_{\boldsymbol{\beta}} X^T Q(\rho)(\sigma^2 I + \Omega)^{-1} Q^T(\rho) X \Sigma_{\boldsymbol{\beta}},$$

facilitating sampling from (A.15). Finally, if $\mu_{\boldsymbol{\beta}}$ and $\Sigma_{\boldsymbol{\beta}}$ were viewed as unknown we could introduce hyperparameters. In this case $\Sigma_{\boldsymbol{\beta}}$ would typically be diagonal and $\mu_{\boldsymbol{\beta}}$ might be $\mu_0 \mathbf{1}$, but the simultaneous diagonalization would still simplify the implementation of the slice Gibbs sampler.

We note an alternate strategy that does not marginalize over $\boldsymbol{\beta}$ and does not require simultaneous diagonalization. The likelihood of $(\boldsymbol{\beta}, \sigma^2, \boldsymbol{\rho})$ is given by

$$L(\boldsymbol{\beta}, \sigma^2, \boldsymbol{\rho}; \boldsymbol{Y}) \propto |\sigma^2 H(\rho)|^{-\frac{1}{2}} \exp\left\{-(\boldsymbol{Y} - X\boldsymbol{\beta})^T H(\rho)^{-1}(\boldsymbol{Y} - X\boldsymbol{\beta})/2\sigma^2\right\}. \tag{A.16}$$

Letting $\theta_1 = (\boldsymbol{\beta}, \sigma^2)$ and $\theta_2 = \rho$ with normal and inverse gamma priors on $\boldsymbol{\beta}$ and σ^2, respectively, we can update $\boldsymbol{\beta}$ and σ^2 componentwise conditional on $\theta_2, \boldsymbol{Y}$, since $\boldsymbol{\beta}|\sigma^2, \theta_2, \boldsymbol{Y}$ is normal while $\sigma^2|\boldsymbol{\beta}, \theta_2, \boldsymbol{Y}$ is inverse gamma. $\theta_2|\theta_1, U, \boldsymbol{Y}$ is updated using the slice Gibbs sampler with shrinkage as described earlier.

A.6.3 Mean structure process with nugget

Extending (A.11) and (A.13) we now assume that $Y(\mathbf{s}_1), \ldots, Y(\mathbf{s}_n)$ are observations from a spatial process over $\mathbf{s} \in D$ such that

$$Y(\mathbf{s}_i) = \boldsymbol{X}^T(\mathbf{s}_i)\boldsymbol{\beta} + w(\mathbf{s}_i) + \epsilon(\mathbf{s}_i) \; . \tag{A.17}$$

As above, we adopt a prior of the form $\pi_1(\boldsymbol{\beta})\pi_2(\tau^2)\pi_3(\sigma^2)\pi_4(\boldsymbol{\rho})$, where $\pi_1(\boldsymbol{\beta})$ is $N(\boldsymbol{\mu}_\beta, \Sigma_\beta)$. Note that we could again marginalize over $\boldsymbol{\beta}$ and the $w(s_i)$ as in the previous section, but the resulting marginal covariance matrix is of the form $\sigma^2 H(\boldsymbol{\rho}) + X\Sigma_\beta X^T + \tau^2 I$. The simultaneous diagonalization trick does not help here since $Q(\boldsymbol{\rho})$ is not orthogonal. Instead we just marginalize over the $w(s_i)$, obtaining the joint posterior $p(\boldsymbol{\beta}, \tau^2, \sigma^2, \boldsymbol{\rho}, U | \boldsymbol{Y})$ proportional to

$$\pi_1(\boldsymbol{\beta})\pi_2(\tau^2)\pi_3(\sigma^2)\pi_4(\boldsymbol{\rho}) \, I\left(U < |\sigma^2 H(\boldsymbol{\rho}) + \tau^2 I|^{-\frac{1}{2}}\right.$$
$$\left. \times \exp\{-(\boldsymbol{Y} - X\boldsymbol{\beta})^T(\sigma^2 H(\boldsymbol{\rho}) + \tau^2 I)^{-1}(\boldsymbol{Y} - X\boldsymbol{\beta})/2\}\right) \, .$$

We employ the modified scheme suggested below (A.10) taking $\boldsymbol{\theta_1} = \boldsymbol{\beta}$ and $\boldsymbol{\theta_2} = (\tau^2, \sigma^2, \boldsymbol{\rho})$. The required full conditional distribution $p(\boldsymbol{\beta}|\tau^2, \sigma^2, \boldsymbol{\rho}, \boldsymbol{Y})$ is $N(A\mathbf{a}, A)$, where

$$A^{-1} = X^T(\sigma^2 H(\boldsymbol{\rho}) + \tau^2 I)^{-1} X + \Sigma_\beta^{-1}$$
$$\text{and } \mathbf{a} = X^T(\sigma^2 H(\boldsymbol{\rho}) + \tau^2 I)^{-1}\boldsymbol{Y} + \Sigma_\beta^{-1}\boldsymbol{\mu}_\beta \, .$$

A.7 Structured MCMC sampling for areal model fitting

Structured Markov chain Monte Carlo (SMCMC) was introduced by Sargent, Hodges, Carlin (2000) as a general method for Bayesian computing in richly parameterized models. Here, "richly parameterized" refers to hierarchical and other multilevel models. SMCMC (pronounced "smick-mick") provides a simple, general, and flexible framework for accelerating convergence in an MCMC sampler by providing a systematic way to update groups of similar parameters in blocks while taking full advantage of the posterior correlation structure induced by the model and data. Sargent (2000) apply SMCMC to several different models, including a hierarchical linear model with normal errors and a hierarchical Cox proportional hazards model.

Blocking, i.e., simultaneously updating multivariate blocks of (typically highly correlated) parameters, is a general approach to accelerating MCMC convergence. Liu (1994) and Liu et al. (1994) confirm its good performance for a broad class of models, though Liu et al. (1994, Sec. 5) and Roberts and Sahu (1997, Sec. 2.4) give examples where blocking slows a sampler's convergence. In this section, we show that spatial models of the kind proposed by Besag, York, and Mollié (1991) using nonstationary "intrinsic autoregressions" are richly parameterized and lend themselves to the SMCMC algorithm. Bayesian inference via MCMC for these models has generally used single-parameter updating algorithms with often poor convergence and mixing properties. There have been some recent attempts to use blocking schemes for similar models. Cowles (2002, 2003) uses SMCMC blocking strategies for geostatistical and areal data models with normal likelihoods, while Knorr-Held and Rue (2002) implement blocking schemes using algorithms that exploit the sparse matrices that arise out of the areal data model.

We study several strategies for block-sampling parameters in the posterior distribution when the likelihood is Poisson. Among the SMCMC strategies we consider here are blocking using different-sized blocks (grouping by geographical region), updating jointly with and without model hyperparameters, "oversampling" some of the model parameters, reparameterization via hierarchical centering and "pilot adaptation" of the transition kernel. Our results suggest that our techniques will generally be far more accurate (produce less correlated samples) and often more efficient (produce more effective samples per second) than univariate sampling procedures.

SMCMC algorithm basics

Following Hodges (1998), we consider a hierarchical model expressed in the general form,

$$
\begin{bmatrix} y \\ \hline 0 \\ \hline M \end{bmatrix} = \begin{bmatrix} X_1 \mid 0 \\ \hline H_1 \mid H_2 \\ \hline G_1 \mid G_2 \end{bmatrix} \begin{bmatrix} \theta_1 \\ \hline \theta_2 \end{bmatrix} + \begin{bmatrix} \epsilon \\ \hline \delta \\ \hline \xi \end{bmatrix} . \tag{A.18}
$$

The first row of this layout is actually a collection of rows corresponding to the "data cases," or the terms in the joint posterior into which the response, the data y, enters directly. The terms in the second row (corresponding to the H_i) are called "constraint cases" since they place stochastic constraints on possible values of θ_1 and θ_2. The terms in the third row, the "prior cases" for the model parameters, have known (specified) error variances for these parameters. Equation (A.18) can be expressed as $Y = X\Theta + E$, where X and Y are known, Θ is unknown, and E is an error term with block diagonal covariance matrix $\Gamma = \mathrm{Diag}(\mathrm{Cov}(\epsilon), \mathrm{Cov}(\delta), \mathrm{Cov}(\xi))$. If the error structure for the data is normal, i.e., if the ϵ vector in the constraint case formulation (A.18) is normally distributed, then the conditional posterior density of Θ is

$$
\Theta | Y, \Gamma \sim N((X^T\Gamma^{-1}X)^{-1}(X^T\Gamma^{-1}Y) , (X^T\Gamma^{-1}X)^{-1}) . \tag{A.19}
$$

The basic SMCMC algorithm is then nothing but the following two-block Gibbs sampler :

(a) Sample Θ as a single block from the above normal distribution, using the current value of Γ.

(b) Update Γ using the conditional distribution of the variance components, using the current value of Θ.

In our spatial model setting, the errors are not normally distributed, so the normal density described above is not the correct conditional posterior distribution for Θ. Still, a SMCMC algorithm with a Metropolis-Hastings implementation can be used, with the normal density in (A.19) taken as the candidate density.

A.7.1 Applying structured MCMC to areal data

Consider again the Poisson-CAR model of Subsection 5.4.3, with no covariates so that $\mu_i = \theta_i + \phi_i$, $i = 1, \ldots, N$, where N is the total number of regions, and $\{\theta_1, .., \theta_N\}, \{\phi_1, .., \phi_N\}$ are vectors of random effects. The θ_i's are independent and identically distributed Gaussian normal variables with precision parameter τ_h, while the ϕ_i's are assumed to follow a $CAR(\tau_c)$ distribution. We place conjugate gamma hyperpriors on the precision parameters, namely $\tau_h \sim G(\alpha_h, \beta_h)$ and $\tau_c \sim G(\alpha_c, \beta_c)$ with $\alpha_h = 1.0$, $\beta_h = 100.0$, $\alpha_c = 1.0$ and $\beta_c = 50.0$ (these hyperpriors have means of 100 and 50, and standard deviations of 10,000 and 2,500, respectively, a specification recommended by Bernardinelli et al., 1995).

There is a total of $2N + 2$ model parameters: $\{\theta_i : i = 1, \ldots N\}$, $\{\phi_i : i = 1, \ldots N\}$, τ_h and τ_c. The SMCMC algorithm requires that we transform the Y_i data points to $\hat{\mu}_i = \log(Y_i/E_i)$, which can be conveniently thought of as the response since they should be roughly linear in the model parameters (the θ_i's and ϕ_i's). For the constraint case formulation, the different levels of the model are written down case by case. The data cases are $\hat{\mu}_i$, $i = 1, \ldots, N$. The constraint cases for the θ_i's are $\theta_i \sim N(0, 1/\tau_h)$, $i = 1, \ldots, N$. For the constraint cases involving the ϕ_i's, the differences between the neighboring ϕ_i's can be used to get an unconditional distribution for the ϕ_i's using pairwise differences (Besag et al., 1995). Thus the constraint cases can be written as

$$(\phi_i - \phi_j)|\tau_c \sim N(0, 1/\tau_c) \tag{A.20}$$

for each pair of adjacent regions (i, j).

To obtain an estimate of Γ, we need estimates of the variance-covariance matrix corresponding to the $\hat{\mu}_i$'s (the data cases) and initial estimates of the variance-covariance matrix for the constraint cases (the rows corresponding to the θ_i's and ϕ_i's). Using the delta method, we can obtain an approximation as follows: assume $Y_i \sim N(E_i e^{\mu_i}, E_i e^{\mu_i})$ (roughly), so invoking the delta method we can see that $\mathrm{Var}(log(Y_i/E_i))$ is approximately $1/Y_i$. A reasonably good starting value is particularly important here since we never update these variance estimates (the data variance section of Γ stays the same throughout the algorithm). For initial estimates of the variance components corresponding to the θ_i's and the ϕ_i's, we can use the mean of the hyperprior densities on τ_h and τ_c, and substitute these values into Γ.

As a result, the SMCMC candidate generating distribution is thus of the form (A.19), with the Y_i's replaced by $\hat{\mu}$. To compute the Hastings ratio, the distribution of the ϕ_i's is rewritten in the joint pairwise difference form with the appropriate exponent for τ_c (Hodges, Carlin, and Fan, 2003):

$$p(\phi_1, \phi_2, ..., \phi_N | \tau_c) \propto \tau_c^{(N-1)/2} \exp\left\{ -\frac{\tau_c}{2} \sum_{i \sim j} (\phi_i - \phi_j)^2 \right\}, \qquad (A.21)$$

where $i \sim j$ if i and j are neighboring regions. Finally, the joint distribution of the θ_i's is given by

$$p(\theta_1, \theta_2, ..., \theta_N | \tau_h) \propto \tau_h^{N/2} \exp\left\{ -\frac{\tau_h}{2} \sum_{i=1}^{N} \theta_i^2 \right\}. \qquad (A.22)$$

As above, the response vector is $\hat{\mu}^T = \{\log(Y_1/E_1), \ldots, \log(Y_N/E_N)\}$. The $(2N + C) \times 2N$ design matrix for the spatial model is defined by

$$X = \left[\begin{array}{c|c} I_{N \times N} & I_{N \times N} \\ \hline -I_{N \times N} & 0_{N \times N} \\ \hline 0_{C \times N} & A_{C \times N} \end{array} \right]. \qquad (A.23)$$

The design matrix is divided into two halves, the left half corresponding to the N θ_i's and the right half referring to the N ϕ_i's. The top section of this design matrix is an $N \times 2N$ matrix relating $\hat{\mu}_i$ to the model parameters θ_i and ϕ_i. In the ith row, a 1 appears in the ith and $(N + i)$th columns while 0's appear elsewhere. Thus the ith row corresponds to $\mu_i = \theta_i + \phi_i$. The middle section of the design matrix is an $N \times 2N$ matrix that imposes a stochastic constraint on each θ_i separately (θ_i's are i.i.d normal). The bottom section of the design matrix is a $C \times 2N$ matrix with each row having a -1 and 1 in the $(N + k)$th and $(N + l)$th columns, respectively, corresponding to a stochastic constraint being imposed on $\phi_l - \phi_k$ (using the pairwise difference form of the prior on the ϕ_i's as described in (A.20) with regions l and k being neighbors). The variance-covariance matrix Γ is a diagonal matrix with the top left section corresponding to the variances of the data cases, i.e., the $\hat{\mu}_i$'s. Using the variance approximations described above, the $(2N + C) \times (2N + C)$ block diagonal variance-covariance matrix is

$$\Gamma = \left[\begin{array}{c|c|c} Diag(1/Y_1, 1/Y_2, \ldots, 1/Y_N) & 0_{N \times N} & 0_{N \times C} \\ \hline 0_{N \times N} & \frac{1}{\tau_h} I_{N \times N} & 0_{N \times C} \\ \hline 0_{C \times N} & 0_{C \times N} & \frac{1}{\tau_c} I_{C \times C} \end{array} \right]. \qquad (A.24)$$

Note that the exponent on τ_c in (A.21) would actually be $C/2$ (instead of $(N-1)/2$) if obtained by taking the product of the terms in (A.20). Thus, (A.20) is merely a form we use to describe the distribution of the ϕ_is for our constraint case specification. The formal way to incorporate the distribution of the ϕ_is in the constraint case formulation is by using an alternate specification of the joint distribution of the ϕ_i's, as described in Besag and Kooperberg (1995). This form is a $N \times N$ Gaussian density with precision matrix, Q,

$$ p(\phi_1, \phi_2, ..., \phi_N | \tau_c) \propto \exp\left(-\frac{\tau_c}{2} \boldsymbol{\phi}^T Q \boldsymbol{\phi} \right), \text{ where } \boldsymbol{\phi}^T = (\phi_1, \phi_2, ..., \phi_N), $$
(A.25)

and

$$ Q_{ij} = \begin{cases} c & \text{if } i = j \text{ where } c = \text{ number of neighbors of region } i \\ 0 & \text{if } i \text{ is not adjacent to } j \\ -1 & \text{if } i \text{ is not adjacent to } j \end{cases} . $$

However, it is possible to show that this alternate formulation (using the corresponding design and Γ matrices) results in the same SMCMC candidate mean and covariance matrix for $\boldsymbol{\Theta}$ given τ_h and τ_c as the one described in (A.19); see Haran, Hodges, and Carlin (2003) for details.

A.7.2 Algorithmic schemes

Univariate MCMC (UMCMC): For the purpose of comparing the different blocking schemes, one might begin with a univariate (updating one variable at a time) sampler. This can be done by sampling τ_h and τ_c from their gamma full conditional distributions, and then, for each i, sampling each θ_i and ϕ_i from its full conditional distribution, the latter using a Metropolis step with univariate Gaussian random walk proposals.

Reparameterized Univariate MCMC (RUMCMC): One can also reparameterize from $(\theta_1, \ldots, \theta_N, \phi_1, \ldots, \phi_N)$ to $(\mu_1, \ldots, \mu_N, \phi_1, \ldots, \phi_N)$, where $\mu_i = \theta_i + \phi_i$. The (new) model parameters and the precision parameters can be sampled in a similar manner as for UMCMC. This "hierarchical centering" was suggested by (Besag et al. (1995) and Waller et al. (1997) for the spatial model, and discussed in general by Gelfand et al. (1995, 1996).

Structured MCMC (SMCMC): A first step here is *pilot adaptation*, which involves sampling (τ_h, τ_c) from their gamma full conditionals, updating the Γ matrix using the averaged (τ_h, τ_c) sampled so far, updating the SMCMC candidate covariance matrix and mean vector using the Γ matrix, and then sampling $(\boldsymbol{\theta}, \boldsymbol{\phi})$ using the SMCMC candidate in a Metropolis-Hastings step. We may run the above steps for a "tuning" period, after which we fix the SMCMC candidate mean and covariance, sampled (τ_h, τ_c) as before, and use the Metropolis-Hastings to sample $(\boldsymbol{\theta}, \boldsymbol{\phi})$ using SMCMC proposals. Some related strategies include adaptation of the Γ matrix more or less frequently, adaptation over shorter and longer periods of time, and pilot adaptation while blocking on groups of regions.

Our experience with pilot adaptation schemes indicates that a single proposal, regardless of adaptation period length, will probably be unable to provide a reasonable acceptance rate for the many different values of (τ_h, τ_c) that will be drawn in realistic problems. As such, we typically turn to oversampling Θ relative to (τ_h, τ_c); that is, the SMCMC proposal is always based on the current (τ_h, τ_c) value. In this algorithm, we sample τ_h and τ_c from their gamma full conditionals, then compute the SMCMC proposal based on the Γ matrix using the generated τ_h and τ_c. For each (τ_h, τ_c) pair, we run a Hastings independence subchain by sampling a sequence of length 100 (say) of Θ's using the SMCMC proposal. Further implementational details for this algorithm are given in Haran (2003).

Reparameterized Structured MCMC (RSMCMC): This final algorithm is the SMCMC analogue of the reparametrized univariate algorithm (RUMCMC). It follows exactly the same steps as the SMCMC algorithm, with the only difference being that Θ is now $(\boldsymbol{\mu}, \boldsymbol{\phi})$ instead of $(\boldsymbol{\theta}, \boldsymbol{\phi})$, and the proposal distribution is adjusted according to the new parameterization.

Haran, Hodges, and Carlin (2003) compare these schemes in the context of two areal data examples, using the notion of *effective sample size*, or ESS (Kass et al., 1998). ESS is defined for each parameter as the number of MCMC samples drawn divided by the parameter's so-called autocorrelation time, $\kappa = 1 + 2\sum_{k=1}^{\infty} \rho(k)$, where $\rho(k)$ is the autocorrelation at lag k. One can estimate κ from the MCMC chain, using the initial monotone positive sequence estimator as given by Geyer (1992). Haran et al. (2003) find UMCMC to perform poorly, though the reparameterized univariate algorithm (RUMCMC) does provide a significant improvement in this case. However, SMCMC and RSMCMC still perform better than both univariate algorithms. Even when accounting for the amount of time taken by the SMCMC algorithm (in terms of effective samples per second), the SMCMC scheme results in a far more efficient sampler than the univariate algorithm; for some parameters, SMCMC produced as much as 64 times more effective samples per second.

Overall, experience with applying several SMCMC blocking schemes to real data sets suggests that SMCMC provides a standard, systematic technique for producing samplers with far superior mixing properties than simple univariate Metropolis-Hastings samplers. The SMCMC and RSMCMC schemes appear to be reliable ways of producing good ESSs, irrespective of the data sets and parameterizations. In many cases, the SMCMC algorithms are also competitive in terms of ES/s. In addition since the blocked SMCMC algorithms mix better, their convergence should be easier to diagnose and thus lead to final parameter estimates that are less biased.

These estimates should also have smaller associated Monte Carlo variance estimates.

APPENDIX B

Answers to selected exercises

Chapter 1

3. As hinted in the problem statement, level of urbanicity might well explain the poverty pattern evident in Figure 1.2. Other regional spatially oriented covariates to consider might include percent of minority residents, percent with high school diploma, unemployment rate, and average age of the housing stock. The point here is that spatial patterns can often be explained by patterns in existing covariate data. Accounting for such covariates in a statistical model may result in residuals that show little or no spatial pattern, thus obviating the need for formal spatial modeling.

7.(a) The appropriate R code is as follows:

```
#  R program to compute geodesic distance
#  see also www.auslig.gov.au/geodesy/datums/distance.htm

#  input:   point1=(long,lat) and point2=(long,lat)
#               in degrees
#  output:  distance in km between the two points
#  example:
point1 <- c(87.65,41.90)  # Chicago (downtown)
point2 <- c(87.90,41.98)  # Chicago (O'Hare airport)
point3 <- c(93.22,44.88)  # Minneapolis (airport)
#  geodesic(point1,point3) returns 558.6867

geodesic <- function(point1, point2){
  R <- 6371
    point1 <- point1 * pi/180
    point2 <- point2 * pi/180
      d <- sin(point1[2]) * sin(point2[2]) +
            cos(point1[2]) * cos(point2[2]) *
            cos(abs(point1[1] - point2[1]))
    R*acos(d)
}
```

406 ANSWERS TO SELECTED EXERCISES

(b) Chicago to Minneapolis, 562 km; New York to New Orleans, 1897.2 km.

8. Chicago to Minneapolis, 706 km; New York to New Orleans, 2172.4 km. This overestimation is expected since the approach stretches the meridians and parallels, or equivalently, presses the curved domain onto a plane, thereby stretching the domain (and hence the distances). As the geodesic distance increases, the quality of the naive estimates deteriorates.

9. Chicago to Minneapolis, 561.8 km; New York to New Orleans, 1890.2 km. Here, the slight underestimation is expected, since it finds the straight line by penetrating (burrowing through) the spatial domain. Still, this approximation seems quite good even for distances close to 2000 km (e.g., New York to New Orleans).

10.(a) Chicago to Minneapolis, 562.2 km; New York to New Orleans, 1901.5 km.

(b) Whenever all of the points are located along a parallel or a meridian, this projection will not be defined.

Chapter 2

4.(a) Conditional on $\mathbf{u}$, finite realizations of Y are clearly Gaussian since $(Y(\mathbf{s}_i))_{i=1}^n = (W(x_i))_{i=1}^n$, where $x_i = \mathbf{s}_i^T \mathbf{u}$, and W is Gaussian. The covariance function is given by $Cov(Y(\mathbf{s}), Y(\mathbf{s}+\mathbf{h})) = c(\mathbf{h}^T \mathbf{u})$, where c is the (stationary) covariance function of W.

(b) For the marginal process, we need to take expectation over the distribution of $\mathbf{u}$, which is uniform over the n-dimensional sphere. Note that

$$Cov(Y(\mathbf{s}), Y(\mathbf{s}+\mathbf{h})) = E_{\mathbf{u}}\left[Cov\left(W(\mathbf{s}^T\mathbf{u}), W((\mathbf{s}+\mathbf{h})^T\mathbf{u})\right)\right]$$
$$= E_{\mathbf{u}}\left[c(\mathbf{h}^T\mathbf{u})\right].$$

Then, we need to show that $E_{\mathbf{u}}\left[c(\mathbf{h}^T\mathbf{u})\right]$ is a function of $\|\mathbf{h}\|$. Now, $\mathbf{h}^T\mathbf{u} = \|\mathbf{h}\|\cos\theta$, so $E_{\mathbf{u}}\left[c(\mathbf{h}^T\mathbf{u})\right] = E_\theta\left[c(\|\mathbf{h}\|\cos\theta)\right]$. But θ, being the angle made by a uniformly distributed random vector $\mathbf{u}$, has a distribution that is invariant over the choice of $\mathbf{h}$. Thus, the marginal process $Y(\mathbf{s})$ has isotropic covariance function $K(r) = E_\theta\left[c(r\cos\theta)\right]$. Note: The above covariance function (in $\Re^n$) can be computed using spherical integrals as

$$K(r) = \frac{2\Gamma(n/2)}{\sqrt{\pi}\Gamma((n-1)/2)} \int_0^1 c(r\nu)\left(1-\nu^2\right)^{(n-3)/2} d\nu.$$

10. If $\tau^2 = 0$, then $\Sigma = \sigma^2 H(\phi)$. If $\mathbf{s}_0 = \mathbf{s}_k$, where $\mathbf{s}_k$ is a monitored site,

we have $\boldsymbol{\gamma}^T = \sigma^2 [H(\phi)]_{k*}$, the kth row of $\sigma^2 H(\phi)$. Thus, $\mathbf{e}_k^T H(\phi) = (1/\sigma^2)\boldsymbol{\gamma}^T$, where $\mathbf{e}_k = (0,\ldots,1,\ldots,0)^T$ is the kth coordinate vector. So $\mathbf{e}_k^T = (1/\sigma^2)\boldsymbol{\gamma}^T H^{-1}(\phi)$. Substituting this into equation (2.18), we get

$$E[Y(\mathbf{s}_k)|\mathbf{y}] = \mathbf{x}_k^T\boldsymbol{\beta} + \mathbf{e}_k^T(\mathbf{y} - X\boldsymbol{\beta}) = \mathbf{x}_k^T\boldsymbol{\beta} + y(\mathbf{s}_k) - \mathbf{x}_k^T\boldsymbol{\beta} = y(\mathbf{s}_k)\ .$$

When $\tau^2 > 0$, $\Sigma = \sigma^2 H(\phi) + \tau^2 I$, so the Σ^{-1} in equation (2.18) does not simplify, and we do not have the above result.

Chapter 3

1. Brook's Lemma, equation (3.7), is easily verified as follows: Starting with the extreme right-hand side, observe that

$$\frac{p(y_{10},\ldots,y_{n0})}{p(y_{n0}|y_{10},\ldots,y_{n-1,0})} = p(y_{10},\ldots,y_{n-1,0})\ .$$

Now observe that

$$p(y_n|y_{10},\ldots,y_{n-1,0})p(y_{10},\ldots,y_{n-1,0}) = p(y_{10},\ldots,y_{n-1,0},y_n)\ .$$

The result follows by simply repeating these two steps, steadily moving leftward through (3.7).

3. We provide two different approaches to solving the problem. The first approach is a direct manipulative approach, relying upon elementary algebraic simplifications, and might seem a bit tedious. The second approach relies upon some relatively advanced concepts in matrix analysis, yet does away with most of the manipulations of the first approach.

Method 1: In the first method we derive the following identity:

$$\mathbf{u}^T D^{-1}(I - B)\mathbf{u} = \sum_{i=1}^n \frac{u_i^2}{\tau_i^2}\left(1 - \sum_{j=1}^n b_{ij}\right) + \sum_{i<j} \frac{b_{ij}}{\tau_i^2}(u_i - u_j)^2\ ,\quad\text{(B.1)}$$

where $\mathbf{u} = (u_1,\ldots,u_n)^T$. Note that if this identity is indeed true, the right-hand side must be strictly positive; all the terms in the r.h.s. are strictly positive by virtue of the conditions on the elements of the B matrix unless $\mathbf{u} = \mathbf{0}$. This would imply the required positive definiteness.

We may derive the above identity either by starting with the l.h.s. and eventually obtaining the r.h.s., or vice versa. We adopt the former. So,

$$\begin{aligned}\mathbf{u}^T D^{-1}(I - B)\mathbf{u} &= \sum_i \frac{u_i^2}{\tau_i^2} - \sum_i\sum_j \frac{b_{ij}}{\tau_i^2}u_i u_j \\ &= \sum_i \frac{u_i^2}{\tau_i^2} - \sum_i \frac{b_{ii}}{\tau_i^2}u_i^2 - \sum_i\sum_{j\neq i} \frac{b_{ij}}{\tau_i^2}u_i u_j\end{aligned}$$

$$= \sum_i \frac{u_i^2}{\tau_i^2}(1 - b_{ii}) - \sum_i \sum_{j \neq i} \frac{b_{ij}}{\tau_i^2} u_i u_j.$$

Adding and subtracting $\sum_i \sum_{j \neq i} (u_i^2/\tau_i^2) b_{ij}$ to the last line of the r.h.s., we write

$$\mathbf{u}^T D^{-1}(I - B)\mathbf{u} = \sum_i \frac{u_i^2}{\tau_i^2}\left(1 - \sum_j b_{ij}\right) + \sum_i \sum_{j \neq i} \frac{u_i^2}{\tau_i^2} b_{ij}$$

$$- \sum_i \sum_{j \neq i} \frac{b_{ij}}{\tau_i^2} u_i u_j$$

$$= \sum_i \frac{u_i^2}{\tau_i^2}\left(1 - \sum_j b_{ij}\right) + \sum_i \sum_{j \neq i} \frac{b_{ij}}{\tau_i^2}\left(u_i^2 - u_i u_j\right)$$

$$= \sum_i \frac{u_i^2}{\tau_i^2}\left(1 - \sum_j b_{ij}\right) + \sum_{i<j} \frac{b_{ij}}{\tau_i^2}\left(u_i - u_j\right)^2.$$

To explain the last manipulation,

$$\sum_i \sum_{j \neq i} \frac{b_{ij}}{\tau_i^2}\left(u_i^2 - u_i u_j\right) = \sum_{i<j} \frac{b_{ij}}{\tau_i^2}\left(u_i - u_j\right)^2, \qquad \text{(B.2)}$$

note that the sum on the l.h.s. of (B.2) extends over the $2 \times \binom{n}{2}$ (unordered) pairs of (i, j). Consider any particular pair, say, (k, l) with $k < l$, and its "reflection" (l, k). Using the symmetry condition, $b_{kl}/\tau_k^2 = b_{lk}/\tau_l^2$, we may combine the two terms from this pair as

$$\frac{b_{kl}}{\tau_k^2}\left(u_k^2 - u_k u_l\right) + \frac{b_{lk}}{\tau_l^2}\left(u_l^2 - u_l u_k\right) = \frac{b_{kl}}{\tau_k^2}\left(u_k - u_l\right)^2.$$

Performing the above trick for each of the $\binom{n}{2}$ pairs, immediately results in (B.2).

Method 2: The algebra above may be skipped using the following argument, based on eigenanalysis. First, note that, with the given conditions on B, the matrix $D^{-1}(I - B)$ is (weakly) diagonally dominant. This means that, if $A = D^{-1}(I - B)$, and $R_i(A) = \sum_{j \neq i} |a_{ij}|$ (the sum of the absolute values of the ith row less that of the diagonal element), then $|a_{ii}| \geq R_i(A)$, for all i, with strict inequality for at least one i. Now, using the Gershgorin Circle Theorem (see, e.g., Theorem 7.2.1 in Golub and Van Loan, p. 320), we immediately see that 0 cannot be an interior point of Gershgorin circle. Therefore, all the eigenvalues of A must be nonnegative. But note that all the elements of B are strictly positive. This means that all the elements of A are nonzero, which means that 0 cannot be a boundary point of the Gershgorin circle. Therefore, 0 must be an exterior point of the circle, proving that all the eigenvalues of A must be strictly positive. So A, being symmetric, must be positive definite.

Note: It is important that the matrix D be chosen so as to ensure $D^{-1}(I - B)$ is symmetric. To see that this condition cannot be relaxed, consider the following example. Let us take $B = \begin{pmatrix} 0.3 & 0.5 \\ 0.1 & 0.9 \end{pmatrix}$. Clearly the matrix satisfies the conditions laid down in the problem statement. If we are allowed to choose an arbitrary D, we may take $D = I_2$, the 2×2 identity matrix, and so $D^{-1}(I - B) = \begin{pmatrix} 0.7 & -0.5 \\ -0.1 & 0.1 \end{pmatrix}$. But this is not positive definite, as is easily seen by noting that with $\mathbf{u}^T = (1, 2)$, we obtain $\mathbf{u}^T D^{-1}(I - B)\mathbf{u} = -0.1 < 0$.

4. Using the identity in (B.1), it is immediately seen that, taking B to be the scaled proximity matrix (as in the text just above equation (3.15)), we have $\sum_{j=1}^{n} b_{ij} = 1$, for each i. This shows that the first term on the r.h.s. of (B.1) vanishes, leading to the second term, which is a pairwise difference prior.

Chapter 4

1. The complete `WinBUGS` code to fit this model is given below. Recall "#" is a comment in `WinBUGS`, so this version actually corresponds model for part (c).

```
model
{
    for (i in 1:N) {
        y[i] ~ dbern(p[i])
#        logit(p[i]) <- b0 + b1*kieger[i] + b2*team[i]
#        logit(p[i]) <- b0 + b2*(team[i]-mean(team[]))
        logit(p[i]) <- b0 + b1*(pct[i]-mean(pct[]))
        pct[i] <- kieger[i]/(kieger[i]+team[i])
    }
    b0 ~ dnorm(0, 1.E-3)
    b1 ~ dnorm(0, 1.E-3)
    b2 ~ dnorm(0, 1.E-3)
}

HERE ARE INITS:
list(b0=0, b1=0, b2=0)
```

	95% Credible intervals			
	β_1	β_2	DIC	p_D
Model				
(a)	(−3.68, 1.21)	(.152, 2.61)	8.82	1.69
(b)	—	(.108, 1.93)	9.08	1.61
(c)	(−70.8, −3.65)	—	8.07	1.59

Table B.1 *Posterior summaries, Carolyn Kieger prep basketball logit model.*

```
HERE ARE THE DATA:
list(N = 9,                            # number of observations
  y = c(1,1,1,1,0,1,1,1,0),            # team win/loss
  kieger = c(31,31,36,30,32,33,31,33,32),  # Kieger points
  team = c(31,16,35,42,19,37,29,23,15))    # team points
```

Running a single Gibbs sampling chain for 20,000 iterations after a 1,000-iteration burn-in period, Table B.1 gives the resulting 95% equal tail posterior credible intervals for β_1 and β_2 for each model, as well as the corresponding DIC and p_D scores.

(a) Running this model produces MCMC chains with slowly moving sample traces and very high autocorrelations and cross-correlations (especially between β_0 and β_1, since Kieger's uncentered scores are nearly identical). The 95% equal-tail confidence interval for β_1 includes 0, suggesting Kieger's score is not a significant predictor of game outcome; the p_D score of just 1.69 also suggests there are not 3 "effective" parameters in the model (although none of these posterior summaries are very trustworthy due to the high autocorrelations, hence low effective sample MCMC sample size). Thus, the model is not acceptable either numerically (poor convergence; unstable estimates due to low effective sample size) or statistically (model is overparametrized).

(b) Since "kieger" was not a significant predictor in part (a), we delete it, and center the remaining covariate ("team") around its own mean. This helps matters immensely: numerically, convergence is much better and parameter and other estimates are much more stable. Statistically, the DIC score is not improved (slightly higher), but the p_D is virtually unchanged at 1.6 (so both of the remaining parameters in the model are needed), and β_2 is more precisely estimated.

(c) Again convergence is improved, and now the DIC score is also better. β_1 is significant and negative, since the higher the proportion of points scored by Kieger (i.e., the lower the output by the rest of the team), the less likely a victory becomes.

(d) The p_i themselves have posteriors implied by the β_j posteriors and reveal that the team was virtually certain to win Games 1, 3, 4, 6, and 7 (where Kieger scored a lower percentage of the points), but could well have lost the others, especially Games 2 and 9 (the former of which the team was fortunate to win anyway). This implies that the only thing that might still be missing from our model is some measure of how *few* points the opponent scores, which is of course governed by how well Kieger and the other team members play on *defense*. But fitting such a model would obviously require defensive statistics (blocked shots, etc.) that we currently lack.

5.(a) In our implementation of WinBUGS, we obtained a slightly better DIC score with Model XI (7548.3, versus 7625.2 for Model XII), suggesting that the full time-varying complexity is not required in the survival model. The fact that the 95% posterior credible interval for γ_3, (−0.43, .26), includes 0 supports this conclusion.

(b) We obtained point and 95% interval estimates of −0.20 and (−0.25, −0.14) for γ_1, and −1.61 and (−2.13, −1.08) for γ_2.

(c) Figure B.1 plots the estimated posteriors (smoothed histograms of WinBUGS output). In both the separate (panel a) and joint (panel b) analyses, this patient's survival is clearly better if he receives ddC instead of ddI. However, the joint analysis increases the estimated median survival times by roughly 50% in both groups.

(d) Estimation of the random effects in NLMIXED is via empirical Bayes, with associated standard errors obtained by the delta method. Approximate 95% prediction intervals can then be obtained by assuming asymptotic normality. We obtained point and interval estimates in rough agreement with the above WinBUGS results, and for broadly comparable computer runtimes (if anything, our NLMIXED code ran slower). However, the asymmetry of some of the posteriors in Figure B.1 (recall they are truncated at 0) suggests traditional confidence intervals based on asymptotic normality and approximate standard errors will not be very accurate. Only the fully Bayesian-MCMC (WinBUGS) approach can produce exact results and corresponding full posterior inference.

Chapter 5

1. The calculations for the full conditionals for β and $\mathbf{W}$ follow from the results of the general linear model given in Example 4.2. Thus, with a $N(A\alpha, V)$ prior on β, (i.e., $p(\beta) = N(A\alpha, V)$) the full conditional for

$\boldsymbol{\beta}$ is $N\left(D\mathbf{d}, D\right)$, where

$$D^{-1} = \left(\frac{1}{\tau^2}X^T X + V^{-1}\right)^{-1}$$

$$\text{and } \mathbf{d} = \frac{1}{\tau^2}X^T\left(\mathbf{Y} - \mathbf{W}\right) + V^{-1}A\boldsymbol{\alpha}.$$

Note that with a flat prior on $\boldsymbol{\beta}$, we set $V^{-1} = 0$ to get

$$\boldsymbol{\beta}|\mathbf{Y}, \mathbf{W}, X, \tau^2 \sim N\left(\left(X^T X\right)^{-1}X^T\left(\mathbf{Y} - \mathbf{W}\right), \tau^2\left(X^T X\right)^{-1}\right).$$

Similarly for $\mathbf{W}$, since $p\left(\mathbf{W}\right) = N\left(\mathbf{0}, \sigma^2 H\left(\phi\right)\right)$, the full conditional distribution is again of the form $N\left(D\mathbf{d}, D\right)$, but where this time

$$D^{-1} = \left(\frac{1}{\tau^2}I + \frac{1}{\sigma^2}H^{-1}\left(\phi\right)\right)^{-1}$$

$$\text{and } \mathbf{d} = \frac{1}{\tau^2}\left(\mathbf{Y} - X\boldsymbol{\beta}\right).$$

Next, with $p\left(\tau^2\right) = IG\left(a_\tau, b_\tau\right)$, we compute the full conditional distribution for τ^2, $p\left(\tau^2|\mathbf{Y}, X, \boldsymbol{\beta}, \mathbf{W}\right)$, as proportional to

$$\frac{1}{(\tau^2)^{a_\tau+1}}\exp\left(-b_\tau/\tau^2\right)$$
$$\times \frac{1}{(\tau^2)^{n/2}}\exp\left(-\frac{1}{2\tau^2}\left(\mathbf{Y} - X\boldsymbol{\beta} - \mathbf{W}\right)^T\left(\mathbf{Y} - X\boldsymbol{\beta} - \mathbf{W}\right)\right)$$
$$\propto \frac{1}{(\tau^2)^{a_\tau+n/2}}\exp\left(-\frac{1}{\tau^2}\left(b_\tau + \frac{1}{2}\left(\mathbf{Y} - X\boldsymbol{\beta} - \mathbf{W}\right)^T\left(\mathbf{Y} - X\boldsymbol{\beta} - \mathbf{W}\right)\right)\right),$$

where n is the number of sites. Thus we have the conjugate distribution

$$IG\left(a_\tau + \frac{n}{2}, \, b_\tau + \frac{1}{2}\left(\mathbf{Y} - X\boldsymbol{\beta} - \mathbf{W}\right)^T\left(\mathbf{Y} - X\boldsymbol{\beta} - \mathbf{W}\right)\right).$$

Similar calculations for the spatial variance parameter, σ^2, yield a conjugate full conditional when $p\left(\sigma^2\right) = IG\left(a_\sigma, b_\sigma\right)$, namely

$$\sigma^2 \mid \mathbf{W}, \phi \sim IG\left(a_\sigma + \frac{n}{2}, \, b_\sigma + \frac{1}{2}\mathbf{W}^T H^{-1}\left(\phi\right)\mathbf{W}\right).$$

Finally, for the spatial correlation function parameter ϕ, no closed form solution is available, and one must resort to Metropolis-Hastings or slice sampling for updating. Here we would need to compute

$$p\left(\phi|\mathbf{W}, \sigma^2\right) \propto p\left(\phi\right) \times \exp\left(-\frac{1}{2\sigma^2}\mathbf{W}^T H^{-1}\left(\phi\right)\mathbf{W}\right).$$

Typically the prior $p\left(\phi\right)$ is taken to be uniform or gamma.

5.(a) These relationships follow directly from the definition of $w\left(\mathbf{s}\right)$ in equation (5.23):

$$Cov\left(w\left(\mathbf{s}\right), w\left(\mathbf{s}'\right)\right)$$

$$= Cov\left(\int_{\Re^2} k\left(\mathbf{s}-\mathbf{t}\right) z\left(\mathbf{t}\right) dt \, , \, \int_{\Re^2} k\left(\mathbf{s}'-\mathbf{t}\right) z\left(\mathbf{t}\right) dt\right)$$

$$= \sigma^2 \int_{\Re^2} k\left(\mathbf{s}-\mathbf{t}\right) k\left(\mathbf{s}'-\mathbf{t}\right) dt \, .$$

and

$$var\left(w\left(\mathbf{s}\right)\right) \;=\; \sigma^2 \int_{R^2} k^2\left(\mathbf{s}-\mathbf{t}\right) dt,$$

obtained by setting $\mathbf{s}=\mathbf{s}'$ above.

(b) This follows exactly as above, except that we adjust for the covariance in the stationary $z(\mathbf{t})$ process:

$$Cov\left(w\left(\mathbf{s}\right), w\left(\mathbf{s}'\right)\right) \;=\; \int_{\Re^2}\int_{\Re^2} k\left(\mathbf{s}-\mathbf{t}\right) k\left(\mathbf{s}'-\mathbf{t}\right)$$
$$\times Cov\left(z\left(\mathbf{t}\right), z\left(\mathbf{t}'\right)\right) dtdt'$$
$$=\; \sigma^2 \int_{\Re^2}\int_{\Re^2} k\left(\mathbf{s}-\mathbf{t}\right) k\left(\mathbf{s}'-\mathbf{t}\right) \rho\left(\mathbf{t}-\mathbf{t}'\right) dtdt'$$
$$\text{and } var\left(w\left(\mathbf{s}\right)\right) \;=\; \sigma^2 \int_{\Re^2}\int_{\Re^2} k\left(\mathbf{s}-\mathbf{t}\right) k\left(\mathbf{s}-\mathbf{t}'\right) \rho\left(\mathbf{t}-\mathbf{t}'\right) dtdt',$$

obtained by setting $\mathbf{s}=\mathbf{s}'$ above.

10. From (5.47), the full conditional $p(\phi_i|\boldsymbol{\phi}_{j\neq i}, \boldsymbol{\theta}, \boldsymbol{\beta}, \mathbf{y})$ is proportional to the product of a Poisson and a normal density. On the log scale we have

$$\log p(\phi_i|\boldsymbol{\phi}_{j\neq i}, \boldsymbol{\theta}, \boldsymbol{\beta}, \mathbf{y}) \propto -E_i e^{\mathbf{x}_i'\boldsymbol{\beta}+\theta_i+\phi_i} + \phi_i y_i - \frac{\tau_c m_i}{2}(\phi_i - \bar{\phi}_i)^2 \, .$$

Taking two derivatives of this expression, it is easy to show that in fact $(\partial^2/\partial\phi_i^2)\log p(\phi_i|\boldsymbol{\phi}_{j\neq i}, \boldsymbol{\theta}, \boldsymbol{\beta}, \mathbf{y}) < 0$, meaning that the log of the full conditional is a concave function, as required for ARS sampling.

Chapter 6

6.(a) Denoting the likelihood by L, the prior by p, and writing $\mathbf{y} = (y_1, y_2)$, the joint posterior distribution of m_1 and m_2 is given as

$$p(m_1, m_2|\mathbf{y}) \;\propto\; L(m_1, m_2; \mathbf{y})p(m_1, m_2)$$
$$\propto\; (7m_1 + 5m_2)^{y_1} e^{-(7m_1+5m_2)}$$
$$\times (6m_1 + 2m_2)^{y_2} e^{-(6m_1+2m_2)}$$
$$\times m_1^{a-1} e^{-m_1/b} m_2^{a-1} e^{-m_2/b} \, ,$$

so that the resulting full conditional distributions for m_1 and m_2 are

$$p(m_1|m_2, \mathbf{y}) \;\propto\; (7m_1 + 5m_2)^{y_1}(6m_1 + 2m_2)^{y_2} m_1^{a-1} e^{-m_1(13+b^{-1})};$$
$$p(m_2|m_1, \mathbf{y}) \;\propto\; (7m_1 + 5m_2)^{y_1}(6m_1 + 2m_2)^{y_2} m_2^{a-1} e^{-m_2(7+b^{-1})} \, .$$

We see immediately that conjugacy is absent; these two expressions are not proportional to any standard distributional form. As such, one might think of univariate Metropolis updating to obtain samples from the joint posterior distribution $p(m_1, m_2 | \mathbf{y})$, though since this is a very low-dimensional problem, the use of MCMC methods here probably constitutes overkill!

Drawing our Metropolis candidates from Gaussian distributions with means equal to the current chain value and variances $(0.3)^2$ and $(0.1)^2$ for δ_1 and δ_2, respectively, for each parameter we ran five independent sampling chains with starting points overdispersed with respect to the suspected target distribution for 2000 iterations. The observed Metropolis acceptance rates were 45.4% and 46.4%, respectively, near the 50% rate suggested by Gelman et al. (1996) as well as years of Metropolis "folklore." The vagueness of the prior distributions coupled with the paucity of the data in this simple example (in which we are estimating two parameters from just two data points, y_1 and y_2) leads to substantial autocorrelation in the observed chains. However, plots of the observed chains as well as the convergence diagnostic of Gelman and Rubin (1992) suggested that a suitable degree of algorithm convergence obtains after 500 iterations. The histograms of the remaining $5 \times 1500 = 7500$ iterations shown in Figures B.2(a) and (b) provide estimates of the marginal posterior distributions $p(m_1 | \mathbf{y})$ and $p(m_2 | \mathbf{y})$. We see that point estimates for m_1 and m_2 are 18.5 and 100.4, respectively, implying best guesses for $7m_1 + 5m_2$ and $6m_1 + 2m_2$ of 631.5 and 311.8, respectively, quite consistent with the observed data values $y_1 = 632$ and $y_2 = 311$. Also shown are 95% Bayesian credible intervals (denoted "95% BCI" in the figure legends), available simply as the 2.5 and 97.5 empirical percentiles in the ordered samples.

(b) By the Law of Iterated Expectation, $E(Y_{3a} | \mathbf{y}) = E[E(Y_{3a} | \mathbf{m}, \mathbf{y})]$. Now we need the following well-known result from distribution theory:

Lemma: If $X_1 \sim Po(\lambda_1)$, $X_2 \sim Po(\lambda_2)$, and X_1 and X_2 are independent, then

$$X_1 \mid (X_1 + X_2 = n) \sim Bin\left(n, \frac{\lambda_1}{\lambda_1 + \lambda_2}\right). \quad \blacksquare$$

We apply this lemma in our setting with Y_{3a} playing the role of X_1, y_1 playing the role of n, and the calculation conditional on $\mathbf{m}$. The result is

$$
\begin{aligned}
E(Y_{3a} | \mathbf{y}) &= E[E(Y_{3a} | \mathbf{m}, \mathbf{y})] = E[E(Y_{3a} | m_1, y_1)] \\
&= E\left[y_1 \left(\frac{2m_1 + 2m_2}{7m_1 + 5m_2} \right) \Big| y_1 \right]
\end{aligned}
$$

$$\approx \frac{y_1}{G} \sum_{g=1}^{G} \frac{2m_1^{(g)} + 2m_2^{(g)}}{7m_1^{(g)} + 5m_2^{(g)}} \equiv \hat{E}(Y_{3a}|\mathbf{y}) , \quad (B.3)$$

where $\{(m_1^{(g)}, m_2^{(g)}), g = 1, \ldots, G\}$ are the Metropolis samples drawn above. A similar calculation produces a Monte Carlo estimate of $E(Y_{3b}|\mathbf{y})$, so that our final estimate of $E(Y_3|\mathbf{y})$ is the sum of these two quantities. In our problem this turns out to be $\hat{E}(Y_3|\mathbf{y}) = 357.0$.

(c) Again using Monte Carlo integration, we write

$$p(y_3|\mathbf{y}) = \int p(y_3|\mathbf{m}, \mathbf{y}) p(\mathbf{m}|\mathbf{y}) d\mathbf{m} \approx \frac{1}{G} \sum_{g=1}^{G} p(y_3|\mathbf{m}^{(g)}, \mathbf{y}) .$$

Using the lemma again, $p(y_3|\mathbf{m}, \mathbf{y})$ is the convolution of two independent binomials,

$$Y_{3a}|\mathbf{m}, \mathbf{y} \sim Bin\left(y_1 , \frac{2m_1 + 2m_2}{7m_1 + 5m_2}\right) , \quad (B.4)$$

$$\text{and} \quad Y_{3b}|\mathbf{m}, \mathbf{y} \sim Bin\left(y_2 , \frac{m_1 + m_2}{6m_1 + 2m_2}\right) . \quad (B.5)$$

Since these two binomials do not have equal success probabilities, this convolution is a complicated (though straightforward) calculation that unfortunately will not emerge as another binomial distribution. However, we may perform the sampling analog of this calculation simply by drawing $Y_{3a}^{(g)}$ from $p(y_{3a}|\mathbf{m}^{(g)}, y_1)$ in (B.4), $Y_{3b}^{(g)}$ from $p(y_{3b}|\mathbf{m}^{(g)}, y_2)$ in (B.5), and defining $Y_3^{(g)} = Y_{3a}^{(g)} + Y_{3b}^{(g)}$. The resulting pairs $\{(Y_3^{(g)}, \mathbf{m}^{(g)}), g = 1, \ldots, G\}$ are distributed according to the joint posterior distribution $p(y_3, \mathbf{m}|\mathbf{y})$, so that marginally, the $\{Y_3^{(g)}, g = 1, \ldots, G\}$ values have the desired distribution, $p(y_3|\mathbf{y})$.

In our setting, we actually drew 25 $Y_{3a}^{(g)}$ and $Y_{3b}^{(g)}$ samples for each $\mathbf{m}^{(g)}$ value, resulting in $25(7500) = 187,500 \, Y_3^{(g)}$ draws from the convolution distribution. A histogram of these values (and a corresponding kernel density estimate) is shown in Figure B.3. The mean of these samples is 357.2, which agrees quite well with our earlier mean estimate of 357.0 calculated just below equation (B.3).

Chapter 7

2.(a) This setup closely follows that below equation (2.17), so we imitate this argument in the case of a bivariate process, where now $\mathbf{Y}_1 = Y_1(\mathbf{s}_0)$ and $\mathbf{Y}_2 = \mathbf{y}$. Then, as in equation (2.18),

$$E[Y_1(\mathbf{s}_0)|\mathbf{y}] = \mathbf{x}^T(\mathbf{s}_0)\boldsymbol{\beta} + \boldsymbol{\gamma}^T \Sigma^{-1}(\mathbf{y} - X\boldsymbol{\beta}) ,$$

parameter	2.5%	50%	97.5%
θ_1	-0.437	-0.326	-0.216
β_1	3.851	5.394	6.406
β_2	-2.169	2.641	7.518
σ_1	0.449	0.593	2.553
σ_2	0.101	1.530	6.545
ϕ_1	0.167	0.651	0.980
ϕ_2	0.008	0.087	0.276
τ	4.135	5.640	7.176

Table B.2 *Posterior quantiles for the conditional LMC model.*

where $\gamma^T = \left(\gamma_1^T, \gamma_2^T\right)$, where $\gamma_1^T = (c_{11}(\mathbf{s}_0 - \mathbf{s}_1), \ldots, c_{11}(\mathbf{s}_0 - \mathbf{s}_n))$ and $\gamma_2^T = (c_{12}(\mathbf{s}_0 - \mathbf{s}_1), \ldots, c_{12}(\mathbf{s}_0 - \mathbf{s}_n))$. Also,

$$\Sigma_{2n \times 2n} = \begin{pmatrix} C_{11} & C_{12} \\ C_{12}^T & C_{22} \end{pmatrix} + \begin{pmatrix} \tau_1^2 I_n & 0 \\ 0 & \tau_2^2 I_n \end{pmatrix},$$

with $C_{lm} = (c_{lm}(\mathbf{s}_i - \mathbf{s}_j))_{i,j=1,\ldots,n}$ with $l, m = 1, 2$.

(b) The approach in this part is analogous to that of Chapter 2, Exercise 10. Observe that with $\mathbf{s}_0 = \mathbf{s}_k$, $\left(\mathbf{e}_k^T : \mathbf{0}\right) \Sigma = \gamma^T$ if and only if $\tau_1^2 = 0$, where $\mathbf{e}_k^T = (0, \ldots, 1, \ldots, 0)$ is the n-dimensional kth coordinate vector. This immediately leads to $E[Y_1(\mathbf{s}_k)|\mathbf{y}] = y_1(\mathbf{s}_k)$; $E[Y_2(\mathbf{s}_k)|\mathbf{y}] = y_2(\mathbf{s}_k)$ is shown analogously.

4.(a) Let $Y_1(\mathbf{s})$ be the temperature at location $\mathbf{s}$, $Y_2(\mathbf{s})$ be the precipitation at location $\mathbf{s}$, and $X(\mathbf{s})$ be the elevation at location $\mathbf{s}$. We then fit the following conditional LMC, as in equation (7.28):

$$Y_1(\mathbf{s}) = \theta_1 X(\mathbf{s}) + \sigma_1 w_1(\mathbf{s})$$
$$Y_2(\mathbf{s}) | Y_1(\mathbf{s}) = \beta_1 X(\mathbf{s}) + \beta_2 Y_1(\mathbf{s}) + \sigma_2 w_2(\mathbf{s}) + \epsilon(\mathbf{s}),$$

where $\epsilon(\mathbf{s}) \sim N(0, \tau^2)$, $w_i(\mathbf{s}) \sim GP(0, \rho(\cdot, \phi_i))$, for $i = 1, 2$.
The file www.biostat.umn.edu/~brad/data/ColoradoLMCa.bug on the web contains the WinBUGS code for this problem. Table B.2 gives a brief summary of the results. The results are more or less as expected: temperature is negatively associated with elevation, while precipitation is positively associated. Temperature and precipitation do not seem to be significantly associated with each other. The spatial smoothing parameters ϕ_1 and ϕ_2 were both assigned $U(0, 1)$ priors for this analysis, but it would likely be worth investigating alternate choices in order to check prior robustness.

(b) These results can be obtained simply by switching Y_1 and Y_2 in the data labels for the model and computer code of part (a).

6.(a) This follows directly by noting $var(\mathbf{v}_1) = \lambda_{11}I_n$, $var(\mathbf{v}_2) = \lambda_{22}I_n$, and $cov(\mathbf{v}_1, \mathbf{v}_2) = \lambda_{12}I$.

(b) Note that $var(\boldsymbol{\phi}_1) = \lambda_{11}A_1A_1^T$, $var(\boldsymbol{\phi}_2) = \lambda_{22}A_2A_2^T$, and also that $cov(\boldsymbol{\phi}_1, \boldsymbol{\phi}_2) = \lambda_{12}A_1A_2^T$. So with $A_1 = A_2$, the dispersion of $\boldsymbol{\phi}$ is given by

$$\Sigma_{(\boldsymbol{\phi})} = \begin{pmatrix} \lambda_{11}AA^T & \lambda_{12}AA^T \\ \lambda_{12}A^TA & \lambda_{12}AA^T \end{pmatrix} = \Lambda \otimes AA^T.$$

Taking A as the square root of $(D_W - \rho W)^{-1}$ yields $\Sigma_{(\boldsymbol{\phi})} = \Lambda \otimes (D_W - \rho W)^{-1}$. Note that the order of the Kronecker product is different from equation (7.34), since we have blocked the $\boldsymbol{\phi}$ vector by components rather than by areal units.

(c) In general, with $A_1 \neq A_2$, we have

$$\Sigma_{(\boldsymbol{\phi})} = \begin{pmatrix} \lambda_{11}A_1A_1^T & \lambda_{12}A_1A_2^T \\ \lambda_{12}A_2A_1^T & \lambda_{12}A_2A_2^T \end{pmatrix} = \mathcal{A}(\Lambda \otimes I)\mathcal{A}^T,$$

where $\mathcal{A} = BlockDiag(A_1, A_2)$. For the generalized MCAR, with different spatial smoothness parameters ρ_1 and ρ_2 for the different components, take A_i as the Cholesky square root of $(D_W - \rho_i W)^{-1}$ for $i = 1, 2$.

Chapter 8

2. The code in www.biostat.umn.edu/~brad/data/ColoradoS-T1.bug fits model (8.6), the additive space-time model. This is a "direct" solution, where we explicitly construct the temporal process. By contrast, the file www.biostat.umn.edu/~brad/data/ColoradoS-T2.bug uses the spatial.exp function, tricking it to handle temporal correlations by setting the y-coordinates to 0.

4.(a) Running five chains of an MCMC algorithm, we obtained point and 95% interval estimates of –0.01 and [–0.20, 0.18] for β; using the same reparametrization under the chosen model (10) in Waller et al. (1997), the point and interval estimates instead are –0.20 and [–0.26, –0.15]. Thus, using this reparametrization shows that age adjusting has eliminated the statistical significance of the difference between the two female groups.

(b) Figure B.4 shows the fitted age-adjusted lung cancer death rates per 1000 population for nonwhite females for the years 1968, 1978, and 1988. The scales of the three figures show that lung cancer death rates are increasing over time. For 1968, we see a strong spatial pattern of increasing rates as we move from northwest to southeast, perhaps the result of an unmeasured occupational covariate (farming versus mining). Except for persistent low rates in the northwest corner, however,

node (unit)	Mean	sd	MC error	2.5%	Median	97.5%
W_1 (A)	−0.0491	0.835	0.0210	−1.775	−0.0460	1.639
W_3 (C)	−0.183	0.9173	0.0178	−2.2	−0.136	1.52
W_5 (E)	−0.0320	0.8107	0.0319	−1.682	−0.0265	1.572
W_6 (F)	0.417	0.8277	0.0407	−1.066	0.359	2.227
W_9 (I)	0.255	0.7969	0.0369	−1.241	0.216	1.968
W_{11} (K)	−0.195	0.9093	0.0209	−2.139	−0.164	1.502
ρ_1 (A)	1.086	0.1922	0.0072	0.7044	1.083	1.474
ρ_3 (C)	0.901	0.2487	0.0063	0.4663	0.882	1.431
ρ_5 (E)	1.14	0.1887	0.0096	0.7904	1.139	1.521
ρ_6 (F)	0.935	0.1597	0.0084	0.6321	0.931	1.265
ρ_9 (I)	0.979	0.1683	0.0087	0.6652	0.971	1.339
ρ_{11} (K)	0.881	0.2392	0.0103	0.4558	0.861	1.394
τ	1.73	1.181	0.0372	0.3042	1.468	4.819
β_0	−7.11	0.689	0.0447	−8.552	−7.073	−5.874
β_1	0.596	0.2964	0.0105	0.0610	0.578	1.245
RR	3.98	2.951	0.1122	1.13	3.179	12.05

Table B.3 *Posterior summaries, MAC survival model (10,000 samples, after a burn-in of 1,000).*

this trend largely disappears over time, perhaps due to increased mixing of the population or improved access to quality health care and health education.

Chapter 9

1.(a) Table B.3 summarizes the results from the nonspatial model, which are based on 10,000 posterior samples obtained from a single MCMC chain after a burn-in of 1,000 iterations. Looking at this table and the raw data in Table 9.14, basic conclusions are as follows:

- Units A and E have moderate overall risk $(W_i \approx 0)$ but increasing hazards $(\rho > 1)$: few deaths, but they occur late.
- Units F and I have high overall risk $(W_i > 0)$ but decreasing hazards $(\rho < 1)$: several early deaths, many long-term survivors.
- Units C and K have low overall risk $(W_i < 0)$ and decreasing hazards $(\rho < 1)$: no deaths at all; a few survivors.
- The two drugs differ significantly: CI for β_1 (RR) excludes 0 (1).

2.(b) The appropriate interval censored `WinBUGS` code is as follows:

```
model
{
for (i in 1:N) {
  TimeSmoking[i] <- Age[i] - AgeStart[i]
  RelapseT[i] ~ dweib(rho[i],mu[i])I(censored.time1[i],
    censored.time2[i])
  log(mu[i]) <- beta0 + beta[1]*TimeSmoking[i]
    + beta[2]*SexF[i] + beta[3]*SIUC[i]
    + beta[4]*F10Cigs[i] + W[County[i]]
  rho[i] <- exp(lrho[County[i]])
  }

# for (i in 1:regions) {W[i] ~ dnorm(0.0, tau_W)}
# for (i in 1:regions) {lrho[i] ~ dnorm(0.0, tau_rho)}

for (i in 1:sumnum) {weights[i] <- 1}

W[1:regions] ~ car.normal(adj[], weights[], num[], tau_W)
lrho[1:regions] ~ car.normal(adj[], weights[], num[],
  tau_rho)

for (i in 1:4) { beta[i] ~ dnorm(0.0, 0.0001)}
beta0 ~ dnorm(0.0,0.0001)
tau_W ~ dgamma(0.1,0.1)
tau_rho ~ dgamma(0.1,0.1)
}
```

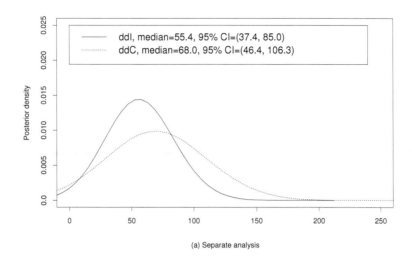

(a) Separate analysis

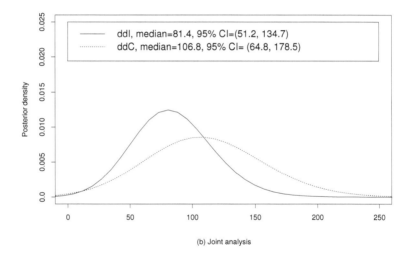

(b) Joint analysis

Figure B.1 Median survival time for a hypothetical patient (male, negative AIDS diagnosis at study entry, intolerant of AZT): (a) estimated posterior density of median survival time of the patient from separate analysis; (b) estimated posterior density of median survival time of the patient from joint analysis.

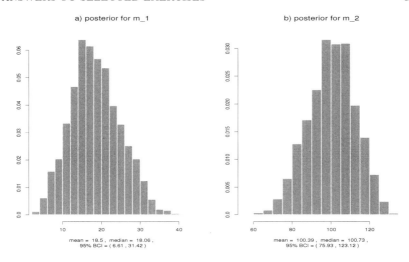

Figure B.2 *Posterior histograms of sampled* **m** *values, motivating example.*

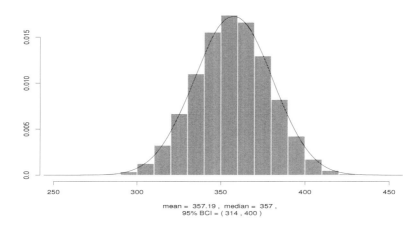

Figure B.3 *Posterior histogram and kernel density estimate, sampled* Y_3 *values, motivating example.*

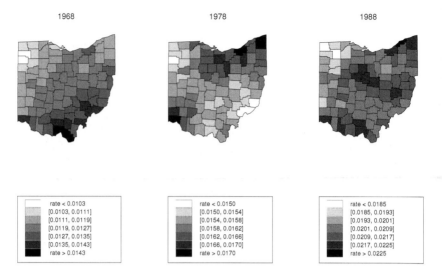

Figure B.4 *Fitted median lung cancer death rates per 1000 population, nonwhite females.*

References

Abrahamsen, N. (1993). Bayesian kriging for seismic depth conversion of a multi-layer reservoir. In *Geostatistics Troia, '92*, ed. A. Soares, Boston: Kluwer Academic Publishers, pp. 385–398.

Abramowitz, M. and Stegun, I.A. (1965). *Handbook of Mathematical Functions*. New York: Dover.

Agarwal, D.K. and Gelfand, A.E. (2002). Slice Gibbs sampling for simulation based fitting of spatial data models. Technical report, Institute for Statistics and Decision Sciences, Duke University.

Agarwal, D.K., Gelfand, A.E., and Silander, J.A. (2002). Investigating tropical deforestation using two-stage spatially misaligned regression models. *J. Agric. Biol. Environ. Statist.*, **7**, 420–439.

Agarwal, D.K., Gelfand, A.E., Sirmans, C.F. and Thibadeau, T.G. (2005). Non-stationary spatial house price models. To appear *J. Statist. Plann. Inf.*

Agresti, A. (2002). *Categorical Data Analysis*, 2nd ed. New York: Wiley.

Aitken, M., Anderson, D., Francis, B., and Hinde, J. (1989). *Statistical Modelling in GLIM*. Oxford: Oxford Statistical Science.

Akima, H. (1978). A method of bivariate interpolation and smooth surface fitting for irregularly distributed data points. *ACM Transactions on Mathematical Software*, **4**, 148-164.

Andersen, P.K. and Gill, R.D. (1982). Cox's regression model for counting processes: A large sample study. *Ann. Statist.*, **10**, 1100–1120.

Anselin, L. (1988). *Spatial Econometrics: Models and Methods*. Dordrecht: Kluwer Academic Publishers.

Anton, H. (1984). *Calculus with Analytic Geometry*, 2nd ed. New York: Wiley.

Armstrong, M. and Diamond, P. (1984). Testing variograms for positive definiteness. *Mathematical Geology*, **24**, 135–147.

Armstrong, M. and Jabin, R. (1981). Variogram models must be positive definite. *Mathematical Geology*, **13**, 455–459.

Arnold, B.C. and Strauss, D.J. (1991). Bivariate distributions with conditionals in prescribed exponential families. *J. Roy. Statist. Soc., Ser. B*, **53**, 365–375.

Assunção, R.M. (2003). Space-varying coefficient models for small area data. *Environmetrics*, **14**, 453–473.

Assunção, R.M., Potter, J.E., and Cavenaghi, S.M. (2002). A Bayesian space varying parameter model applied to estimating fertility schedules. *Statistics in Medicine*, **21**, 2057–2075.

Assunção, R.M., Reis, I.A., and Oliveira, C.D.L. (2001). Diffusion and prediction of Leishmaniasis in a large metropolitan area in Brazil with a Bayesian space-time model. *Statistics in Medicine*, **20**, 2319–2335.

Bailey, T.C. and Gatrell, A.C. (1995). *Interactive Spatial Data Analysis*. Essex: Addison Wesley Longman.

Bailey, M.J., Muth, R.F., and Nourse, H.O. (1963). A regression method for real estate price index construction. *J. Amer. Statist. Assoc.*, **58**, 933–942.

Banerjee, S. (2000). On multivariate spatial modelling in a Bayesian setting. Unpublished Ph.D. dissertation, Department of Statistics, University of Connecticut.

Banerjee, S. (2005). On geodetic distance computations in spatial modelling. To appear *Biometrics*.

Banerjee, S. and Carlin, B.P. (2002). Spatial semiparametric proportional hazards models for analyzing infant mortality rates in Minnesota counties. In *Case Studies in Bayesian Statistics, Volume VI*, eds. C. Gatsonis et al. New York: Springer-Verlag, pp. 137–151.

Banerjee, S. and Carlin, B.P. (2003). Semiparametric spatio-temporal frailty modeling. *Environmetrics*, **14**, 523–535.

Banerjee, S., Gamerman, D., and Gelfand, A.E. (2003). Spatial process modelling for univariate and multivariate dynamic spatial data. Technical report, Division of Biostatistics, University of Minnesota.

Banerjee, S. and Gelfand, A.E. (2002). Prediction, interpolation and regression for spatially misaligned data. *Sankhya, Ser. A*, **64**, 227–245.

Banerjee, S. and Gelfand, A.E. (2003). On smoothness properties of spatial processes. *J. Mult. Anal.*, **84**, 85–100.

Banerjee, S., Gelfand, A.E., Knight, J., and Sirmans, C.F. (2004). Spatial modelling of house prices using normalized distance-weighted sums of stationary processes. *J. Bus. Econ. Statist.*, **22**, 206–213.

Banerjee, S., Gelfand, A.E., and Polasek, W. (2000). Geostatistical modelling for spatial interaction data with application to postal service performance. *J. Statist. Plann. Inf.*, **90**, 87–105.

Banerjee, S., Gelfand, A.E., Sirmans, C.F. (2004). Directional rates of change under spatial process models. *J. Amer. Statist. Assoc.*, **98**, 946–954.

Banerjee, S., Wall, M.M., and Carlin, B.P. (2003). Frailty modeling for spatially correlated survival data, with application to infant mortality in Minnesota. *Biostatistics*, **4**, 123–142.

Barry, R.P. and Ver Hoef, J.M. (1996). Blackbox kriging: Spatial prediction without specifying variogram models. *J. Agric. Biol. Environ. Statist.*, **1**, 297–322.

Bayes, T. (1763). An essay towards solving a problem in the doctrine of chances. *Philos. Trans. Roy. Soc. London*, **53**, 370–418. Reprinted, with an introduction by George Barnard, in 1958 in *Biometrika*, **45**, 293–315.

Becker, R.A. and Wilks, A.R. (1993). Maps in S. Technical report, AT&T Bell Laboratories; website www.research.att.com/areas/stat/doc/93.2.ps.

Berger, J.O. (1985). *Statistical Decision Theory and Bayesian Analysis*, 2nd ed. New York: Springer-Verlag.

Berger, J.O. and Pericchi, L.R. (1996). The intrinsic Bayes factor for linear models. In *Bayesian Statistics 5*, eds. J.M. Bernardo, J.O. Berger, A.P. Dawid, and A.F.M. Smith. Oxford: Oxford University Press, pp. 25–44.

Berkson, J. and Gage, R.P. (1952). Survival curve for cancer patients following treatment. *J. Amer. Statist. Assoc.*, **47**, 501–515.

Berliner, L.M. (2000). Hierarchical Bayesian modeling in the environmental sciences. *Allgemeines Statistisches Archiv (Journal of the German Statistical Society)*, **84**, 141–153.

Bernardinelli, L., Clayton, D., and Montomoli, C. (1995). Bayesian estimates of disease maps: how important are priors? *Statistics in Medicine*, **14**, 2411–2431.

Bernardinelli, L. and Montomoli, C. (1992). Empirical Bayes versus fully Bayesian analysis of geographical variation in disease risk. *Statistics in Medicine*, **11**, 983–1007.

Bernardinelli, L., Pascutto, C., Best, N.G. and Gilks, W.R. (1997). Disease mapping with errors in covariates. *Statistics in Medicine*, **16**, 741–752.

Bernardo, J.M. and Smith, A.F.M. (1994). *Bayesian Theory.* New York: Wiley.

Besag, J. (1974). Spatial interaction and the statistical analysis of lattice systems (with discussion). *J. Roy. Statist. Soc., Ser. B*, **36**, 192–236.

Besag, J. and Green, P.J. (1993). Spatial statistics and Bayesian computation (with discussion). *J. Roy. Statist. Soc., Ser. B*, **55**, 25–37.

Besag, J., Green, P., Higdon, D., and Mengersen, K. (1995). Bayesian computation and stochastic systems (with discussion). *Statistical Science*, **10**, 3–66.

Besag, J. and Kooperberg, C. (1995). On conditional and intrinsic autoregressions. *Biometrika*, **82**, 733–746.

Besag, J., York, J.C., and Mollié, A. (1991). Bayesian image restoration, with two applications in spatial statistics (with discussion). *Annals of the Institute of Statistical Mathematics*, **43**, 1–59.

Best, N.G., Ickstadt, K., and Wolpert, R.L. (2000). Spatial Poisson regression for health and exposure data measured at disparate resolutions. *J. Amer. Statist. Assoc.*, **95**, 1076–1088.

Best, N.G., Waller, L.A., Thomas, A., Conlon, E.M. and Arnold, R.A. (1999). Bayesian models for spatially correlated diseases and exposure data. In *Bayesian Statistics 6*, eds. J.M. Bernardo et al. Oxford: Oxford University Press, pp. 131–156.

Billheimer, D. and Guttorp, P. (1996). Spatial models for discrete compositional data. Technical Report, Department of Statistics, University of Washington.

Breslow, N.E. and Clayton, D.G. (1993). Approximate inference in generalized linear mixed models. *J. Amer. Statist. Assoc.*, **88**, 9–25.

Breslow, N.E. and Day, N.E. (1987). *Statistical Methods in Cancer Research, Volume II – The Design and Analysis of Cohort Studies.* Lyon: International Agency for Research on Cancer.

Brook, D. (1964). On the distinction between the conditional probability and the joint probability approaches in the specification of nearest-neighbour systems. *Biometrika*, **51**, 481–483.

Brooks, S.P. and Gelman, A. (1998). General methods for monitoring convergence of iterative simulations. *J. Comp. Graph. Statist.*, **7**, 434–455.

Brown, P.E., Kåresen, K.F., Roberts, G.O., and Tonellato, S. (2000). Blur-generated nonseparable space-time models. *J. Roy. Statist. Soc., Ser. B*, **62**, 847–860.

Brown, P., Le, N. and Zidek, J. (1994). Multivariate spatial interpolation and exposure to air pollutants. *The Canadian Journal of Statistics*, **22**, 489–509.

California Office of Statewide Health Planning and Development (1997). *Hospital Patient Discharge Data (Public Use Version)*. Sacramento, CA: State of California.

Cancer Surveillance and Control Program (1997). Case completeness and data quality audit: Minnesota Cancer Surveillance System 1994–1995. Technical report, Minnesota Department of Health.

Carlin, B.P. and Banerjee, S. (2003). Hierarchical multivariate CAR models for spatio-temporally correlated survival data (with discussion). In *Bayesian Statistics 7*, eds. J.M. Bernardo, M.J. Bayarri, J.O. Berger, A.P. Dawid, D. Heckerman, A.F.M. Smith, and M. West. Oxford: Oxford University Press, pp. 45–63.

Carlin, B.P., Chaloner, K., Church, T., Louis, T.A., and Matts, J.P. (1993). Bayesian approaches for monitoring clinical trials with an application to toxoplasmic encephalitis prophylaxis. *The Statistician*, **42**, 355–367.

Carlin, B.P. and Hodges, J.S. (1999). Hierarchical proportional hazards regression models for highly stratified data. *Biometrics*, **55**, 1162–1170.

Carlin, B.P. and Louis, T.A. (2000). *Bayes and Empirical Bayes Methods for Data Analysis*, 2nd ed. Boca Raton, FL: Chapman and Hall/CRC Press.

Carlin, B.P. and Pérez, M.-E. (2000). Robust Bayesian analysis in medical and epidemiological settings. In *Robust Bayesian Analysis (Lecture Notes in Statistics, Vol. 152)*, eds. D.R. Insua and F. Ruggeri. New York: Springer-Verlag, pp. 351–372.

Carlin, B.P., Xia, H., Devine, O., Tolbert, P., and Mulholland, J. (1999). Spatio-temporal hierarchical models for analyzing Atlanta pediatric asthma ER visit rates. In *Case Studies in Bayesian Statistics, Volume IV*, eds. C. Gatsonis et al. New York: Springer-Verlag, pp. 303–320.

Carlin, B.P., Zhu, L., and Gelfand, A.E. (2001). Accommodating scale misalignment in spatio-temporal data. In *Bayesian Methods with Applications to Science, Policy and Official Statistics*, eds. E.I. George et al. Luxembourg: Office for Official Publications of the European Communities (Eurostat), pp. 41–50.

Case, K.E. and Shiller R.J. (1989). The efficiency of the market for single family homes. *American Economic Review*, **79**, 125–137.

Casella, G. and George, E. (1992). Explaining the Gibbs sampler. *The American Statistician*, **46**, 167–174.

Casella, G., Lavine, M., and Robert, C.P. (2001). Explaining the perfect sampler. *The American Statistician*, **55**, 299–305.

Centers for Disease Control and Prevention (1987). Underreporting of alcohol-related mortality on death certificates of young U.S. Army veterans. In *Morbidity and Mortality Weekly Report*, U.S. Department of Health and Human Services, Vol. 36, No. 27 (July 1987), pp. 437–440.

Centers for Disease Control and Prevention (1996). Mortality trends for Alzheimer's disease, 1979–1991. In *Vital and Health Statistics*, U.S. Department of Health and Human Services, Series 20, No. 28 (January 1996), p. 3.

Chambers, J.A., Cleveland, W.S., Kleiner, B., and Tukey, P.A. (1983). *Graphical Methods for Data Analysis*. Belmont, CA: Wadsworth.

Cherry, S., Banfield, J., and Quimby, W.F. (1996). An evaluation of a nonparametric method of estimating semi-variograms of isotropic spatial processes. *Journal of Applied Statistics*, **23**, 435–449.

Chen, M.-H., Ibrahim, J.G., and Sinha, D. (1999). A new Bayesian model for survival data with a surviving fraction. *J. Amer. Statist. Assoc.*, **94**, 909–919.

Chen, M.-H., Shao, Q.-M., and Ibrahim, J.G. (2000). *Monte Carlo Methods in Bayesian Computation*. New York: Springer-Verlag.

Chib, S. and Greenberg, E. (1998). Analysis of multivariate probit models. *Biometrika*, **85**, 347–361.

Chiles, J.P. and Delfiner, P. (1999). *Geostatistics: Modeling Spatial Uncertainty*. New York: Wiley.

Christakos, G. (1984). On the problem of permissible covariance and variogram models. *Water Resources Research*, **20**, 251–265.

Christakos, G. (1992). *Random Field Models in Earth Sciences*. New York: Academic Press.

Clayton, D. (1991). A Monte Carlo method for Bayesian inference in frailty models. *Biometrics*, **47**, 467–485.

Clayton, D. (1994). Some approaches to the analysis of recurrent event data. *Statistics in Medical Research*, **3**, 244–262.

Clayton, D.G. and Kaldor, J.M. (1987). Empirical Bayes estimates of age-standardized relative risks for use in disease mapping. *Biometrics*, **43**, 671-681.

Cliff, A.D. and Ord, J.K. (1973). *Spatial Autocorrelation*. London: Pion.

Clifford, P. (1990). Markov random fields in statistics. In *Disorder in Physical Systems*. Oxford: Oxford University Press, pp. 20–32.

Congdon, P. (2001). *Bayesian Statistical Modelling*. Chichester: Wiley.

Congdon, P. (2003). *Applied Bayesian Modelling*. Chichester: Wiley.

Congdon, P. and Best, N.G. (2000). Small area variation in hospital admission rates: Adjusting for referral and provider variation. *J. Roy. Statist. Soc. Ser. C (Applied Statistics)*, **49**, 207–226.

Cohn, D.L., Fisher, E., Peng, G., Hodges, J., Chesnutt, J., Child, C., Franchino, B., Gibert, C., El-Sadr, W., Hafner, R., Korvick, J., Ropka, M., Heifets, L., Clotfelter, J., Munroe, D., and Horsburgh, R. (1999). A prospective randomized trial of four three-drug regimens in the treatment of disseminated *Mycobacterium avium* complex disease in AIDS patients: Excess mortality associated with high-dose clarithromycin. *Clinical Infectious Diseases*, **29**, 125–133.

Cole, T.J. and Green, P.J. (1992). Smoothing reference centiles; The LMS method and penalized likelihood. *Statistics in Medicine*, **11**, 1305–1319.

Conlon, E.M. and Waller, L.A. (1999). Flexible spatial hierarchical models for mapping disease rates. *Proceedings of the Statistics and the Environment Section of the American Statistical Association*, pp. 82–87.

Cooley, J.W. and Tukey, J.W. (1965). An algorithm for the machine computation of complex Fourier series. *Mathematics of Computation*, **19**, 297–301.

Corsten, L.C.A. (1989). Interpolation and optimal linear prediction. *Statistica Neerlandica*, **43**, 69–84.

Cowles, M.K. (2002). MCMC sampler convergence rates for hierarchical normal linear models: A simulation approach. *Statistics and Computing*, **12**, 377–389.

Cowles, M.K. (2003). Efficient model-fitting and model-comparison for high-dimensional Bayesian geostatistical models. *Journal of Statistical Planning and*

Inference, **112**, 221–239.

Cowles, M.K. and Carlin, B.P. (1996). Markov chain Monte Carlo convergence diagnostics: A comparative review. *J. Amer. Statist. Assoc.*, **91**, 883–904.

Cox, D.R. and Oakes, D. (1984). *Analysis of Survival Data*. London: Chapman and Hall.

Cramér, H. and Leadbetter M.R. (1967). *Stationary and Related Stochastic Processes*. New York: Wiley.

Cressie, N.A.C. (1993). *Statistics for Spatial Data*, 2nd ed. New York: Wiley.

Cressie, N.A.C. (1996). Change of support and the modifiable areal unit problem. *Geographical Systems*, **3**, 159–180.

Cressie, N. and Huang, H.-C. (1999). Classes of nonseparable spatio-temporal stationary covariance functions. *J. Amer. Statist. Assoc.*, **94**, 1330–1340.

Damian, D., Sampson, P.D., and Guttorp, P. (2001). Bayesian estimation of semiparametric non-stationary spatial covariance structures. *Environmetrics*, **12**, 161–178.

Damien, P., Wakefield, J., and Walker, S. (1999). Gibbs sampling for Bayesian non-conjugate and hierarchical models by using auxiliary variables. *J. Roy. Statist. Soc., Ser. B*, **61**, 331–344.

Daniels, M.J. and Kass, R.E. (1999). Nonconjugate Bayesian estimation of covariance matrices and its use in hierarchical models. *J. Amer. Statist. Assoc.*, **94**, 1254–1263.

DeGroot, M.H. (1970). *Optimal Statistical Decisions*. New York: McGraw-Hill.

Dempster, A.M. (1972). Covariance selection. *Biometrics*, **28**, 157–175.

Dempster, A.P., Laird, N.M., and Rubin, D.B. (1977). Maximum likelihood estimation from incomplete data via the EM algorithm (with discussion). *J. Roy. Statist. Soc., Ser. B*, **39**, 1–38.

DeOliveira, V. (2000). Bayesian prediction of clipped Gaussian random fields. *Computational Statistics and Data Analysis*, **34**, 299–314.

DeOliveira, V., Kedem, B., and Short, D.A. (1997). Bayesian prediction of transformed Gaussian random fields. *J. Amer. Statist. Assoc.*, **92**, 1422–1433.

Devesa, S.S., Grauman, D.J., Blot, W.J., Pennello, G.A., Hoover, R.N., and Fraumeni, J.F., Jr. (1999). *Atlas of Cancer Mortality in the United States, 1950–94*. NIH Publ. No. 99-4564, Bethesda, MD: National Institutes of Health; website `www-dceg.ims.nci.nih.gov/atlas/index.html`.

Devine, O.J., Qualters, J.R., Morrissey, J.L., and Wall, P.A. (1998). Estimation of the impact of the former Feed Materials Production Center (FMPC) on lung cancer mortality in the surrounding community. Technical report, Radiation Studies Branch, Division of Environmental Hazards and Health Effects, National Center for Environmental Health, Centers for Disease Control and Prevention.

Diggle, P.J. (2003). *Statistical Analysis of Spatial Point Patterns*, 2nd ed. London: Arnold.

Diggle, P.J. and Ribeiro, P.J. (2002). Bayesian inference in Gaussian model-based geostatistics. *Geographical and Environmental Modelling*, **6**, 129–146.

Diggle, P.J., Tawn, J.A., and Moyeed, R.A. (1998). Model-based geostatistics (with discussion). *J. Roy. Statist. Soc., Ser. C (Applied Statistics)*, **47**, 299–350.

Doksum, K.A. and Gasko, M. (1990). On a correspondence between models in binary regression analysis and in survival analysis. *International Statistical Review*, **58**, 243–252.

Duan, J. and Gelfand, A.E. (2003). Finite mixture model of nonstationary spatial data. Technical report, Institute for Statistics and Decision Sciences, Duke University.

Eberly, L.E. and Carlin, B.P. (2000). Identifiability and convergence issues for Markov chain Monte Carlo fitting of spatial models. *Statistics in Medicine*, **19**, 2279–2294.

Ecker, M.D. and Gelfand, A.E. (1997). Bayesian variogram modeling for an isotropic spatial process. *J. Agric. Biol. Environ. Statist.*, **2**, 347–369.

Ecker, M.D. and Gelfand, A.E. (1999). Bayesian modeling and inference for geometrically anisotropic spatial data. *Mathematical Geology*, **31**, 67–83.

Ecker, M.D. and Gelfand, A.E. (2003). Spatial modeling and prediction under stationary non-geometric range anisotropy. *Environmental and Ecological Statistics*, **10**, 165–178.

Ecker, M.D. and Heltshe, J.F. (1994). Geostatistical estimates of scallop abundance. In *Case Studies in Biometry*, eds. N. Lange, L. Ryan, L. Billard, D. Brillinger, L. Conquest, and J. Greenhouse. New York: Wiley, pp. 107–124.

Elliott, P., Wakefield, J.C., Best, N.G., and Briggs, D.J., eds. (2000). *Spatial Epidemiology: Methods and Applications*. Oxford: Oxford University Press.

Ewell, M. and Ibrahim, J.G. (1997). The large sample distribution of the weighted log-rank statistic under general local alternatives. *Lifetime Data Analysis*, **3**, 5–12.

Farewell, V.T. (1982). The use of mixture models for the analysis of survival data with long term survivors. *Biometrics*, **38**, 1041–1046.

Farewell, V.T. (1986). Mixture models in survival analysis: Are they worth the risk? *Canadian Journal of Statistics*, **14**, 257–262.

Finkelstein, D.M. (1986). A proportional hazards model for interval-censored failure time data. *Biometrics*, **42**, 845–854.

Flowerdew, R. and Green, M. (1989). Statistical methods for inference between incompatible zonal systems. In *Accuracy of Spatial Databases*, eds. M. Goodchild and S. Gopal. London: Taylor and Francis, pp. 239–247.

Flowerdew, R. and Green, M. (1992). Developments in areal interpolating methods and GIS. *Annals of Regional Science*, **26**, 67–78.

Flowerdew, R. and Green, M. (1994). Areal interpolation and yypes of data. In *Spatial Analysis and GIS*, S. Fotheringham and P. Rogerson, eds., London: Taylor and Francis, pp. 121–145.

Fotheringham, A.S. and Rogerson, P., eds. (1994). *Spatial Analysis and GIS*. London: Taylor and Francis.

Fuentes, M. (2001). A high frequency kriging approach for non-stationary environmental processes. *Environmetrics*, **12**, 469–483.

Fuentes, M. (2002a). Spectral methods for nonstationary spatial processes. *Biometrika*, **89**, 197–210.

Fuentes, M. (2002b). Modeling and prediction of non-stationary spatial processes. *Statistical Modeling*, **2**, 281–298.

Fuentes, M. and Smith, R.L. (2001) Modeling nonstationary processes as a convolution of local stationary processes. Technical report, Department of Statistics, North Carolina State University.

Fuentes, M. and Smith, R.L. (2003). A new class of models for nonstationary processes. Technical report, Department of Statistics, North Carolina State University.

Gamerman, D. (1997). *Markov Chain Monte Carlo: Stochastic Simulation for Bayesian Inference*. Boca Raton, FL: Chapman and Hall/CRC Press.

Gamerman, D., Moreira, A.R.B., and Rue, H. (2005). Space-varying regression models: Specifications and simulation. To appear *Computational Statistics and Data Analysis*.

Gatsonis, C., Hodges, J.S., Kass, R.E., and Singpurwalla, N.D., eds. (1993). *Case Studies in Bayesian Statistics*. New York: Springer-Verlag.

Gatsonis, C., Hodges, J.S., Kass, R.E., and Singpurwalla, N.D., eds. (1995). *Case Studies in Bayesian Statistics, Volume II*. New York: Springer-Verlag.

Gatsonis, C., Hodges, J.S., Kass, R.E., McCulloch, R.E., Rossi, P., and Singpurwalla, N.D., eds. (1997). *Case Studies in Bayesian Statistics, Volume III*. New York: Springer-Verlag.

Gatsonis, C., Kass, R.E., Carlin, B.P., Carriquiry, A.L., Gelman, A., Verdinelli, I., and West, M., eds. (1999). *Case Studies in Bayesian Statistics, Volume IV*. New York: Springer-Verlag.

Gatsonis, C., Kass, R.E., Carlin, B.P., Carriquiry, A.L., Gelman, A., Verdinelli, I., and West, M., eds. (2002). *Case Studies in Bayesian Statistics, Volume V*. New York: Springer-Verlag.

Gatsonis, C., Kass, R.E., Carriquiry, A.L., Gelman, A., Higdon, D., Pauler, D., and Verdinelli, I., eds. (2003). *Case Studies in Bayesian Statistics, Volume VI*. New York: Springer-Verlag.

Gelfand, A.E., Ecker, M.D., Knight, J.R., and Sirmans, C.F. (2005). The dynamics of location in home price. To appear *Journal of Real Estate Finance and Economics*.

Gelfand, A.E. and Ghosh, S.K. (1998). Model choice: a minimum posterior predictive loss approach. *Biometrika*, **85**, 1–11.

Gelfand, A.E., Kim, H.-J., Sirmans, C.F., and Banerjee, S. (2003). Spatial modeling with spatially-varying coefficient processes. *J. Amer. Statist. Assoc.*, **98**, 387–396.

Gelfand, A.E., Kottas, A., and MacEachern, S.N. (2003). Nonparametric Bayesian spatial modeling using dependent Dirichlet processes. Technical report, Institute for Statistics and Decision Sciences, Duke University.

Gelfand A.E. and Mallick, B.K. (1995). Bayesian analysis of proportional hazards models built from monotone functions. *Biometrics*, **51**, 843–852.

Gelfand, A.E., Sahu, S.K., and Carlin, B.P. (1995). Efficient parametrizations for normal linear mixed models. *Biometrika*, **82**, 479–488.

Gelfand, A.E., Sahu, S.K., and Carlin, B.P. (1996). Efficient parametrizations for generalized linear mixed models (with discussion). In *Bayesian Statistics 5*, eds. J.M. Bernardo, J.O. Berger, A.P. Dawid, and A.F.M. Smith. Oxford: Oxford University Press, pp. 165–180.

Gelfand, A.E., Schmidt, A.M., Banerjee, S., and Sirmans, C.F. (2004). Nonstationary multivariate process modeling through spatially varying coregionalization (with discussion). *Test*, **13**, 1–50.

Gelfand, A.E. and Smith, A.F.M. (1990). Sampling-based approaches to calculating marginal densities. *J. Amer. Statist. Assoc.*, **85**, 398–409.

Gelfand, A.E. and Vounatsou, P. (2003). Proper multivariate conditional autoregressive models for spatial data analysis. *Biostatistics*, **4**, 11–25.

Gelfand, A.E., Zhu, L., and Carlin, B.P. (2001). On the change of support problem for spatio-temporal data. *Biostatistics*, **2**, 31–45.

Gelman, A., Carlin, J.B., Stern, H.S., and Rubin, D.B. (2004). *Bayesian Data Analysis*, 2nd ed. Boca Raton, FL: Chapman and Hall/CRC Press.

Gelman, A., Roberts, G.O., and Gilks, W.R. (1996). Efficient Metropolis jumping rules. In *Bayesian Statistics 5*, eds. J.M. Bernardo, J.O. Berger, A.P. Dawid, and A.F.M. Smith. Oxford: Oxford University Press, pp. 599–607.

Gelman, A. and Rubin, D.B. (1992). Inference from iterative simulation using multiple sequences (with discussion). *Statistical Science*, **7**, 457–511.

Geman, S. and Geman, D. (1984). Stochastic relaxation, Gibbs distributions and the Bayesian restoration of images. *IEEE Trans. on Pattern Analysis and Machine Intelligence*, **6**, 721–741.

Geyer, C.J. (1992). Practical Markov Chain Monte Carlo (with discussion). *Statistical Science*, **7**, 473–511.

Ghosh, M. (1992). Constrained Bayes estimates with applications. *J. Amer. Statist. Assoc.*, **87**, 533–540.

Gikhman, I.I. and Skorokhod, A.V. (1974). *Stochastic Differential Equations*. Berlin: Springer-Verlag.

Gilks, W.R., Richardson, S., and Spiegelhalter, D.J., eds. (1996). *Markov Chain Monte Carlo in Practice*. London: Chapman and Hall.

Gilks, W.R. and Wild, P. (1992). Adaptive rejection sampling for Gibbs sampling. *J. Roy. Statist. Soc., Ser. C (Applied Statistics)*, **41**, 337–348.

Gneiting, T. (2002). Nonseparable, stationary covariance functions for space-time data. *J. Amer. Statist. Assoc.*, **97**, 590–600.

Goldman, A.I. (1984). Survivorship analysis when cure is a possibility: A Monte Carlo study. *Statistics in Medicine* **3**, 153–163.

Goldman, A.I., Carlin, B.P., Crane, L.R., Launer, C., Korvick, J.A., Deyton, L., and Abrams, D.I. (1996). Response of CD4$^+$ and clinical consequences to treatment using ddI or ddC in patients with advanced HIV infection. *J. Acquired Immune Deficiency Syndromes and Human Retrovirology*, **11**, 161–169.

Goldstein, H. (1995). *Kendall's Library of Statistics 3: Multilevel Statistical Models*, 2nd ed. London: Arnold.

Golub, G.H. and van Loan, C.F. (1996). *Matrix Computations*, 3rd ed. Baltimore, MD: Johns Hopkins University Press.

Gotway, C.A. and Young, L.J. (2002). Combining incompatible spatial data. *J. Amer. Statist. Assoc.*, **97**, 632–648.

Green, P.J. and Richardson, S. (2002). Hidden Markov models and disease mapping. *J. Amer. Statist. Assoc.*, **97**, 1055–1070.

Greenwood, J.A. (1984). A unified theory of surface roughness. *Proceedings of the Royal Society of London. Series A, Mathematical and Physical Sciences*, **393**, 133–157.

Griffith, D.A. (1988). *Advanced Spatial Statistics*. Dordrecht, the Netherlands: Kluwer.

Grzebyk, M. and Wackernagel, H. (1994). Multivariate analysis and spatial/temporal scales: real and complex models. In *Proceedings of the XVIIth International Biometrics Conference*, Hamilton, Ontario, Canada: International Biometric Society, pp. 19–33.

Guggenheimer, H.W. (1977). *Differential Geometry*. New York: Dover Publications.

Guo, X. and Carlin, B.P. (2004). Separate and joint modeling of longitudinal and event time data using standard computer packages. *The American Statistician*, **58**, 16–24.

Guttman, I. (1982). *Linear Models: An Introduction*. New York: Wiley.

Guyon, X. (1995). *Random Fields on a Network: Modeling, Statistics, and Applications*. New York: Springer-Verlag.

Haining, R. (1990). *Spatial Data Analysis in the Social and Environmental Sciences*. Cambridge: Cambridge University Press.

Hall, P., Fisher, N.I., and Hoffmann, B. (1994). On the nonparametric estimation of covariance functions. *Ann. Statist.*, **22**, 2115–2134.

Handcock, M.S. (1999). Comment on "Prediction of spatial cumulative distribution functions using subsampling." *J. Amer. Statist. Assoc.*, **94**, 100–102.

Handcock, M.S. and Stein, M.L. (1993). A Bayesian analysis of kriging. *Technometrics*, **35**, 403–410.

Handcock, M.S. and Wallis, J. (1994). An approach to statistical spatial-temporal modeling of meteorological fields (with discussion). *J. Amer. Statist. Assoc.*, **89**, 368–390.

Haran, M. (2003). Efficient perfect and MCMC sampling methods for Bayesian spatial and components of variance models. Unpublished Ph.D. dissertation, School of Statistics and Division of Biostatistics, University of Minnesota.

Haran, M., Hodges, J.S., and Carlin, B.P. (2003). Accelerating computation in Markov random field models for spatial data via structured MCMC. *Journal of Computational and Graphical Statistics*, **12**, 249–264.

Harville, D.A. (1997). *Matrix Algebra from a Statistician's Perspective*. New York: Springer-Verlag.

Hastings, W.K. (1970). Monte Carlo sampling methods using Markov chains and their applications. *Biometrika*, **57**, 97–109.

Heikkinen, J. and Högmander, H. (1994). Fully Bayesian approach to image restoration with an application in biogeography. *Applied Statistics*, **43**, 569–582.

Henderson, R., Diggle, P.J., and Dobson, A. (2000). Joint modelling of longitudinal measurements and event time data. *Biostatistics*, **1**, 465–480.

Heywood, P.F., Singleton, N., and Ross, J. (1988). Nutritional status of young children – The 1982/3 National Nutrition Survey. *Papua New Guinea Medical Journal*, **31**, 91–101.

Higdon, D.M. (1998a). Auxiliary variable methods for Markov chain Monte Carlo with applications. *J. Amer. Statist. Assoc.*, **93**, 585–595.

Higdon, D.M. (1998b). A process-convolution approach to modeling temperatures in the north Atlantic Ocean. *Journal of Environmental and Ecological*

Statistics, **5**, 173–190.

Higdon, D.M. (2002). Space and space-time modeling using process convolutions. In *Quantitative Methods for Current Environmental Issues*, eds. C. Anderson, V. Barnett, P.C. Chatwin, and A.H. El-Shaarawi. London: Springer-Verlag, pp. 37–56.

Higdon, D., Lee, H., and Holloman, C. (2003). Markov chain Monte Carlo-based approaches for inference in computationally intensive inverse problems (with discussion). In *Bayesian Statistics 7*, eds. J.M. Bernardo, M.J. Bayarri, J.O. Berger, A.P. Dawid, D. Heckerman, A.F.M. Smith, and M. West. Oxford: Oxford University Press, pp. 181–197.

Higdon, D., Swall, J., and Kern, J. (1999). Non-stationary spatial modeling. In *Bayesian Statistics 6*, eds. J.M. Bernardo, J.O. Berger, A.P. Dawid, and A.F.M. Smith. Oxford: Oxford University Press, pp. 761–768.

Hill, R.C., Knight, J.R., and Sirmans, C.F. (1997). Estimating capital asset price indexes. *The Review of Economics and Statistics*, **79**, 226–233.

Hill, R.C., Sirmans, C.F., and Knight, J.R. (1999). A random walk down main street. *Regional Science and Urban Economics*, **29**, 89–103.

Hjort, N. and Omre, H. (1994). Topics in spatial statistics (with discussion). *Scand. J. Statist.*, **21**, 289–357.

Hoaglin, D.C., Mosteller, F., and Tukey, J.W. (1983). *Understanding Robust and Exploratory Data Analysis*. New York: Wiley.

Hoaglin, D.C., Mosteller, F., and Tukey, J.W. (1985). *Exploring Data Tables, Trends, and Shapes*. New York: Wiley.

Hobert, J.P., Jones, G.L., Presnell, B. and Rosenthal, J.S. (2002). On the applicability of regenerative simulation in Markov chain Monte Carlo. *Biometrika*, **89**, 731–743.

Hodges, J.S. (1998). Some algebra and geometry for hierarchical models, applied to diagnostics (with discussion). *J. Roy. Statist. Soc., Series B*, **60**, 497–536.

Hodges, J.S., Carlin, B.P., and Fan, Q. (2003). On the precision of the conditionally autoregressive prior in spatial models. *Biometrics*, **59**, 317–322.

Hoel, P.G., Port, S.C., and Stone, C.J. (1972). *Introduction to Stochastic Processes*. Boston: Houghton Mifflin.

Hogmander, H. and Møller, J. (1995). Estimating distribution maps from atlas data using methods of statistical image analysis. *Biometrics*, **51**, 393–404.

Hoeting, J.A., Leecaster, M., and Bowden, D. (2000). An improved model for spatially correlated binary responses. *J. Agr. Biol. Env. Statist.*, **5**, 102–114.

Hrafnkelsson, B. and Cressie, N. (2003). Hierarchical modeling of count data with application to nuclear fall-out. *Environmental and Ecological Statistics*, **10**, 179–200.

Huang, H.-C. and Cressie, N.A.C. (1996). Spatio-temporal prediction of snow water equivalent using the Kalman filter. *Computational Statistics and Data Analysis*, **22**, 159–175.

Huerta, G., Sanso, B., and Stroud, J.R. (2003). A spatiotemporal model for Mexico city ozone levels. *J. Roy. Statist. Soc., Ser. C (Applied Statistics)*, **53**, 231–248.

Ingram, D.D. and Kleinman, J.C. (1989). Empirical comparisons of proportional hazards and logistic regression models. *Statistics in Medicine*, **8**, 525–538.

Isaaks, E.H. and Srivastava, R.M. (1989). *An Introduction to Applied Geostatistics.* Oxford: Oxford University Press.

Jin, X. and Carlin, B.P. (2003). Multivariate parametric spatio-temporal models for county level breast cancer survival data. Research Report 2003–002, Division of Biostatistics, University of Minnesota.

Jones, C.B. (1997). *Geographical Information Systems and Computer Cartography.* Harlow, Essex, UK: Addison Wesley Longman.

Journel, A.G. and Froidevaux, R. (1982). Anisotropic hole-effect modelling. *Math. Geology*, **14**, 217–239.

Journel, A.G. and Huijbregts, C.J. (1978). *Mining Geostatistics.* New York: Academic Press.

Kaiser, M.S. and Cressie, N. (2000). The construction of multivariate distributions from Markov random fields. *J. Mult. Anal.*, **73**, 199–220.

Kaluzny, S.P., Vega, S.C., Cardoso, T.P., and Shelly, A.A. (1998). *S+SpatialStats: User's Manual for Windows and UNIX.* New York: Springer-Verlag.

Karson, M.J., Gaudard, M., Linder, E. and Sinha, D. (1999). Bayesian analysis and computations for spatial prediction (with discussion). *Environmental and Ecological Statistics*, **6**, 147–182.

Kashyap, R. and Chellappa, R. (1983). Estimation and choice of neighbors in spatial interaction models of images. *IEEE Transactions on Information Theory*, IT-29, 60–72.

Kass, R.E., Carlin, B.P., Gelman, A., and Neal, R. (1998). Markov chain Monte Carlo in practice: A roundtable discussion. *The American Statistician*, **52**, 93–100.

Kass, R.E. and Raftery, A.E. (1995). Bayes factors. *J. Amer. Statist. Assoc.*, **90**, 773–795.

Kent, J.T. (1989). Continuity properties for random fields. *Annals of Probability*, **17**, 1432–1440.

Killough, G.G., Case, M.J., Meyer, K.R., Moore, R.E., Rope, S.K., Schmidt, D.W., Schleien, B., Sinclair, W.K., Voillequé, P.G., and Till, J.E. (1996). Task 6: Radiation doses and risk to residents from FMPC operations from 1951–1988. Draft report, Radiological Assessments Corporation, Neeses, SC.

Kim, H., Sun, D., and Tsutakawa, R.K. (2001). A bivariate Bayes method for improving the estimates of mortality rates with a twofold conditional autoregressive model. *J. Amer. Statist. Assoc.*, **96**, 1506–1521.

Kinnard, W.N. (1971). *Income Property Valuation.* Lexington, MA: Heath-Lexington Books.

Knight, J.R., Dombrow, J., and Sirmans, C.F. (1995). A varying parameters approach to constructing house price indexes. *Real Estate Economics*, **23**, 87–105.

Knorr-Held, L. (2002). Some remarks on Gaussian Markov random field models for disease mapping. In *Highly Structured Stochastic Systems*, eds. N. Hjort, P. Green and S. Richardson. Oxford: Oxford University Press.

Knorr-Held, L. and Best, N.G. (2001). A shared component model for detecting joint and selective clustering of two diseases. *J. Roy. Statist. Soc. Ser. A*, **164**, 73–85.

Knorr-Held, L. and Rue, H. (2002). On block updating in Markov random field models for disease mapping. *Scand. J. Statist.*, **29**, 597–614.

Krige, D.G. (1951). A statistical approach to some basic mine valuation problems on the Witwatersrand. *J. Chemical, Metallurgical and Mining Society of South Africa*, **52**, 119–139.

Lahiri, S.N., Kaiser, M.S., Cressie, N., and Hsu, N.-J. (1999). Prediction of spatial cumulative distribution functions using subsampling (with discussion). *J. Amer. Statist. Assoc.*, **94**, 86–110.

Laird, N.M. and Ware, J.H. (1982). Random-effects models for longitudinal data. *Biometrics*, **38**, 963–974.

Langford, I.H., Leyland, A.H., Rasbash, J., and Goldstein, H. (1999). Multilevel modelling of the geographical distributions of diseases. *Applied Statistics*, **48**, 253–268.

Lawson, A.B. (2001). *Statistical Methods in Spatial Epidemiology*. New York: Wiley.

Lawson, A.B. and Denison, D.G.T., eds. (2002). *Spatial Cluster Modelling*. Boca Raton, FL: Chapman and Hall/CRC Press.

Le, C.T. (1997). *Applied Survival Analysis*. New York: Wiley.

Le, N. and Zidek, J. (1992). Interpolation with uncertain spatial covariances: A Bayesian alternative to kriging. *J. Mult. Anal.*, **43**, 351–374.

Le, N.D., Sun, W., and Zidek, J.V. (1997). Bayesian multivariate spatial interpolation with data missing by design. *J. Roy. Statist. Soc., Ser. B*, **59**, 501–510.

Leecaster, M.K. (2002). Geostatistic modeling of subsurface characteristics in the Radioactive Waste Management Complex Region, Operable Unit 7-13/14. Technical report, Idaho National Engineering and Environmental Laboratory, Idaho Falls, ID.

Lele, S. (1995). Inner product matrices, kriging, and nonparametric estimation of the variogram. *Mathematical Geology*, **27**, 673–692.

Li, Y. and Ryan, L. (2002). Modeling spatial survival data using semiparametric frailty models. *Biometrics*, **58**, 287–297.

Liu, J.S. (1994). The collapsed Gibbs sampler in Bayesian computations with applications to a gene regulation problem. *J. Amer. Statist. Assoc.*, **89**, 958–966.

Liu, J.S. (2001). *Monte Carlo Strategies in Scientific Computing*. New York: Springer-Verlag.

Liu, J.S., Wong, W.H., and Kong, A. (1994). Covariance structure of the Gibbs sampler with applications to the comparisons of estimators and augmentation schemes. *Biometrika*, **81**, 27–40.

Louis, T.A. (1982). Finding the observed information matrix when using the EM algorithm. *J. Roy. Statist. Soc., Ser. B*, **44**, 226–233.

Louis, T.A. (1984). Estimating a population of parameter values using Bayes and empirical Bayes methods. *J. Amer. Statist. Assoc.*, **79**, 393–398.

Luo, Z. and Wahba, G. (1998). Spatio-temporal analogues of temperature using smoothing spline ANOVA. *Journal of Climatology*, **11**, 18–28.

Lusht, K.M. (1997). *Real Estate Valuation*. Chicago: Irwin.

MacEachern, S.N. and Berliner, L.M. (1994). Subsampling the Gibbs sampler. *The American Statistician*, **48**, 188–190.

Majumdar, A. and Gelfand, A.E. (2003). Convolution methods for developing cross-covariance functions. Technical report, Institute for Statistics and Decision Sciences, Duke University.

Mardia, K.V. (1988). Multi-dimensional multivariate Gaussian Markov random fields with application to image processing. *Journal of Multivariate Analysis*, **24**, 265–284.

Mardia, K.V. and Goodall, C. (1993). Spatio-temporal analyses of multivariate environmental monitoring data. In *Multivariate Environmental Statistics*, eds. G.P. Patil and C.R. Rao. Amsterdam: Elsevier, pp. 347–386.

Mardia, K.V., Goodall, C., Redfern, E.J., and Alonso, F.J. (1998). The kriged Kalman filter (with discussion). *Test*, **7**, 217–285.

Mardia, K.V., Kent, J.T., and Bibby, J.M. (1979). *Multivariate Analysis*. New York: Academic Press.

Mardia, K.V. and Marshall, R.J. (1984). Maximum likelihood estimation of models for residual covariance in spatial regression. *Biometrika*, **71**, 135–146.

Matérn, B. (1960; reprinted 1986). *Spatial Variation, 2nd ed.* Berlin: Springer-Verlag.

Matheron, G. (1963). Principles of geostatistics. *Economic Geology*, **58**, 1246–1266.

Matheron, G. (1982). Pour une analyse krigeante des données regionaliées. Technical report, Ecole Nationale Supérieure des Mines de Paris.

McBratney, A. and Webster, R. (1986). Choosing functions for semi-variograms of soil properties and fitting them to sampling estimates. *Journal of Soil Science*, **37**, 617–639.

Mengersen, K.L., Robert, C.P., and Guihenneuc-Jouyaux, C. (1999). MCMC convergence diagnostics: A reviewww (with discussion). In *Bayesian Statistics 6*, eds. J.M. Bernardo, J.O. Berger, A.P. Dawid, and A.F.M. Smith. Oxford: Oxford University Press, pp. 415–440.

Metropolis, N., Rosenbluth, A.W., Rosenbluth, M.N., Teller, A.H., and Teller, E. (1953). Equations of state calculations by fast computing machines. *J. Chemical Physics*, **21**, 1087–1091.

Meyer, T.H., Ericksson, M., and Maggio, R.C. (2001). Gradient estimation from irregularly spaced datasets. *Mathematical Geology*, **33**, 693–717.

Mira, A., Møller, J., and Roberts, G.O. (2001). Perfect slice samplers. *J. Roy. Statist. Soc., Ser. B*, **63**, 593–606.

Mira, A. and Sargent, D.J. (2005). Strategies for speeding Markov Chain Monte Carlo algorithms. To appear *Statistical Methods and Applications*.

Mira, A. and Tierney, L. (2001). Efficiency and convergence properties of slice samplers. *Scandinavian Journal of Statistics*, **29**, 1–12.

Møller, J. and Waagepetersen, R. (2004). *Statistical Inference and Simulation for Spatial Point Processes*. Boca Raton, FL: Chapman and Hall/CRC Press.

Monahan, J.F. (2001). *Numerical Methods of Statistics*. Cambridge: Cambridge University Press.

Mueller, I., Vounatsou, P., Allen, B.J., and Smith, T. (2001). Spatial patterns of child growth in Papua New Guinea and their relation to environment, diet, socio-economic status and subsistence activities. *Annals of Human Biology*, **28**, 263–280.

Mugglin, A.S. and Carlin, B.P. (1998). Hierarchical modeling in Geographic Information Systems: population interpolation over incompatible zones. *J. Agric. Biol. Environ. Statist.*, **3**, 111–130.

Mugglin, A.S., Carlin, B.P., and Gelfand, A.E. (2000). Fully model based approaches for spatially misaligned data. *J. Amer. Statist. Assoc.*, **95**, 877–887.

Mugglin, A.S., Carlin, B.P., Zhu, L., and Conlon, E. (1999). Bayesian areal interpolation, estimation, and smoothing: An inferential approach for geographic information systems. *Environment and Planning A*, **31**, 1337–1352.

Murray, R., Anthonisen, N.R., Connett, J.E., Wise, R.A., Lindgren, P.G., Greene, P.G., and Nides, M.A. for the Lung Health Study Research Group (1998). Effects of multiple attempts to quit smoking and relapses to smoking on pulmonary function. *J. Clin. Epidemiol.*, **51**, 1317–1326.

Mykland, P., Tierney, L., and Yu, B. (1995). Regeneration in Markov chain samplers. *J. Amer. Statist. Assoc.*, **90**, 233–241.

Neal, R.M. (2003). Slice sampling (with discussion). *Annals of Statistics*, **31**, 705–767.

O'Hagan, A. (1994). *Kendall's Advanced Theory of Statistics Volume 2b: Bayesian Inference*. London: Edward Arnold.

O'Hagan, A. (1995). Fractional Bayes factors for model comparison (with discussion). *J. Roy. Statist. Soc., Ser. B*, **57**, 99–138.

Omre, H. (1987). Bayesian kriging – merging observations and qualified guesses in kriging. *Math. Geology*, **19**, 25–39.

Omre, H. (1988). A Bayesian approach to surface estimation. In *Quantitative Analysis of Mineral and Energy Resources*, eds. C.F. Chung et al., Boston: D. Reidel Publishing Co., pp. 289–306.

Omre, H. and Halvorsen, K.B. (1989). The Bayesian bridge between simple and universal kriging. *Math. Geology*, **21**, 767–786.

Omre, H., Halvorsen, K.B. and Berteig, V. (1989). A Bayesian approach to kriging. In *Geostatistics*, ed. M. Armstrong, Boston: Kluwer Academic Publishers, pp. 109–126.

Overton, W.S. (1989). Effects of measurements and other extraneous errors on estimated distribution functions in the National Surface Water Surveys. Technical Report 129, Department of Statistics, Oregon State University.

Pace, R.K. and Barry, R. (1997a). Sparse spatial autoregressions. *Statistics and Probability Letters*, **33**, 291–297.

Pace, R.K. and Barry, R. (1997b). Fast spatial estimation. *Applied Economics Letters*, **4**, 337–341.

Pace, R.K., Barry, R., Gilley, O.W., and Sirmans, C.F. (2000). A method for spatial-temporal forecasting with an application to real estate prices. *International J. Forecasting*, **16**, 229–246.

Pardo-Igúzquiza, E. and Dowd, P.A. (1997). AMLE3D: A computer program for the inference of spatial covariance parameters by approximate maximum likelihood estimation. *Computers and Geosciences*, **23**, 793–805.

Pearson, F. (1990). *Map Projections: Theory and Applications*. Boca Raton, FL: CRC Press.

Penberthy, L. and Stivers, C. (2000). Analysis of cancer risk in district health departments in Virginia, 1992–1995. Technical report, Cancer Prevention and Control Project, Virginia Department of Health.

Pickle, L.W., Mungiole, M., Jones G.K., and White, A.A. (1996). *Atlas of United States Mortality*. Hyattsville, MD: National Center for Health Statistics.

Press, W.H., Teukolsky, S.A, Vetterling, W.T., and Flannery, B.P. (1992). *Numerical Recipes in C*. Cambridge: Cambridge University Press.

Rao, C.R. (1973). *Linear Statistical Inference and its Applications*, 2nd ed. New York: Wiley.

Raudenbush, S.W. and Bryk, A.S. (2002). *Hierarchical Linear Models: Applications and Data Analysis Methods*, 2nd ed. Newbury Park, CA: Sage Press.

Rehman, S.U. and Shapiro, A. (1996). An integral transform approach to cross-variograms modeling. *Computational Statistics and Data Analysis*, **22**, 213–233.

Ribeiro, P.J. and Diggle, P.J. (2001). geoR: a package for geostatistical analysis. *R News*, **1**, 14–18.

Ripley, B.D. (1981). *Spatial Statistics*. New York: Wiley.

Robert, C.P. (1994). *The Bayesian Choice: A Decision-Theoretic Motivation*. New York: Springer-Verlag.

Robert, C.P. and Casella, G. (1999). *Monte Carlo Statistical Methods*. New York: Springer-Verlag.

Roberts, G.O. and Rosenthal, J.S. (1999). Convergence of slice sampler Markov chains. *J. Roy. Statist. Soc., Ser. B*, **61**, 643–660.

Roberts, G.O. and Sahu, S.K. (1997). Updating schemes, correlation structure, blocking and parameterization for the Gibbs sampler. *J. Roy. Statist. Soc., Ser. B*, **59**, 291–317.

Roberts, G.O. and Smith, A.F.M. (1993). Simple conditions for the convergence of the Gibbs sampler and Metropolis-Hastings algorithms. *Stochastic Processes and their Applications*, **49**, 207–216.

Robinson, W.S. (1950). Ecological correlations and the behavior of individuals. *American Sociological Review*, **15**, 351–357.

Rogers, J.F. and Killough, G.G. (1997). Historical dose reconstruction project: estimating the population at risk. *Health Physics*, **72**, 186–194.

Royle, J.A. and Berliner, L.M. (1999). A hierarchical approach to multivariate spatial modeling and prediction. *Journal of Agricultural, Biological and Environmental Statistics*, **4**, 1–28.

Rue, H. and Tjelmeland, H. (2002). Fitting Gaussian Markov random fields to Gaussian fields. *Scand. J. Statist.*, **29**, 31–49.

Sain, S.R. and Cressie, N. (2002). Multivariate lattice models for spatial environmental data. In *Proc. A.S.A. Section on Statistics and the Environment*. Alexandria, VA: American Statistical Association, pp. 2820–2825.

Sampson, P.D. and Guttorp, P. (1992). Nonparametric estimation of nonstationary spatial covariance structure. *J. Amer. Statist. Assoc.*, **87**, 108–119.

Sanso, B. and Guenni, L. (1999). Venezuelan rainfall data analysed using a Bayesian space-time model. *Applied Statistics*, **48**, 345–362.

Sargent, D.J., Hodges, J.S., and Carlin, B.P. (2000). Structured Markov chain Monte Carlo. *Journal of Computational and Graphical Statistics*, **9**, 217–234.

Scalf, R. and English, P. (1996). Border Health GIS Project: Documentation for inter-censal zip code population estimates. Technical report, Impact Assessment Inc., Environmental Health Investigations Branch, California Department of Health Services.

Schafer, J.L. (2000). *Analysis of Incomplete Multivariate Data*. London: Chapman and Hall/CRC Press.

Schervish, M.J. and Carlin, B.P. (1992). On the convergence of successive substitution sampling. *J. Computational and Graphical Statistics*, **1**, 111–127.

Schmidt, A.M. and O'Hagan, A. (2005). Bayesian inference for nonstationary spatial covariance structure via spatial deformations. To appear *J. Roy. Statist. Soc., Ser. B*.

Shapiro, A. and Botha, J. (1991). Variogram fitting with a general class of conditionally nonnegative definite functions. *Computational Statistics and Data Analysis*, **11**, 87–96.

Sheehan, T.J., Gershman, S.T., MacDougall, L.A., Danley, R.A., Mroszczyk, M., Sorensen, A.M., and Kulldorff, M. (2000). Geographic assessment of breast cancer screening by towns, zip codes and census tracts. *Journal of Public Health Management Practice*, **6(6)**, 48–57.

Short, M., Carlin, B.P., and Bushhouse, S. (2002). Using hierarchical spatial models for cancer control planning in Minnesota. *Cancer Causes and Control*, **13**, 903–916.

Silverman, B. (1986). *Density Estimation for Statistics and Data Analysis*. Boca Raton, FL: Chapman and Hall/CRC Press.

Skeel, R.D. (1980). Iterative refinement implies numerical stability for Gaussian elimination. *Mathematics of Computation*, **35**, 817–832.

Smith, R.L. (1996). Estimating nonstationary spatial correlations. Technical report, Department of Statistics, Cambridge University, UK.

Smith, R.L. (2001). *Environmental Statistics*. Lecture notes for CBMS course at the University of Washington, under revision for publication; website www.unc.edu/depts/statistics/postscript/rs/envnotes.pdf.

Snyder, J.P. (1987). *Map Projections: A Working Manual*. Professional Paper 1395, United States Geological Survey.

Solow, A.R. (1986). Mapping by simple indicator kriging. *Mathematical Geology*, **18**, 335–354.

Spiegelhalter, D.J., Best, N., Carlin, B.P., and van der Linde, A. (2002). Bayesian measures of model complexity and fit (with discussion). *J. Roy. Statist. Soc., Ser. B*, **64**, 583–639.

Spiegelhalter, D.J., Thomas, A., Best, N., and Gilks, W.R. (1995a). BUGS: Bayesian inference using Gibbs sampling, Version 0.50. Technical report, Medical Research Council Biostatistics Unit, Institute of Public Health, Cambridge University.

Spiegelhalter, D.J., Thomas, A., Best, N., and Gilks, W.R. (1995b). BUGS examples, Version 0.50. Technical report, Medical Research Council Biostatistics Unit, Institute of Public Health, Cambridge University.

Stein, M.L. (1999a). *Interpolation of Spatial Data: Some Theory for Kriging*. New York: Springer-Verlag.

Stein, M.L. (1999b). Predicting random fields with increasingly dense observations. *Annals of Applied Probability*, **9**, 242–273.

Stein, M.L. (2003). Space-time covariance functions. Technical Report #4, CISES, University of Chicago.

Stein, A. and Corsten, L.C.A. (1991). Universal kriging and cokriging as a regression procedure. *Biometrics*, **47**, 575–587.

Stein, A., Van Eijnbergen, A.C., and Barendregt, L.G. (1991). Cokriging nonstationary data. *Mathematical Geology*, **23**, 703–719.

Stern, H.S. and Cressie, N. (1999). Inference for extremes in disease mapping. In *Disease Mapping and Risk Assessment for Public Health*, eds. A. Lawson, A. Biggeri, D. Böhning, E. Lesaffre, J.-F. Viel, and R. Bertollini. Chichester: Wiley, pp. 63–84.

Stigler, S. (1999). *Statistics on the Table: The History of Statistical Concepts and Methods*. Boston: Harvard University Press.

Stroud, J.R., Müller, P., and Sanso, B. (2001). Dynamic models for spatiotemporal data. *J. Roy. Statist. Soc., Ser. B*, **63**, 673–689.

Supramaniam, R., Smith, D., Coates, M., and Armstrong, B. (1998). *Survival from Cancer in New South Wales in 1980 to 1995*. Sydney: New South Wales Cancer Council.

Tanner, M.A. (1996). *Tools for Statistical Inference: Methods for the Exploration of Posterior Distributions and Likelihood Functions*, 3rd ed. New York: Springer-Verlag.

Tanner, M.A. and Wong, W.H. (1987). The calculation of posterior distributions by data augmentation (with discussion). *J. Amer. Statist. Assoc.*, **82**, 528–550.

Thomas, A.J. and Carlin, B.P. (2003). Late detection of breast and colorectal cancer in Minnesota counties: An application of spatial smoothing and clustering. *Statistics in Medicine*, **22**, 113–127.

Tobler, W.R. (1979). Smooth pycnophylactic interpolation for geographical regions (with discussion). *J. Amer. Statist. Assoc.*, **74**, 519–536.

Tolbert, P., Mulholland, J., MacIntosh, D., Xu, F., Daniels, D., Devine, O., Carlin, B.P., Klein, M., Dorley, J., Butler, A., Nordenberg, D., Frumkin, H., Ryan, P.B., and White, M. (2000). Air pollution and pediatric emergency room visits for asthma in Atlanta. *Amer. J. Epidemiology*, **151:8**, 798–810.

Tonellato, S. (1997). Bayesian dynamic linear models for spatial time series. Technical report (Rapporto di riceria 5/1997), Dipartimento di Statistica, Universita CaFoscari di Venezia, Venice, Italy.

United States Department of Health and Human Services (1989). *International Classification of Diseases*, 9th revision. Washington, D.C.: DHHS, U.S. Public Health Service.

Vargas-Guzmán, J.A., Warrick, A.W., and Myers, D.E. (2002). Coregionalization by linear combination of nonorthogonal components. *Mathematical Geology*, **34**, 405–419.

Vaupel, J.W., Manton, K.G., and Stallard, E. (1979). The impact of heterogeneity in individual frailty on the dynamics of mortality. *Demography*, **16**, 439–454.

Vecchia, A.V. (1988). Estimation and model identification for continuous spatial processes. *J. Roy. Statist. Soc., Ser. B*, **50**, 297–312.

Ver Hoef, J.M. and Barry, R.P. (1998). Constructing and fitting models for cokriging and multivariable spatial prediction. *Journal of Statistical Planning and Inference*, **69**, 275–294.

Wackernagel, H. (1998). *Multivariate Geostatistics: An Introduction with Applications*, 2nd ed. New York: Springer-Verlag.

Wakefield, J. (2001). A critique of ecological studies. *Biostatistics*, **1**, 1–20.

Wakefield, J. (2003). Sensitivity analyses for ecological regression. *Biometrics*, **59**, 9–17.

Wakefield, J. (2005). Ecological inference for 2×2 tables. To appear (with discussion) *J. Roy. Statist. Soc., Ser. B*.

Wakefield, J. and Morris, S. (2001). The Bayesian modeling of disease risk in relation to a point source. *J. Amer. Statist. Assoc.*, **96**, 77–91.

Wakefield, J. and Salway, R. (2001). A statistical framework for ecological and aggregate studies. *J. Roy. Statist. Soc., Ser. A*, **164**, 119–137.

Wall, M.M. (2004). A close look at the spatial structure implied by the CAR and SAR models. *J. Statist. Plann. Inf.*, **121**, 311–324.

Waller, L.A. and Gotway, C.A. (2004). *Applied Spatial Statistics for Public Health Data*. New York: Wiley.

Waller, L.A., Carlin, B.P., and Xia, H. (1997). Structuring correlation within hierarchical spatio-temporal models for disease rates. In *Modelling Longitudinal and Spatially Correlated Data*, eds. T.G. Gregoire, D.R. Brillinger, P.J. Diggle, E. Russek-Cohen, W.G. Warren, and R.D. Wolfinger, New York: Springer-Verlag, pp. 308–319.

Waller, L.A., Carlin, B.P., Xia, H., and Gelfand, A.E. (1997). Hierarchical spatio-temporal mapping of disease rates. *J. Amer. Statist. Assoc.*, **92**, 607–617.

Waller, L.A., Turnbull, B.W., Clark, L.C., and Nasca, P. (1994). Spatial pattern analyses to detect rare disease clusters. In *Case Studies in Biometry*, eds. N. Lange, L. Ryan, L. Billard, D. Brillinger, L. Conquest, and J. Greenhouse. New York: Wiley, pp. 3–23.

Wei, W.W.S. (1990). *Time Series Analysis: Univariate and Multivariate Methods*. Menlo Park, CA: Addison-Wesley.

Weiss, R.E. (1996). Bayesian model checking with applications to hierarchical models. Technical report, Department of Biostatistics, UCLA School of Public Health. Available online at rem.ph.ucla.edu/~rob/papers/index.html

West, M. and Harrison, P.J. (1997). *Bayesian Forecasting and Dynamic Models*, 2nd ed. New York: Springer-Verlag.

Whittaker, J. (1990). *Graphical Models in Applied Multivariate Statistics*. Chichester: Wiley.

Whittle, P. (1954). On stationary processes in the plane. *Biometrika*, **41**, 434–449.

Wikle, C. and Cressie, N. (1999). A dimension reduced approach to space-time Kalman filtering. *Biometrika*, **86**, 815–829.

Wikle, C.K., Milliff, R.F., Nychka, D., and Berliner, L.M. (2001). Spatiotemporal hierarchical Bayesian modeling: Tropical ocean surface winds. *J. Amer. Statist. Assoc.*, **96**, 382–397.

Wilkinson, J.H. (1965). *The Algebraic Eigenvalue Problem*. Oxford: Clarendon Press.

Woodbury, A. (1989). Bayesian updating revisited. *Math. Geology*, **21**, 285–308.

Xia, H. and Carlin, B.P. (1998). Spatio-temporal models with errors in covariates: mapping Ohio lung cancer mortality. *Statistics in Medicine*, **17**, 2025–2043.

Yaglom, A.M. (1962). *An Introduction to the Theory of Stationary Random Functions*. New York: Dover Publications.

Yakovlev, A.Y. and Tsodikov, A.D. (1996). *Stochastic Models of Tumor Latency and their Biostatistical Applications.* New Jersey: World Scientific.

Zhu, J., Lahiri, S.N., and Cressie, N. (2005). Asymptotic inference for spatial CDFs over time. To appear *Statistica Sinica.*

Zhu, L. and Carlin, B.P. (2000). Comparing hierarchical models for spatio-temporally misaligned data using the Deviance Information Criterion. *Statistics in Medicine,* **19**, 2265–2278.

Zhu, L., Carlin, B.P., English, P. and Scalf, R. (2000). Hierarchical modeling of spatio-temporally misaligned data: Relating traffic density to pediatric asthma hospitalizations. *Environmetrics,* **11**, 43–61.

Zhu, L., Carlin, B.P., and Gelfand, A.E. (2003). Hierarchical regression with misaligned spatial data: Relating ambient ozone and pediatric asthma ER visits in Atlanta. *Environmetrics,* **14**, 537–557.

Zimmerman, D.L. (1993). Another look at anisotropy in geostatistics. *Mathematical Geology,* **25**, 453–470.

Zimmerman, D.L. and Cressie, N. (1992). Mean squared prediction error in the spatial linear model with estimated covariance parameters. *Ann. Inst. Stastist. Math.,* **44**, 27–43.

Author index

Abrahamsen, N., 129
Abramowitz, M., 28
Agarwal, D., 73, 74, 86, 206
Agresti, A., 72
Aitken, J., 39
Akima, H., 47
Allen, B., 246, 247
Alonso, F., 276
Andersen, P., 313
Anderson, D., 39
Anselin, L., 92, 96
Anthonisen, N., 335
Anton, H., 143
Armstrong, M., 33, 34
Arnold, B., 76
Arnold, R., 166, 285, 317
Assunção, R., 246, 321

Bailey, M., 260
Bailey, T., 2
Banerjee, S., 17, 36, 81, 155, 156, 163, 164, 218–220, 247, 275, 304, 311, 315, 318, 344, 346
Banfield, J., 137
Barendregt, L., 222
Barry, R., 34, 86, 88, 137, 153, 169, 243
Bayes, T., 101
Becker, R., 92
Berger, J., 99, 100, 103, 107
Berkson, J., 329, 331
Berliner, M., 119, 233
Bernardinelli, L., 164, 197, 201, 285, 286, 400
Bernardo, J., 99
Berteig, V., 129

Besag, J., 7, 8, 76–79, 82, 164, 169, 188, 197, 389, 390, 392, 398, 400, 402
Best, N., 107–109, 166, 245, 285, 286, 288, 295, 317, 335
Bibby, J., 151
Botha, J., 34, 137
Bowden, D., 83
Breslow, N., 70, 147, 148, 295
Brook, D., 76, 86
Brooks, S., 118
Brown, P., 129, 222, 275
Bryk, A., 361
Burnett, R., 288

Cardoso, T., 50, 67, 89, 90, 93, 138, 144
Carlin, B., xvi, 1, 73, 81, 99, 100, 103, 105, 107–110, 113, 115, 117, 119, 123, 127, 128, 163, 164, 166, 173, 176, 182, 184, 188, 190, 192, 200, 203, 211, 212, 247, 284, 285, 288, 293, 295–297, 302, 304, 310, 311, 315, 317, 318, 329, 335, 339, 362, 366, 369, 379, 398, 401–403, 417
Carlin, J., xvi, 99
Carriquiry, A., 123
Case, K., 260
Casella, G., 77, 110
Cavenaghi, S., 246
Chambers, J., 39
Chen, M.-H., 110, 329, 331, 332
Cherry, S., 137
Chib, S., 225
Chiles, J., 2
Christakos, G., 33, 66
Clark, L., 185, 187